00 SPR 7122

Agricultural
and Food Policy

Fourth Edition

Agricultural and Food Policy

RONALD D. KNUTSON
Texas A&M University

J. B. PENN
Sparks Commodities, Inc.

BARRY L. FLINCHBAUGH
Kansas State University

Prentice Hall
Upper Saddle River, New Jersey 07458

Library of Congress Cataloging-in-Publication Data
Knutson, Ronald D.
    Agricultural and food policy / Ronald D. Knutson, J. B. Penn,
Barry L. Flinchbaugh. — 4th ed.
        p.   cm.
    Includes bibliographical references and index.
    ISBN 0-13-753989-4
    1. Agriculture and state—United States.   2. Food supply—
Government policy—United States.   3. United States—Commercial
policy.   4. United States—Economic policy—1981–1993.   I. Penn, J.
B.   II. Flinchbaugh, Barry L.   III. Title.
HD1761.K65 1998
338.1'873—dc21   97-23274
                                                                       CIP

Acquisitions Editor: *Charles Stewart*
Production Editor: *Michael Jennings*
Production Liaison: *Eileen M. O'Sullivan*
Managing Editor: *Mary Carnis*
Director of Manufacturing and Production: *Bruce Johnson*
Manufacturing Manager: *Ed O'Dougherty*
Production Manager: *Marc Bove*
Marketing Manager: *Melissa Bruner*
Editorial Assistant: *Kimberley Yehle*
Page make-up/formatting: *Carlisle Communications*
Printer/Binder: *RR Donnelley & Sons Company*

© 1998 by Prentice-Hall, Inc.
A Simon & Schuster Company
Upper Saddle River, New Jersey 07458

Printed in the United States of America
10 9 8 7 6 5 4 3 2 1

ISBN: 0-13-753989-4

Prentice-Hall International (UK) Limited, *London*
Prentice-Hall of Australia Pty. Limited, *Sydney*
Prentice-Hall Canada Inc., *Toronto*
Prentice-Hall Hispanoamericana, S.A., *Mexico*
Prentice-Hall of India Private Limited, *New Delhi*
Prentice-Hall of Japan, Inc., *Tokyo*
Simon & Schuster Asia Pte. Ltd., *Singapore*
Editora Prentice-Hall do Brasil, Ltda., *Rio de Janeiro*

Clancy Knut O'Bear, C.D., a delightful Lhasa Apso, is an important member of the Knuston family. As a puppy, Clancy once became so enraged with the poor exam of a student that he chewed it into hundreds of tiny pieces. Fortunately, Clancy moved on to more scholarly activities. He became a serious student and enthusiastically attended eight semesters of doggie obedience school. Sometimes he graduated near the top of his class, and sometimes he did not quite graduate. In spite of his academic inconsistencies, Clancy eventually earned his Companion Dog Certificate from the American Kennel Club. This book is dedicated to all students who, like Clancy, keep trying to succeed.

Ronald D. Knutson

Contents

PART IV   Consumers, Rural Development, and Agribusiness

17 Impact of Agricultural and Food Policy on Agribusiness   488

Index   513

Preface

This book is designed and written primarily as an undergraduate textbook on agricultural and food policy. It recognizes that policy formulation involves a blending of economics and politics. It also recognizes that the government policies and programs that are uniquely important to agriculture today are more than the traditional domestic farm programs. In fact, as one looks back on the past two decades, the policy decisions having the greatest impact on agriculture may arguably have been in the international, consumer, and general economic policy arenas.

Understanding contemporary domestic farm policy decisions requires knowledge of the process of policy formulation, the macroeconomics of agriculture, the international agricultural economic and policy environment, and the fundamental economic relationships and principles that affect today's agriculture. These topics, therefore, are treated before discussions of more traditional policy instruments such as target prices, loan rates, grain reserves, and production controls. Subsequent chapters describe and analyze contemporary issues such as the structure of agriculture, price controls, nutrition policy, food safety, farm labor, and the use of finite resources.

The issues are treated in a current context intended to capture the interest of students. The book does not prescribe solutions to problems. Instead, it emphasizes developing an understanding of the problems, policy alternatives, and their consequences.

Recognition is given to the fact that the nature of agricultural and food problems changes over time. This is due to the inherent volatility of agriculture, food production, and the domestic and international factors affecting the politics of food. While policymakers in the 1970s worried about production deficits, their attention in the 1980s returned to surpluses and low prices that have tended to characterize agriculture since the first major farm programs of the 1930s.

In the 1990s, attention shifted rather abruptly to greatly increased concern about environmental issues and related public interest concerns. In addition, the international scene received greater attention as the second world of centrally planned economies disintegrated and as the United States attempted to aid in the establishment of viable market economies internationally. The development of a stable food supply pipeline was an integral part of that strategy that all farmers could actively support.

The problems of agriculture, or their symptoms, have a tendency to reappear periodically. Agricultural and food policy goals shift as the nature of the problem and related priorities change. These apparent policy cycles underscore the need to review the lessons of past policy experience. Yet agricultural policy tools, over time, evolve as old programs and are modified to fit changing agricultural circumstances and demands of policymakers.

In 1996 those circumstances and demands changed abruptly as less government expenditures and involvement was demanded by the Congress. While several farm organizations resisted, they eventually fell in line behind the Freedom to Farm proposal. The fact of passage of a farm bill that decouples payments from production represents an historic turning point in history.

Therefore, in this the fourth edition the major change involves a restructuring of Part III on domestic farm and resource policy. Chapter 10 has been converted to an historical perspective chapter. A new chapter 11 has been added to cover the 1996 farm bill changes and to discuss future policy options.

Likewise Chapter 15 on food and nutrition policy has undergone substantial updating for food safety and welfare reform policy changes.

Economic principles are introduced throughout the book where they are particularly relevant to analyzing a particular problem, policy, program, or consequence. This is done to develop an understanding of how the tools of analysis can be used to provide insight into the economic impact of particular policies. Ultimately, our goal is to provide students with a framework and the tools to evaluate policies on which they, as the leaders of the future, will have to make decisions.

Many people at Purdue University, the University of Minnesota, Pennsylvania State University, Texas A&M University, Kansas State University, and in the U.S. Department of Agriculture had a unique influence on the content of this book through their contributions to the education of the authors. Primary among these were Don Paarlberg, G. Edward Schuh, George Brandow, J. C. Bottum, Willard W. Cochrane, G. Art Barnaby, Edward G. Smith, John Penson, Dennis Fisher, and James W. Richardson. The first edition benefited from the detailed comments of B. L. Flinchbaugh, William H. Meyer, and G. Edward Schuh. Dr. Boehm leaves as an author after three editions and turns over the responsibility to Dr. Barry Flinchbaugh, an authority on the 1996 farm bill and the politics of agriculture. In addition, Flinchbaugh deserves special credit for his tenacious admonitions to remove personal prejudices, biases, and value judgments from the book.

This fourth edition benefited from the many comments from students and professors who used the first three editions. Dawne Hicks has been our loyal processor of drafts and redrafts. Likewise, Sharron Knutson continues to correct her husband's errors in writing, grammar, and reasoning. David P. Ernstes was in charge of graphics production. He is credited with having corrected and updated previous graphics, with the help of many USDA contributors. All are to be credited for their effort and patience with the authors.

Ronald D. Knutson
J. B. Penn
B. L. Flinchbaugh

About the Authors

Ron Knutson grew up in rural Minnesota and received his B.S. from the University of Minnesota. He has a M.S. from Pennsylvania State University and a Ph.D. from the University of Minnesota. He is currently a professor, extension policy educator, and director of the Agricultural and Food Policy Center (AFPC) at Texas A&M University. AFPC performs analyses of the impacts of policy changes for the agriculture committees of the Congress. Dr. Knutson was formerly a professor at Purdue University, the staff economist for the Agricultural Marketing Service of USDA, and Administrator of the Farmer Cooperative Service of USDA. He is on the board of directors of the Farm Foundation whose mission is policy education. He is chairman of the University Study Committee for Milk Marketing Order Reform mandated by the 1996 farm bill. Dr. Knutson also developed the Texas A&M Agricultural Policy Congressional Intern Program.

JB Penn grew up in rural Arkansas and received his B.S. from Arkansas State University. He has a M.S. from Louisiana State University and a Ph.D. from Purdue University. Dr. Penn is one of a class of outstanding Purdue graduate students, recruited by Dr. Charles French in the late 1960s, who now enjoys high level government and private sector positions in policy and agribusiness. This group of professionals is often referred to as the Purdue Mafia. Dr. Penn is Senior Vice President of Sparks Companies, Inc., an economic information and consulting company serving the food and agriculture industry for over twenty years. Past positions enable Dr. Penn to bring a unique perspective to policy issues and analysis. He has held the position of administrator of the Economic Research Service/USDA and was also a member of the Council of Economic Advisors to the President.

Barry Flinchbaugh grew up in rural Pennsylvania and received his B.S. from Pennsylvania State University. He has a M.S. from Pennsylvania State University and a Ph.D. from Purdue University. Dr. Flinchbaugh is also part of the Purdue Mafia. He is a professor and extension economist at Kansas State University and has also been a county agent in Pennsylvania. He is a nationally known policy educator and speaker on policy issues. Throughout the 1996 farm bill debate, Dr. Flinchbaugh worked closely with Congressman (now Senator) Roberts in crafting its Freedom to Farm provisions. At the same time, he has enjoyed a close working relationship with Secretary Glickman. He currently chairs the Commission on 21st Century Production Agriculture authorized in the 1996 farm bill. Dr. Flinchbaugh is an ardent Harry Truman fan and is a fellow of the Harry S. Truman Institute.

Agricultural and Food Policy

PROCESS

PART I

Economic policy is a course of action pursued by the government in the management of national economic affairs. It is a product of both economics and politics. It is implemented by government programs enacted into law by elected political representatives on behalf of the citizenry—or special-interest segments of the population.

Farm policy is sectoral economic policy that deals with agriculture and food. Understanding farm policy requires a knowledge of both the political process by which laws are enacted and administered and the economic origin and consequences of those laws. Developing an appreciation for the process of policy development is the purpose of Part I.

Chapter 1 is designed to provide insight into the factors influencing people's attitudes or feelings about policies relating to agriculture. Why is support for farm programs eroding? Does it mean an end to agricultural and food policy?

Chapter 2 provides a condensed overview of the problems agriculture and food policies are designed to address. These problems are discussed in greater depth later in the book.

Chapter 3 describes how policy is made. It is a minicourse in how government makes decisions regarding farm and food programs. Emphasis is placed on the political actors.

Chapter 4 explains how agriculture and food policy interest groups organize to affect the policy outcome. Specific information is provided on the typical policy positions taken by the major farm, agribusiness, and public interest groups that influence farm and food policy decisions.

1 THE POLICY SETTING

The translation of values into public policy is what politics is about.
                                                    —Willard Graylin

The constitution of the United States stipulates that government exists to ensure domestic tranquility, provide for the common defense, establish justice, protect individual liberties, and promote the general welfare. Historically, one of the major policy issues faced by government has been the expanding size and role of government, particularly as it relates to the function of promoting the general welfare. A wide philosophical gap separates public opinions regarding the extent to which the powers of government should be utilized in solving economic and social problems. This is particularly true of agriculture, where the extent of government involvement continues to be a major controversial issue.

What should government do to treat the problems of low farm prices and low farm incomes? Some believe that government should not get involved. The free market will solve the problem. Low farm prices, they suggest, are a consequence of excess supplies. Low farm prices will, if allowed to exist, provide the incentive for reduced production and expanded consumption. Less production will bring higher prices and higher farm incomes. The problem, therefore, is self-correcting if the market is allowed to operate.

Some might argue that the decline of communism demonstrates the superiority of the unfettered market as an allocation tool. Others suggest that such a market remedy is too harsh, that food production is too important for a completely laissez-faire approach, and that without assistance, only the largest and most efficient farms will survive. Government, they suggest, should provide a level of assistance that allows all farmers an opportunity to survive, compete, and earn an income comparable to their nonfarm counterparts. It should, at the same time, ensure that consumers at home and abroad have access to an adequate diet.

This same line of argument goes on around the world—not just in the United States. The performance of the market system is as much of an issue in trade policy, market policy, food policy, rural development policy, and resource policy as it is in farm policy. The issue is one of whether conditions exist under which one might reasonably anticipate that government will perform better than free markets in allocating goods and services. This issue is becoming increasingly complex as, for example, cities bid water supplies away from farmers. This raises questions as to the future direction of food production in states such as California, the largest agricultural state in terms of value of sales.

Adding to the complexity is the erosion of individual rights, particularly property rights, associated with increased regulation by government. Although curbs on freedom have always been concerns associated with farm program restrictions, environmental regulatory activity is viewed by many farmers and landowners as undue infringement and "taking" of property value in violation of the Constitution. While all of this is happening, rural American cities and the related infrastructure have deteriorated, much like the central cities of metropolitan America.

POLICY

Policy is a course of action or guiding principle pursued by the government. It influences or determines the actions and decisions of government. Economic policy involves principles or actions related to the management of the national economy. For example, free trade in international markets is an economic policy. An administration that embraces a free-trade policy is opposed to restrictions on product imports and to subsidies for export. It actively pursues international actions that will reduce barriers to trade.

Agriculture and food policy involves the principles that guide government involvement in the production; the resources utilized in production; the marketing and consumption of food; and the conditions under which people live in rural America. Production is defined broadly enough to include the resources used in the production process. Marketing includes both the domestic and international aspects of the agricultural economy. Consumption encompasses the retail price, distribution, nutrition, and safety aspects of food. Although the emphasis in this book is on economic policy affecting production, marketing, and consumption, it is impossible to discuss these issues without considering broader issues. That is, foreign policy, nutrition policy, and environmental policy often have such pervasive economic implications that these effects become major considerations in agriculture and food policy decisions. Likewise, the status of the rural economy affects the conditions under which farmers and other rural residents live. Thus, the agriculture committees of the Congress have a major impact on government policy affecting all rural residents.

Agriculture policy involves numerous interrelated and highly controversial issues. Exhibited in the Smithsonian Institution's National Museum of American History is a tractor that was driven from the wheat-producing areas of the Texas High Plains to Washington, D.C., by protesting farmers in the 1970s. In the 1980s, farm foreclosure sales were stopped by farmers who protested government policies regarding farm prices and credit, just as had occurred 50 years earlier. Such protests are by no means limited to the United States. Both European and Japanese governments have experienced demonstrations by farmers against efforts to reduce assistance from government. In 1996 cows were herded through the streets of Paris, France, to protest low beef prices. In the 1990s, farmers are expressing concern over limitations on their right to farm in deference to the impacts of their management practices on endangered species, wetlands, or veal calves.

Yet farmers are not of one mind regarding appropriate policies for food and agriculture. A 1989 survey of farmers in 21 states revealed that 33 percent favored the current voluntary commodity programs and 35 percent wanted to eliminate all government programs. [1]Only 11 percent favored mandatory production controls. The views appear to be shifting against mandatory programs. A pre-1985 farm bill poll indicated an even split between voluntary programs, mandatory controls, and elimination of all programs.[2] A post-1985 farm bill poll of wheat producers indicated that a majority favored mandatory controls.[3] The decline to 6 percent support prior to the 1996 bill appears to be a marked retreat from mandatory controls. In a 1994 survey of farmers in 15 states, 37 percent of the respondents favored retaining a voluntary commodity program with target prices and support prices while 41 percent favored a phaseout of these programs. Six percent favored mandatory production controls. Moreover, there appears to be increased support among farmers for program phaseout. Farmers' views on issues of sustainability and pesticide use in 1994 are by no means uniform, with 39 percent agreeing in regulation to reduce pollution-contributing farming practices and 43 percent disagreeing.

FACTS, BELIEFS, VALUES, AND GOALS

A **policy position** *indicates a conclusion as to what the role of government ought to be with respect to a particular problem or a set of circumstances.* Policy positions are derived from the interaction of facts, beliefs, values, and goals that are held by individuals (Figure 1.1). In a firm or organization, differences among individuals in facts, beliefs, values, and goals should be discussed and rationalized before a policy position is developed. This generally involves a process of education and compromise.

---

[1]Harold D. Guither, Bob F. Jones, Marshall A. Martin, and Robert G. F. Spitze, *U.S. Farmers' Preferences for Agricultural and Food Policy After 1995*, N.C. 545 (Urbana, Ill.: University of Illinois, November 1994), p. 24.

[2]Harold D. Guither, Bob F. Jones, Marshall A. Martin, and Robert G. F. Spitze, *U.S. Farmers' Views on Agricultural and Food Policy*, N.C. 227 (Urbana, Ill.: University of Illinois, December 1984), p. 15.

[3]USDA Press Release, *USDA Releases Wheat Poll Results* (Washington, D.C.: Office of Information, August 15, 1986).

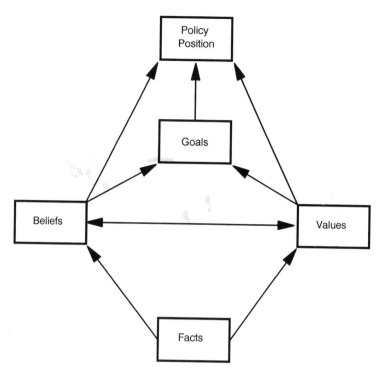

FIGURE 1.1   **Factors influencing one's policy position.**

Facts

A **fact** *is something known with certainty.* Facts describe what is. If it can be objectively verified, rational people will tend to agree on a fact. In the physical or biological sciences, facts are more readily determined and agreed on than in the social sciences. Facts are more nebulous in social sciences such as economics. The definition of farm income is an example. When comparing incomes of farmers and nonfarmers, should farm income include income earned from an off-farm job? Should it include changes in the value of the farmer's land and other assets?

Causal relationships are also more definitive in the physical and biological sciences. A specific herbicide kills certain weeds. Causal relationships in social sciences are less precise, less measurable, less readily agreed on, and almost always subject to qualification. For example, economists disagree over whether government support of farm prices and incomes aids the survival of the family farm or hastens its demise. They also disagree over whether the inheritance tax exemption helps to preserve the family farm from generation to generation, or simply attracts outside investors, or both.

The inability to be definitive does not mean that economics is useless, that there are no observable facts, that economic explanations are useless, or that a knowledge of economic implications is meaningless. It does mean that a need exists to identify, analyze, weigh, and evaluate economic facts, relationships, and impacts. Different perspectives on facts need to be understood and evaluated when analyzing a policy issue. In addition, factual knowledge is important to objectivity in making policy decisions. Thus, research and education regarding facts and relationships are very important to the policy process.

Beliefs

Beliefs *describe what people think is reality.* A belief involves mental conviction, acceptance, confidence, or faith that a proposition is true. Beliefs are not dependent on the intrinsic, objective truth of the proposition. There are true beliefs, partially true beliefs, and false beliefs. Beliefs, therefore, can be based on fact, partially based on fact, or have no basis in fact. It is generally possible to sort out beliefs that have a factual basis from those that do not.

Many policy disagreements arise when beliefs are based only partially on facts. Such beliefs are not only a source of disagreement; they can also be deceptive. Averages frequently fall into this category. For example, during the 1970s, the income of the farm population from all sources averaged 90.2 percent of nonfarmers' average income. But in two of those years, farm income was higher than nonfarm income. In the mid-1980s, farm income once again fell substantially below nonfarm income. Yet farmers who gross more than $100,000 in sales had consistently higher average incomes than the nonfarm population. It might legitimately be argued that domestic farm policy cannot be expected to significantly help farmers with less than $100,000 in sales.

Policy disagreements frequently have their roots in mythology or notions that are based more on tradition, values, or convenience than on fact. For example, many farmers ascribe to the myth that land is the source of all wealth. This myth has its roots in eighteenth century economic thought developed by the physiocrats. It fails to recognize that land is only one economic factor of production—the others generally being labor, capital, management, and water, which are as important to productivity as land. Extensions of this physiocratic doctrine lead to other myths, such as the notion that recessions or depressions in agriculture lead to recessions or depressions in the overall economy. Reality suggests that agriculture's impact on the economy is no greater than its share of overall economic activity.

Values

Values *are conceptions of what should be.* They provide an image of what is good and right and thus specify that some things are better than others. Values indicate what is desirable. They provide justification for proposed or actual behavior.

Values are influenced by beliefs and by facts. Values also influence beliefs. For example, farmers value individual initiative. This value arises in part from the belief that individuals are responsible for their own fate through their own initiative or lack thereof. Thus, farmers frequently believe that many people receiving public assistance could earn a living if they were willing to work.

Historically, many of the values attributed to farmers have been associated with the concept of Jeffersonian agrarianism. The agrarian ideology has three basic tenets:

- Agriculture is the basic occupation of humankind.
- Rural life is morally superior to urban life.
- A nation of small independent farmers is the proper basis for a democratic society.[4]

The agrarian ideology, with its declaration of moral superiority and its blueprint for democracy, was highly acceptable to the American people of the nineteenth and early twentieth centuries. Out of this ideology grew a body of rhetoric, known as the agricultural creed, that has garnered widespread support for farm programs. The articles of the **agricultural creed,** as explained by Paarlberg, include the following:

- Farmers are good citizens, and a high percentage of the population should be on farms.
- Farming is not only a business but a way of life.
- Farming should be a family enterprise.
- The land should be owned by the person who tills it.
- It is good to make two blades of grass grow where one grew before.
- Anyone who wants to farm should be able to do so.
- A farmer should be his own boss.[5]

Even today, despite monumental changes in U.S. society, it would be a mistake to suggest that the agrarian ideology and its associated agricultural creed are dead. Its application can still be seen in political campaign rhetoric extolling the family farm and lauding the farmer as the backbone of democracy. It can be seen in the tendency to view farmers as a homogeneous body having similar problems, justifying the need for a single national farm policy. Agrarianism thus continues to serve as one of the justifications for farm programs. Closely related, agrarianism serves as the foundation for many of the values still held by farmers and their organizations.[6]

Despite such campaign rhetoric and organizational dogma, substantial disagreement exists over whether rural-urban differences in values exist any longer.

---

[4]Edward W. Hassinger, *The Rural Component of American Society* (Danville, Ill.: The Interstate Printers and Publishers, 1978), pp. 83–85.

[5]Don Paarlberg, *American Farm Policy* (New York: John Wiley & Sons, 1964), p. 3.

[6]Hassinger, *Rural Component*, p. 95.

One school of thought holds that they do. These proponents point to studies that suggest the following values are held in high esteem by farmers:

- Quality education is viewed as the means to occupational achievement and success. Technology, being a product of education and research, has traditionally been looked on favorably by farmers and ranchers.
- Work and proficiency in one's job is a key to success. The work ethic is generally believed to be held in stronger esteem by farmers than by urban people.
- Puritan ethical standards are stronger in rural America. Farmers are, in general, more religious and express greater opposition to divorce, premarital sex, abortion, drugs, and consumption of alcoholic beverages.
- Conservation has always been a major concern of farmers because of their ties to the soil.
- Property rights associated with land and water are more sacred than those associated with other forms of property or business operations. Property rights seem to be more important to farmers than to other businesspeople.
- Personal freedom, patriotism, and support of the democratic system are strongly held values that are consistent with the agrarian ideal.

The high value placed on personal freedom can be associated with the desires of farmers and ranchers to be their own bosses. However, studies have also shown a tendency for farmers to conform to typical patterns of behavior and commonly held beliefs and values in rural America.[7] Despite this trend toward conformity, no consensus exists among farmers on any value, related belief, or behavior. This lack of agreement on values could be a source of disagreement on policy remedies to problems.[8]

Although farmers desire equal status for themselves in society, studies have consistently shown a very conservative attitude toward movements giving equal rights to racial minorities and women. These attitudes are consistent with findings of farmers' willingness to lend a helping hand in time of need, tempered by considerably less support for food stamp programs, which ironically increase the demand for farm products. It appears that farmers attribute many of the problems of minorities and the poor to a lack of willingness to work.[9]

The opposing school of thought holds that rural and urban values have changed and blended over time so that they are now so similar that significant differences no longer exist. Copp, for example, notes that "rural society as we

[7]Olaf F. Larson, "Values and Beliefs of Rural People," in T. R. Ford, ed., *Rural U.S.A.: Persistence and Change* (Ames, Iowa: Iowa State University Press, 1978), p. 93.
[8]Ibid., p. 111.
[9]Ibid., pp. 98–99.

New York City or Lubbock: Which Is Which?

While there [New York City] I stumbled upon a "country-western" bar where—to my sincere surprise—I witnessed students from CCNY doing a passable rendition of a country dance called the cotton-eyed Joe. They drank Pearl and Lone Star beer and vigorously applauded star-spangled cowboy musicians who played the latest Willie Nelson hits. They wore cowboy hats, boots, and belts with buckles as big as bulls. Meanwhile, in Lubbock [Texas], the "city" fathers were bemoaning problems previously associated with urbanism—the local drug crisis and the finding of two bodies in the trunk of an abandoned car. And, to the astonishment of many, the once fogyish Texas Tech *University Daily* rather matter-of-factly reported that daughters of farmers from across the South Texas plains were posing nude for *Playboy* magazine. At the same time, the "kickers" at Coldwater—a local cowboy watering hole—did the same dances displayed in New York City; they also drink the same beer, listen to the same kinds of music, and ride the same mechanical bulls with the same gusto that might be seen at any tavern in any community of any size across the United States. Simply put, after a beer or three in either a Lubbock or New York bar, even the most sensitive anthropologist could become confused about his or her geographical whereabouts.

Source: Thomas K. Pinhey, "Two Chickens: A Response to Bealer's Question," *The Rural Sociologist,* Vol. 1 (January 1981), p. 26.

used to know it is virtually nonexistent."[10] Even Larson, a proponent of the existence of value differences, admits that studies show a surprisingly uniform ranking of values between rural and urban people.[11] The major motivating forces facilitating this change have been increased mobility, school consolidation, television, improved education, and increased off-farm employment.

Whether significant differences in rural–urban values, in fact, exist has become a major source of controversy among sociologists. Pinhey charges that the existence of value differences is assumed, not real.[12] Bealer wants to believe that such differences exist but admits that the evidence supporting them suggests the

[10]Paul McKay, "Modern America Brings Changes to Rural Life," *The Bryan-College Station Eagle* (Bryan, Tex.), November 4, 1981, p. 10D; and William P. Kuvlesky and James H. Copp, "Rural America: The Present Realities and Future Prospects," in *Toward an American Rural Renaissance,* unpublished manuscript (College Station, Tex.: Texas A&M University, 1982), pp. 16, 25.

[11]Larson, "Values and Beliefs," p. 94.

[12]Thomas K. Pinhey, "Two Chickens: A Response to Bealer's Question," *Rural Sociol.* (January 1981), pp. 26–30.

need for considerable caution.[13] Their verification, Bealer argues, must await detailed study of values held by the operators of different sizes and types of farms.

If farmers' values are no different from those of the general population, it removes the agrarian arguments for preserving the family farm. It does not mean that farm programs are no longer justified. It does, however, shift the burden for justification of domestic farm policy primarily to economic differences between farm and nonfarm sectors as opposed to a combination of economic differences and preserving the ideals of Jeffersonian agrarianism and the agricultural creed.

Agriculture's ability to secure public support and legislation favorable to it is influenced by the beliefs and values of the urban and suburban majority toward farm people. The image of rural people held by their city cousins has been very favorable. Rural people are considered friendlier, healthier, more honest and hardworking, as getting more enjoyment out of life, and as having fewer tensions and fewer pressures. But even though rural areas were viewed as the best place to raise children, they were considered to offer the least opportunity for a young person.[14] Available evidence suggests that the general population has had a very favorable image of farmers. A survey of a random sample of U.S. adults concerning their attitudes toward agriculture, farming, and farmers indicates strong support for farming as a way of life, for family farms, and for maintaining an agriculture where farmers can make economic decisions independently (Table 1.1).

This favorable image of farmers leads to widespread public support for government programs that support farm prices and incomes. For example, a public opinion poll conducted early in 1985 revealed that only 24 percent of the voting population favored cutting government farm subsidies and crop controls as a means of reducing the federal deficit.[15] The vast majority, 71 percent, favored cutting the deficit by some other means, and only 5 percent were not sure. A higher proportion (38 percent) of the voting population favored cutting food stamps, although a clear majority (58 percent) still favored finding some other way. Farm and food stamp programs enjoyed considerably more support than loans to college students (which 56 percent favored ending) but less support than social security, for which only 12 percent favored canceling cost-of-living adjustments.

In June 1987, when the government was spending more than $25 billion on farm subsidies, 40 percent said that the government should be spending more to help farmers.[16] Interestingly, urban people appear to be more willing to support farm subsidies than rural people. While 51 percent of the people polled in cities of over 500,000 population said spending on farm programs should rise, only 41 percent of those in rural areas supported higher subsidies.[17] (*Note:* All of these

---

[13]Robert C. Bealer, "On Policy Matters and Rural-Urban Differences," *Rural Sociol.* (January 1981), pp. 19–25.

[14]Larson, "Values and Beliefs," pp. 93–94.

[15]"High Demand for Government Services" in "Opinion Roundup," *Public Opinion* (February-March 1985), p. 19.

[16]Kenneth E. John, "Top Priorities," *The Washington Post National Weekly Edition*, June 8, 1987, p. 38.

[17]*U.S. News and World Report*, March 24, 1986, p. 82.

TABLE 1.1   Attitudes of U.S. Adults Concerning Agriculture, Farming, and Farmers (in percentages)

	Agree	Undecided	Disagree
Agriculture is the most basic occupation in our society, and almost all other occupations depend on it.	80.0	11.6	8.4
Farming involves understanding and working with nature; therefore, it's a much more satisfying occupation than others.	57.4	22.4	20.1
We hear so much about crime and corruption today because our nation is becoming so urbanized.	53.6	25.2	21.1
Farming should be an occupation where farmers can make their economic decisions independently.	76.8	18.1	5.1
Farmers ought to appreciate farming as a good way of life and be less concerned about their cash income.	14.3	11.8	74.0
The family farm must be preserved because it is a vital part of our heritage.	82.1	8.6	9.3
Government should have a special policy to ensure that family farms survive.	67.3	19.3	13.5
Obtaining greater efficiency in food production is more important than preserving the family farm.	22.3	22.4	55.4
Most consumers would be willing to have food prices raised to help preserve the family farm.	24.1	23.1	52.8
Corporate farms should pay more taxes than family farms.	66.8	16.6	16.6

Source: Brenda Jordan and Luther Tweeten, *Public Perceptions of Farm Problems*, Res. Rep. P-894 (Stillwater, Okla.: Oklahoma Agricultural Experiment Station, June 1987), p. 3.

surveys were done at a time when the financial crisis in rural America regularly made the news.)

A favorable image of farmers is consistent with a nonfarm perspective of farmers as not being the main source of problems such as rising food prices and food safety. Instead, consumer activists have tended to focus their criticisms on bad decisions by government, on suppliers of inputs to farmers, or on the excessive market concentration in food processing and distribution.

Yet there has been an increasing tendency for certain interest groups to be critical of farmers' production practices such as confinement systems for livestock and poultry production. Support for farmers could be eroding as agriculture becomes more industrialized and farmers are viewed as being no different than other businesspeople. Support may erode as fewer Americans have relatives who currently live on farms. There are expectations that farmers should be even better

TABLE 1.2   Attitudes of U.S. Adults Concerning Environmental and Food Issues Relating to Agriculture (in percentages)

	Agree	Undecided	Disagree
Use of antibiotics in animals' feed is a threat to human health.	50.8	34.2	14.9
Laws regulating excess soil erosion are badly needed.	56.9	35.5	7.6
Farmers who fail to adopt needed soil conservation practices should be financially penalized.	37.7	33.5	28.8
There should be government policy to protect prime farmland from urban growth.	81.7	12.5	5.7
Farm products should be sold only to countries that support the United States in world affairs.	60.0	15.8	24.1
Today's food is safer than it has ever been.	35.8	24.8	39.3

Source: Brenda Jordan and Luther Tweeten, *Public Perceptions of Farm Problems*, Res. Rep. P-894 (Stillwater, Okla.: Oklahoma Agricultural Experiment Station, June 1987), p. 3.

citizens and curb some of the negative externalities associated with their operations, such as water pollution and soil erosion.

The clear implication is that there are values other than those toward farmers that are important to agriculture and food policy. Table 1.2 indicates that public perceptions of a number of environmental and food policy issues is not nearly as favorable to farmers. Antibiotics used in feeds to produce the vast majority of livestock and poultry are suspect. A majority favor laws regulating erosion. A plurality favor financial penalties for those who fail to adopt soil conservation practices. The vast majority would protect prime farmland—presumably, even if the land could be sold at a higher price for development, denying the farmer the capital appreciation benefits. The majority would presumably embargo exports to countries that do not support us in world affairs. A plurality believe that food is less safe today.

Because all of the responses in Tables 1.1 and 1.2 came from the same survey and set of respondents, the differences indicate the complexity of the values affecting agriculture and food policy. Each issue, or each group of issues, must be evaluated from the perspective of the person being most directly impacted. An alternative perspective is that the negative responses to the questions in Table 1.2 signal a tear in society's image of the fabric of agriculture. There was some evidence of a tear in the 1996 farm bill debate. Defenders of farmers, particularly from the leadership of the new Republican majority, appeared not to exist. Whether this was a reflection of their constituency, perhaps, will need to await the next farm policy debate.

Goals

Goals *are desired ultimate end results or objectives.* A goal is the purpose toward which an endeavor is directed. Goals are long term in nature. The inability of groups of individuals to achieve their goals may lead to visible dissatisfaction, agitation, and the eventual need to turn to government for assistance in goal achievement.

The choice of goals is influenced by a person's values and beliefs. Whether the inability to achieve a goal becomes a public issue depends on the importance attached to it, the influence of the group identifying with the goal, and the extent to which the goal is not being achieved under current government and private initiatives. Farmers hold a wide variety of goals. Some more important ones include the following:

- Self-preservation and survival are goals of every human being. While the values of many farmers favor individual as opposed to government initiative to solve problems, they usually turn to government when farm prices and incomes plummet and their survival is threatened. They are not, however, as likely to turn to government for a solution to incremental problems or less visible action that could, in the long term, lead to their self-destruction. For example, the political pressures for government assistance accelerated sharply in the 1980s as an increasingly large number of farmers faced financial failure and bankruptcy. Sociological studies revealed that conditions of severe financial and/or mental stress were evident in much of rural America.[18] In contrast, most farmers have never become sufficiently disturbed by progressive trends toward large-scale farming and corporate-integrated production marketing systems to advocate policy steps that effectively curb these trends. The unwillingness of family farmers to support policies that would inhibit encroachment by corporate and nonfarm interests is cited by Breimyer as a case in which the self-preservation instinct has not prevailed.[19]

- Raising the standard of living has traditionally been a goal of farming and public policy toward farming. The traditional standard used for this goal has been the attainment of the same level of income, housing, and personal possessions as the nonfarm population. Once this standard is achieved, there is no evidence that the goal of continuous increases in the standard of living is any less important.

- Ownership of farmland and the related private property rights satisfy farmers' values favoring freedom and independence. Farmland is so closely tied to agriculture that its ownership becomes a natural status symbol and key to raising the standard of living and long-term survival. To most farmers, ownership implies a right to farm in a manner deemed appropriate by the farmer that is not to be infringed on by societal values favoring endangered species or animal rights.

---

[18]William D. Heffernan and Judith Bortner Heffernan, "Impact of the Farm Crisis on Rural Families and Communities," *Rural Sociol.* (May 1986), p. 160.

[19]Harold F. Breimyer, *Farm Policy: 13 Essays* (Ames, Iowa: Iowa State University Press, 1977), pp. 67–78.

- Progress, efficiency, and productivity goals are consistent with farmers' faith in the <u>work ethic</u> as a key to survival. Since farmers as individuals or even groups have little or no influence on price, improved efficiency and productivity have become keys to progress. This helps to explain their traditional strong support for education and research.

The goals of food and agriculture policy are not solely the province of farmers. The goals of individuals and groups other than farmers have an increasingly important role in policy decisions. Such goals can be characterized as having a public interest orientation. Some examples are listed:

- Producing an ample supply of food at reasonable prices is a goal that has traditionally been used as a public interest justification for policies subsidizing farm prices and incomes. Farmers from time to time characterize this goal as an integral part of a policy designed to ensure that consumers will enjoy low food prices—the so-called cheap food policy.
- Expanding agricultural exports has been an increasingly important policy goal since the 1970s. Arguably, it has become the most important goal because agriculture is one of the few segments of the American economy that consistently generates a positive balance of trade—an excess of the value of exports over imports.
- Removing hunger and malnutrition as a goal applies to both American and foreign consumers. It is clear, however, that the domestic priority for hunger relief ranks significantly above foreign priorities unless there is a specific foreign policy objective to be served, such as Somalia.
- Maintaining health and reducing health hazards is a goal that gives rise to policies and programs designed to protect the environment and ensure the safety of the food supply. The resulting regulations are frequently viewed by farmers and agribusiness as conflicting with their goals of increased income and their values supporting personal freedom.
- Preserving and allocating resources such as land and water for future generations has become a major concern. Although the goal of soil conservation has long been a concern of farmers, resource policy has taken on a new public interest dimension. Resource policy is a battleground between farmers and environmentalists that is rapidly coming into the mainstream of agriculture and food policy. In the process, farmers are finding themselves subject to more regulations and consider their property rights to be in jeopardy.

THE IMPORTANCE OF COMPROMISE

Persons belonging to the same organization tend to have common goals, values, and beliefs. However, even within an organization, the goals, values, and beliefs of all individuals are not the same, nor are they held with the same intensity. To arrive at a cohesive policy position, compromise among the members of a group with respect to goals, values, or beliefs is frequently necessary. The willingness of the members to compromise is a source of strength.

Without compromise, constant friction among members of the group is possible. Sufficient friction results in an inability of the group to arrive at a policy position. Historically, considerable attention has been focused on differences that exist among farm groups in their attitudes toward government involvement in agriculture and toward specific policies and programs. These differences appear to have their roots less in goals, such as the need to increase farm prices and incomes, than in values favoring personal freedom, opposition to government involvement, and beliefs in the market system as a means of solving problems.

More recently, the major conflicts have been between farmers and agribusiness on the one hand and environmentalists on the other. These conflicts become considerably more divisive than those among farmers and could eventually jeopardize the base of government support for agriculture. In the 1996 farm bill debate, the effects of this divisiveness were clearly seen in dairy policy where conflict among producers and with processors led to a congressional decision to drop the dairy price support program and mandate reform of the marketing order system.

THE IMPACT OF TIME

Goals, values, and beliefs change over time. Such changes may result from improved communication, exposure to new ideas, improved education, or a change in the nature of the problem. For example, the decline in values associated with Jeffersonian agrarianism has been attributed to improved roads, improved schools, improved communication, and higher farm incomes.[20]

The goals of policy may also change over time because of the relative importance of individuals or groups influencing policy. For example, increased consumer and environment activism has made the goals related to conservation, food safety, nutrition, and the preservation of endangered species more important in the policy process.

WHY GOVERNMENT BECOMES INVOLVED

The specific reasons for government involvement in agriculture have changed as the nature of the farm problem and the overall political, social, and economic environment within which agriculture operates has changed. Five most frequently mentioned reasons for government involvement in agriculture are indicated first. These are followed by two more basic economic concepts used to explain government involvement in agriculture.

[20]McKay, "Modern America" and Kuvlesky and Copp, "Rural America."

- Low farm income has traditionally been the major justification for programs that support farm prices and incomes. These programs, however, have become increasingly controversial as government costs have risen, farm numbers have declined, farm income has increased, and farm size has become more diverse.

- The need to stabilize farm prices and incomes has also provided an important justification for farm programs. Stability is desired both to reduce the incidence of mistakes in production decisions and to reduce economic stress on farm families. Stability has been used as a justification both for raising prices and for lowering them.

- The importance of an adequate supply of food historically has been used to justify government programs that expand farm production, such as irrigation projects, agricultural research, and extension. U.S. agricultural abundance has been seized on as a source of export earnings and as a diplomatic weapon of foreign policy. Food has thus become recognized as having a value that extends beyond nutrition.

- The safety of the food supply became an important issue once the ability of the U.S. farmer to produce an adequate supply of food was demonstrated. Over time, the safety issue evolved from a concern with sanitation to the contemporary controversy over additives, pesticides, residues, and nutrition.

- Protecting the capacity of agriculture to produce in future generations has led to programs that conserve the soil. This concern has since spread to conservation of limited water supplies, to protection of water quality, to preservation of prime farmland in populous areas, and to the more recent concern about sustainability.

The two underlying economic rationales that explain government involvement in agriculture are:

Externalities *are benefits or costs accruing to some individuals or groups apart from a market transaction.* For example, a farmer experiencing cash flow problems may decide to avoid conservation practices and thus incur higher rates of soil erosion. The result is excessive runoff and pollution of streams. That pollution is an externality to people who fish, cities that utilize surface water for their water supply, and people who simply enjoy clean streams. Externalities lead to regulations and sometimes to subsidies. Regulations may be enacted to prohibit or limit pollution created by livestock farms. Farmers may be required to engage in certain conservation practices in return for government subsidies. Ironically, subsidies encourage more intensive farming, which, in the absence of conservation requirements, would result in a higher level of erosion. Where supply and demand do not accurately reflect all the benefits and costs of production, the price system cannot be expected to bring about an allocation of resources that best satisfies the wants of society. Externalities, therefore, create market failures—*the failure of markets to allocate resources in society's interest.*

Market failures occur when market systems do not supply goods or services in the manner desired by society. Market failures extend beyond the concept of externalities. Many other cases exist in which government has stepped in with services, subsidies, or regulations designed to fill the gap with **public goods.** Where public benefits are large and the market would not otherwise provide them, public goods result. For example, the land-grant university system was set up because market rewards were not sufficient for either farmers or agribusiness to invest in large-scale programs of agriculture research and education. Similarly, market information, crop and livestock production reports, rural electrification, and farm credit are programs for markets that were judged to have failed to provide sufficient services to agriculture and rural America. Government programs filled the void. Many of the disagreements that exist over the role of government in agriculture result from questions of whether the market has failed—that is, whether it can be relied on to perform certain functions better than government. The answer could also be expected to change over time as agriculture becomes better (or worse) in managing programs. Some would argue that low farm incomes are the result of market failures due to the slowness with which resources adjust out of agriculture.

ECONOMICS, ECONOMISTS, AND PUBLIC POLICY

Economics plays an important role in the development of agriculture and food policies. The traditional farm problems of surplus production and instability are rooted in economics. Agriculture and food policies, in turn, have a direct impact on the production and marketing decisions of farmers and ranchers. Economic variables have a major impact on the world's ability to satisfy future food needs with limited resources. Many of the contemporary food concerns have a direct impact on food availability, cost of production, and prices.

This is not to imply that economics has exclusive jurisdiction over agricultural and food policy issues. Values, beliefs, facts, and goals, as well as individual and group behavior, make sociology and psychology an integral part of policy development and implementation. Policy is determined in the political arena, and policy decisions are fundamentally political decisions.

The Role of Economics

Within this interdisciplinary setting, economics has three main functions:

 ■ It provides insight into the origin of economic problems. This insight can be traced from the aggregate or macrolevel to the individual, firm, or microlevel. An understanding of the origin of problems is crucial to the development of solutions.

2 ■ It assists in developing policy and program alternatives for solving problems.

3. ■ It can be used to analyze the consequences of policies. It is an understanding of the consequences, more than anything else, that is critical to informed public policy decision making.

The greatest need of the policymaker is for an objective analysis of the consequences of the alternative courses of action available. Consequences may be analyzed for a number of interest groups, the most important of which includes farmers, input suppliers, market intermediaries (e.g., brokers), marketing firms (e.g., processors and retailers), consumers, taxpayers, and environmentalists. In other words, almost everyone should be concerned about agriculture and food policy, although the consequences may be quite different for each interest group. As a result, economic analysts exist throughout the public and private sectors. Such analysts are employed by interest groups such as farm organizations, businesses, consulting firms, and the government. Interest groups are more effective in getting their policy proposals implemented when they are based on factual economic analysis. Their arguments are more persuasive, and policymakers are more likely to listen to a case based on facts.

Within government, it is not unusual to have two or three separate economic studies of a controversial policy proposal. Such studies may be carried out from the perspective of a member of Congress who is concerned about being reelected; by the secretary of agriculture, who is concerned about the proposal's impact on farm income and on U.S. Department of Agriculture (USDA) spending; or by the president's economic advisers, who are concerned about its impact on inflation and the budget.

All these analyses will bear on the final public policy outcome. However, the final decision is political. Yet, even here, the politician is often faced with a decision that involves weighing the impact of the policy alternatives on various constituencies—farmers, businesspeople, consumers, environmentalists, and taxpayers.

The role of economics does not end with the decision of the policymaker. Once a policy or program has been implemented, its economic impact must be evaluated as a means of detecting problems, identifying the magnitude of specific relationships, fine-tuning programs, and guiding future decisions.

Throughout this whole process, there is a very important role to be played by those supplying information on policy issues and the policy process. This includes the work of journalists, reporters, and analysts for the media, interest groups, and private firms. Once a bill is enacted into law, an extensive amount of work is involved in explaining its consequences, including how it is to be implemented.

Some General Economic Principles

Throughout this book, relevant economic principles are introduced and discussed as they apply to specific policy topics. There are, however, three general concepts that are of overriding concern to understanding the origin of economic problems in agriculture. Discussion of these economic concepts at this point not only helps us to understand subsequent discussions of food and agricultural problems but also provides readers an opportunity to review and sharpen their theoretical tools of economic analysis.

Supply and Demand

Many agriculture and food policies are designed to overtly influence the supply and demand for farm products. For example, international market development programs are designed to expand demand and, thereby, raise prices. Supply control programs raise prices by restricting production. Other programs are designed to avoid some of the consequences of shifts in supply or demand. For example, the target price program guarantees farmers a specific return per unit of commodity produced, regardless of market price. The government makes up the difference between the market price and the target price in the form of a direct payment. Farmers make decisions on what to produce based on the target price, not the market price. An understanding of the supply and demand for farm products and the factors affecting them becomes critical to understanding policy.

Figure 1.2 contains a representation of typical supply and demand schedules that are used extensively throughout this book. The demand schedule slopes downward to the right, indicating that as price falls, consumers will buy more of a product. Every consumer has an individual demand schedule. The market demand (in Figure 1.2) is the sum of all individual consumer demands. The slope of the demand schedule indicates the responsiveness of consumers to price changes. Economists refer to this sensitivity as the *elasticity of demand. The elasticity of demand relates the percentage change in quantity demanded resulting from a 1 percent change in the price of the product.*

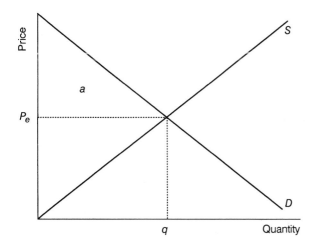

FIGURE 1.2
Supply and demand schedule.

The demand for farm products generally comes from two segments—domestic and export markets. Domestic demand is generally more inelastic than export demand. That is, domestic buyers are less responsive to increases in U.S. farm prices than are foreign buyers. This more elastic foreign demand has important policy implications. For example, U.S. producer returns can be increased by charging a higher price in the domestic market than in the foreign market—if the two markets can be effectively separated. Demand elasticities vary from product to product. For example, the demand for a staple such as flour is less price responsive (more inelastic) than demand for meat, which has several substitutes. In addition, demand for all foods is more inelastic than it is for any individual food.

The magnitude of the elasticity of demand is very important from a policy perspective. If the elasticity is less than -1.0, then farmers' income can be raised by increasing the price. That is, the percentage increase in price is greater than the percentage reduction in quantity demanded, yielding an increase in revenue to farmers. The more inelastic the demand, the greater the benefit from a price increase to farmers.

Demand shifts in response to income, population, and consumer preferences. When income rises, demand generally increases. *The relationship between changes in income and corresponding changes in demand is referred to as the income elasticity of demand.* Income elasticities vary from food to food and consumer to consumer. Low-income consumers generally are more income elastic than high-income consumers, meaning that poorer consumers increase food consumption more in response to an income increase than do the wealthy. Similarly, export demand is more income elastic than domestic demand; that is, foreign consumers purchase more food products in response to an increase in income than do domestic consumers.

The supply schedule slopes upward and to the right. An individual farm's supply schedule is the marginal cost curve (Figure 1.3). Farmers maximize profits by equating the returns they receive for a unit of the product with the marginal cost. As price per unit increases, the quantity supplied increases. Each farm has its own supply curve, which in the short run is the segment of its marginal cost curve above average variable cost. If the firm cannot cover its average variable cost in the short run, it stops producing. In the long run, the supply schedule is the segment of the marginal cost curve above average cost. The market supply schedule is the sum of all individual farm marginal cost schedules.

For most of an undergraduate student's training in economics, the point at which the supply and demand schedule crosses is the magic competitive equilibrium price. However, for many agricultural commodities, the equilibrium price is not politically acceptable, giving rise to government involvement in agriculture. The policy becomes one of trying to change that price, to the benefit of farmers. When government becomes involved, the market price may no longer be at the competitive equilibrium, *P,* in Figure 1.2. Getting used to the price not being at equilibrium is a problem for many students studying agriculture and food policy.

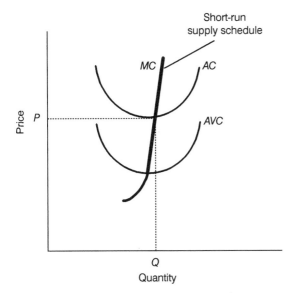

FIGURE 1.3
Short-run supply schedule for a farm.

Fallacy of Composition

Individuals and firms, by and large, pursue production and consumption decisions that enhance their own welfare, whether measured in terms of profits, wealth, or satisfaction. Yet, as a result of pursuit of individual goals, the group as a whole may be worse off. *This inverse relationship between the pursuit of individual goals and group results is referred to as the fallacy of composition.*

Many different illustrations of the fallacy of composition exist. One, which also provides the opportunity to introduce the concept of market equilibrium, involves incentives for firms operating in a purely competitive market to increase production. Figure 1.4 provides a representation of a typical farmer's marginal cost (MC) and average cost (AC) of producing different quantities of corn on a given number of acres of land. There are enough of these typical farmers that no single farmer can have a significant influence on the marketplace.

Figure 1.4 also contains a market supply (S_1) and demand (D) schedule. The market price in Figure 1.4 is P_1, and Q_1 is the quantity supplied. At this price, the typical farm in the market produces quantity q_2, at the point where marginal cost is equal to marginal revenue. That is, the extra cost of producing the last unit of output is equal to the extra revenue from selling that unit of output. At this quantity, the typical farm's average cost of producing corn is P_2. It is thus making a pure profit of $(P_1 - P_2)q_2$, the shaded area. A pure profit is one that more than covers the farmer's normal return on capital contributed to the farm operation.

With the farmer in a pure profit position, there is incentive for the typical farmer to plant a larger number of acres of corn and for new producers to begin producing corn. As this occurs, the supply schedule shifts to the right. In pure competition, one farmer can expand the number of acres planted without having

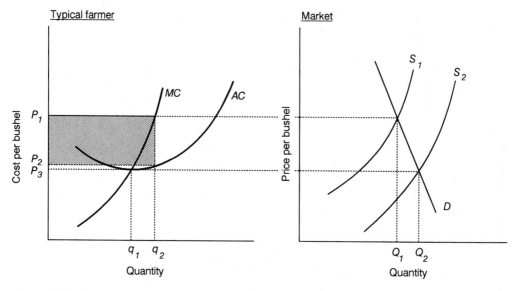

FIGURE 1.4   Representation of firm and market equilibrium concepts.

a significant price effect. However, if all farmers expand their output, price falls and profits fall. This overall reduction in profits illustrates the fallacy of composition concept—even though it is profitable for any individual farm to expand production, if all farms expand, profits decline for all farms. Incentives will exist for individual farms to increase production until the supply schedule shifts to S_2. At S_2, the market is in equilibrium. That is, supply and demand are equal at a price (P_3) where only normal profits exist. Consequently, profits provide no more than a normal return on capital invested.

The fallacy of composition is a source of many errors in agricultural policy decisions. For example, lowering the inheritance tax has been a major policy objective for many farm organizations. This action is favored when viewed from the perspective of individual family farms. However, when viewed from a broader and longer term economic perspective, it encourages outside investment in agriculture, raises land prices, and, in turn, raises the cost of production. The objective of promoting intergeneration transfer is thus thwarted by the fallacy of composition. To avoid falling into the fallacy of composition trap, policy consequences need to be viewed both from the perspective of the individual farmer and from the perspective of the market, as well as from the economy as a whole. Closely related, they must be viewed both from a long-run perspective after market forces have worked themselves out as well as from a short-run perspective.

Consumer and Producer Surplus

Who gains and who loses from farm policy is one of the most interesting and important issues confronting economists, interest groups, and policymakers.

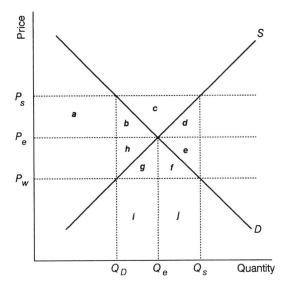

FIGURE 1.5
Impact of a price support program on consumer and producer surplus for wheat.

Making definitive judgments regarding this issue is not easy and requires several assumptions. Yet within the context of the supply and demand schedules, certain quantitative comparisons can be made.

In indicating the quantities of products that will be purchased by consumers at alternative prices, the demand schedule provides insight into the amount of satisfaction obtained from a change in government policy. The area above the price line in Figure 1.2 (area *a*) is the consumer surplus that consumers gain from purchasing quantity *q* of the product. Starting at the first unit of product purchased, the consumer surplus is large, gradually declining until, at unit *q*, the consumer surplus is zero—the satisfaction gained is just equal to the price. If a higher price were charged, the consumer surplus would logically decline because consumers would be less well off.

The producer surplus is measured by the area above the supply schedule and under the equilibrium price. At any price above P_e, the producer realizes a return (surplus) of the area bounded by the equilibrium price line, the new price line, and the supply curve.

A specific example is illustrated in Figure 1.5, which involves supply (*S*) and demand (*D*) schedules for wheat and a free-market equilibrium price (P_e) and quantity (Q_e). Suppose that the government, at the urging of wheat producer interests, decided to establish a wheat price support at price P_S. What would be the impact on producer and consumer surplus?

Logically, consumer surplus would decrease because of the higher price paid for wheat. The specific reduction in consumer surplus (*CS*) is indicated by the sum of area *a* and area *b* (*a* + *b* = CS loss). Looked at differently, areas (*a* + *b*) is the amount consumers would have to be compensated to make them as well off under the wheat price support P_S as under the free-market policy and P_e.

Producer surplus would increase with price support P_S by area (*a* + *b* + *c*). Note that producers benefit more than consumers lose. Accordingly, producers could more than compensate consumers for the higher price support and still be better off by area *c*. Does this mean that society is better off with the higher price support?

Not necessarily. What happened to the wheat surplus of the difference between Q_S and Q_D? To maintain the price at P_S, the government had to buy $Q_S - Q_D$. It had to pay $P_S (Q_S - Q_D)$ or area ($b + c + d + e + f + g + h + i + j$). What does the government do with the wheat that it buys? It either sells it or stores it. Assume that, in this case, it sells the wheat to the beleaguered countries of the former Soviet republic (FSR) at price P_W. What then is the net loss associated with the higher price support? The answer is area ($d + e$). Economists refer to the area *d* and *e* net loss as a deadweight loss. No one receives the benefits from these areas. Area *c* is not included because it is a wash benefit to producers against the taxpayer cost. Likewise, areas *i + j* are not included because they are a wash against revenue from the export sale. Areas *b, h, g,* and *f* represent a misallocation of resources because they are a loss to consumers, a gain to producers, and a cost to taxpayers for a net cost to society.

Limits of Economics

Although the tools of economic theory are useful in providing insight into the origin of economic problems and the consequences of alternative solutions to those problems, they have a limited usefulness in deciding which alternative is "best."

At least two important assumptions are encompassed in economic analysis involving producer and consumer surplus. First, a dollar to every consumer, taxpayer, and producer has the same satisfaction (utility) associated with it. That is, a dollar of subsidy to producers means exactly the same as a dollar extra cost of bread to a poor central city consumer, or each and every taxpayer. Of course, many value judgments must be made to satisfy this important assumption. The only case for which economists can say that one policy option is "better" than the current policies is one in which someone or some group is made better off with no one else, including taxpayers, being made worse off. Such changes are referred to as *movements in the direction of Pareto optimality. A* **Pareto optimum** *exists when it is impossible to make anyone better off without making someone else worse off.*

Because such situations are rarely found, the Pareto criteria are not very useful in most policy decisions. As a result, some economists suggest using the **compensation principle** as a basis for policy decisions. *This principle suggests that as long as those who are made better off by a policy change are able to more than compensate those who are made worse off, the change is justified.* Considerable debate exists, however, over whether compensation, in fact, has to be paid to those who are made worse off.

Resolving this debate inherently involves making interpersonal comparisons, which again illustrates our first assumption that each person or group values the dollar similarly. For example, increased satisfaction to farmers from higher price supports and reduced satisfaction to consumers and taxpayers must be directly proportional to dollar costs and revenues involved. Such an assumption is, of course, a value judgment. Although economists can be useful in helping to evaluate the trade-off in terms of the magnitude of the economic costs and benefits, politicians, whether elected or appointed, are responsible for making such judgments. If politicians make such value judgments incorrectly a sufficient proportion of the time, the electorate can be expected to remove them from office. The second assumption is that there must be no externalities associated with production. That is, as in the wheat example, there must be no erosion that adversely affects water quality and no residues from the use of extra chemicals in production. Such nonmarket values are discussed extensively in Chapter 13.

SYNOPSIS

Most of the remainder of this book involves explaining the economic conditions that lead to government involvement in agriculture. This includes explaining how specific programs affect agriculture as well as the domestic and world economy. This is accomplished in four major parts or sections:

1. Part I recognizes that agriculture and food policy is developed through the political process at the urging of interest groups. Policy is thus more than economics. It is designed to solve the problems of people through the political process. Understanding policy thus requires a knowledge of the overall nature of the problems being solved, the process of policy formulation, and the role of the various interest groups in the process. This is the focus of the first four chapters.
2. Part II recognizes the increasingly important role that macroeconomics and international policy instruments play in U.S. agriculture. Too often there is a tendency to think about agriculture and food in a narrow state or national context. Chapters 5, 6, and 7 are designed to dispel this tendency—to recognize that U.S. production and policies have world implications but to realize that significant limits exist on America's ability to direct or control agricultural economic events. Chapter 8 then explains the impacts of macroeconomic policies on agriculture and the relationship to international markets.
3. Part III includes four chapters covering the various aspects of domestic policy. These chapters stretch beyond traditional farm price and income policies to include discussions of resource and structure policy. These areas have become so interwoven with a mixture of subsidies, cross-compliance, and regulations that farm resource and structure policies must be rationalized together.
4. Part IV recognizes that domestic agriculture is more than farms and related resources. It includes issues of consumers, food safety, rural development, and the impact of policy on agribusiness. These interests are an integral part of developing a national agricultural food policy.

ADDITIONAL READINGS

1. An extensive compilation of articles on the philosophical and economic issues surrounding the family farm is contained in Gary Comstock, ed., *Is There a Moral Obligation to Save the Family Farm?* (Ames, Iowa: Iowa State University Press, 1987).

2. A good comparison of how farmers, rural nonfarm residents, and urban folks view farm, food, and environmental issues is contained in Brenda Jordan and Luther Tweeten, *Public Perceptions of Farm Problems*, Res. Rep. P-894 (Stillwater, Okla.: Oklahoma Agricultural Experiment Station, June 1987).

3. One of the best discussions of consumer and producer surplus under alternative government policies is contained in Peter G. Helmberger,*Economic Analysis of Farm Programs* (New York: McGraw-Hill, 1991), pp. 15–100.

POLICY PROBLEMS OF FOOD AND AGRICULTURE

A problem well stated is a problem half solved.
—*Charles F. Kettering*

Food and agriculture policies and programs are intended to solve chronic problems encountered in the production and marketing of agricultural products. These policies and programs evolve gradually over the years as the problems become increasingly apparent and change in character. The nature of the problems confronting both producers and consumers of food has changed dramatically in the past 10 years. Policymakers have begun to respond to those changes.

In this chapter, we provide an overview of the major problems confronting the agriculture and food sector and how they have changed. A more detailed discussion of each problem area is contained in subsequent chapters. This chapter concludes with five different philosophical approaches to dealing with the current and evolving mix of problems.

SYMPTOMS, PROBLEMS, AND POLICIES IN TRANSITION

The agriculture policies and programs that exist today evolved from the problems and policies that existed in the past. This happens because problems and policies change gradually and unevenly. As experience is gained with particular programs, adjustments are made. In addition, people differ in their perceptions of the nature of problems and how they have changed. These differences account for a substantial proportion of the disagreement over the appropriateness of current and evolving policies.

From the 1930s through the 1960s, the most pervasive problem confronting agriculture was excess capacity, resulting in low farm prices and incomes. Throughout this period, farm income averaged only 51 percent of nonfarm income. Not until 1965 did average farm income exceed 70 percent of nonfarm income.[1]

The origin of this low price and income problem was excess food and fiber production capacity, combined with too many farmers among whom to divide the available income. Agriculture's production capacity expanded more rapidly than did effective demand, thus holding prices and income down. Over time, low incomes encouraged farmers and their wives, sons, and daughters to seek employment outside farming. Those with smaller and moderate-size farms became more dependent on off-farm income. More often than not, the sons and daughters of farmers ended up living in metropolitan cities where job opportunities were more plentiful than in rural America. A result has been rural economic decline. Despite massive migration from the farm to the city, average farm incomes remained low relative to nonfarm incomes. Improvement in the situation was slow and gradual during the 40-year period.

The chronic nature of this excess capacity problem led to extensive government involvement in agriculture. Public debate centered on how to achieve higher incomes for farmers. Price supports were employed along with efforts to control production.

The existence of low farm prices and income was by no means limited to the U.S. farmer. Farm product prices throughout most of the world were low. American price support policies frequently resulted in domestic prices being above world prices. Under these conditions, U.S. products were not priced to compete in the world market, and any export sales generally required subsidies. Export subsidies were not only expensive but invited retaliation by other countries that desired to protect their own producers from the effects of low prices. The result was the erection of barriers to trade designed to prevent foreign products from entering the United States and undermining domestic programs.

In the mid-1960s, there was increasing recognition that if the United States was to become a competitor in international markets, domestic price levels had to be reduced. But the political unpopularity of this move required at least partial compensation for farmers, and direct payments from the government were instituted.

Economic conditions in agriculture improved marginally in the late 1960s. Then in the early 1970s they improved abruptly when a confluence of forces brought the biggest boom since early in the twentieth century, often referred to as the *golden years* of agriculture. Adverse weather and disease combined to reduce global crop production. Concurrently, demand had begun to increase rapidly as a result of population growth, rising incomes, and changes in foreign policy. This confluence of forces resulted in a growing concern over the ability of the world to feed itself. The problem of supplying expanded international markets was exaggerated by a decision of the Soviet leadership to enter the world markets and purchase large quantities of grain in response to reduced production. Previously, the

[1]Agricultural Statistics (Washington, D.C.: USDA, various issues).

Soviets had responded to production shortages simply by reducing consumption. With strong demand from other importers, notably the developing countries, world agriculture was suddenly in a boom period.

During the early 1970s, attendees at world food conferences expressed concern about the capacity of agriculture to meet present and future food needs. Prices for farm products—sometimes almost panic markets—reached record levels, such as $13 per bushel for soybeans in 1973 (from 1990–1996 soybeans averaged $5.75 per bushel). The mentality of policymakers, so long confronted by chronic surpluses, suddenly shifted to deficits. Farmers were encouraged to expand acreage—to plant from fence row to fence row. They responded and, in the face of strong export demand, were rewarded with higher incomes. They also expanded the size of their farm operations and, in the process, bid land prices far above a level that could be sustained based on past earnings.

These developments created the opportunity for government to take substantial steps to reduce its role in agriculture. New market-oriented farm policies were implemented that placed greater reliance on supply and demand forces to determine prices. At the same time, policy inconsistencies became more apparent, notably, export embargoes to control domestic farm prices and/or to obtain foreign political concessions while attempting to expand markets.

Disagreement arose over whether the newfound prosperity of the 1970s represented a permanent change toward a much tighter world supply–demand balance or was simply an aberration. The answer was soon forthcoming when burdensome supplies once again developed and prices tumbled in the early 1980s. Assuming, or hoping, that these adversities were only temporary, government once again attempted to support farm prices and income and to control production. U.S. farm products once again became priced out of the world market by high domestic price supports and a strong dollar.

In an effort to restore competitiveness in world markets, the 1985 farm bill began to lower support prices sharply while maintaining farm income through direct payments to farmers. In addition, foreign buyers, including the Soviet Union and China, were given subsidies to buy surplus U.S. grain, to the chagrin of exporters competing with the United States, notably, Canada, Australia, Argentina, and Thailand. The result was that the costs of farm programs skyrocketed to $26 billion in the mid-1980s.

The events of the 1960s through the 1980s may be viewed as a **policy cycle**— agriculture going from surpluses to deficits and back to surpluses, from bust to boom and back to bust. Agriculture policy had cycled from supporting prices and incomes and controlling production to market-oriented policies and then back to supporting prices and incomes and controlling production.

The concept of policy cycles is a useful way to review history. It provides lessons on the mistakes and successes of the past. Yet the concept can easily be oversimplified and overdrawn. Agriculture has changed in several profound ways:

- It has become export and world-market dependent. Agriculture's production plant expanded to serve a world market with the crops from about two of every five acres being exported. Reducing production to serve only domestic markets became a much less feasible policy option after the 1970s.

- The face of world trade has been changed by developments in Eastern Europe and the former Soviet Union. The potential entry of Eastern European countries into the European Union has long-term implications for competitive relationships between these two blocs. Concurrently, the former Soviet republics are striving to make the transition to market-based economies.

- The fragility of agriculture has become more evident. Complacency regarding the importance of an abundant food supply was removed by the 1970s experience. In the 1980s, complacency about the durability of agriculture was shaken once again as the price of farm real estate plummeted, many rural banks failed, and the cooperative Farm Credit System required federal assistance to remain solvent.

- Public interest in agriculture and its policies accelerated sharply. Consumer interests extended beyond the price of food to the safety, nutritional content, and quality of the food supply. Foreign policy interests in agriculture mounted as food once again became an international negotiating tool. Environmentalists became more interested in agriculture's impact on the quality of water, wildlife habitat, and the future productive capacity of the soil resource.

- The face of agriculture itself changed as large farms became more dominant and moderate-sized farms began to disappear. The inevitability of a bimodal small farm/large farm distribution of farm sizes began to gain general acceptance. The efficiency and resilience of moderate-sized family farms were in doubt.

- As agriculture entered the 1990s, the resource constraints on agriculture became increasingly apparent. These constraints must be viewed in terms of the externalities inherent in agriculture. They also reflect society's efforts to rationalize often conflicting interests in a cheap food supply, a cleaner environment, reduced resource abundance, farmers' perceived private property rights, and the desire for prosperity in rural communities. These conflicting goals appear to be on a collision course—one that is not easily resolved. Farmers, their organizations, and agribusinesses are not in a very good political position to fight these battles as the balance of political representation continues to shift decisively toward urban areas with each reapportionment of the House of Representatives.

- As resource constraints become more apparent, the length of the policy cycles can be expected to decrease or become more erratic in length. Arguably, the next food crisis arose in 1996 when carryover stocks of grain in the United States reached a record low. Shorter or more erratic cycles provide the potential for more rapid changes in the policy agenda—the thrust of items debated in the Congress.

TODAY'S FOOD AND AGRICULTURAL PROBLEMS

Food and agricultural problems can no longer be characterized in simple terms such as low prices and incomes. Today, there are several problems, including

- The world food problem
- The farm and resource problem
- The consumer food problem
- The rural development problem

These problems are interrelated and have overlapping dimensions (Figure 2.1). Their importance shifts over time with the emergence of specific issues. All dimensions of the problems affect farmers and the conditions under which food is produced, marketed, and consumed. Major dimensions of each of the problems are briefly reviewed here. More detailed discussion is presented in later chapters.

The World Food Problem

The world food problem is perhaps the most complex of the major problems. This is true not only because its solution involves satisfying the nutritional needs of more than six billion people but also because governmental systems, policies, and programs of the many nations of the world must be rationalized. It has at least three major dimensions:

- **Distribution.** Reducing hunger and malnutrition is primarily a distribution problem. Properly distributed, there is enough food to feed the approximately 500 million malnourished people in the world. This was true even during

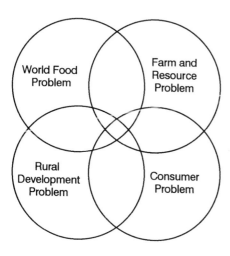

FIGURE 2.1
Food and agricultural policy problems are interrelated.

the world food crisis of the 1970s. The cause of the distribution problem is low incomes—a lack of economic growth and development. The solution frequently lies with trade. Open trade channels facilitate economic growth and development. With higher incomes, food assistance or aid is a less frequent need. Yet massive food assistance programs are sometimes necessary to avoid widespread starvation and malnutrition (e.g., sub-Saharan Africa).

■ **Trade.** Keeping trade channels open has been a perpetual problem. Although the benefits of free trade to society as a whole are abundantly evident, from the perspective of an individual farmer—a milk producer in Germany or a peanut farmer in Virginia—the benefits of protectionism are equally clear. It is generally believed that U.S. agriculture, broadly defined to include farmers and agribusiness, stands to gain more from free trade than it would lose. This explains why the United States typically is a strong advocate of reduced trade barriers. Yet while advocating free trade, many countries' farm policies are protectionist. Policies designed to encourage domestic production frequently mean domestic prices above world prices, thus requiring imports to be restricted. Some developing countries have policies that discourage farm production by placing ceilings on prices at which products may be sold, to the benefit of consumers. Open trade channels would ameliorate many of these problems, albeit at the expense of some producers and countries.

■ **Capacity.** The adequacy of global production capacity was the subject of several studies during the 1970s.[2] These studies have proven too conservative for the times. Given favorable weather, agriculture has proved to have tremendous capacity to respond to market needs, but that response is to economic incentives. In the 1990s, increased concern arose over competition for water supplies, particularly in the West, and, in the longer run, the warming of the earth's atmosphere. Likewise, public support for agricultural research and education, as well as public investments in the infrastructure that serves agriculture, including irrigation systems and roads, has been critically important to its development. As the public's willingness to invest in agriculture declines, agribusiness becomes a more important component of agriculture's resource base.

Expanding production is dependent on the existence of incentives or rewards to producers. In market-oriented economies, these incentives are higher prices and profits, and appreciation of asset values. Farmers respond to profit incentives. Richard Lyng, secretary of agriculture for President Reagan, put it this way: "When farm prices rise to profitable levels, even the roosters begin to lay eggs."[3]

---

[2]See particularly the *World Food and Nutrition Study* (Washington, D.C.: National Academy of Science, 1977); and *The Global 2000 Report to the President* (Washington, D.C.: Council on Environmental Quality and Department of State, 1981).

[3]Richard Lyng, speech presented at the 1980 Agricultural Outlook Conference (Washington, D.C.: USDA, November 19, 1980).

It would be a mistake to leave the impression that the world food problem is simply a matter of supply and demand for food. It is, in fact, much more complex. Many of the complexities result from governments whose decision makers respond to immediate political pressures. Decisions of governments around the world affect American farmers, related businesses, and consumers. For example, assurance of an adequate food supply at reasonable prices is such a high priority that most countries see that there is enough food to feed their people.

The Farm and Resource Problem

The farm problem, like the world food problem, is also multidimensional, making it difficult to treat from a policy perspective. Of particular importance is the increasing interrelationship between farm issues and resource issues. In previous editions of this book, the resource problem has been treated separately from the farm problem. However, soil conservation issues have always been interwoven with farm policy issues. Some conservation programs might more appropriately have been considered to have been production adjustment (control) programs.

In a contemporary context, the resource issues are an important dimension of the farm problem. Conservation compliance as a condition for receiving farm program benefits, the conservation reserve program, wetland reserves, and potential water allocation systems are all excellent examples. Moreover, virtually all resource issues directly impact farmer costs and, therefore, become part of the farm problem. In addition, the resolution of several of the resource issues may have considerable impact on agriculture. Some of the primary dimensions of the farm and resource problem are discussed in the following subsections.

Instability. Agricultural production is inherently unstable, largely because of relatively stable domestic needs interacting with production variability due to weather conditions, pests, and diseases. This instability leads not only to variation in domestic production but also to abrupt changes in export demand as weather, pests, and disease affect output in other countries. Export demand is also affected by changes in the value of the dollar or policies of importing countries. The supply and demand for farm products is sufficiently inelastic so that changes in either can result in proportionately larger changes in prices. Unless buffered by government programs, farmers typically face a high degree of price and income risk.

Government policies themselves can be a major source of instability in agriculture. Sharp changes in policy, such as imposition of an embargo on exports, can have devastating effects on farm prices. Sharp changes in farm program spending make it difficult for farmers to plan their investment and output. This is true not only for the United States but also for agriculture throughout the world.[4]

---

[4]D. Gale Johnson, "The World Food Situation: Recent and Prospective Developments," in *The Role of Markets in the World Food Economy* (Boulder, Colo.: Westview Press, 1983), pp. 1–33.

Excess capacity and fixed assets.   Throughout much of the period from the 1920s through the 1960s, agriculture was characterized by chronic excess capacity. This capacity problem was the result of relatively high price supports and resource fixity. Since 1970, agriculture has entered a period of what appears to be cyclical or periodic excess capacity. Contributing to the appearance of periodic tight supplies was the fact that about 35 million acres of highly erodible cropland were retired from production in the conservation reserve program. One of the major issues under the 1996 farm bill involves the quantity of highly erosive land that should be kept out of production. Increased concern has been expressed about the need to put more land into production to satisfy record levels of export demand.

Assets, such as land and machinery, have traditionally been slow to adjust out of agriculture because many production costs tend to be fixed. Production thus has tended to continue as long as variable costs are recovered and there is some residual contribution to fixed costs. Also, agricultural labor and management skills often are not well adapted to move into other occupations. However, with animal agriculture becoming more vertically coordinated, there may be some tendency toward greater responsiveness to market needs in this sector. In other words, the more business-oriented farmers and segments of agriculture appear to be more responsive to market signals.

Diversity.   Agriculture has become highly diverse and is in transition from a family farm-based structure to a business-based structure. This transition increases the diversity of agriculture. Wide differences in efficiency are evident between smaller farms and larger, often vertically coordinated farms. Differences in efficiency lead to income disparities. The result is that considerable stress is placed on smaller and moderate-sized family farms that try to make a living from farming alone.

Survival in agriculture has become dependent on technical agronomic skills combined with business and financial management skills. A majority of the production is produced on farms where college-level management skills are absolutely essential. An increasing proportion of the technical skills are being supplied by consultants whether they be pest management specialists, nutrition specialists, or financial management specialists.

These changes have converted or are converting the family farms that have dominated agriculture into family business operations or even into nonfamily-based agribusiness operations. Some find it very difficult to adapt to or accept these changes. This issue is part of a larger issue known as the *structure of agriculture*.[5] This issue is concerned with the trends toward large-scale integrated agriculture, the potential development of a land-holding class, the decline of rural communities, and the ability of government programs to effectively adjust and serve the needs of a diverse and changing clientele.

---

[5]Harold F. Breimyer, "Perspectives on a Time to Choose, the USDA Report on Structure of Agriculture," in *Structure of Agriculture* (Washington, D.C.: Subcommittee on Forests, Family Farms, and Energy, Committee on Agriculture, U.S. House of Representatives, February 27, 1981), pp. 123–126.

Resource scarcity.   Competition for limited natural resources is increasing. This competition is most apparent on the East and West Coasts where population pressures claim farmland for residential and industrial uses. In the West and Southwest, this competition extends to the availability of limited water resources where irrigated agriculture utilizes more than 70 percent of the available water. Near every metropolitan area, there is competition for land and, in many cases, water.

Traditionally, most of these competitive relationships have been handled by the market, by common law rights, by local government, or by state government. The trend in the past two decades was to pass regulatory responsibilities from the federal to the state and local levels, but now questions are arising as to whether it will be necessary for the federal government to become more involved in these resource allocation disputes.

Environmental externalities.   The days when agriculture is treated differently than industry from an environmental perspective are rapidly coming to an end. Once thought to be a source of nonpoint pollution, agriculture experienced less stringent environmental regulations than did other industries. Certain confined animal feeding operations (dairy, hogs, poultry, and cattle), however, are now being treated as point pollution, requiring zero discharge. Then again, the requirements for nonpoint pollution are becoming more like those of the zero-discharge point regulations. Closely related, the endangered species regulations appear to be having more influence on agriculture and forestry than on other industries.

The worst may be yet to come. These increased regulatory requirements make farmers feel that they are losing their rights to farm—that the government is "taking" their private property rights from them without compensation. These most interesting conflicts are likely to be played out in the courts during the next several years.

The Consumer Food Problem

Historically, the influence of consumers on food and agriculture policy has ebbed and flowed. From 1900 through 1970, the major emphasis of consumer interests was assurance that food was processed under sanitary conditions and protected against food poisoning or outbreaks of diseases. This changed abruptly in the early 1970s when consumer interests expanded to major food and farm policy issues such as food prices, nutrition, food labeling, pesticides, animal rights, marketing orders, and certain commodity support programs. The primary dimensions of consumer food issues include the following:

Nutrition.   Contemporary consumer involvement in food policy likely had its roots in the concern about poverty in America that emerged in the early 1960s, out of which evolved the food stamp program. It and related family nutrition programs have since expanded to include well over 25 million people at a cost of more than $29 billion. Food programs expanded during the Reagan–Bush years despite efforts to reduce the scope and costs of the programs. Growth then accelerated during the Clinton years. The child nutrition (school lunch) programs feed more than 25 million students at a cost to the federal government of nearly $6 billion. Total food program costs, therefore, have risen to over $37 billion. This is a considerably larger component of USDA's budget than are farm subsidy costs.

With the U.S. population increasing at an annual rate of less than 1 percent and one-third of the farm production exported, the ability of the United States to feed itself is not an issue. There is no reason, with proper governmental assistance and education, that hunger and malnutrition should be a problem. Major policy changes have been implemented in making food labels more useful from a consumer nutrition perspective. Revamping of food assistance programs to make them more responsive to needs and less subject to abuse remains on the agenda. Increased questions arise as to whether nutrition should be moved from the USDA to the Department of Health and Human Resources.

Price.   There is a danger that, in a particular year, foreign demand will be so strong that unrestricted exports could lead to rapid increases in food prices. Such a situation occurred in 1974 when soybean prices became sufficiently high that an export embargo was imposed. The combination of sharply increased world demand for food and general inflationary pressures increased the proportion of income spent on food in the United States from a low 16.2 percent in 1973 to 16.8 percent in 1980.[6] That proportion since has declined to about 14 percent. While the proportion of income spent on food in the United States is lower than in most other countries in the world, real increases in food expenditures relative to income place visible pressure on consumers' budgets. Thus, despite a low share of income spent on food, questions arise as to what can be done to reduce the price of food. These questions could become more complex as the costs of meeting environmental regulations are imposed on agriculture.

Safety.   After many years of agonizing debate, two significant actions were taken in the design of food safety policy during the Clinton administration. For decades the United States operated under an antiquated zero-tolerance requirement for cancer-causing substances. This law was modified to provide reasonable certainty of no harm but broadened coverage to include a range of potential health problems resulting from chemical use in agriculture and food. Of equal significance was the adoption of new meat and poultry inspection procedures including the Hazard Analysis Critical Control Point procedures that are widely used in other food industries.

---

[6]Data from the Bureau of Labor Statistics (Washington, D.C.: U.S. Department of Commerce, 1981) and from Julie Kurland, "Food Spending and Income," *Natl. Food Rev.* (Washington, D.C.: ERS, USDA, Winter 1986), p. 30.

The Rural Development Problem

Rural development has returned to the agriculture and food policy agenda with unknown strength and vigor. This topic has not received a priority position on the agenda since the New Deal infrastructure policies extended telephone and electrical service to every farm and rural resident in the 1930s. Rural decline is most apparent in the South and on the Great Plains in areas isolated from the benefits of metropolitan growth. The linkage to the farm problem lies in the decline in off-farm employment opportunities, the deterioration in the quality of education and health services, and the impacts on the quality of services available to farmers from the agribusiness sector. The following list describes the major dimensions of the rural development problem:

- **Infrastructure.** Deterioration has occurred in the rural infrastructure of available services. This infrastructure includes deterioration in the quality of roads, bridges, and telephone service in truly remote areas.
- **Health and education.** Deterioration in the quality of rural health and education has become a major concern as the per capita cost of providing these services has increased.
- **Selective outmigration.** The lack of jobs in rural communities has encouraged migration of youth from rural areas to the city over several generations. The result has been a persistent decline in rural human resource capacity and leadership. The result is a deterioration in the ability of rural governmental systems to solve problems.
- **Poverty.** Some of the highest levels of poverty are found in rural America. Delivery systems for federal social programs are geared to urban constituencies.
- **Farm labor.** The treatment of migratory farm labor has been the subject of much federal and state policy. Issues that have tended to drive a wedge between farmers and farm labor relate to wage rates, the ability of farm laborers to organize, and the conditions under which they can strike.

While farmers live in an environment of rural, economic, and social decline, farm organizations, typically, have shown little interest in policies that foster rural development. As rural representation declines in the Congress, political collaboration between farm and rural interests makes more sense. Yet few apparently view such a political alliance as having potential—even though 25 percent of the population is rural and only 2 percent are farmers. The only item consistently on the agenda of President Clinton's two secretaries of agriculture has been rural development.

PHILOSOPHIES OF PROBLEM SOLVING

As noted in Chapter 1, one's attitude toward policy is influenced by facts, values, beliefs, and goals. That is, facts, values, beliefs, and goals influence one's conclusions concerning the perception of the problem, the economic relationships that relate to solving problems, the appropriate role of government in agriculture, and the effectiveness of government in solving problems. Specific combinations of facts, values, beliefs, and goals, therefore, lead to identifiable philosophies toward government's role in agriculture. Five such philosophical approaches discussed here include *free market, humanitarian, agricultural fundamentalist, stabilizer,* and *regulator.* These approaches are discussed in their pure form. Individuals may prefer combinations or variations from the pure form because of differences in the values or beliefs they hold. No attempt is made to evaluate the merits of any of these philosophical approaches.

1 **The Free Market** — *Republicans, Farm Bureau*

In the **free-market approach,** *the forces of supply and demand determine product prices and also allocate and ration available supplies.* This approach normally places a high value on the role of profits, private enterprise, initiative, and hard work. Little confidence is placed in the ability of government to solve or even ameliorate problems.

To the free-market advocate, agriculture would be better off if government programs—particularly price and income programs—were eliminated. At most, the role of government should be limited to research, education, provision of production and market information, and actions to reduce barriers to free trade in foreign markets. Even here, however, an increased role for private sector research, information, and analysis is seen.

Supporters of this view were once mostly Republicans, agribusinesspeople, and members of the American Farm Bureau. More recently, the base of support appears to have expanded. This has led to a greater bipartisan consensus on farm issues such as the General Agreement on Tariffs and Trade (GATT) and the North American Free Trade Agreement (NAFTA). Bipartisan support for freer markets also means less shift in domestic farm policy between Republican and Democratic administrations.

2. **The Humanitarian** — *Production Stimulator*

The **humanitarian** believes that *the major agricultural problem is that of adequately feeding an ever-expanding world population and solving issues of malnutrition at home and abroad.* Government's role in this context is to provide the basis for increased production and food assistance.

Government policies consistent with the philosophy of a humanitarian include increased foreign food aid and development assistance, expanded food programs designed to reach the needy, producer income supplements to provide continuous production incentives in the event of low farm prices, and government-held stocks of grain to guard against a production shortfall. The agenda of the humanitarian also includes that of environmentalists who seek greater food safety, improved water quality, and the protection of endangered species. Potential conflicts between the production and environmental goals of humanitarians go largely unrecognized.

The humanitarians have a basic belief in the right to food. The **right-to-food resolution,** adopted by the 1974 World Food Conference, *holds that every person in the world has a right to an adequate diet.* Associated with this right are government programs designed to encourage production and get food to those who are in the greatest need.

Religious groups such as the Interreligious Task Force on U.S. Food Policy can be directly associated with the humanitarian philosophy. Reports of the National Academy of Sciences, environmentalists, and greenhouse effect advocates also tend to support this point of view.[7]

The Agricultural Fundamentalist

The **agricultural fundamentalist** believes that the *root of all wealth lies in agriculture and the soil, that agriculture has "moral properties—the capacity to engender in human beings an elevated behavior,"*[8] and that it creates values in human beings that are generally recognized as being good. The USDA was created with a fundamentalist philosophy. Its seal is inscribed: "Agriculture is the foundation of manufacture and commerce."

The policy prescription of the fundamentalist is to maintain agriculture's economic health and thereby maintain the economy's health. High farm prices give farmers more money to spend in rural communities, which in turn works its way through the whole economy by stimulating employment and investment in new plants and equipment. The fundamentalist philosophy also holds that high farm prices are necessary to preserve the values that are instilled in farm people. The basic policy prescription of the fundamentalist is government establishment of price floors for agricultural commodities at cost of production plus a reasonable profit.

Although there are those who would like to believe that farmers are the backbone of the economy, the fundamentalist movement appears to be slowly dying, but it is not dead. Fundamentalism is part of the history of farm organizations. It last appeared as the American Agriculture Movement onslaught of Washington, D.C., in 1978–1979. The remnant is a tractor in the Smithsonian that was driven to the capitol by farmers as a protest against low farm prices and a call for parity.

The fundamentalists who exist today may find friends and allies among the sustainable agriculture advocates, the humanitarians, and the environmentalists—all of whom seem to believe that large farms are bad and that smaller family farms are good. Yet there is probably still a bit of fundamentalism in every

---

[7]World Food and Nutrition Study.
[8]Harold F. Breimyer, *Farm Policy: 13 Essays* (Ames, Iowa: Iowa State University Press, 1977), p. 6.

farmer and farm organization. It is not unusual for the calm of a conservative meeting to be broken by fundamentalists' pleas for government assistance that is said to generate benefits that extend throughout the economy.

The Stabilizer

The **stabilization** philosophy holds that *the major problem in agriculture is instability*. Instability, it is suggested, undermines the family farm structure, results in errors in production and marketing decisions, and fosters inflation. Government policy, to the stabilizer, should ensure that farm prices move over a relatively narrow range and that supplies are always available. This can be accomplished through government control of a portion of the grain stocks and through the establishment of floor prices.

Stabilizers have friends in humanitarians and sustainable agriculture advocates. Both seem to believe in stocks and smaller farms. Agricultural fundamentalists could be enticed to be stabilizers with high price supports.

Other segments of agriculture and agribusiness appear to have moved beyond the stabilizer philosophy to the free-market stance by recognizing that tools such as futures, options, and contracts can be used to reduce the risk of price instability.

The Regulator

The **regulator** believes that *market incentives cannot be relied on to make decisions on the appropriate combination of inputs to be used in production, the allocation of land and water resources, or the processing and marketing practices of food processors and retailers.* They view regulation as being essential to agriculture operating in the public interest. There are economic reasons supporting this view, such as the existence of profit motives, externalities, and public goods. On the other hand, there are those who believe that many of these regulations unduly infringe on property rights. The main advocates of the regulatory approach are environmentalists and food safety advocates.

THE SINGLE-PROBLEM TRAP

It is all too easy to fall into the trap of treating each of the four agriculture and food problem areas separately. Policymakers do this regularly by responding only to the issues of the moment. Reality suggests that the four areas are not mutually exclusive but, in fact, are interrelated. Increasing world food demand affects the prices received by farmers, the cost of food, its availability to consumers, the availability of resources for future food production, and the economic health of rural America. These interrelationships must always be kept in mind. In other words, students of policy must be continuously looking at causes and effects that extend beyond the immediate problem or policy alternative. That is the approach taken in this book.

ADDITIONAL READING

1. In a contemporary context, one of the best ways to follow farm policy issues as they evolve is by reading the magazine *Choices*, published by the American Agricultural Economics Association. This quarterly publication is designed to communicate to policymakers and other economists views on the economic dimensions of evolving agricultural issues, the policy options, and their consequences.

3 THE POLICY PROCESS

Congress is so strange. A man gets up to speak and says nothing. Nobody listens—and then everybody disagrees.

—Boris Marskaloa

Implementation of a cohesive agricultural and food policy requires the establishment of a broad base of support in the Congress as well as in the executive branch of government. Establishing such support depends on more than just numbers. Strength is in organization, coupled with knowledge of how government works, how decisions are made, and how to mobilize broad support for specific causes. The enactment of a new government program into law is by no means the end. Programs may never be implemented because of a lack of appropriations. Even if funds are provided, decisions must be made on the rules and regulations for implementation. Specific administrative procedures exist for the issuance of such rules and regulations, including the opportunity for public input.

The essence of the policy process is politics. Politics has been defined as the art of the possible, the art of compromise, and the art of determining who gets what. The policy formulation process is often a key determinant of the content of policy. Thus, it is important to have an understanding of the nature of the policy process and of the central actors who attempt to influence the direction and substance of policies for food and agriculture.

Policymakers are the actors who influence and determine policymaking. These actors include more than the key congresspeople, senators, their staff, the president, the secretary of agriculture, and other U.S. Department of Argriculture (USDA) political appointees. They may include individuals in other agencies of government, such as the Office of Management and Budget, Department of State, Food and Drug Administration, and Environmental Protection Agency (EPA). They include the organizations that influence all of the key people and agencies. These may be hired lobbyists, farm organization directors, or just well-wired farmers. The actors also include comparable leaders of environmental, food safety, agribusiness, trade, and food assistance interests.

When these interests reach an irreconcilable conflict among themselves or with a government agency, the actors include the lawyers and judges in the legal system. In other words, there are many policymakers, each with various types and levels of interest and influence.

In this chapter, we describe the structure of the federal policy process in terms of how the various branches and agencies of government interact to develop and implement food and agricultural policies and programs. An attempt is also made to provide some insight into the dynamics of the policy process. In the next chapter, we examine the role of special-interest groups in influencing food and agricultural policy.

The American democratic form of government is intended to reflect the will of the people—government of the people, by the people, and for the people. In a representative government, those charged with making laws, determining national policies, and being responsible for the execution of laws and the implementation of policy are selected primarily through popular elections. Although many of the people involved in the policy process are appointed, elected persons usually appoint the top decision makers and, ultimately, are responsible for decisions regardless of who makes them.

The framers of the Constitution embodied in the American system of government the principle of "separation of powers" by establishing three equal branches of government: legislative, executive, and judicial. These three branches provide a system of checks and balances whereby no branch becomes clearly dominant. Through the interaction of these branches of government, policies and programs are proposed, developed, adopted, interpreted, and carried out (Figure 3.1).

THE LEGISLATIVE BRANCH

The very first provision of the Constitution (Article I, Section 1) creates the legislative branch by providing that "all legislative powers herein created shall be vested in a Congress of the United States, which shall consist of a Senate and a House of Representatives." **In this sense, the Congress truly is the policymaking body.**

The Senate is composed of 100 members: 2 from each state, irrespective of population or area. The six-year term of office is arranged so that it does not terminate for both senators from a particular state at the same time.

The House of Representatives is composed of 435 members elected every two years from among the 50 states, apportioned according to their total population.[1] The membership of the House is reapportioned among and within the states every 10 years to reflect changes in population. This has proved important

---

[1]In addition to the representatives from the 50 states, there is a resident commissioner from the Commonwealth of Puerto Rico and one delegate each from American Samoa, the District of Columbia, Guam, and the Virgin Islands. They have most of the prerogatives of representatives, with the important exception of the right to vote on matters before the House.

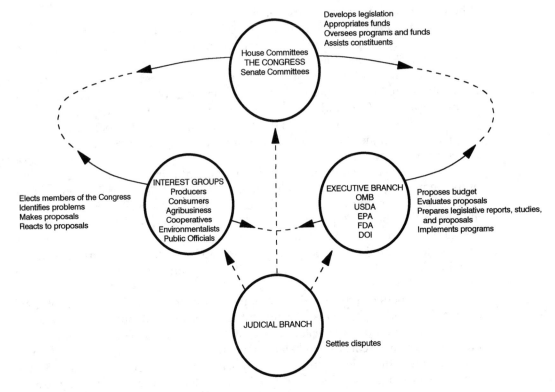

FIGURE 3.1   Model of the agricultural food policymaking process.

to agricultural policy because population has shifted over time from rural to urban areas. As a result, farm and rural groups have experienced a decline in their representation and have become minority groups. The number of rural districts and farm-oriented representatives has gradually declined.

The dramatic impacts of the 1990 census on reapportionment is indicated in Figure 3.2. Not only has rural representation declined, but shifts have also occurred geographically toward the South. According to the census, none of the traditional farm states gained representation. Note that rural states such as South Dakota, North Dakota, and Vermont have only one congressperson. Montana declined to only one. Iowa, Kansas, Kentucky, and Louisiana each lost one congressperson. These relatively small changes mask the reality that, in every state, there are fewer truly rural congresspersons. Thus, while Texas gained 3 congresspersons (from 27 to 30), most of the representation flowed in the direction of the metropolitan centers of Dallas, Houston, Austin, and San Antonio. The same phenomenon occurred in every state.

The impacts of reapportionment on representation can be evaluated from a longer run perspective in Table 3.1. After the dust settled on the 1992 election, rural representation declined to an estimated 71 congresspersons—down 54 percent

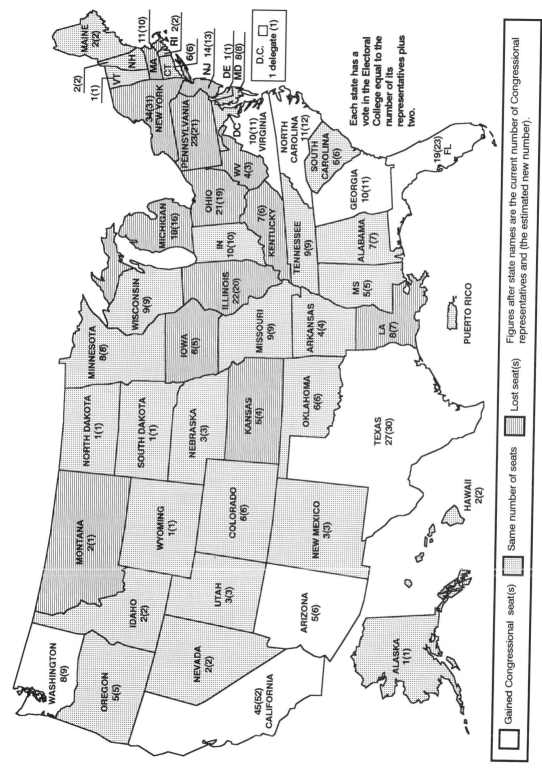

FIGURE 3.2 Population shifts and their political effect.

45

TABLE 3.1   Makeup of U.S. House of Representatives and Districts by Residence of Population, 1968, 1973, 1985, 1992

	1968	*1973*	*1985*	*1992*	Percent Change *1968–1992*
Urban and suburban	280	305	347	364	+30
Rural	155	130	88	71	−54
Total	435	435	435	435	

since 1968. Little change has occurred in rural representation since 1992 because reapportionment will not occur again until 2002.

The important message from this decline is that it takes 218 votes to pass a farm bill in the House. Rural districts, agriculture's powerbase, account for only 71 votes. Where do the rest come from? That issue faces every farm constituency and is the subject of further discussion later in this chapter, as well as the next.

As significant as the decline in rural representation is the unprecedented degree of turnover that has been occurring in the Congress. Since the 1992 elections, a trend can be noted of increased dissatisfaction by members with the institution of Congress (members not seeking reelection) and increased accountability at the polls (incumbents being defeated). In the 1992 elections, 110 house seats and 11 senate seats changed hands. In the 1994 elections, 87 house seats and 9 senate seats changed hands. In the 1996 elections, 74 house seats and 15 senate seats changed hands. Consequently, in 1997, 16 of the 50 members of the House Agriculture Committee were freshmen and 38 had begun their service since 1992. Only 5 members were present when the previous farm bill was debated and passed in 1989. Only two members of the minority and not a single member of the majority were present when President Carter imposed the most recent embargo on agricultural exports in 1980.

Agriculture policymaking has historically been bipartisan. In the 1980s the government was split—Republican administration and Democrat Congress. The 1981, 1985 and 1990 farm bills, therefore, had to be developed through partisan cooperation. In 1994 the Republicans gained control of Congress for the first time in 40 years but a Democrat was in the White House. This divided government was maintained in the 1996 elections. The 1996 farm bill was, therefore, proposed, debated, and passed under distinctly different circumstances than previous farm bills, especially with more partisan friction. While the agricultural committee was controlled by rural Republicans, the House leadership was less rural.

The Constitution stipulates that the Congress must assemble at least once every year, beginning at noon on the third day of January. A Congress lasts for two years, commencing in January of the year following the biennial election of members.

The chief function of the Congress is to make laws that establish U.S. policy. Definite lines of jurisdiction exist between the House and Senate on certain matters. For example, only the Senate may "advise and consent" on treaties and certain nominations by the president. The House initiates revenue bills (taxes, etc.) and usually takes the leadership on appropriation of funds to carry out the functions of government.

Leadership Control[2]

The leadership of the Congress is vested in the party steering committees in both the House and Senate. These committees, made up of ranking leaders of each party, formulate party rules, develop strategies, and make committee assignments. The majority party's committee also nominates the chair of the committees and proposes rules to govern committee and floor action. All members of each house belong to their respective party caucuses, which approve major rules proposed by the steering committees. Since the majority party controls each chamber, its rules usually govern committee and floor action. The composition of each committee reflects the proportionate party membership of each chamber. Thus, a majority of each committee is from the majority party. This is the reason that a united majority party can have a high degree of control over actions of the Congress. It is also why a president whose party also controls the Congress tends to be particularly powerful.

Committee Power Center

Committees are centers of power for determining the fate of particular pieces of legislation, and the person chairing usually is the focal point for power within the committee. The Senate has 16 standing committees, and the House has 19. Each committee has jurisdiction over legislation in specific areas, and all measures affecting a particular area of law are referred to that committee. For example, proposed sugar legislation containing provisions for a tariff on imported sugar would be referred to the House Ways and Means Committee, which has jurisdiction over all revenue bills. Sugar legislation without such a provision, focusing only on a domestic price support program for growers, would be referred to the agriculture committee.[3] A proposal containing both provisions would be referred to both committees.

Senators and representatives usually seek membership on a committee that has jurisdiction in a field most relevant to their constituents or in which they are most qualified and interested. The committees still operate largely by the seniority system, although this was substantially modified by the reform-minded 93rd Congress of 1974. Members rank in seniority according to the order of the year

---

[2]This section draws heavily on B. L. Flinchbaugh and Mark A. Edelman, "The Changing Politics of the Farm and Food System," in *The Farm and Food System in Transition* (East Lansing, Mich.: Michigan State University, 1984).

[3]The term *agriculture committee* refers to both the Senate Committee on Agriculture, Nutrition, and Forestry and the House Committee on Agriculture. Both committees have essentially the same jurisdiction.

they were appointed to the committee. The ranking member of the committee from the majority party is usually elected chair.

Most of the committees have two or more subcommittees that have initial jurisdiction in the consideration of bills in particular subject areas. Subcommittee chairs are also members of the majority party. It should be apparent that the power of the majority party to elect the committee and subcommittee chairs gives them tremendous power since all bills originate in the committees. Each committee is provided with a professional staff to assist in the consideration of bills. The staff are appointed by the chair on a permanent basis without regard, in theory, to political affiliations and then must be agreed to by majority vote of the committee.

The Policymaking Process

Generally, the broad national policies of this country are embodied in statutes enacted by the Congress. The policymaking process, thus, is a major component of the formal legislative process. This is especially true for food and agricultural policy for which the major legislation is the omnibus "farm bill," which expires and is reconsidered by the Congress every four or five years. The 1996 farm bill is a seven-year bill expiring in 2002. The term *farm bill,* however, is really a misnomer. In recent times, farm bills have become truly omnibus, *encompassing not only the traditional farm programs but also the domestic food assistance, foreign aid* (P.L. 480), *rural development, environment, research, extension, and various other program authorizations.* The rationale for such an omnibus bill is to *broaden the base of support for farm legislation*—to get the 218 required votes in the House and 51 in the Senate.

Sources of Legislation

The sources of ideas for legislation are unlimited, and draft bills originate in many diverse quarters. The first source is from members of Congress who may have campaigned for election on the promise of introducing specific legislation. Constituents—either as individuals or groups such as bar associations, labor unions, farm organizations, and the like—may suggest proposals and present draft bills to members.

Proposed legislation may also emanate from the executive branch through *executive communication.* In recent years, this has become a prolific source of proposals, usually in the form of a letter from the president, a member of the president's cabinet, or the head of an independent agency transmitting a draft of a proposed bill to Congress. Such bills may be introduced by a member of the president's party.

However, getting unpopular bills introduced is not always easy. For example, President Reagan's proposed 1985 farm bill to withdraw government support for agriculture was sufficiently unpopular that no member of the Senate wanted to be associated with it. Finally, Senator Helms, chairman of the Senate's agriculture committee, introduced it for discussion and debate, but without his support. Of course, the power of the executive to influence legislation is stronger when the same party controls the White House and the Congress, as happened when President Clinton was elected or previously when President Johnson implemented the Great Society programs. With a Republican Congress in 1995 the Clinton administration did not even attempt to draft a farm bill.

Consideration by Committee

Members of the committee considering a bill have considerably more influence on its content and fate than do noncommittee members. Members participate in hearings, engage in discussions, vote on detailed provisions and specific language, and approve the final form of the bill as it leaves committee.

The functions of the congressional committees are distinctly separated as to authorizations, budget, and appropriations. *The subject matter committees,* such as agriculture, *are* **authorizing committees.** Farm bills originate with the agriculture committees. Farm bills authorize expenditures of funds for specific purposes. The **budget committees** *set broad limits on how much can be spent. Authorizations must be within these limits. The actual appropriation of funds for carrying out the legislation is the function of the* **appropriations committees** in the House and Senate. Since the appropriations committees must approve funds for agriculture and related programs, most of which are administered by the USDA, they are very influential in determining how and when programs will be implemented.

The biggest hurdle for any bill is getting out of committee with a favorable vote. Immediately after its introduction, a bill is referred to a committee. The chair may then refer the bill to a particular subcommittee for consideration. At the same time, copies of the bill are sent to the executive branch cabinet departments or agencies concerned with the subject matter and frequently to the General Accounting Office with a request for an official report of views on the necessity or desirability of enacting the bill into law.[4]

The executive branch's report is referred to as a **legislative report,** *which states whether the bill is favored or opposed by the administration.* The legislative report generally contains specific recommendations for change. Before being returned from the executive branch to the committee, legislative reports are submitted to the Office of Management and Budget in the Executive Office of the President to determine whether the bill is consistent with the president's programs. The committee, of course, is not bound by the recommendation of the executive branch but desires that additional information for its consideration.

---

[4]The General Accounting Office (GAO) was established in 1921 as part of the legislative branch. It is an independent, nonpolitical agency operating under the direction of the comptroller general. GAO's main function is to conduct investigations for the Congress. Such investigations are not limited to audits but may include research to improve government effectiveness, investigations to determine if federal money is being appropriately spent, legality investigations, and assistance in the settlement of financial disputes between the government and the public.

Hearings, by a committee or subcommittee, are held on bills considered to be of significant importance by the leadership. Hearings provide an opportunity for both proponents and opponents to express their views and have them become a part of the bill's record. In the case of food and agricultural legislation, the secretary of agriculture or a designated representative usually appears at such hearings to provide the administration's views on the pending proposal. While the hearing is theoretically a process open to anyone who desires to appear, the chairperson of the committee or subcommittee exercises considerable control over who appears. In other words, the chair can control the agenda within the committee. As a result, some committees will be more open than others.

Upon completion of the hearing, a subcommittee drafts a bill or "marks up" a proposal before it and advances the draft to the committee. Then the committee considers the bill in executive or **markup session,** *where its final form is determined.* All proposals and amendments are considered and decided by members of the committee. High-level politically appointed representatives of USDA attend markup sessions to explain the administration's position on the bill's provisions and to provide technical assistance as requested by the committee. Generally speaking, both the members of the committee and the administration are interested in developing bills that the president will sign rather than veto. Interest-group lobbyists communicate informally with the members and their personal staffs throughout the markup process to influence specific provisions. Decisions are tentative until the entire committee renders final judgment.

A final vote is taken to determine the committee's action, which usually is to report the bill favorably or to table it. Since tabling a bill is normally tantamount to preventing further consideration, adverse reports by the committee usually are not made. On rare occasions, a committee may report a bill without recommendation, which means, typically, that the bill has little chance of passage.

Agriculture committees.   The agriculture committees of the two houses have long been a focal point for food and agricultural policy. The jurisdiction of the agriculture committees is broad, encompassing policies and programs for production agriculture, pesticides, marketing, research, rural development, forestry, domestic food assistance, and some aspects of foreign trade. Generally, the agriculture committees do not have jurisdiction on matters relating to international relations, endangered species, wetlands, clean water, clean air, and tax policy. But in areas such as pesticides, wetlands, and environment, there would be claims of nonexclusive jurisdiction by any single committee.

The character and composition of the agriculture committees have changed markedly over time. The chairperson and a few senior members once wielded almost total control over the committees. They could bargain with the executive branch and interest groups from a position of strength and "deliver the votes" on deals that were made. There was relatively little turnover in agriculture committee membership until after the 1970s. The leadership remained stable and long term until the elections of 1994 and 1996. Congressman Kika de la Garza held his position from 1981 to 1994. Representative Bob Poage (Dem.–Tex.) chaired the

House committee for 8 years and Representative Harold Cooley (Dem.–N.C.) for 16 years before him. Senator Patrick Leahy (Dem.–Vt.) held his chair position from 1987 to 1994. Herman Talmadge (Dem.–Ga.) chaired the Senate committee for 10 years and his predecessor, Allen Ellender (Dem.–La.), for 18 years.

For the first time in 40 years, the 1994 election turned both houses of Congress over to the Republicans. Pat Roberts (Rep.–Kans.) became chair of the House Committee on Agriculture. Ironically the last Republican to chair the committee was from the same Congressional district; Clifford Hope of Kansas. Senator Lugar (Rep.–Ind.) became chair of the Senate committee. The 1996 election maintained Republican control in both houses. Chairman Roberts was elected to the Senate and sought a position on the Senate Appropriations Committee and the chairmanship of the Agriculture Appropriations Subcommittee. Newly elected Congressman Bob Smith (Rep.–Oreg.) was promised the House chairmanship even though he "retired" in 1994 and sought reelection in 1996. The leadership reinstated his seniority and passed over Congressman Larry Combest (Rep.–Tex.) who voted against Freedom to Farm[5] which had leadership support. Senator Lugar remained chair in the Senate.

The control of the agriculture committees has generally rested in the South. This resulted, in large part, from the combination of the one-party Democratic South and the seniority system. This southern control began to change in the late 1960s and early 1970s as several southern members retired, the seniority system became less entrenched, and reelection became less secure. The emergence of the two-party system in the South also reduced its control of the agriculture committees.

When the Senate reorganized its committee structure in 1977, the Select Committee on Nutrition and Human Needs was terminated and the Agriculture Committee was expanded to include its functions. The name was changed to the Committee on Agriculture, Nutrition, and Forestry to reflect the broader responsibility. The Senate committee has 18 members. When the Republicans gained control of the Senate in 1995, the committee and subcommittees came under Republican leadership. The four Senate agriculture subcommittees tend to be organized along functional lines:

- Forestry, Conservation, and Rural Revitalization
- Marketing, Inspection, and Product Promotion
- Production and Price Competitiveness
- Research, Nutrition, and General Legislation

Of these, the Production and Price Competitiveness Subcommittee has the most impact on major farm price and income programs. However, subcommittees are not as important in the Senate agriculture committee because more of the committee's

---

[5]Freedom to Farm is the popular title of the legislation separating government payments from the production of specific program crops, which became the 1996 farm bill, the Federal Agriculture Improvement and Reform Act of 1996 (FAIR).

business tends to be conducted in the full committee. This is possible with fewer committee members.

A larger and more diverse membership (50 members) makes the House Committee on Agriculture more unwieldy than the Senate committee. The membership normally is divided by party, based on the two-to-one-plus-one rule. The 103rd Congress reduced the number of subcommittees from eight to six, giving it more of a cross-commodity focus like the Senate. With Republican control in 1995, the subcommittees were further reduced to five:

- General Farm Commodities
- Livestock, Dairy, and Poultry
- Risk Management and Specialty Crops
- Department Operations, Nutrition and Foreign Agriculture
- Resource Conservation, Research and Forestry

Some have interpreted this subcommittee reorganization as a loss of political influence for commodity groups. Because of the larger number of committee members, the focal point for action in the House agriculture committee tends to be the subcommittees.

For years, the so-called agricultural establishment controlled the agricultural policy agenda in the Congress. The **agricultural establishment** *is composed of the agriculture committees of Congress, farm organizations, and the USDA.* Their control meant that not only could they determine what issues were brought up for action, but they could also prevent consideration of matters that could be harmful to agriculture's interests.[6] Farm subsidies escaped the "welfare" connotation. Payments to farmers controlling large acreages were not raised as a real issue. Much of this control was exercised within the agriculture committees. Issues were brought up that the agriculture committees wanted brought up. The unfavorable issues that surfaced from time to time were tabled. They seldom got beyond the subcommittee and never reached the floor of the House and the Senate.

The precise turning point for the establishment's failure to control the farm and food policy agenda is difficult to pinpoint (see box). It can probably be traced to the constitutional requirement that House seats be apportioned on the basis of shifts in population from rural to urban areas. The establishment's loss of control became particularly apparent in the 93rd Congress in 1974 when a rapid turnover of membership began to occur. With a large number of new members, the seniority system came under attack. Bob Poage, after eight years, was ousted as chairman of the House Committee on Agriculture. In subsequent years, a balancing of agricultural, environmental, and consumer interests became increasingly necessary.

---

[6]Don Paarlberg, *Farm and Food Policy: Issues of the 1980s* (Lincoln, Neb.: University of Nebraska Press, 1980), pp. 10–12.

Who's Got the Ball?

The biggest issue in agricultural policy is: Who is going to control the farm policy agenda and what subjects will be on it?

I like to watch football on television. The first question I ask myself when I switch on the set is, who's got the ball?

The agricultural establishment had the ball for a hundred years, but sometime during the last ten years there was a turnover. It was not rapid, or clean-cut, or dramatic, as in a football game. In fact, it has been so gradual that we have not fully realized it. But the initiative has changed hands nonetheless.

Source: Don Paarlberg, "The Farm Policy Agenda," in *Increasing Understanding of Public Problems and Policies* (Chicago: Farm Foundation, 1975), pp. 95–96.

Another change of particular significance occurs whenever the party that is in the majority changes in either house of Congress. All committee and subcommittee chairpersons change. Thus, in 1980, the chair of the Senate Committee on Agriculture, Nutrition, and Forestry shifted from Senator Talmadge (Dem.–Ga.) to Senator Jesse Helms (Rep.–N.C.). Then, in 1987, Senator Leahy assumed the chair when the Democrats regained control of the Senate. In 1995 a complete shift occurred when the Republicans gained control of both houses. Senator Lugar (Rep.–Ind.) took over from Senator Leahy. Congressman Roberts (Rep.–Kans.) took over from Congressman de la Garza (Dem.–Tex.). In 1997 Congressman Smith (Rep.–Oreg.) became chairman in the House when Congressman Roberts was elected to the Senate.

Foreign relations and foreign affairs.   As agricultural trade has increased, trade negotiations have experienced breakthroughs such as the North American Free Trade Agreement (NAFTA). Agriculture's role in various facets of foreign policy, such as food diplomacy, trade, food aid, and development assistance, has resulted in the House International Relations and Senate Foreign Relations Committee members claiming jurisdiction in foreign agricultural issues. One example is NAFTA, which flowed through the Senate Foreign Relations Committee and the House International Relations Committees in the approval process. Another example is the foreign food aid (P.L. 480) legislation. The House Agriculture and International Relations Committees share jurisdiction on this legislation. This reflects the increasing tendency of committees to share jurisdiction, although one committee is given primary jurisdiction. As exports increased and trade agreements became more common, the Foreign Relations and International Relations Committees assumed greater importance relative to the agriculture committees.

Environment.   Environmental issues have been a major congressional concern since the 1970s and the creation of the EPA. More than one committee has jurisdiction in this arena. On the Senate side, the Committee on the Environment and Public Works has primary jurisdiction on pollution issues while the Committee on Labor and Human Resources has jurisdiction on pesticide issues. In the House, the Commerce and Resources Committee have primary jurisdiction on environmental issues. The agriculture committees have jurisdiction over the Federal Insecticide, Fungicide, and Rodenticide Act (FIFRA). Resources control endangered species legislation on the House side.

Food safety.   Food safety programs largely reside within the jurisdiction of the Science Committee on the House side and the Committee on Labor and Human Resources on the Senate side. This includes oversight of the powerful Food and Drug Administration (FDA), which administers the interpretation of pesticide and food additive tolerances. Interestingly, food assistance programs such as food stamps and meat inspection are controlled by the agriculture committees, although Clinton's reinventing government proposal would have shifted meat inspection jurisdiction to the Energy and Labor committees.

Other committees.   Several other committees influence agriculture, food, and rural policy. For example, the agriculture committees have limited jurisdiction on issues such as rural health and rural education. These issues fall within the jurisdiction of the Economic and Educational Opportunities Committee on the House side and the Committee on Labor and Human Resources on the Senate side.

The Small Business Committee of the House has a large constituency in rural areas and, therefore, has shown great interest in problems of agriculture. The House Judiciary Committee has jurisdiction over antitrust matters relating to agricultural cooperatives. The House Ways and Means Committee and Senate Finance Committee have a major impact on how farmers and ranchers are taxed. Trade matters involving tariffs come under their jurisdiction as well. Grazing fees on federal lands are normally decided by the Ways and Means and Finance Committees, but the appropriation committees have also taken increased jurisdiction.

Reported Bills

If a committee votes to report a bill favorably to the parent body, it is accompanied by a committee report, which usually describes the purpose and scope of the bill and the reasons for its recommended approval. Committee reports, perhaps, are the single most valuable element of the legislative history of a law. The reports are used by the courts, executive departments, and agencies, as well as the public, as a source of information on the purpose, meaning, and intended interpretation of the law.

Floor Action by the House and Senate

Democratic tradition demands that bills be given consideration by the entire membership of both the House and Senate with ample opportunity for debate and amendment. With heightened interest in food and agricultural policy issues, amendments are more frequent and the debate more intense than when the agricultural establishment was in control. For example, the honey, sugar, and dairy provisions of the 1990 farm bill were strongly challenged by Congressmen Richard Armey (Rep.–Tex.) and Charles Schumer (Dem.–N.Y.) on the grounds that they were not needed, were a waste of taxpayer money, and the public interest could better be served by the free market.

The 1994 budget reconciliation process resulted in floor action that terminated funding for the wool, mohair, and honey programs. Coalition leaders work hard to forge compromises and gain the necessary votes to pass, defeat, or amend a bill to gain more acceptable form. Administration appointees work behind the scenes trying to secure preferred provisions and support for their position. Interest groups lobby to gain support for their preferences. After all the debate has taken place and amendments are voted on, a final vote is taken.

Conference Committee

Typically, the House and Senate work on a bill at the same time. Sometimes one house may wait until the other completes action to minimize differences, but more often than not, some conflicting terms or provisions need to be resolved. Sometimes these differences may be substantial. A conference committee is used to resolve these differences. The conference committee proceedings, although among the most influential, are the least understood. Their importance in the legislative process has led to their being called the "third house."

The conference committee normally includes an equal number of members from both houses, with party representation in proportion to membership in both houses. The conferees include the committee chair, relevant subcommittee chairs, and senior members of the committee most interested in the legislation. The actions the conferees may take are governed by a complex set of rules, which, in practice, are frequently circumvented or waived. The general rules constrain the conferees to treat only areas of disagreement between versions of the two bills, with the result being within the bounds defined by the bills.

However, when the Senate and House bills are widely different, what constitutes middle ground is the subject of considerable discretion. Such differences have been particularly prevalent in areas affecting the environment where the House and Senate appear to have had different priorities. Committee staff play a very important role in writing the final language of a bill. Since that language is put together quickly, it is not unusual for the Congress to enact "technical corrections" after final passage. In this case, the conference committee's job becomes one of forging a compromise bill that will be acceptable to the House, the Senate, and the president.

Conference decisions normally cannot be amended when returned to the parent bodies. If both bodies accept the report of the conferees, the final passage of the bill is complete. The bill then becomes an *enrolled bill* and is sent to the White House for presidential action.

Presidential Signature or Veto

Once an enrolled bill is delivered to the White House, the president has 10 days (excepting Sundays) to take action on it. The enrolled bill is handled by the Office of Management and Budget (OMB) in the Executive Office of the President. OMB circulates the bill to all concerned agencies, asking for their views and recommendations to the president as to whether the bill should be signed or vetoed. These statements are then summarized and included in an *enrolled bill memorandum,* which is sent to the president for final decision.

The following options are available to the chief executive [7]:

- Accept the legislation by signing the bill.
- Do nothing when the Congress is in session. The bill becomes law after the tenth day if no action is taken. This is sometimes done to indicate a negative view of the bill when it is politically untenable to embrace it, and it is sometimes done when serious constitutional questions may surround the legality of the bill.
- Do nothing when the Congress is not in session. This "pocket veto" can occur when the president fails to act on the bill after adjournment at the end of the second session.
- Veto the bill, returning it to the Congress with a "veto message" citing reasons for the veto. A vetoed bill can still become law if two-thirds of the members of both houses vote to override the veto.

The Budget Process

Budget concerns now play an increasingly important role in farm policy development and implementation. Cost considerations are important in framing provisions of a farm bill, and, once enacted, the funds must be appropriated to operate the programs.

The basic provisions of a farm bill, such as price supports and direct farmer payments, are known as *entitlements.* An entitlement means that any farmer who

---

[7]Beginning in 1997, the president has been granted permanent legislative line-item veto authority on appropriations bills and targeted tax provisions (providing special treatment to a limited class of taxpayers). After presidential signature, the president has five days to send a message to Congress rescinding specific items. Congress can pass a "disapproved bill" to overturn a presidential recision. If the president vetoes the "disapproved bill," two-thirds votes are required in both houses to override the veto. The line-item veto legislation was part of the Republicans' Contract with America. It was passed in an effort to control deficit spending.

qualifies for the program is entitled to the authorized benefits. About the only time that Congress has control over entitlement expenditures is when the legislation is enacted.[8] Many other farm bill provisions, such as soil conservation, export subsidies, and research and extension programs require direct annual appropriations to implement them. Since Freedom to Farm was enacted, however, market transaction payments are fixed and require annual appropriations so it can be argued that they are no longer entitlements. The budget process is composed of five basic steps:

1. Proposed budget by the president
2. Congressional budget resolution
3. Authorizing committee reconciliation
4. Appropriation process
5. Revenue raising

Presidential budget proposal. The president is required by the Budget and Accounting Act of 1921 to prepare and submit a budget to the Congress. This process, centralized in the Office of Management and Budget, begins almost two years before the fiscal year begins. In the spring of each year, executive agencies begin preparing their budget requests under some instructions from OMB. Informal within-department hearings are held, with OMB representatives generally in attendance. These initial hearings are for the departments to determine their priorities and the overall amount of their requests.

Adjustments for specific programs and total requests frequently are mandated by OMB. When serious disagreements exist between OMB and departments, their settlement is left to the president. The proposed budget is put into final form in the fall and winter and presented to the Congress in the president's annual budget message.

Budget resolution and reconciliation process.   Over time, Congress has become increasingly concerned about the size and complexity of federal spending activities, the growing importance of federal spending in relation to national income, the deficit, and the amount of relatively "uncontrollable" spending.[9] In the 1970s, concern about the ability of Congress to control spending culminated in passage of the 1974 Budget and Impoundment Act, which gave the Congress a means to establish and enforce spending and debt ceilings and revenue floors.

---

[8]An exception to this rule has occurred when Congress has pursued balanced budget initiatives when entitlements may be reduced by the legally required percentage to bring the budget into balance.

[9]One of the major factors resulting in spending being "uncontrollable" has been the enactment of entitlement programs, which, for example, tie welfare benefits to specified eligibility standards and changes in the consumer price index, or agricultural subsidies to changes in the costs of production and market prices.

The 1974 Budget and Impoundment Act created two new budget committees[10] and a staff office, the Congressional Budget Office (CBO), to support the process. These tools enable the Congress to develop and manage national economic and fiscal policy through actions taken by joint resolutions and, thus, not subject to veto by the president. The act links both the authorization and appropriation processes to the budget.

Almost a full year before a new fiscal year begins, the president must provide the budget committees with an estimate of the cost of current programs for the next fiscal year. Then the president's proposed budget is required 15 days after the Congress convenes. The appropriations committees begin work on appropriation bills; authorizing committees begin work on proposals coming before them; and budget committees begin work on the first concurrent budget resolution.

The budget process contains two closely related steps:

1. **Budget resolution** sets revenue estimates, sets a ceiling on the level of spending, and establishes guidelines for program expenditures. These guidelines are sufficiently broad that the authorizing committees have latitude within which they can adjust programs.
2. **Budget reconciliation** is the process by which the authorizing committees adjust programs to fit within the budget resolution.

Budget resolution and reconciliation activities proceed on a very tight schedule. All authorizing committees, including the agriculture committees, must present recommendations on proposed spending levels to the budget committees by March 15. By April 15, the budget committees must report a budget resolution to the floor. By May 15, the authorizing committees must report any bills that include spending authority for the upcoming fiscal year, and the Congress must complete work on the first budget resolution for the coming fiscal year. The 1996 farm bill no longer ties agricultural subsidies to changes in the costs of production and market prices. Farm program spending is now controllable.

Once the first budget resolution is complete and adopted, spending ceilings are then known to the appropriations committees. Any new spending authorized by newly reported bills is estimated. To the extent that the revenue floors and the spending ceilings in the first resolution are within total spending proposals, the process is virtually complete. Only modifications caused by changing economic conditions must be provided for in the second and final budget resolution, which must be completed in September before the beginning of the new fiscal year in October.

In the event that proposed spending exceeds the amount in the budget resolution, the authorizing committees are directed to submit legislation to reconcile spending, revenues, or debts (budget reconciliation). The Congress must complete

---

[10]The creation of the budget committees actually was triggered by confrontations between the Congress and President Nixon over whether monies appropriated could be impounded by the president and not spent.

action on reconciliation bills by September 25. Although these dates are somewhat flexible, the steps in the process are specified and the Congress adheres to them. If spending threatens to exceed the budget resolution, either spending programs must be scaled back or the resolution itself must be changed with necessary supplemental appropriations.

The budget process has had the important impact of constraining spending in the 1981, 1985, 1990, and 1996 farm bills. Ceilings imposed by the budget committees were used by congressional leadership and the administration to bring the bills' costs within the guidelines. For example, budget ceilings were used as a bargaining lever in the 1985 farm bill to lower target prices and to implement a 15 percent reduction in deficiency payments in the Omnibus Budget Reconciliation Act of 1990. The impetus behind fixed market transaction payments in the 1996 farm bill was budget control and restrained spending. The budget resolution/reconciliation process has a major impact on proposals for new spending because, for example, if the agriculture committee authorizes a market transition payment increase, it must find the money to pay for that increase from another program, with the potential for upsetting another interest group.

Recently, the budget process has also impacted the way farm programs are implemented on an annual basis in the intervening years between farm bills. That is, each year the budget resolution has sufficiently reduced farm program spending such that the agricultural committees have been forced to make adjustments, for example, in the number of acres on which farm subsidies are paid. In 1993, the budget reconciliation process included a decision to end the wool, mohair, and honey subsidies, as noted earlier. The fixed payments on the 1996 farm bill should minimize this problem.

Appropriations process. As indicated previously, appropriations are required before authorizations can be implemented. Many bills and provisions of bills are never implemented because appropriations are not forthcoming.

Within the House and Senate appropriations committees exist the highly influential subcommittees for agriculture. For 40 years, the Subcommittee on Agriculture and Related Agencies of the House Committee on Appropriations was chaired by Representative Jamie L. Whitten (Dem.–Miss.) until he became ill in 1992 and relinquished his chair in 1993. Whitten became chair of the House Committee on Appropriations in 1978 while retaining leadership of the Subcommittee on Agriculture and Related Agencies. By virtue of very nearly controlling the budget of the USDA, he was one of the most influential people in determining USDA decisions on agricultural policy and often was referred to as the "permanent secretary of agriculture." The new chair of the Appropriations Committee is Congressman David Livingston (Rep.–La.); Congressman Joe Sheen (Rep.–N.M.) is the Agriculture Appropriations Subcommittee chair.

A situation similar to that of Congressman Whitten existed on the Senate side where Senator Quentin Burdick (Dem.–N.D.) chaired the Senate Agriculture Appropriations Subcommittee. He became ill and died in 1992. The new subcommittee chair is Senator Thad Cochrane (Rep.–Miss.).

Revenue raising process.   Revenue decisions are made by **tax committees;** namely, the Senate Finance Committee and the House Ways and Means Committee. In the 1990s, the willingness of Congress to appropriate money for domestic programs such as agriculture will be determined, in part, by the national attitude toward taxation and the desire to bring the budget deficit under control. In addition to cuts in subsidies, user fees have replaced tax revenue as a means to pay for most government services, many of which are mandatory. For example, meat packers pay a user's fee for mandated inspection by the Food Safety and Inspection Service in USDA.

Resistance to new taxes, as one of the most enduring political issues, and the eagerness of both parties in the 1996 elections to cut taxes create constant pressure to hold down federal spending. In the 1980s, political pressures focused on reducing federal expenditures, including farm program costs. But, as the budget deficit has become larger and more burdensome, taxes have become a national political issue. These same forces continue to operate in the 1990s and can be expected to have an equally great, or even greater, impact on agriculture program spending.

THE EXECUTIVE BRANCH

While the role of the Congress is to enact legislation that represents the broad will of the people, the role of the executive branch is to implement and administer these laws. This means carrying out programs to effectuate the intent of the laws. The president, of course, has a political agenda that he attempts to have the Congress enact. The executive branch influences legislation through its proposals to the Congress, support of or opposition to bills, lobbying efforts, and use of the veto.

The executive branch is structured with the chief executive and staff, operating within the Executive Office of the President (EOP), and cabinet departments charged with specific areas of responsibility. In modern times, the structure of the executive branch has become increasingly complex with the addition of several offices within the EOP, as well as several new cabinet departments.

The Executive Office of the President

A general distinction is made between the White House—the president, vice president, and their personal staffs and advisers—and the agencies that EOP comprises. Although each president operates somewhat differently, the EOP includes the following agencies that influence food and agricultural policy issues:

- The *Office of Management and Budget* is responsible for development of budget proposals and oversight of executive branch expenditures as appropriated by the Congress. In addition, OMB supervises executive branch development of legislative reports and coordinates policy and procedures for all executive branch procurement. It has several staff that specifically oversee the USDA. Because of its strategic position in the executive branch policy-making apparatus, with all cabinet heads effectively reporting through OMB, this agency is probably the most powerful in the domestic policy side of government. Its power is magnified when the president and the majority on the Hill are of the same party. Sometimes OMB directors extend their influence to higher levels within the administration. For example, President Clinton's former Chief of Staff, Leon Panetta, was previously the OMB director. When he was in the Congress, Panetta was a member of the House Agriculture Committee and the Budget Committee. His knowledge of the farm bill and the budget process made him both an asset and a liability to the agriculture establishment.

- The *Council of Economic Advisers (CEA)* is responsible for assessing economic conditions and reporting annually to the Congress. The three-member council also advises the president and assists in development of the administration's national economic policy. The degree of the CEA's influence depends heavily on the chairperson's relationship with the director of OMB and the secretary of the treasury, the combination of which is often referred to as the troika because of their potentially combined supreme influence on domestic policy decisions of which agriculture is one dimension.

- The *National Security Council (NSC)* integrates domestic, foreign, and military policy relating to national security. The NSC also advises the president on the development of foreign policy and coordinates foreign policy development and execution. It has become involved in food diplomacy, as indicated by the Somalian and Rwandan humanitarian military assistance missions and in embargo issues.

- The *Central Intelligence Agency (CIA)* is responsible for foreign intelligence and counterintelligence activities outside the United States. The CIA director has cabinet rank. It maintains an intelligence system on crop production conditions around the world that is independent of USDA's World Agricultural Outlook Board and Foreign Agriculture Service.

- The *Office of the U.S. Trade Representative (USTR)* advises the president on trade matters, coordinates executive branch trade policy, and is responsible for directing U.S. participation in trade negotiations. It has several staff members who handle agricultural trade policy issues.

- The *Office of Science and Technology Policy (OSTP)* is responsible for advising the president on matters relating to science policy. It is involved in analyzing all areas of national concern for which policy will have scientific implications. These areas include agricultural, food safety, and environmental research.

- The *Council on Environmental Quality (CEQ)* advises the president on environmental matters and assists in policy development. Agriculture has a major stake in presidential initiatives such as those affecting water quality and air pollution.

The growing complexity and interdependence of a modern society mean that virtually any public concern today cuts across the assigned responsibilities of several of the cabinet departments and interdependent regulatory agencies. Therefore, a major function of the EOP entities is to assimilate and coordinate the actions and decisions of the many departments, agencies, and offices to ensure a coherent national policy. For example, the Council of Economic Advisers reviews proposed USDA actions to ensure their consistency with the president's overall national economic policy, while OMB assesses USDA actions to make sure that they conform with administration budget policy.

The appropriate role of the EOP versus that of the cabinet departments in determining policy has long been a point of contention, but an especially visible one in the past few administrations. Increasing concern has developed in the agriculture establishment that officials in the EOP, such as the director of OMB, might be making the critical decisions on farm policy. Although this concerns farmers, it is argued that OMB and other agency involvement is necessary to develop a consistent national policy. While the power of OMB is not a new concern, its influence has become greater as budget pressures have grown.

The essential nature of the problem is that as the executive branch has grown more vast and complex, increasingly greater coordination is required to harmonize the policies and programs of each cabinet department with the overall philosophy and broad national policies emanating from the chief executive. Numerous examples exist of cabinet departments catering to constituent interests and striking off in policy directions at odds with the broad policies enunciated by the White House. Bureaucratic infighting between departments, including USDA and the White House has, at such times, been intense.[11]

The issue of cabinet autonomy has been particularly controversial, with much attention given to reported discord and conflict between the secretaries of agriculture and the EOP and the White House during the past two decades. A degree of disorder and disharmony in the executive branch policy decision process for food and agriculture has always existed. This is inevitable to some extent; it quite naturally arises that the favored policy course of USDA is often inconsistent with broader national policy objectives being pursued by the administration as overseen by the EOP. However, an adversarial element to USDA–EOP relations, properly harnessed, is generally regarded as important and useful to reaching better decisions by ensuring close scrutiny and broad consideration of the many aspects of an issue.

One of the factors contributing to USDA–EOP conflict has been increased interest in agriculture and food issues by other cabinet agencies such as the Treasury and State Departments, EPA, and FDA. The links of agriculture to other

---

[11]An excellent discussion of White House cabinet pressure is contained in Joseph Califano, Jr., *Governing America: An Insider's Report from the White House and the Cabinet* (New York: Simon and Schuster, (1981). David A. Stockman's *The Triumph of Politics* (New York: Harper & Row Publishers, 1986) also provides an excellent OMB perspective on the political aspects of economic decisions, with several illustrations from agriculture.

departments became stronger as agriculture's contribution to the economy's balance of trade increased as exports increased, as rising food prices put inflationary pressures on the economy, as commodity subsidies and food aid programs resulted in larger budget outlays, as trade and aid agreements became tools of foreign diplomacy, and as questions arose about agriculture's contribution to environmental problems and the safety of the food supply. These changes have created problems for the agricultural establishment in controlling its farm, resource, and food policy agenda.

The U.S. Department of Agriculture

Shortly after its creation in 1862, President Lincoln stated his intention for the Department of Agriculture: "It is precisely the people's department, in which they feel more directly concerned than any other."[12] Although the mission and focus of the USDA are sources of continuing controversy, the scope and responsibilities of the department have significantly expanded and contracted over time. For example, many of the food inspection and safety programs now handled by FDA were once a part of USDA. USDA now plays a secondary role to EPA on issues of pesticide policy and its implementation although, on the Hill, the agriculture committees still have substantial control over pesticide policy. USDA was once the regulator of the futures market—a function that has since been moved to the Commodity Future Trading Commission. Suggestions have been made that USDA's food assistance programs could more appropriately be handled by the Department of Health and Human Services (DHHS).

From time to time, major changes in the structure of USDA are suggested. Less frequently, such organizational changes actually take place. Such a watershed occurrence took place in 1993 as part of President Clinton's reinventing government task force headed by Vice President Gore. Structural changes in USDA had been discussed previously by Senator Richard Lugar (Rep.–Ind.), the Republican chairman of the Senate Agriculture Committee. He persuaded former Secretary Edward G. Madigan to initiate a comprehensive study of the structure of USDA during the Bush administration. The Madigan study recommended sweeping changes in the organization of USDA, particularly in terms of consolidation of field offices for administration of producer-oriented programs. Vice President Gore's task force adopted most of the recommendations of the Madigan study and added to them.

This confluence of forces gave bipartisan support to the USDA structural changes, virtually ensuring their implementation. As the reader reviews the following brief description of the USDA agencies as conceived by the reinventing government task force, the diversity of USDA becomes apparent. While there are political reasons for this diversity, questions are increasingly arising as to whether

---

[12]Abraham Lincoln, Fourth Annual Message to the Congress (Washington, D.C., December 6, 1864).

Mixing Oranges and Oranges

If there was any thin sliver of the welfare state where the Reagan Administration might have raised the free enterprise and anti-spending banner, it was against the socialistic enterprises of U.S. Agriculture. But by 1984, we had accommodated to the political facts of life here, too. As I contemplated the task of formulating a strategy to deal with the nation's massive deficit after the election, two White House episodes regarding agriculture stood out in my mind vividly. They were the smoking gun which proved that the White House couldn't even tackle the fabulous excesses of the farm pork barrel, and that was the very bottom of the whole spending barrel.

The first episode occurred in the summer of 1982. The issue was agriculture marketing orders, an out-and-out socialist relic from the New Deal that tells every California orange and lemon grower how many of these little fruits can be marketed each week.

The established growers like this kind of lemon socialism because it keeps prices up, supplies down, and new competition out of the market.

So, I'd located some photographs of this lemon socialism at work. They showed gargantuan mountains—bigger than the White House—of California oranges rotting in the field. The reason for all this deliberate garbage creation was that the USDA orange commissar had cut back the weekly marketing quotas, fearing that a bumper crop would drop the price and give consumers too good a deal on oranges.

Since we'd also just talked about a free food program for the homeless, my pictures did seem to suggest something rather ludicrous, and everyone around the cabinet table began to laugh. But then the California politicians swung into action.

Dick Lyng, an old California Reaganaut and Undersecretary of Agriculture, said I was fibbing. "The USDA had nothing to do with this. The growers elect their own committees to stabilize the market."

"You remember, Mr. President," he added, "that a lot of our friends out there depend on these marketing orders."

Well, okay. Some of our friends are members of the Navel Orange Growers Soviet. It wasn't a compelling argument. . .

I asked him how about year-round Florida oranges that come right off the free market, with no supply control by a Florida Orange Grower Soviet at all. He said my point wasn't valid because I was mixing oranges and oranges.

Jim Lake, the Reagan campaign press secretary and paid lobbyist in the off season, had another point. He just went up to the Hill and got a law passed making it illegal for the director of OMB even to read the marketing orders before they were stamped by USDA. That was that for free enterprise in California. Needless to say, there remained equally compelling cases for other variations of Big Government in the other forty-nine states.

Source: Excerpt from *The Triumph of Politics* by David A. Stockman, Copyright © 1986 by David A. Stockman. Reprinted by permission of Harper & Row, Publishers, Inc.

several programs are properly located in terms of government effectiveness and efficiency. This will be seen particularly in the discussion of whether the Food Safety Inspection Service (FSIS) should be located in USDA or in DHHS, as suggested in the reinventing government report.

USDA normally has 15 to 20 agencies; the number varies from time to time as each new administration makes reorganizations. It employs more than 100,000 people, making it one of the largest cabinet departments. Organizationally, USDA's functions are grouped into broad place categories with an undersecretary or an assistant secretary having direct responsibility for the agencies carrying out programs in closely related areas (Figure 3.3). The secretary of agriculture is aided by a deputy secretary and by an office staff composed of the general council, inspector general, a chief economist, a chief financial officer, and a group of budget analysts. In performing the policy advisory function, the chief economist has a group of key policy analysts directly tied to this office who, in turn, draw on the expertise of the Economic Research Service (ERS).

USDA's programs extend to virtually every aspect of food and agriculture, the full scope generally being indicated by the following brief descriptions of the various agencies reflecting the reinventing government changes. We will return to the issue of the appropriate USDA scope later in this chapter.

Commodity, Farm, and International Trade Services

The **Farm Service Agency (FSA)** represents the most substantial and substantive change in the reinventing government package for USDA. It represents one-stop shopping where the farmer can sign up for farm program participation, obtain operating loans, and obtain assistance on conservation issues. This very large and powerful agency is a result of the consolidation of the former Agriculture Stabilization and Conservation Service (ASCS), the former lending programs of the former Farmers Home Administration (FmHA), and the former Federal Crop Insurance Corporation. The structural change appears to have been designed primarily to consolidate the county offices of the ASCS and FmHA into a one-stop farm program center. It also handles county-level functions for conservation programs. In doing so, it retains the "grassroots" committee structure of elected farmer input into farm program administration at the county level. FSA state committees advise USDA farm program administrators on policy and handle appeals from the county level. In contrast with the elected county committees, state committee people are nominated by the highest ranking elected senator from the majority party, appointed by the secretary and, ultimately, by the president. If there is no majority party senator for the state, the highest ranking majority party congressperson makes the nomination to the secretary of agriculture.

The **Foreign Agriculture Service (FAS)** has the responsibility of administering USDA international trade and development activities. FAS's principal missions include expansion of the foreign markets for U.S. farm products; maintaining

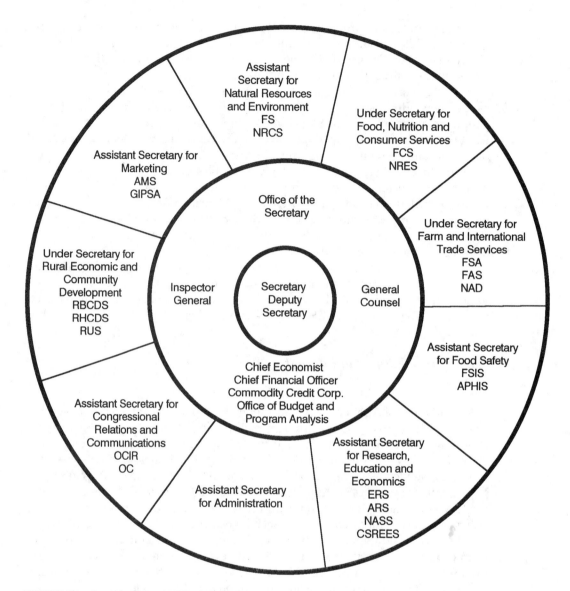

FIGURE 3.3   Organization of USDA by functional areas and agencies.

USDA's intelligence system on crop production and policy changes in other countries; and facilitating international educational, training, and scientific exchanges. It stations more than 150 attachés and counselors in American embassies in major food-producing and consuming countries around the world to monitor conditions and facilitate market development and sales.

The **National Appeals Division (NAD)** operates as the clearing house for appeals by farmers from decisions made by state committees on both farm programs and farm loans. Operating by legally binding administrative procedures, it operates as a "court" within the USDA to which farmers may appeal their cases when they feel mistreated.

Marketing Services

The **Agricultural Marketing Service (AMS)** directs the market order programs for fruits, vegetables, and milk; manages product-grading activities; and collects and disseminates market news information. All of these programs are producer and/or marketing-firm initiated.

The **Grain Inspection, Packers, and Stockyards Administration (GIPSA)** is a regulatory agency that monitors livestock markets for potential unfair trade practices and is responsible for determining grades of grain, testing, and quality assurance for grain moving in interstate commerce and international markets.

The **Food Safety and Inspection Service (FSIS)** conducts the federal meat and poultry inspection programs. The reinventing government task force recommended that this agency be transferred to the Food and Drug Administration in the Department of Health and Human Services. This recommendation made sense because all food safety and inspection activities, except fish, would then be in a single government agency. However, from a political perspective, it made no sense to the agriculture establishment. USDA would lose significant components of the meat industry as constituents. Moreover, it was feared that FDA would be less sympathetic to, and understanding of, the problems of animal agriculture. As a defensive strategy, the agriculture committees created an assistant secretary for food safety position in USDA.

The **Animal and Plant Health Inspection Service (APHIS)** is responsible for the inspection and certification of animals and plants moving in interstate and international commerce. It handles decisions involving the protection of new plant varieties, which has become more important with the advent of biotechnology.

Rural Economic and Community Development

The **Rural Housing and Community Development Service (RHCDS)** operates credit programs for rural communities and rural housing. The rural housing program of its predecessor agency came under fire for making a disproportionately large share of its loans to residents in counties immediately adjacent to metropolitan areas rather than to residents of truly rural America.[13]

The **Rural Utilities Service (RUS)** combines the rural electric, telephone, water, and sewer lending and grants programs of USDA. This agency was instrumental in the extension of electric power, water, telephone, and sewer systems

---

[13]General Accounting Office, *Concentration of Rural Home Loans,* GAO/RCED-93-57 (Washington, D.C.: U.S. Congress, June 1993).

located in remote rural areas of the country. It continues to supply electric and telephone cooperatives with large amounts of frequently subsidized credit. While these credit subsidies have been on the "hit list" of both Republican and Democratic administrations, they have survived. The only agriculture cut explicitly mentioned in President Clinton's first address to the Congress was the elimination of Rural Electrification Administration subsidies.

The **Rural Business and Cooperative Development Service (RBCDS)** is the center for rural business and cooperative development in USDA. As such, it provides technical assistance to rural businesses with a focal point on cooperatives. It also serves as the commercialization center for nontraditional (organic) farming systems.

Natural Resources and Environment

The **Natural Resources Conservation Service (NRCS)** administers several resource-related programs of soil and water conservation, water quality enhancement, and environmental improvement. These programs provide technical assistance to farmers, frequently on a cost-share basis. Farmers feared that the formation of NRCS as a sequel to the Soil Conservation Service (SCS), under an assistant secretary for natural resources and environment, would open the door for a USDA niche controlled by environmentalists. Adding fuel to the fire is the liaison role that the NRCS and its assistant secretary have with EPA.

The **Forest Service (FS)** is responsible for the management of the national forests and federal wilderness lands that encompass nearly 200 million acres. Its Department of Interior counterparts are the Bureau of Land Management, which manages other federal lands, and the National Park Service, which manages national parks.

Food, Nutrition, and Consumer Service Programs

The **Food and Consumer Service (FCS)** administers food assistance and nutrition programs that account for 40 to 70 percent of the USDA budget, depending on the size of farm program expenditure. The largest of these programs, in terms of people served, is the school lunch program, which provides more than 25 million students in the nation's schools with subsidized lunches. In 1992, 1 of every 10 Americans (approximately 25 million people) received benefits from the food stamp program—the most expensive of the nutrition programs.[14] More than 5 million of these people also are reached through specially targeted programs such as the Women, Infant, and Children (WIC) Division program, which provides direct commodity distribution, prenatal care, and nutrition education.

The **Nutrition Research and Education Service (NRES)** focuses on the goal of merging nutrition research and education with USDA's food assistance pro-

---

[14]Masao Matsumoto, personal correspondence (Washington, D.C.: ERS, USDA, January 26, 1993).

grams. This agency has tried to stake out a lead role in all USDA nutrition education, including that of the extension service.

Research, Education, and Economics

The **Agricultural Research Service (ARS)** conducts and coordinates the research programs of USDA. Its problem-oriented biological research and library facilities are headquartered in Beltsville, Maryland, and include regional laboratories located throughout the United States.

The **Cooperative State Research Education and Extension Service (CSREES)** coordinates and directs funding for joint federal–state research and education programs conducted primarily in agricultural experiment stations and extension services located at land-grant universities. The formation of this joint research–education agency was viewed as a reduction in status for extension activities, which previously had their own agency. However, Secretary Dan Glickman, a former congressman from Kansas, long-time member of the House Agriculture Committee, and supporter of land-grant research and extension, has often been quoted that he intends to be known as the research and education secretary.

The **Economic Research Service (ERS)** employs the largest single group of agricultural economists in the world. ERS is the social science research and analysis agency for the secretary of agriculture. It monitors the global food and agriculture system and conducts research to enhance understanding of the functioning and performance of the food and agriculture economy, domestic and international. These studies often touch on sensitive political issues for the administration, such as the impact of free trade while international trade negotiations are in process. As a result of these sensitivities, questions have arisen regarding the release of ERS research results to the political opposition. If ERS is to exist as a respected research agency, these issues must be directly confronted with objectivity.

The **National Agricultural Statistical Service (NASS)** is responsible for collecting primary data from the nation's farmers and agribusiness firms as a means of improving and monitoring the performance of the farm sector, including markets for farm products. The work of NASS is critically important to the smooth functioning of markets, but crop and livestock production reports are often criticized by farmers and their organizations for being biased—particularly when the effect of an anticipated large crop is to drive the market down. In reality, NASS has a very good record—the best in the world—for accurately reporting agricultural statistics.

ORGANIZING FOR POLICY DECISIONS IN USDA

It becomes rather obvious after this brief review that the functions and programs of USDA are broad in scope. It is equally obvious that this breadth and diversity hold great potential for conflicts within USDA itself and that unanimity on any policy position would be rare indeed. Each of the agencies has a particular clientele and objectives that from, time to time, conflict. Thus, reaching a departmental policy position, let alone an executive branch position, has become increasingly difficult.

Political Appointments

One of the privileges of a president is to appoint people to decision-making positions within the executive branch. In addition, numerous committee appointments are made at the state level. This power of appointment is necessary for the president to have trusted individuals of the same political and philosophical persuasion to make and implement decisions consistent with the platform on which the president was elected.

This power of appointment is limited by the civil service laws to persons in "politically sensitive" positions. This involves a large number of people ranging from the top White House staff and cabinet secretaries, such as the secretary of agriculture, to most agency heads, and even their personal secretaries.

A new president has the power to make appointments to more than 3,000 political jobs across all departments of the federal government.[15] The president also has the power to move members of the senior executive service, the top government professionals, from one position to another.

Within USDA, the president has the power to make appointments to 290 high-level supervisory and staff jobs. These appointments include the secretary, deputy secretary, all undersecretaries, and assistant secretaries, and many agency administrators and their staffs. In addition, directors of state offices of USDA for program implementation and the related state committees are politically sensitive. Even appointments to county USDA farm program implementing committees are politically sensitive.

Appointments to these positions realistically involve political consideration as well as people's qualifications. Financial support, work for the campaign, and support from key politicians—particularly from senior members of the agriculture committees—are certainly important factors. The existence of this support may be influenced by the attitudes of special-interest groups, such as farm and agribusiness organizations, toward particular candidates for political appointments. In terms of qualifications, past experience in government plays a role. It is not unusual to see appointments to higher level positions in the same party from one administration to the next. For example, Secretary Clayton Yeutter for the Reagan administration served as an Assistant USDA Secretary in the Nixon administration. Many Clinton administration appointees worked in the previous Carter administration. Secretary Bob Bergland served in the Carter administration, Madigan in the Bush administration, and Mike Espy and Dan Glickman in the Clinton administration came from the Congress.

---

[15]"Pick of the Crop: 'Plums' that Could Make a Difference," *The Washington Post*, November 4, 1992.

It is not difficult to find people who meet the qualifications for political positions, although it may be more difficult to decide who exactly should get the job. When a new party comes into office, an agriculture transition team can easily expect to receive and review the resumes of more than 5,000 individuals who are suggested for political appointments. Needless to say, the competition for the 290 available positions is keen and highly political.

Decision Process in USDA

Decision making is handled differently by each secretary of agriculture. The demands on a secretary's time—for public relations and speech making and other such functions—in addition to the administrative and management functions are enormous. As a result, there is a tendency for the secretary to rely very heavily on a few people for guidance in decision making. In the Bergland–Carter years, that person was the chief economist. For the Block–Reagan years, it was the deputy secretary. Some particularly strong secretaries, such as Secretary Earl Butz (Nixon/Ford administrations) and Glickman (Clinton administration) tended to use the undersecretaries and assistant secretaries more as an advisory body, with the secretary being the clear decision focal point.

Economic impacts play an important role in policy decisions. Three centers exist in USDA for economic analysis:

1. The economic analysis staff in the office of the chief economist handles the daily policy analysis needs of the chief economic adviser to the secretary.
2. The policy analysis division of the ERS analyzes longer term policy issues.
3. The agency responsible for program administration in the decision area frequently employs its own policy analysts.

For example, a farm program decision under previous legislation prior to the 1996 farm bill involving the proportion of land to remove from production each year would normally be analyzed by two groups.[16] The Farm Service Agency will have a position based on its analyses and experience in managing the program and its commodity inventory levels. The chief economist's economic analysis staff would have its own evaluation, which would be broader in nature and would perhaps consider the impacts on exports.

Once these analyses have been completed, they will flow through channels of the chief economist, the undersecretary for farm and international trade services, the deputy secretary, and, ultimately, the secretary of agriculture. At any one of these levels, more analyses may be called for on additional program impacts or options.

ERS would tend to get involved in only longer term studies, for example, those mandated by the Congress. When the secretary of agriculture has made the

---

[16]The 1996 farm bill eliminates annual land set-asides.

decision, it flows to the White House where it will normally be routed to OMB, the Council of Economic Advisers, and other affected agencies. Once again, the USDA recommendations will be analyzed, this time largely in terms of their impact on inflation and government spending, as well as their consistency with other programs and the overall objectives of the president.

The Changing Role of USDA

USDA was created in 1862 to conduct agricultural research and disseminate its results. Its role expanded dramatically in the 1930s under the leadership of Secretary Henry A. Wallace. For the next three decades, the emphasis was placed on supporting farm prices and incomes—often under the guise of conservation—and improving rural electrical and telephone services.

The Great Society programs dramatically expanded USDA's role in assisting the poor with food stamps, school lunches, and improved nutrition. These programs have continued to expand despite periodic political opposition extending back to the 1970s. The Great Society was also the origin of dramatically expanded concern about food safety and environmental quality.

Beginning in the 1970s, the emphasis in USDA on trade and market development began to mushroom. Support for initiatives in trade expansion received broad-based support. Budget crunch conditions led to questions of whether public monies should be spent supporting market development activities when the main beneficiaries were multinational exporters and large-scale farm operators.

Beginning in the 1970s, questions began to arise regarding the appropriate organizational structure for the government, including USDA. In the early 1970s, President Nixon proposed a massive reorganization of the executive branch that would have combined USDA with the Department of Interior and other functions into a giant natural resources department. It was at this time that Secretary Butz was appointed and many allege that his acceptance of the appointment was made conditional on USDA remaining a separate entity, but that he would not oppose the transfer of some major food assistance programs to other departments.[17]

The Carter administration sought to broaden USDA's concerns for many diverse groups making it, in essence, a "department of food" serving farmers, consumers, and people in rural areas. Secretary Bergland, to a considerable extent, was able to achieve this objective, but there was considerable criticism, especially from farm groups, that perceived a diminution of their interests. The appointment of Carol T. Foreman, a vocal consumer advocate, as assistant secretary for consumer affairs was singled out by farm groups as evidence supporting allegations that consumers were more important than farmers to USDA. Secretary Glickman has sought to emphasize the "food" role of USDA. He had similar difficulties with another consumer advocate, former undersecretary Ellen Haas, who resigned in February 1997.

---

[17]See Weldon Barton, "Food, Agriculture, and Administrative Adaptation to Political Change," *Public Admin. Rev.* (March–April 1976), pp. 148–154.

In the 1980 campaign, the Republicans attempted to capitalize politically on this dissatisfaction, indicating that they would return USDA to a more narrow farmer-interest role. President Reagan's initial agriculture secretary, John Block, sought to do this by purging USDA of consumer elements and cutting back on consumer programs. These efforts were largely unsuccessful, although consumers clearly had a smaller role in USDA under Reagan than under Carter.

Beginning in 1989, the GAO initiated a series of eight reports evaluating the structure and management of USDA. This series culminated in a September 1991 report sharply criticizing USDA for being overly production oriented and commodity based—not changing with the times.[18] Capitalizing on the GAO initiative, the *Kansas City Star* published a series of articles with headlines indicating that USDA had grown into a "monster" with an abundance of programs that "produces agency gridlock." The series hardly left a stone unturned in terms of investigating all dimensions of the USDA program—although its food assistance and nutrition programs went relatively unscathed (see box).

Senators Lugar and Leahy (respectively, ranking minority member and chair of the Senate Agriculture Committee at the time) capitalized on the GAO initiative to demand change. Secretary Madigan ordered a special commission to study the GAO report with an emphasis on the consolidation of county offices. At the time, ASCS and FmHA each delivered program benefits to farmers through often separate offices located at the county level. The idea to consolidate these offices was not new. It had been unsuccessfully tried before and failed due to political pressure. Upon his departure, Madigan issued an order closing and consolidating many USDA county offices. Secretary Espy delayed the implementation of this order pending further study of the organization of USDA, including a proposal by Secretary Madigan to reduce the number of high-level officials reporting to the secretary. However, the ultimate USDA reorganization completed by Secretary Glickman was very close to that proposed by Madigan. Even if these reorganizations are successful, they hardly address the broader issues raised by the *Kansas City Star* and by the GAO regarding the nature, scope, and abuse of USDA programs—a subject that is discussed further in later chapters.

The difficulty of narrowly defining the role of USDA in modern times has been evident since the early 1970s when the economic environment for agriculture changed significantly. As agriculture's importance to the national economy grew, so did the executive branch's interest in its activities. Any agricultural and food policy issue attracts the interest of other agencies that feel their responsibilities are affected by the handling of the issue. The Department of State demands a say about food aid going to particular countries as a part of overall foreign policy strategy. The Department of Defense has become involved in delivering food aid as was done in Somalia and Rwanda. The National Security Council and the Department of State are concerned about the negotiation of major bilateral commodity agreements. OMB is concerned about the cost of any policy action or program change. The

---

[18]General Accounting Office, *U.S. Department of Agriculture: Revitalizing Structure, Systems, and Strategies,* GAO/RCED-91-168 (September 3, 1991).

Failing the Grade: Betrayals and Blunders at the Department of Agriculture

The following is a series of lead paragraphs taken from the *Kansas City Star*. Although taken out of context, they are presented to provide a flavor for the nature and origin of the controversy surrounding USDA and the functions it performs:

- "It's not for consumers. In grocery stores nationwide, thousands of USDA labels on meat products are misleading or dead wrong."
- "Eddie Tank grew cotton, wheat, and a little barley here in the fertile flatlands of the San Joaquin Valley. But he specialized in farming the federal government."
- "For dishonest farmers and those who regulate them, opportunity is boundless in the byzantine world of farm programs run by the U.S. Department of Agriculture."
- "In the summer of 1987, a shipment of tainted beef killed four retarded patients at two Utah mental institutions. . . . Government meat inspectors had a problem."
- "Bob Frame doesn't believe in paying good money for a bunch of Hollywood actors to go on television and tell people to eat more beef. . . . It's just a big money scheme for actors and advertisers, Frame said."
- "Tasty as his fruit may be, Burt knows some of it will never end up in grocery produce sections. Instead, the U.S. government will force him to sell some of it for juice, at a loss."
- "Under the department's watch, the number of Black farmers has plummeted 97 percent from the 1920s, from 925,000 to fewer than 23,000 today."
- "The Agriculture Department's civil rights failures begin in its own house."
- "Farmers here and across North Dakota aren't the only ones trying to evade U.S. Department of Agriculture rules forbidding them to drain wetlands."
- "An increasing number of foods labeled 'precooked' or 'ready to eat' by the U.S. Department of Agriculture may not be ready to eat at all."

Source: Kansas City Star, "Failing the Grade, Betrayals and Blunders at the Department of Agriculture" (Kansas City, Mo.: December 8–14, 1991).

Treasury Department wants USDA programs to prevent or at least not exacerbate inflation, improve the balance of trade, and play a positive role in leading to a realistic value of the dollar in foreign currency markets. The Interior Department and the Council of Environmental Quality are concerned about the manner of operation of the Forest Service, with its 92 million acres of national forests, and USDA's stance on environmental matters and water policy. The Department of Health and Human Services is concerned about USDA's operation of the nutrition programs including the food stamp, school lunch, dietary guidelines, and other programs.

Over time, USDA has acquired responsibility for a broad range of program areas that take it far beyond matters of production agriculture. The extent to which these other functions have assumed importance is reflected in its 1995 budget of about $60 billion with 108,000 employees. Of the total federal budget, USDA programs are extremely diverse, falling into 10 of 17 budget categories—more than any other federal agency[19] (Figure 3.4). Its clients include all consumers,

FIGURE 3.4   USDA outlays by major program category, 1962–1996.
Source: USDA Historical Budget Outlays electronic database (Washington, D.C.: USDA/ERS, March 1995).

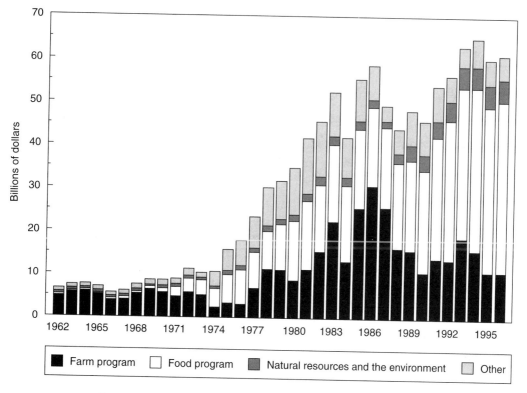

[19]Edward G. Magidan, "Testimony before the Committee on Agriculture," selected speeches and news releases (Washington, D.C.: USDA, June 18–24, 1992), p. 1.

about 1.8 million farmers, about 25 million food stamp recipients, about 25 million schoolchildren, 5 million single parents, timber companies, agribusiness firms, environmentalists, and rural communities.

It is obvious that with this diverse clientele and their frequently conflicting interests, any policy position taken by USDA officials will be controversial. The pressures on USDA are becoming more familiar and frustrating to its traditional agricultural clientele. They see these pressures as complicating the policy process. On the other hand, they recognize the political realities of the need to develop coalitions of diverse interest to pass bills in the Congress. They only grudgingly accept the reality that the secretary of agriculture is not the farmers' representative in Washington. He is the president's man. When he ceases to be the president's man, he becomes an ex-secretary.[20]

THE JUDICIAL BRANCH

The courts have the responsibility for interpreting the Constitution and settling disputes among parties within the framework of existing laws. The Constitution makes the judiciary an equal and coordinate branch of government. American society, during the last decade or two, has increasingly looked to the courts for resolution of some of its most difficult problems. Courts have been asked, for example, to devise rules for reapportionment of legislatures, integration of schools, protection of the environment, and improving the lot of the poor. Complex as they are, these new challenges have taxed the traditional methods of the courts and promoted substantial controversy.

Although settlement of disputes among individuals, determination of violations of the law, and carrying out legal procedures for settlement of estates, bankruptcies, and so on still occupy most of the courts' time, the scope of judicial business has broadened. This broadening has been particularly pronounced in the area of the rights of individuals, groups, and firms to pursue legal remedies against government. The result has been the involvement of courts in decisions that earlier would have been thought unsuited for adjudication. Judicial activity has extended to welfare, prison, and mental hospital administration; to education and employment policy; to road and bridge building; to automotive standards; and to natural resource management.

The impacts have clearly extended to food and agricultural issues. Courts have struck down laws requiring a period of in-state residence as a condition of eligibility for welfare. Federal district courts have set elaborate standards for food handling, ordered the Forest Service to stop clear-cutting timber, and required the Congress to straighten out the conflicting tolerances for carcinogens in the food supply.

---

[20]B. L. Flinchbaugh, "It's Easy to Be Ignored If You Don't Have Your Act Together," speech presented at the National Institute on Cooperative Education, Colorado State University, Fort Collins, Colo., July 28, 1981, p. 1.

What the judges have been doing is new in a special sense. Although no single feature of most of this litigation constitutes an abrupt departure from the past, the aggregate of decisions distinguishes it sharply from the traditional exercise of the judicial function. Many wholly new areas of adjudication have been opened. In the process, the rights of food stamp recipients, consumers, and farmers have been clarified under the law. To some extent, this judicial activity is a response to invitations from Congress and from state legislatures. By enacting legislation that is broad and vague, the Congress has, in essence, passed many of the problems to the courts. The courts then have to deal with the inevitable litigation to determine the "intent of the Congress," which, in some statutes, is far from clear. State legislatures have similarly enacted broad, vague legislation with similar results. Moreover, as federal subsidies to agriculture decline, states are pressured to adopt their own programs and/or interstate compacts designed to accomplish some of the same objectives as the federal subsidies.

To an important extent, increased litigation arises from the realities of increased government regulation and expanded social programs. In this sense, the expansion of judicial activity is a mere concomitant of the growth of government. As governmental activity in general expands, it is very likely that judicial activity will also expand.[21]

USDA, THE ESTABLISHMENT, AND THE FUTURE

In the 1970s, the agricultural establishment began losing control of the farm policy agenda. Agricultural and food policymaking has become more complex and difficult to predict. Although initially the establishment may have thought it was in control, this perception has since given way to these stark realizations:

- Consumers and environmentalists do have an interest in the decisions of USDA.
- The secretary of state and trade representative may have as big a stake in international food issues as the secretary of agriculture.
- The secretary of agriculture cannot consider only farm interests in decisions on agricultural and food policy.

The problem of agriculture being unable to control its agenda becomes most acute when the establishment itself cannot agree on what should be done. Disagreement is fostered by a budget process, which says that any new spending for one USDA program must come at the expense of another USDA program. In dairy, this was carried even further when it was suggested that if dairy program changes increase the cost of supplying dairy products to the WIC program, that cost had to be made up by dairy farmers through an assessment.

[21]Donald L. Horowitz, *The Courts and Social Policy* (Washington, D.C.: The Brookings Institution, 1977).

The interest of the farm constituency in the issues encompassed in food programs was perhaps indicated most clearly when Congressman Charles Stenholm (a farmer; Dem.–Tex.) chose the chair of the Department Operations and Nutrition Subcommittee, which manages USDA's food programs, over the livestock committee. Well over 50 percent of farmers' income in Stenholm's district is from beef and dairy. Well over 50 percent of USDA's budget is food programs!

Prior to the 1970s, the agriculture establishment was consistently able to manage the political process in its favor. Without being in a position to manage the process any longer, the agricultural establishment has been on the defensive. Clearly, the agricultural establishment in the future will have to share control of the agenda with other interest groups.

ADDITIONAL READINGS

1. For a detailed study of the legislative process, see J. B. Penn, "The Federal Policy Process in Developing the Food and Agriculture Act of 1977," in *Agricultural-Food Policy Review,* AFPR-3 (Washington, D.C.: ESCS, USDA, February 1980), pp. 9–46.

2. An excellent discussion of the relationships between the White House and the cabinet is contained in Joseph Califano, Jr., *Governing America: An Insider's Report from the White House and the Cabinet* (New York: Simon & Schuster, 1981). Also see David A. Stockman, *The Triumph of Politics* (New York: Harper & Row, Publishers, 1986).

4

FOOD AND AGRICULTURAL POLICY INTEREST GROUPS

No class of Americans, so far as I know, has ever objected . . . to any amount of government meddling if it appeared to benefit that particular class.

—*Carl Becker*

Interest groups provide the motivating force for the policy process. They identify problems and advance proposals for solving them. They organize to affect the results of elections, to influence the position of candidates, and to influence appointments to key positions as well as the ultimate votes or decisions on issues that affect their members. Few individuals, acting alone, are in a position to have the same amount of influence as an organized interest group.

Food and agricultural policy interest groups include any organization, association, or firm seeking in some way to influence agricultural, resource, food, and rural development policy decisions of the Congress or the executive branch. Interest groups generally employ lobbyists to represent them in the political process and to advise them on political strategy.

Interest groups are active at all levels of government decision making. In the executive branch, interest groups directly contact top USDA appointees, the secretary of agriculture, or the president. They may also attempt to influence decisions of middle-management career employees who make program or policy recommendations to the secretary of agriculture and ultimately to the president. In the Congress, interest groups actively seek to influence congressional staff, the reports of congressional research and information agencies, and, ultimately, the positions of the senators and representatives.

INFLUENCING EFFECTIVENESS

Substantial differences exist in the abilities of interest groups to influence policy. How effective a group is at influencing policy is largely determined by

- The priority of the particular policy problem being considered
- The consequences of alternative means of dealing with the problem
- The number of individuals or firms represented by the interest group
- The economic importance and political influence of the individuals or firms represented by the interest group
- How well the interest group has analyzed the issue
- The strategies employed by the interest group

Priority of the Problem

The starting point for influencing policymakers' decisions is convincing them that the problem confronting the interest group's members deserves high enough priority to devote the time required to resolve it. Members of Congress will often introduce a bill proposed by an interest group, but substantial time and effort are involved in getting the bill seriously considered by the body and ultimately enacted into law. To command this time and effort, the problem must be substantial, must affect people who have political influence, and must not have an apparent private action solution.

Consequences of Alternatives

For the democratic process to work effectively, as much information needs to be brought to bear on a policy decision as possible. This requires an analysis of the consequences of all policy alternatives, including the status quo. These consequences must be analyzed in terms of those groups affected—producers, agribusiness, consumers, and taxpayers. Generally, a decision that benefits one of these groups adversely affects another. Therefore, one interest group frequently has a different perspective on the merits of a particular policy alternative than another group. Policymakers are then required to balance the merits, views, and impacts of one group's proposals against those of others. Therefore, they must know the consequences for each group, perhaps even for subgroups within each group.

Number of Individuals or Firms Represented

Power exists in numbers. Numbers represent the votes required to be elected and reelected. Farmers traditionally have relied heavily on numbers to influence political decisions. In former times, they tended to vote as a bloc and were influenced largely by "pocketbook issues" and their conservative nature. As farm numbers have declined, the political power of farmers and their interest groups has been affected. Increased consumer interest in food and agricultural policy has, at times, been an asset to farmers and, at other times, a liability. For example, the food stamp program has been a factor in holding the interest of many urban members of Congress in support of the farm bill. However, that program also competes with other USDA programs for increasingly limited federal dollars. Since the 1985 farm bill, coalitions involving environmental interest groups have been important to getting the votes needed to enact a farm bill.

Influence of Individuals or Firms Represented

In the political process, economic importance and political influence are good substitutes for numbers. Agribusiness firms, such as the major grain-exporting companies, derive much of their influence from the economic importance of the business and the political influence of the people who manage and represent them. Their managers generally are politically active, and they frequently have their own lobbyists and support trade association lobbying and consulting firm studies designed to place the best light on the alternative they favor. Agribusiness can either be potent political allies or foes to the farm lobby, depending on the issue. Likewise, environmentalists can be allies or adversaries of farmers on particular farm program issues.

Supporting Analyses

At one time, farmers may have had sufficient political clout to obtain support of the Congress or the executive branch of government by virtue of their numbers alone. Those days are past. An interest group's success today is determined by its effectiveness in presenting to policymakers complete and objective analyses of the problem, its causes, the policy options, and their consequences. Most of these analyses are prepared by the staff of interest groups, by industry consulting firms, and by university faculty who serve as industry consultants. As indicated in the previous chapter, USDA does its own analyses of the issues and options. Likewise, university faculty often conduct independent analyses of policy options. Support for a particular proposal is enhanced if analyses from more than one source indicate that a particular position is preferable. Of course, performing such analyses offers employment opportunities for agricultural economists and other scientists having expertise on the substance of specific policy issues.

Strategies Employed

The lobbying strategies employed to influence policymakers' decisions cover a wide range. These include communicating in the form of letters, newsletters, or telephone calls concerning the organization's position; providing information and undertaking special studies; answering questions; "wining and dining"; contributing to cooperative or potentially cooperative elected officials; withholding political support from uncooperative elected officials; promoting individuals for political appointments; and making elective candidates' voting records or positions on relevant issues known to particular interest groups and population segments. Interest groups, as a general rule, are cautious in taking overt public positions on the election of particular candidates from a single

Agriculture's Political Power Base Is Not Declining

Conventional wisdom suggests that declining farm population results in an erosion of agriculture's political power base. Politicians and economists alike have proclaimed the demise of agriculture's power base and the necessity for coalition building with other special-interest groups if traditional farm programs are to be maintained. Reality suggests that the decline of agriculture's political power base is more rhetoric than fact.

In 1996, a conservative Republican authorized a seven year farm program costing more than $46 billion in support of the nation's farmers at a time when money could have been saved by simply reauthorizing the 1990 farm bill. A myriad of editorials chronicled the massive expenditures of farm subsidies; yet polls often reveal general public support for spending even more on farm programs. In fact, polls show that urban dwellers are more sympathetic to farmers' problems than are rural people in rural areas. Yes, farm numbers are declining, but agriculture is continuing to maintain its base of support from both urban and rural politicians and their constituents.

The reasons for this continued support are many:

- Although U.S. agriculture is plagued by surplus production, the American public is constantly reminded, almost nightly, on the six o'clock news of the problems of shortages and famine in various countries around the world.
- Food is one of the basic necessities of life, and with each visit to the grocery store, the public is constantly, although perhaps subliminally, reminded of its direct dependence on an efficient agricultural industry. The choice, therefore, is clear.
- To err on the side of surplus food production is more politically attractive in the body politic than the alternative, even with a $20 to $30 billion price tag.
- It may be that while farm numbers are declining, farmers are becoming much more sophisticated at lobbying. The operators of large farms have enough political muscle to compensate for reduced numbers.
- Consumers, particularly the urban poor, benefit from both food programs and farm programs.

It is not surprising, therefore, that agricultural committees, USDA, and farm organizations capitalize on these strengths. It is also not surprising that the urban members of Congress support farm programs. Agriculture, like defense, sells well whether in New York City or Des Moines.

Source: Edward G. Smith, Department of Agricultural Economics (College Station, Tex.: Texas A&M University, January 1992).

political party. They realize that, over the long term, they must be able to work with any candidate who is elected.

Political action committees (PACs) facilitate the consolidation of political contributions around particular organizations, causes, and issues. PAC funds are obtained through ad hoc contributions; donations based on the gross sales of agribusiness firms; or in the case of agricultural producers, an amount for each unit of product marketed. Participation in a PAC is voluntary, and decisions on how the funds are used are made by a committee of contributors. Agriculture PACs have become an important factor in the political strategy of the sector. Although certainly not the largest U.S. special-interest group, some agriculture PACs—in particular, the dairy cooperative PACs—have ranked among the largest in the United States.

Aside from PACs, interest groups attempt to increase their effectiveness by broadening their base of support and influence through recruiting support from other interest groups. This approach, known as **networking**, involves compromise, horse trading, forming of coalitions, and logrolling.

Compromise, in a political setting, is the willingness of each side to make concessions on a particular issue so that a unified position may be reached. The willingness to compromise is an integral part of effective lobbying. This realization frequently causes opposing parties initially to adopt an extreme position, giving latitude for subsequent compromise. For example, a farm organization may initially support a 30 percent increase in price supports when actually hoping to obtain at least a 20 percent increase.

Horse trading involves the exchange of support for particular proposals among interest groups as a means of broadening the base of support. For example, a food service interest group might support a higher milk price support if, in return, a milk producers' interest group supports increased subsidies for school lunch programs.

Coalitions are alliances of groups or factions formed to attain a particular political end. They frequently result from compromise and horse trading among interest groups. Coalitions frequently are temporary, formed to help a particular piece of legislation get passed, but some tend to be enduring. The coalition of greatest importance to farm policy is Food Chain, a loose but formal alliance of some 130 farm organizations,[1] often also referred to as the *farm bloc.*

Logrolling involves sequentially building congressional support for a particular piece of legislation to the point where it cannot be resisted. Logrolling generally involves compromise and the exchange of political favors—including influence and votes—among legislators to achieve legislation and decisions of interest to one another.

SUBGOVERNMENTS

The political process is built on power relationships among members of Congress, the administration, and interest groups. *When a congressional committee,*

[1]Graeme Browning, "Sagging Aggies," *National Journal* (February 22, 1992), pp. 452–455.

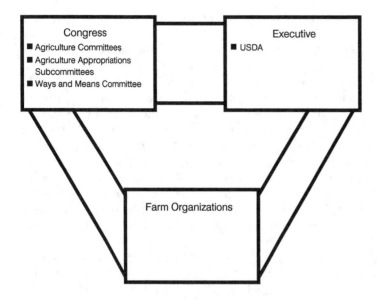

FIGURE 4.1
Iron triangle representing the
agricultural establishment.

administration agency staff, and an interest group become locked in a highly effective, reciprocal, mutually supporting arrangement, it is referred to as a **subgovernment.**[2] In a subgovernment, all three groups find their particular goals supported by the others with the result that policy is produced by this tripartite interaction, often referred to as an iron triangle.

All iron triangle members gain from the subgovernment. For the government agency, a subgovernment offers congressional support for its program and budget, less harassment in hearings or investigations, regular and predictable daily bureaucratic routine, and friendly relations with the interest group. For committee members, a subgovernment offers the chance to represent the interests of their constituents (the interest groups), to reward campaign contributions, and to secure immediate attention from bureaucrats in servicing their constituents' interests. The interest group gains a subsidy, favorable administrative rules, and an avenue for direct input into both congressional and administration decisions.

As indicated in Chapter 3, the original agriculture subgovernment was composed of the agriculture committees; the Ways and Means Committee, controlled by protectionist Congressman Wilbur Mills (Dem.–Ark.);[3] agriculture appropriation subcommittees of the Congress; the USDA; and the farm organizations (Figure 4.1). The agriculture establishment was highly protective of the farmer, promoted high price supports, and kept other interests such as the environment off the agenda.

The nature of this subgovernment has gradually changed over time. A particularly pervasive upheaval occurred in the 1972 post-Watergate election when

[2]Barbara Hinckley, *Stability and Change in Congress* (New York: Harper & Row Publishers, 1983), pp. 234–236.
[3]The power of the Ways and Means Committee was derived from its control over tariff policy.

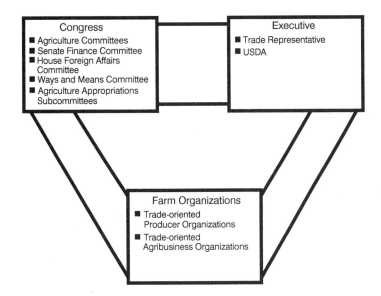

FIGURE 4.2
Iron triangle representing the agricultural trade subgovernment.

powerful representatives such as Mills were unseated. Another blow came in 1993 when Congressman Jamie Whitten (Dem.–Miss.) was replaced as chair of the House Appropriations Committee and the Agriculture Appropriations Subcommittee. However, the largest single factor influencing the policy position of the agriculture establishment subgovernment has been the increased importance of trade, international economic interdependence, and market economies. This gave rise to a reorientation and recomposition of the agriculture establishment in support of policies designed to expand trade.[4] The new trade-oriented subgovernment is illustrated in Figure 4.2.

The Senate Finance and House Foreign Affairs Committees became important actors in support of trade. In the absence of Mills, the Ways and Means Committee became less protectionist. In the executive branch, USDA took on a number of new trade advocacy faces during the 1970s and the 1980s such as Secretary Clayton Yeutter, Carroll Brunthaven (Cook Industries and Sparks Commodities), Richard Bell (Riceland), and Richard Crowder (formerly Pillsbury). The Trade Representative became more agriculture oriented with Yeutter (Chicago Mercantile Exchange) becoming its head. The point is that trade advocates permeated the USDA. As a consequence, the U.S. position in trade negotiations became free trade oriented. In the process, those interest groups having a direct interest in expanded trade became more influential.

Much of agribusiness became an integral part of the agricultural trade-oriented subgovernment. Although the old agriculture establishment lost its political influence, the agricultural trade-oriented subgovernment has gained

---

[4]Ronald D. Knutson, *External Influences on Agriculture: Societal, Environmental and Global Economy,* AFPC Policy Issues Paper 90–2 (College Station, Tex.: Texas A&M University, August 1990).

power. It can be credited with the move to direct payments as a substitute for high price supports, to lower loan rates, to decoupling, to increased flexibility, to the Canadian Free Trade Agreement, and to the North American Free Trade Agreement. In spite of this free-trade orientation, commodity programs remained intact, although they changed their nature from price supports to direct fixed farmer payments. Questions remain as to whether these commodity programs can withstand the pressures to reduce government expenditures. In the 1996 farm bill a schedule of declining fixed decoupled payments was authorized. However, at the time of the analysis, these payments were estimated to cost more than if the provisions of the 1990 farm bill had been reauthorized.

The reelection of President Clinton, the absence of any attention to agriculture during the election campaign, the appointment of Glickman as secretary of agriculture, and the appointment of several environmental and consumer advocates to high-level political positions in USDA have raised questions about the nature and viability of the agriculture iron triangle. Only time will tell whether it can any longer be characterized as such.

1. AGRICULTURE INTEREST GROUPS

Agricultural interests include producers and agribusiness. Each has its own lobbying groups and strategies. In some cases, the distinction between producers and agribusiness interests has become blurred by integration of production and marketing functions—such as in poultry—and by combinations that cut across an entire industry, as in cotton and, to a lesser extent, rice.

a) The Producer Lobby

Every commodity produced in American agriculture tends to have its own lobbying organization. The large number of producer organizations reflects both the geographic scope of U.S. agriculture and the values and beliefs regarding the role of government in agriculture. In general, the producer groups have widely differing views on policy. Despite these differences, the producer lobby has been effective in obtaining support, including substantial government expenditures for farm programs. The strength of the producer lobby lies in the number of farmers it represents; the importance of food and agriculture to particular states, congressional districts, and the nation; and the level of political activism and influence of farmers and farm organizations. In 1990, the farm population represented 1.9 percent of the total U.S. population.[5] However, in some states, the farm population is a substantially higher percentage of total population.

---

[5]Jerome M. Stam *et al., Farm Financial Stress, Farm Exits, and Public Sector Assistance to the Farm Sector in the 1980s,* Agri. Econ. Rpt. 645 (Washington, D.C.: ERS/USDA, April 1991), p. 21.

The political power of farmers tends to be augmented by a sizable number of rural residents who maintain close ties to agriculture and, in many instances, are at least indirectly dependent on agriculture for their jobs and economic well-being. It is also augmented by the economic importance of agriculture relative to other sources of economic activity. The counties where agriculture accounts for more than 20 percent of the income of total labor and business income are indicated in Figure 4.3. It suggests that agriculture's influence is concentrated in the Corn Belt, Great Plains, Northwest, and West.

There should be no doubt that with reduced numbers, farmers' effectiveness as a lobby will increasingly reflect their organization, sophistication, unity, and political contributions, as well as the perceived importance of food to the nation. They must learn to practice the politics of the minority.

For lobbying purposes, producers are organized into three major groups: general farm organizations, commodity groups, and cooperatives.

b) General Farm Organizations

General farm organizations have a producer membership that cuts across commodities. Their lobbying activities cover a wide range of commodity, agricultural, regulatory, rural, and general economic issues. This diversity of member interests sometimes makes it difficult for these organizations to take positions on certain issues and causes splintering within the group. Conflicts often arise, for example, between livestock and feed grain producers over the appropriate government policies with respect to grain price supports. Feed grain producers historically wanted higher price supports, whereas livestock producers preferred lower price supports. Grain producers want to expand exports, whereas beef and dairy producers want to limit imports. Some of the general farm organizations are also involved in the business of selling insurance and farm supplies. Farm supply sales or commodity marketings are frequently made through cooperative organizations having a farm organization identity, such as Farm Bureau and Farmers Union. In these instances, farm organization membership is frequently a condition for the purchase of either insurance or supplies. As a result, it is difficult to distinguish between members who join the farm organization because they embrace its policy philosophy or for supply acquisition reasons. Organization membership statistics, therefore, exaggerate the true policy constituency of farm organizations—particularly general farm organizations.

Over time, new farm organizations tend to be formed whenever adverse economic conditions exist in agriculture for an extended period. Each of the organizations has distinctly different philosophies on at least some aspects of policy to which a majority of its members subscribe.

The American Farm Bureau Federation is the largest of the general farm organizations. Its roots extend to the early 1900s and its activities were closely aligned with Extension Service educational programs. Their ties were shed eventually at the insistence of USDA because the Farm Bureau was actively involved in lobbying for farm programs.

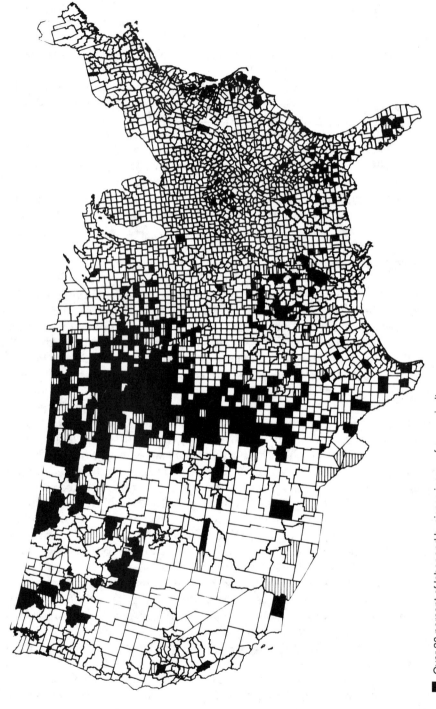

■ Over 20 percent of labor and business income from agriculture.

▦ 15 – 19.99 percent of labor and business income from agriculture.

FIGURE 4.3 Agriculturally dependent counties, 1991.

Source: Compiled from data supplied by the U.S. Department of Commerce, Bureau of Economic Analysis, 1993.

The Farm Bureau is generally recognized as being a voice of conservatism and free enterprise in agriculture. It supports market-oriented farm policies, including lower price supports; unrestricted access to markets; organized farm labor; reduced government regulation; opposition to USDA food programs; and a balanced federal budget. In the 1990s, it endorsed Republican administration proposals in trade negotiations to end all agricultural subsidies by the year 2000. This policy orientation brought the Farm Bureau into close alignment with Republican candidates and officeholders. It is, for example, typical for individuals closely aligned with the Farm Bureau to be on the staff of Republican members of Congress representing rural districts or states, as well as holding high-level positions in USDA during Republican administrations. Of course, when a Democrat becomes president, these Republican ties are downplayed.

Despite its basic conservative nature, the Farm Bureau has not repudiated all farm commodity programs and subsidies. For example, in 1966 it supported Freedom to Farm fixed decoupled payments. Substantial regional differences have traditionally existed in the Farm Bureau. For example, Corn Belt farmers have been less willing to accept production control programs than Southern dairy, peanut, and tobacco farmers. For years, the center of power in the Farm Bureau was in the Corn Belt states of Illinois, Iowa, and Indiana. In the mid-1970s, the balance of power shifted toward the South and West; then in the 1980s, it shifted back to the Corn Belt where it remains today.

The **National Farmers Union** represents the liberal side of the farm policy spectrum. Although organized in 30 states, the Farmers Union's strength lies in the Great Plains and Upper Midwest, with the largest membership in Oklahoma, North Dakota, and Minnesota.

The Farmers Union has traditionally embraced a goal of government-administered parity prices for farmers, including production control programs in times of surpluses. It supports realistic limitations on direct government payments that give preference to family farms. Preservation of the family farm remains a central goal of its policies.

In contrast to the Farm Bureau's free-trade stance, the Farmers Union has advocated pricing and market-sharing agreements with other grain-exporting countries. It has also, from time to time, suggested that grain export decisions be handled by a government-mandated organization such as a marketing board. Also, in contrast with the Farm Bureau, the Farmers Union has become actively involved in marketing milk.

The Farmers Union has a history of joining other liberal causes, including labor union rights, expanded food stamp and child nutrition programs, and increased foreign aid. It has traditionally tended to align itself with Democratic candidates and officeholders. Not surprisingly, it was an ardent opponent of the North American Free Trade Agreement (NAFTA) and the World Trade Organization (WTO).

The **National Grange,** formed in 1867 and once the largest farm organization, is currently the smallest of the three old-line general farm organizations. The Grange is most widely known for leading the fight against the power of railroads and for support of antitrust restraints before the turn of the century. The center of its strength traditionally has been the Northeast and the Northwest.

The **National Farmers Organization (NFO)** grew out of depressed commodity prices and farm incomes in the mid-1950s. The initial theme of the NFO was collective bargaining. To accomplish its objective of raising farm prices, members signed contracts pledging to market their production under NFO direction. The NFO, in turn, attempted to sign supply contracts with both corporate and cooperative processing and marketing firms. These actions met strong resistance from processing and marketing firms. The NFO, attempting to strengthen its bargaining position, initiated holding actions against processors and marketing firms in the 1960s, including dumping milk, burning grain, and killing hogs and cattle. Attempts to prevent marketing by farmers and truckers unsympathetic to the NFO resulted in isolated incidences of violence.

In the early 1970s, support declined for the NFO's militant bargaining strategies. It then augmented its bargaining activities by marketing milk and grain through cooperatives either newly organized by NFO members or by gaining control of existing cooperatives. At this time, NFO also assumed an active role as a Washington lobbyist on behalf of its members. In this role, it typically has taken positions on issues between the Farm Bureau and the Farmers Union. In 1980, the NFO favored the expanded all-risk crop insurance program. The NFO continues to press for favorable marketing arrangements on behalf of its members by signing supply contracts with processors. Like the Farmers Union, some of its strongest programs involve milk marketing in the upper Midwest.

The **American Agriculture Movement (AAM)** was organized in the mid-1970s when farm prices fell sharply from their 1973 peak. In contrast with the NFO, its center of power was in the Great Plains (Texas, Colorado, Kansas, and North and South Dakota) and in Georgia. AAM membership appeared to be composed of disenchanted NFO and Farmers Union members. Many of its leaders were younger farmers, often Vietnam veterans, who had come on rough times after expanding their farm operations and indebtedness during the relatively prosperous years from 1972 to 1975. When the cost-price squeeze became severe in the mid-1970s, these farmers turned to Washington for relief.

AAM's main early tactic for attracting the attention of lawmakers was driving tractors to Washington, D.C., disrupting traffic, and camping out for several weeks on the Capitol Mall until the Congress acted to raise price supports above the levels in the 1977 farm bill. During the demonstrations, goats were turned loose in the Capitol and Secretary Bergland took sanctuary in the White House after his office was invaded by irate AAM members.[6] During this time, extensive discussions occurred between USDA employees and farmers concerning the nature of

---

[6]During this time, Secretary Bergland took time to read a substantial volume of literature published by the USDA and the land-grant universities on the nature of the farm problem. This study led the secretary to identify the changing structure of agriculture and the associated rise of large-scale farms as a central farm policy issue and a source of continuing conflict. Subsequently, major research initiatives and hearings were held on the structure issues. This issue is discussed in Chapter 12.

the farm problem. For many USDA bureaucrats, this may have been their first time to see or talk to a farmer! The AAM also became known for placing substantial pressure on rural businesses to support and contribute financially to support their cause. But such tactics cost the AAM much political support in rural and urban America.

The initial goal of the AAM in Washington was to obtain parity price supports.[7] This was most vigorously pursued in special legislative action following enactment of the 1977 farm bill. However, the price support goals were subsequently reduced substantially. In the debate on the 1985 farm bill, AAM advocated price supports at 70 percent of parity. As farm economic conditions further deteriorated, the AAM backed legislation proposed by Senator Tom Harkin (Dem.–Iowa) and Congressman Richard Gephardt (Dem.–Mo.) to set price supports at 70 percent of parity with escalation to 80 percent, achieved by mandatory production controls and a cartel of grain-exporting nations. Although it still has disruptive elements among its membership, the AAM now appears much more inclined to work within the political process to achieve its goals.

The action of **agricultural women** and the **Family Farm Coalition** warrant discussion as relatively new general farm organization movements influencing policy. Until the mid-1970s, a wife's place in agriculture was generally one of raising children, helping with the farm work, and keeping records. Generally, women were not involved in agricultural interests outside the home farm. The women's rights movement changed this situation for many farm families. Nearly as many farm wives as nonfarm wives now have off-farm jobs.

Farm women's groups have sprung up throughout the major agricultural regions of the United States. Examples include American Agri-Women (AAW), Women Involved in Farm Economics (WIFE), Concerned Farm Wives, Partners in Action for Agriculture, and California Women for Agriculture. Women's organizations generally support maintaining conditions favorable to family farm survival, increased appropriations for extension, work incentives for food stamp recipients, increased ease of estate transfers, less government regulation, expanded export markets, and higher price and income supports.

The Family Farm Coalition has a base of support that is broader than women's farm organizations. They include the NFO, AAM, Farmers Union, some women's groups, some commodity groups, and some public interest groups. Their relatively liberal farm program agenda is aimed at raising farm price supports, excluding large farms from obtaining farm program benefits, and related causes.

Two major organizations having important religious ties are part of the family farm lobby. One is the National Catholic Rural Life Conference (NCRLC). Another is the Interreligious Task Force on U.S. Food Policy, a coalition of 27 denominations and interfaith groups. Although also concerned with hunger-related issues,

---

[7]*Parity* relates the current price of farm products to that which existed in 1910–1914. Since it does not fully take into account changes in technology, the parity price generally is at least double the current farm price. Parity is discussed in greater detail in Chapter 10.

the Interreligious Task Force has consistently supported farm programs designed to maintain an economically sound family farm structure. These religious groups have a definite agricultural fundamentalist orientation. They see the decline of family farms as directly influencing rural church membership, the size of the collection plate, and the health of rural communities. Ties to the more liberal farm organizations such as the Farmers Union, AAM, and NFO are evident.

b) Commodity Organizations

Commodity organizations represent producers of specific agricultural products. They are frequently more effective than the general farm organizations on specific commodity-oriented issues for four major reasons:

- As agricultural production has become more highly specialized, farmers identify more closely with a commodity organization.
- Commodity organizations frequently can speak with a clearer voice on commodity issues. Conflicts among producers of different products, such as dairy farmers and beef ranchers, that arise in general farm organizations are not as apparent in commodity organizations.
- Once underfinanced, most commodity organizations now have access to federal- and/or state-mandated producer check-off funds. A check-off program deducts a small amount per unit or a small percentage of the value of product marketed. Although these funds are intended primarily for research and promotion and cannot be used for lobbying, commodity groups are strengthened financially by the resulting flow of funds and the visibility given them by promotional activities. Federal and state check-off programs now exist for most major commodities as well as for many specialty crops.[8]
- They tied in easily to the commodity orientation of congressional subcommittees, commodity sections of farm bills, and commodity analysis conducted by USDA. The 1996 farm bill was, however, less commodity oriented and the House Agriculture Committee no longer has single commodity subcommittees.

At least 20 commodity organizations are engaged in food and agricultural lobbying in Washington, D.C. Individually, the amount of influence of these organizations changes over time, depending on the strength of their leadership, their cohesiveness, and the nature of their felt needs. From an historical perspective, the most influential of the organizations representing the major commodities include those discussed in the following paragraphs.

---

[8]David S. Cloud, "When Madison Avenue Talks, Farm-Belt Members Listen," *Congressional Quarterly* (November 11, 1989),pp. 3047–3051.

The National Association of Wheat Growers (NAWG) is one of the largest and most influential of the producer commodity organizations. It has traditionally been a leader in overall strategy in farm program development. However, in the 1980s, its relative strength appeared to wane. This further declined with their reluctance to support Freedom to Farm even though several of their state affiliates did so.

Because more than 65 percent of the wheat produced is normally exported, NAWG has a strong interest in export market development and the impact of government policies on the export market. As a result, it has had close working relations with Wheat Associates, a market development check-off funded organization. Accordingly, it has traditionally supported farm programs that provide producers with income support while maintaining the competitive position of wheat in the export market, although some producer segments still appear to favor high loan rates. Since much of the wheat is produced in areas having substantial risk of drought, disaster protection has also been of direct interest.

The **American Soybean Association (ASA)** and the National Cattlemen's Beef Association (discussed later) are probably among the most conservative of the producer organizations. Many of the efforts of the ASA have been in support of domestic and foreign market development programs and in opposition to government interference in export markets. The ASA has been active in promoting the use of soybean meat extenders in the school lunch program.

Not until 1985 did the ASA seek government programs supporting direct payments to soybean producers. In 1985, it sought direct government payments of $50 per acre for soybeans, and in 1987, it sought a marketing loan program similar to that for rice and cotton that would have provided direct income support to its members. In 1990, they received marketing loan authority, which was extended in 1996. The trend appears to be for ASA to support more actively the types of programs that exist for other major crops.

The **National Corn Growers Association (NCGA)** represents producers of a crop that, in volume, is four times as large as wheat. Over time, its political influence has increased decisively to become one of the strongest commodity groups. Although formerly quite conservative, NCGA appears to have become more liberal over time. It has supported substantially higher price and income supports, the grain reserve, subsidized loans to build storage facilities, and efforts to control production. It also supports subsidies for ethanol production (used to increase the octane in gasoline) and high sugar price supports (see box). NCGA also supported decoupled fixed payments in the 1996 farm bill.

The **National Cotton Council (NCC)** is a unique organization in that it attempts to represent the policy interests of the entire cotton industry. Its members include producers, shippers, merchants, exporters, and textile mills. As a result, its policy positions tend to be more closely aligned with agribusiness firms than with producers. Yet its Producers Steering Committee is always a "who's who" of the U.S. cotton farmer kingpins, many of whom individually wield substantial political power. All in all, NCC has to rank high on the list of the most powerful farm organizations. Seldom does it go to the Congress without a unified industry position, and it is always represented on the Hill and in USDA.

Shooting Yourself in the Foot

For years, sugar producer lobbyists have pressed for high price supports for sugar, maintained by controlling sugar imports. An unforeseen side effect of this policy has been development of a large corn sweetener industry. High-fructose corn sweetener (HFCS) developed in this protected market environment. Its lower cost allowed it increasingly to substitute for sugar, to the extent that it has garnered over half of the total domestic caloric sweetener market. As HFCS consumption and domestic sugar production have increased (both due to high price supports), the sugar import quota has been reduced and in the late 1980s was approaching zero. Reductions in import quotas have caused sugar producers in developing countries such as the Philippines to complain loudly.

As HFCS consumption has increased (and sugar consumption declines), corn producers, prodded by HFCS manufacturers, have become more overtly supportive of high sugar price supports. Sugar producers responded to high price supports by increasing the production of sugar while sugar consumers reduced consumption. The 1996 farm bill was a test of the power of sugar. Decoupling support from production was applied to the crops. The support program was phased out in dairy. Peanut quotas became transferable. But sugar lost nothing. In reality, however, the big gainers once again were the HFCS, ADM (a major HFCS processor), and corn producers.

NCC initially opposed decoupled fixed payments in the 1996 farm bill representing primarily the interests of farmers fearing the decline of cotton production under enhanced flexibility. At least two congressmen from cotton country helped to derail Freedom to Farm on its first vote, forcing the chairman to attach it to the Budget Reconciliation Act. However, at the end, Freedom to Farm had cotton support once cotton farmers and their bankers understood it.

The **National Milk Producers Federation (NMPF)** is the voice of dairy farmers who are members of regional dairy cooperatives. Therefore, NMPF could also be classified as a cooperative lobby. The NMPF's major objectives include maintaining milk price supports as high as is politically feasible, maintaining quotas restricting imports of dairy products, protecting the federal milk marketing order system against attack, and maintaining the use of milk products in domestic and foreign food assistance programs. Until the late 1980s when such programs fell out of vogue, the NMPF was very successful in pursuing these objectives. Its success was attributable to the number of producers it represented in every state (accounting for more than 75 percent of total U.S. milk production), strong financial support from its member cooperatives (also politically active), and a substantial number of political action committees associated with its cooperative members. Its reduction in strength resulted from its continued advocacy of unpopular production control programs, its unwillingness to change (compromise), the increasingly divergent regionalism of the dairy industry (particularly in terms of producer size), and its loss of key leadership.

Finally, dairy policy was overhauled under the 'new' leadership of Congressman, and subcommittee chair, Steve Gunderson (Rep.-Wisc.). Price supports were phased out and marketing orders were consolidated.

The **National Cattlemen's Beef Association (NCBA)** is recognized as being an effective lobbyist for cattle raisers and feeders and other agribusiness segments of the industry. The resulting organization is not unlike the NCC, but it does not include the entire industry. The NCBA's objectives include limiting government involvement in its members' operations, limiting beef imports, limiting grazing of acreages idled by farm programs, controlling natural predators, supporting government policies that encourage more beef consumption, opposing the adoption of production controls for milk, and include buying and grazing of grasses and legumes (alfalfa) under the flexibility provisions of the 1996 farm bill. A battle ensued between alfalfa producers and cattlemen that alfalfa producers won. Considerable conflict has developed between the NCBA and public interest advocates over dietary guidelines that suggest moderate beef consumption and reduced cholesterol; the use of hormones in cattle feeding; and issues relating to animal rights, endangered species, property rights, and pollution control.

Until the 1980s, NCBA had shown relatively little interest in policy development for commodities other than beef. However, when the 1983 dairy diversion program resulted in many additional milk cows being slaughtered, beef producers became concerned about the resulting adverse effect on beef prices. NCBA lobbied unsuccessfully against the dairy termination (buyout) program in 1985 and then brought legal action against the secretary of agriculture, alleging that he failed to regulate the flow to market of slaughtered dairy cows, as required by the authorizing legislation. In the early 1990s, it successfully opposed dairy efforts to enact any type of production control program. This flurry of political and legal activity substantially broadened the role of NCBA in commodity program development.

The **Public Lands Council** was formed because NCBA membership could not agree on the appropriate policy regarding such issues as grazing fees on public lands. Western cattle raisers tended to favor low grazing fees on public lands, and Eastern cattle raisers concluded that the resulting competition drove down their prices. The Public Lands Council once dominated Bureau of Land Management decisions on grazing fees and environmental interests. However, with increased influence of environmental interest groups, the Bureau of Land Management and the Public Lands Council have more frequently become forceful adversaries over issues such as grazing fees. These conflicts spread to other areas such as endangered species and water pollution.

c) *Cooperatives*

Cooperatives are becoming some of the more effective producer lobbying forces in agriculture. Although concerned mainly with protecting their members' special organizational privileges, cooperatives were an original advocate of the export-oriented marketing loan concept. Their lobbying strategies take advantage of the combination of the producer numbers they represent, their effectiveness in communicating their needs, and their organization of political action committees.

For some cooperative organizations, lobbying activities relate largely to their legal status. Other cooperatives, such as dairy and rice, combine protection of their legal status with intensive commodity-oriented farm program lobbying. For example, Associated Milk Producers, Inc.; Dairymen, Inc.; and Mid-America Dairymen are cooperative organizations that engage in commodity lobbying activities. Aligned with each of these three organizations is one or more PACs. In recent elections, these PACs have contributed the majority of all contributions made to congressional candidates by agriculture-related PACs.[9] These contributions also helped to fortify the lobbying activities of the National Milk Producers Federation.

Another example of cooperative activity in commodity lobbying involves rice; Riceland Foods, Inc., and Farmers Rice Cooperative attempt directly to maintain a policy environment that facilitates exports while providing their producers income protection in the highly volatile rice market. Rice producers unsuccessfully opposed the enhanced flexibility in the 1996 farm bill.

In addition to such commodity-oriented cooperative lobbying activity, two national cooperative organizations have extensive political activities.

The **National Council of Farmer Cooperatives (NCFC)** is a lobbying organization whose membership is composed basically of large regional cooperatives. It commits a major share of its resources to defending the marketing cooperatives' antitrust exemption provided by the Capper-Volstead Act, maintaining the preferential tax status of cooperatives, and defending market order legislation.[10] In addition, the NCFC directs substantial attention to agricultural chemical and pest control regulations of the Environmental Protection Agency, agricultural transportation issues, and energy problems. Since the mid-1980s, NCFC has actively lobbied for farm bill provisions that support its members, such as the marketing loan concept to be applied to agricultural commodities. It has been an active participant in and supporter of GATT and Canadian and Mexican trade negotiations.

The **National Rural Electric Cooperative Association (NRECA)** is one of the most powerful, yet least noticed, producer lobbying groups. Electrification of rural America occurred through the help of the government. One agency of USDA—the Rural Electrification Administration[11]—was devoted almost exclusively to extending credit to rural electric cooperatives. Much of this credit has been heavily subsidized. Ending or substantially reducing such subsidized cred-

---

[9]Ibid.

[10]Farmer cooperatives have the benefit of a host of exemptions not available to other business firms. Among these are the right to merge with one another and engage in joint pricing activities as long as predatory activity is not involved. Marketing orders help to fortify the effectiveness of these activities that are designed to raise farm prices and incomes. These cooperative benefits are discussed in Chapter 12.

[11]Now the Rural Utilities Service.

its was an objective of the Reagan and Bush administrations. It continues to be advocated by OMB in the Clinton administration.

In a time when budget cutting is a high national political priority, a major share of NRECA's lobbying resources is devoted to defending its existing subsidized credit sources. Other projects include protecting the preferential tax status of cooperatives, assisting in obtaining approval of the location of power plants and transmission lines, and discouraging taxes on energy sources. Under the leadership of former Agriculture Secretary Bergland and former Congressman English, NRECA expanded its concerns beyond rural infrastructure into the arenas of rural health and education.

The Agribusiness Lobby

Agribusiness firms have a large number of food and agricultural lobbyists in Washington that are among the most powerful in the agriculture establishment. Their political clout results from a combination of their knowledge of Capitol Hill, their ability to work with members of Congress and their staffs, their knowledge of how decisions are made in the executive branch, attention to detail, their knowledge of the substance of the issues, and their political contributions through PACs and as individuals. Their effectiveness is enhanced by the fact that they do their job, typically with little public attention.

It is not at all unusual for a person lobbying for the agribusiness sector to move into a high-level post in USDA. For example, Richard Lyng served as the president of the American Meat Institute before becoming secretary of USDA. Conversely, that person might move back to a firm in the same industry. Clayton Yeutter moved from USDA to the Chicago Mercantile Exchange, to Trade Representative, to the secretary of agriculture, to the domestic policy adviser position for President Bush. While in the private sector, he and Lyng sat on many agribusiness and cooperative boards of directors. *This process of changing jobs between private industry and government is referred to as the* **revolving-door syndrome.** While ethics and conflict of interest curbs have been put on revolving door activities, movement between the public and private sector is inherent in the U.S. political process.

Agribusiness firms clearly feel a need to be represented in the political process. They want to protect a market system of privately owned business firms against government encroachment. They have traditionally opposed tax and antitrust preferences for cooperatives. They recognize the role that government plays in promoting economic and market stability. They also recognize government as a major purchaser of processed farm products through the school lunch program, as a storer of grain, as a provider of export credit, as a partner in promoting export sales, and as a provider of economic stability to farmers and of credit for rural development.

Agribusiness and producer interests may well diverge over matters of farm policy. Agribusiness generally favors government programs designed to expand farm production and opposes programs that restrict production. However, the interests of all agribusiness firms are not the same. Those producing farm supplies have a greater stake in farm profitability than do marketing firms. Farm machinery, chemical, and fertilizer sales are lower when farm incomes are reduced. The ability of farmers to pay off farm loans also decreases with income reductions. Lobbyists representing farm supply firms and banks, therefore, are more likely to come to the aid of farmers in times of depressed farm prices and incomes than are marketing firm lobbyists. Yet this aid tends to be more in the form of support for subsidies rather than production controls. Farm input sales occur in proportion to the number of acres farmed or cows milked.

An interesting case in point is the lobbying effort of an agribusiness coalition for the elimination of acreage set-aside programs including its CRP during the debate on the 1996 farm bill. This coalition was considerably weakened when Dwayne Andreas of Archer-Daniels-Midland bolted the coalition and supported continuation of farm programs.

Agribusiness lobbyists may be placed in three groups: (1) general organizations, (2) commodity organizations, and (3) Washington representatives of individual firms. General agribusiness organizations represent both input supply and marketing firms. Commodity organizations focus on an individual input or commodity. Washington representatives of agribusiness may either be employees of a firm, a law firm, or a public relations firm hired to represent them in Washington. Some large agribusiness firms may be represented by general organizations or commodity organizations as well as their own Washington representatives.

q) General Agribusiness Organizations

General agribusiness lobbyists have substantial involvement in a wide range of food and agricultural policy issues.

The **Grocery Manufacturers of America (GMA)** represents most major processors and manufacturers of consumer food products in the United States through its spokesperson, former Agriculture Secretary John Block. Its major policy concerns relate to reducing regulation of nutrition, food safety and quality, advertising, packaging and labeling, antitrust, imports, and retail business practices. Although generally not directly concerned with farm program issues, the GMA regularly joins other groups on commodity-specific issues, such as sugar user organizations, to oppose sugar price supports and import restrictions.

The **Food Marketing Institute (FMI)** represents a large proportion of firms engaged in the retail and wholesale distribution of food. On behalf of its supermarket chain members, FMI traditionally has supported the food stamp program—in part, because it expands grocery sales. It has taken a keen interest in opposing antitrust regulations and related concerns about the competitiveness of the food industry. It has opposed regulations that would require stamping prices on individual grocery items. Since FMI's members handle the products of the GMA and many food retailers are also some of the largest manufacturers of groceries, the FMI has many of the same interests as the GMA.

The **American Frozen Food Institute** represents frozen-food manufacturers across the commodity spectrum. While having major interests in regulations relating to preparation and labeling of frozen foods similar to those of GMA, a primary concern of the Frozen Food Institute has been encouraging expanded sales of frozen prepared foods through the school lunch program. In particular, it has been a strong supporter of providing schools with cash instead of surplus commodities acquired through the farm commodity programs. The schools then would be free to purchase fully prepared meals, of which a substantial share would likely be frozen. It has also had substantial interests in government policy regarding energy and the impacts of refrigeration on the environment.

b) Commodity Agribusiness Organizations

The **National Grain and Feed Association (NGFA)** represents grain marketing firms at all levels of the market channel from county elevator to exporter. The policy positions of the NGFA reflect the market orientation of its major grain-exporting members, unimpeded market operations, and the members' interest in the use of grain as a feed. Although the NGFA favors price supports for farmers, it advocates maintaining them at low levels so as not to interfere unduly with domestic and export uses. Its support for grain reserves reflects the involvement of many of its members in grain storage. It strongly opposes production controls.

The **American Feed Manufacturers Association (AFMA)** has many interests and members in common with the NGFA. For example, it favors not only a market-oriented farm policy but also grain reserves to stabilize feed prices. However, its policy emphasis tends more toward issues of importance to feed production for livestock and poultry, such as government regulation of the use of antibiotics, growth stimulants, and of the amount of residues transmitted from feed to meat, milk, and eggs. It is also directly concerned with protecting the rights of its members to become more involved in livestock and poultry production through various forward integration strategies.

The **North American Grain Export Association (NAGEA)** has as its major purpose the expansion of grain exports. Its members include virtually all of the grain-exporting firms operating in the United States. It has a direct interest in farm, trade, foreign, and general economic policies as they affect exports. Therefore, it was a strong supporter of Bush and Clinton administration proposals to end all agricultural subsidies in the GATT negotiations. It is probably the most conservative and free-market oriented of the agribusiness lobbyists, being generally opposed to higher price supports, production controls, higher inspection standards, embargos, and the use of grain as a tool of foreign policy. In addition, NAGEA has advocated reduced import restrictions on dairy products and beef because such restrictions are inconsistent with the ease of greater market access for its products.

The **American Bakers Association,** as an important grain buyer, at times may have conflicting interests with other agribusiness grain lobbying interests.

Although it shares the market-oriented farm program philosophy, when wheat prices began to rise sharply in the mid-1970s, it asked the secretary of agriculture for relief in terms of either a price freeze or embargo of exports.

The Bakers Association's overall goal is securing policies that hold food price inflation low and encourage full production. Accordingly, it has advocated low price supports not only for grain but also for sugar and dairy products, all main ingredients in the products of its members.

The **American Meat Institute (AMI)** represents meat packers throughout the United States. Its major concerns are increased government regulation of the meat industry, defending sanitary conditions in the meat trade, opposing nutritional labeling, and defending curing additives such as sodium nitrite. AMI has had a continuing interest in the contribution of meat to food price increases. In this regard, it has been concerned about studies of the Small Business Committee, the Department of Justice, and the Federal Trade Commission linking a high market share of the industry's largest packers and packer ownership of feedlots to meat price increases. Its interest in this issue increased in the mid-1980s when the market share of the largest meat packers rose sharply due to several mergers and acquisitions. In the early 1990s, AMI sought to separate meat from the announced changes in requirements for food labeling. It has always been an active advocate of keeping meat inspection activities (FSIS) in USDA rather than in FDA.

The controversy concerning "captive supplies" in the beef industry is also of concern to the AMI. "Captive supplies" include supplies either owned outright by packers or contracted by feedlots with the packers.

The **International Dairy Food Association (IDFA)** represents milk processors and is especially concerned with government regulation of the milk industry. Its activities cover three major areas: (1) government regulation of producer prices under the milk price support and federal marketing order system, (2) regulatory standards for the composition of milk and ice cream products promulgated by the Food and Drug Administration, and (3) government action that influences prices such as sugar for the manufacture of ice cream or the availability of casein for use in cheese or coffee whiteners.

In its lobbying activities, the IDFA has tended to emphasize studies of alternative methods of solving particular milk pricing problems as a means of fostering industry dialogue and providing Congress with information relevant to policy issues. Interestingly, it has not been an active opponent to regulation of producer prices under milk marketing orders. It has been particularly effective in organizing coalitions of user groups for issues relating to milk and sugar. In the 1990s, its effectiveness appeared to increase as that of the milk cooperatives declined.

The **Sweetener Users Association** is a coalition of food firms for which sugar and other sweeteners are an important ingredient. Its major objectives are to eliminate sugar price supports and quotas, which raise the price to its members. Its members include soft-drink bottlers, bakers, ice cream manufacturers, and confectioners. It has lost many battles with the more powerful sugar and HFCS lobbies.

The **Tobacco Institute** has been described by one health advocate as "one of the most lethal trade associations going."[12] Bush administration HHS Secretary Louis Sullivan called tobacco revenue "blood money" and sellers of cigarettes "merchants of death."[13] A large proportion of its efforts have been directed toward defending against the mounting evidence linking smoking and health problems, fighting restrictions on smoking in public places, and supporting studies of methods of reducing potentially harmful substances in tobacco smoke. More recently, it has become a strong advocate of expanded exports. To accomplish its lobbying objectives, it has employed two former representatives, a former senator, and a former governor. Whether or not tobacco is addictive was an issue in the 1996 presidential campaign and the use of tobacco by teenagers is an issue that has put the Tobacco Institute on the defensive.

The **Fertilizer Institute** is the association of firms engaged in both the manufacture and distribution of fertilizers. The institute traditionally has supported programs that improve farm prosperity. It has also traditionally had a special interest in USDA soil conservation programs and has been a supporter of research and education. However, it has been opposed to environmental and sustainable agriculture advocate proposals that would curb fertilizer use by regulation or taxation. Since fertilizer sales occur in proportion to acres farmed, the Fertilizer Institute has been an active opponent of programs that remove farmland from production, especially proposals for mandatory production controls.

The **American Bankers Association (ABA),** agricultural bankers division, protects the interests of bankers in government agricultural lending programs. Although concerned about expanded credit through the Farm Services Agency and the Small Business Administration, public agencies such as the ABA have generally been supportive of expanded government-guaranteed loans. When the Cooperative Farm Credit System required an infusion of federal capital, ABA was mainly concerned with maintaining a level playing field for its banks by such means as securing a secondary market for farm mortgages.

The ABA traditionally has supported government programs that reduce risk in agriculture such as crop insurance, disaster payments, and, in some instances, target prices. Support also has, from time to time, been expressed for higher price and/or income supports on the grounds that it would increase deposits and allow farmers to repay some otherwise insolvent loans. However, it has consistently opposed government subsidies on interest rates for any type of loan. ABA applauded suggestions of the Reagan and Bush administrations to reduce subsidies on rural utility loans and to limit the functions of the Farm Services Agency to guaranteeing loans.

The **National Agricultural Chemical Association (NACA)** is the trade association that represents pesticide manufacturers. It is the focal point for what has

[12]Harold D. Guither *The Food Lobbyists* (Lexington, Mass.: Lexington Books, 1980), pp. 68.
[13]Alyson Pytte, "Tobacco's Clout Stays Strong Through Jobs, Ads,"*Congressional Quarterly*(May 12, 1990).

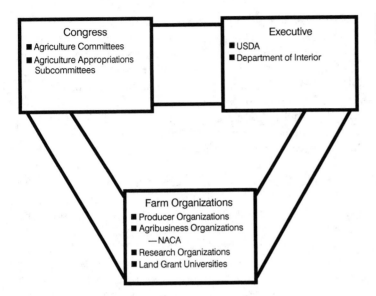

FIGURE 4.4
Iron triangle representing the pesticide subgovernment.

become known as the pesticide subgovernment (Figure 4.4).[14] For years, this subgovernment maintained control of the Federal Insecticide, Fungicide, and Rodenticide Act (FIFRA) within the agriculture committees. NACA is the focus for defense of pesticides, which brings it into direct conflict with environment and food safety advocates.

3 The Public Interest Lobby

It is a prime responsibility of the Congress, the president, and the courts to see that decisions are made in the public interest. The term **public interest** *means to benefit society as a whole as opposed to a particular segment of society.* Other than the special interests of producers and agribusiness firms, there exist a large number of people who are affected by food and agricultural policy and program decisions. These people are generally known as the public. Their interest in policy covers a wide range of values, beliefs, goals, and factual knowledge. Therefore, they belong to many different organizations and express a variety of concerns. *Those groups that attempt to reflect interests other than the special interests of producers and agribusiness groups are generally known as* **public interest groups.**

The public interest in agricultural and food programs differs from farmer and/or agribusiness interest for the following reasons:

- Food is a necessity of life. An adequate supply of food, therefore, is a necessity. Individuals deprived of food suffer serious mental and physiological

[14]Christopher J. Bosso, *Pesticides and Politics: The Life Cycle of a Public Issue* (Pittsburgh, Pa.: University of Pittsburgh Press, 1987).

consequences. Society bears a moral obligation to prevent such consequences. Although both farmers and consumers may agree on this point, questions arise over who should bear the cost of eliminating hunger and malnutrition. Should dairy farmers be required to pay for the higher costs of milk to mothers who receive welfare assistance when the milk price increase resulted from an increase in the price support level?

- Food costs money. The average consumer spends about 11.4 percent of disposable income on food, but lower income consumers spend a considerably higher percentage of income. All consumers are affected by a change in the price of food, and commodity prices do affect retail prices. Also, farm programs can affect farm and food prices. While farmers and consumer advocates may argue over the existence of a "cheap food policy," public interest in the price of food is legitimate and can be expected to continue.

- Food yields externalities. The public is affected by externalities originating from agricultural production such as stream pollution from runoff, soil erosion, pesticide residues, chemicals in the soil and water, and adverse effects on wildlife. The public has a legitimate interest in curbing such externalities.

- Food can be unsafe. Consumers want to be assured of a safe food supply and that farmers, processors, retailers, and the government are protecting it from contamination. Yet even the mention of a safety problem with a particular product has extremely adverse effects on sales, prices, profits, and even the survival of affected firms. Consumers clearly have an interest in food safety.

- Public interest groups represent taxpayers. When reducing the federal budget deficit is a national priority, taxpayers have a legitimate interest in farm program costs, just like the cost of all other government programs.

The question is raised from time to time by producer and agribusiness groups as to the extent to which public interest groups truly represent the public. Each public interest organization does represent a specific set of concerns other than the primary concerns of producers and agribusiness. Each interest group represents the interest of its constituency and its financial contributors whether it is a "public" interest or a special-interest group. The Congress and the executive branch have to gauge the extent of the support of all groups in reaching decisions on various issues. How well they perform this function is determined by the public in subsequent elections.

During the 1970s, a sharp increase occurred in the number of public interest organizations involved in food and agricultural lobbying. The level of their activity and effectiveness likely peaked during the Carter presidency when a substantial flow of public funds supported these groups. The Reagan administration moved quickly to reduce these funds, both because of their Democratic political alignment and as part of overall efforts to trim federal spending. There has been some indication of resurgence during the Clinton administration. Despite their relatively small funding, these interest groups are very often effective in presenting their concerns. They primarily focus on four major areas: consumer concerns, nutrition, hunger, and the environment. Environmentalists, are much better financed than the other three. As a result, they have become one of the most effective lobbies in the United States. A few public interest organizations work in more than one of these areas and are mentioned below more than once.

⚘ The Consumer Food Lobby

The consumer food lobby is led by the Community Nutrition Institute, the Consumer Federation of America, the Consumers Union, Congress Watch, Public Voice for Food and Health Policy, and the Center for Budget and Policy Priorities. These and other public interest organizations frequently form coalitions on specific issues to provide a united consumer lobby.

The **Public Voice for Food and Health Policy** engages in a wide range of agricultural and food policy issues on behalf of consumers. Its participation in farm policy development is built on the formation of coalitions among consumer, nutrition, and environmental groups. Ellen Haas, its founder, has devoted her professional life to representing the public in agricultural and food issues. Before becoming Clinton's undersecretary of agriculture for food programs, Haas was a very effective lobbyist and a nationally recognized figure who had studied and learned all dimensions of agricultural and food policy. Public Voice is relatively small, but its impact is large. Haas was among the most feared of the public interest lobbyists by the agriculture establishment, which lobbied very hard to prevent her from being put in charge of USDA food programs.

The **Consumer Federation of America (CFA)** represents consumers on a wide range of issues, including food and agricultural issues. CFA has a membership composed of credit unions, electric cooperatives, labor unions, state and city consumer associations, and, somewhat surprisingly, the National Farmers Union. CFA once supported changing USDA to a Department of Food responsive to the needs of all affected by the food system. One of its former presidents, Carol T. Foreman, served as assistant secretary of agriculture for food and nutrition programs in the Carter administration. Since its organization in 1968, CFA has been an important factor in enacting stronger federal meat and poultry inspection requirements, eliminating the food stamp purchase requirement, reducing food additives, improving food labeling, encouraging direct farmer marketing programs, and opposing sugar and milk price support increases.

Congress Watch, the lobbying arm of Ralph Nader's **Public Citizen** organization, provides oversight on a wide range of public issues before the Congress. Congress Watch regularly testifies before the agriculture committees on farm programs as well as consumer issues. It also monitors the voting records of members of Congress and their relation to lobbyists that support them with political contributions. For example, it has been critical of large contributions made by dairy cooperatives to members serving on the agriculture committees.

It is noteworthy that Ralph Nader's Public Citizen has been associated with several projects that looked critically at agricultural firms and lobbyists. For example, the Agriculture Accountability Project sponsored a study that was critical of ties among the land-grant universities, large farmers, and agribusiness firms.[15] Another one of its studies charged that most large cooperatives were controlled by their management rather than by their producer members.[16] Both studies, although extensively criticized, contained some elements of truth and had an impact. Some suggest that is why they were criticized so extensively.

Consumers Union has also published several studies critical of farm programs, government regulation of food industries, and the products of food firms. Its main enterprise, however, traditionally has been evaluating quality and performance of particular products, including food products. After the early 1970s, it also became an active advocate for particular causes related to food, especially regulations directly affecting food prices, the use of hormones in raising livestock and poultry, and the safety of the food supply. Its studies, as a general rule, receive extensive publicity and are calculated to have maximum impact.

The **Community Nutrition Institute (CNI)** has been a major consumer force for specific food and agricultural policy issues. The CNI is a nonprofit advocacy agency whose goal is ensuring all people an adequate, safe, nutritious, and affordable diet. It is, to an important extent, a consumer research and lobbying agency, treating a broad scope of domestic consumer concerns. Its activities include studies that are largely supported by contracts with government agencies. Many of its projects also involve assembling experts on particular topics for workshops, seminars, and conferences. CNI publishes the *CNI Weekly Report*, a newsletter that provides an excellent summary of current issues and legislative activity related to food.

CNI has been a leading advocate for a more consumer-oriented USDA. It has been deeply involved in efforts to accomplish a reevaluation of the marketing order program. CNI was active in the 1973 meat boycott and has been involved in all farm bills since 1973. It effectively led opposition to Reagan administration proposed cuts in nutrition programs. However, it declined in influence when Ellen Haas left to form Public Voice.

The **Center on Budget and Policy Priorities (CBPP)** was founded by Robert Greenstein, the administrator of USDA's Food and Nutrition Service during the Carter administration. CBPP was organized to do analytical work on Reagan food policy. Its success in serving the policy analysis needs of both Republicans and Democrats is based on the accuracy of its numbers, timeliness, and objectivity.

Labor unions, including the AFL-CIO, United Food and Commercial Workers, United Autoworkers, Aerospace and Agricultural Implement Workers, and the National Association of Farm Worker Organizations, have been active in promoting consumer causes. They have surfaced concerns of their members in consumer advocacy groups as well as in the Congress. Equally important, they have provided a critical base of financial support for the consumer movement.

---

[15]Jim Hightower, *Hard Tomatoes, Hard Times* (Cambridge, Mass.: Schenkman Publishing Company, 1973).

[16]Linda Kravitz, *Who's Minding the Coops?* (Washington, D.C.: Agriculture Accountability Project, 1976).

The major labor union concerns generally can be classified as being those of middle-class Americans—the cost of food, assurances that the food supply is safe, and nutritional labeling—and, in the 1970s, of strikers.

The center of the agriculture labor movement for many years was in California under the leadership of Cesar Chavez. The major thrust of the movement to organize farm labor initially occurred among grape and lettuce workers. These efforts were actively supported by nonfarm unions. Their success is illustrated by the fact that California became the first state to establish an agricultural labor practices law.

The Nutrition, Food Safety, and Quality Lobby

The nutrition, food safety, and quality lobby is concerned primarily with the impact of diet on health. The relationships among health, diet, nutrition, and the safety of the food supply have spawned some of the most direct conflicts between public interest advocates and producers. Issues such as the establishment of nutrition guidelines, nutritional labeling, the linkage of cholesterol to heart disease, the relationship of nitrites to cancer, the use of growth stimulants in cattle feeding, the approval of bovine somatotropin (rBST), and genetic engineering such as Roundup Ready soybeans have created highly emotional public reactions.

With an origin in issues of sanitation in the 1890s, the food safety lobby made the safety of eggs, broilers, and red meat important concerns that reappeared in the 1990s. The nutrition, food safety, and quality lobby has been a strong supporter of research to examine the nature and extent of the causal effect of diet on health. As a result, it should not be surprising that the major interest groups involved in this area have an applied scientific orientation.

The **Center for Science in the Public Interest** is a highly respected consumer activist organization that investigates food, nutrition, consumer, and environmental concerns. The center has become a focal point for specific issues that are closely related to new scientific evidence. Its activities are closely coordinated with those of other consumer, hunger, and environmental groups. Its monthly magazine, *Nutrition Action,* has become a guiding force for focusing on nutrition and food problems in the United States and other countries.

In addition to its major concerns about the relationship of food to health, the center has spoken out on such diverse issues as low-nutrient processed foods, corporate investment in agriculture, food prices, and conflicts of interest between industry and government. It created substantial controversy within agriculture and the food industry when it created a list of foods that it considered to be good and bad for kids. In one of its lighter but yet serious moments, it raised questions concerning the nutritional example set by President Reagan's Jelly Belly (a jelly bean candy) consumption. More recently, it has become an advocate against the use of biotechnological products in agriculture and the potential introduction of new genetic or transgenic material into the food supply and the environment.

The **Health Research Group** is another Ralph Nader organization related to Public Citizen and Congress Watch. It has been concerned primarily with the safety of food additives, and is extremely critical of existing food safety regulations and of the scientific staff in both the Food and Drug Administration and USDA.

Health Research has advocated a policy of considering all "generally recognized as safe" (GRAS) substances as potentially dangerous unless scientifically proven otherwise. It has also suggested that new food additives not be permitted in food unless their benefit is demonstrated to be more than cosmetic and greater than an existing additive. It has proposed that all safety-related testing of additives be shifted from industry to academic centers. Needless to say, the food industry has become very concerned with such suggestions.

The **Society of Nutrition Education** is an organization of professionals in nutrition education and related fields. Its major mission has been to increase the effectiveness of nutrition education. Out of its ranks, new leadership in nutrition education developed, advocating a more activist role in educating consumers and influencing policy on the relationships between diet and health. Specific concerns include support for the issuance of food consumption guidelines, dietary goals, increased nutrition research, educational activity on the dangers of additives and highly processed foods, and improved food labeling. The society has been a major factor in shifting the thinking of nutrition educators from the basic food groups to those ideals embedded in the dietary goals and guidelines, including those embodied in the food pyramid.

The Hunger Lobby

The **hunger lobby** *is concerned with issues related to hunger, malnutrition, and the adequacy of the food supply worldwide.* Concern about hunger in America became a national policy issue in the early 1960s and more recently as the proportion of poor and homeless has increased. Prior to the 1960s, hunger in America was viewed largely as a problem for the church and local communities rather than government. The concern of the hunger lobby switched from the United States to the world when the hunger problem was substantially reduced in the United States. Thus, in the early 1970s, the combination of hunger, worldwide food shortages, and recurring famines led to the formation of new organizations and a renewed commitment on the part of existing organizations to address the problems of world hunger and malnutrition. The concerns about hunger and malnutrition commanded national attention again in the 1980s and the 1990s, compounded by African civil strife that eventually led to the U.S. humanitarian military initiative in Somalia and Rwanda. International attention was drawn to hunger and malnutrition during the Rome World Food Summit in 1996.

These organizations are active in influencing foreign agricultural development assistance and food aid policies. They also actively attempt to increase the level of awareness and concern by the American people for problems of world hunger and malnutrition. The leadership and the conscience for the hunger lobby lie largely in church-related organizations.

Bread for the World is a nondenominational Christian public interest movement concerned about adequacy of food for the poor in the United States and the rest of the world. It advocates a greater U.S. commitment to problems of hunger and food shortage and to agricultural development policy. Specifically, it has been a strong proponent of domestic and international grain reserves, domestic food programs for those in the greatest need, and appropriate combinations of foreign food aid and development aid. It has expressed concern about U.S. emphasis on surplus disposal and the accomplishment of foreign policy objectives, as opposed to increasing foreign food production, eliminating hunger, and increasing development aid. Bread for the World coordinates its activities with the relief mission-oriented programs of Church World Service, Lutheran World Relief, and Catholic Relief Services.

The **World Watch Institute** raised to national attention the adequacy of the world food supply and associated resource problems. Its founder, Lester Brown, is a prolific author and has exhibited a unique ability to sense the emergence of problems and to draw together facts and concerns in a timely manner. World Watch frequently is the first to address evolving policy issues involving the use of the world's limited resources in its annual *State of the World* books and related publications. Producer groups and "think tanks" such as the Hudson Institute often challenge the pessimistic predictions of Mr. Brown.

The **Food Research and Action Center (FRAC)** is a nonprofit law firm and advocacy center working to end malnutrition in the United States. FRAC has sought to expand the coverage of food assistance programs through both advocacy and litigation. FRAC won cases involving the requirements to issue food stamps retroactively to eligible recipients and a requirement that USDA increase its efforts to inform potential food stamp recipients. Former Congressman Richard Wampler (Rep.–Va.), then senior minority member of the House Agriculture Committee, once expressed frustration with FRAC for "frustrating the whole process with lawsuits."[17] Interestingly, FRAC has received a large share of its financial aid from the federal government, against whom the lawsuits are filed.[18]

The Resource and Environment Lobby

Controversies involving resource use and the environment rival or exceed those in nutrition and food safety for creating conflict among producers, agribusiness, and the public interest lobby. The concerns cover a wide range of issues from endangered species, water quality, and pesticide use to the organizational rights of farm labor. The lobbyists are well organized, enjoy stronger financial support

---

[17]*The Food Stamp Program*(Washington, D.C.: Committee on Agriculture, U.S. House of Representatives, April 17, 1977), p. 349.

[18]Many of the Reagan administration efforts to trim the federal budget severely restricted or eliminated federal funds available to groups such as FRAC.

than many other public interest groups, and, in many instances, raise issues with longer term implications for the ability to produce an adequate supply of food at reasonable prices.

The **Sierra Club** is one of the oldest, largest (with more than 500,000 members), and most effective organizations working to protect and conserve natural resources.[19] Its major concerns include wilderness preservation, air and water pollution, soil conservation, energy conservation, endangered species, elimination of toxic substances, population, and land use planning. Specific concerns involving food and agriculture include availability of water for irrigation, use of public lands for grazing, existence of toxic substances in the food supply, ability to feed future populations, and humane treatment of livestock.[20] Sierra Club proposals for dealing with these issues generally would increase production costs and reduce the level of production. The Sierra Club was an active supporter of the conservation reserve program, which removes erosive land from production, requiring approved soil conservation plans in return for farm program benefits, and prohibiting wetlands from receiving farm program benefits.

The **Environmental Defense Fund** in many respects serves as the scientific backstop for the environmentalists. As such, its mission is finding scientifically sound solutions to environmental problems. Its support comes from a combination of membership dues, contributions, and grants from foundations for specific studies. The fund has done extensive research on the causes of cancer, chemicals in food, integrated pest management, and the impact of Environmental Protection Agency (EPA) policies on the environment. It has been a strong advocate of tighter EPA regulation.

The **Natural Resources Defense Council (NRDC)** is an aggressive resource policy counterpart to the Environmental Defense Fund. It became a savvy and highly controversial advocate/litigant against agricultural chemicals in its aggressive pursuit of Alar-treated apples. The resulting media furor alarmed food producers and retailers.[21] The result was a lawsuit by agricultural interests against NRDC with the charge that its statements about Alar's health and environmental effects could not be substantiated. The Alar controversy made NRDC one of the most feared environmental advocates.

Friends of the Earth is an environmental lobby based in the United States but with affiliations in several foreign countries. It was founded in 1969 when its president moved from the Sierra Club. Its agricultural concerns relate primarily to pesticide issues. It favors less pesticide use and would restrict U.S. corporations from selling or using pesticides internationally that are not approved for use in the United States. It would also place increased restrictions on pesticide use by households and has expressed concern about what it considers to be excess profits earned by the chemical companies.

---

[19]Richard L. Stroup, "The Market—Conservation's Best Friend," *The Wall Street Journal,* April 19, 1990.

[20]Kathryn Ann Utrup, "How Sierra Club Members See Environmental Issues," *Sierra* (March–April 1979), pp. 14–18.

[21]Carol S. Kramer, *Food Safety: The Consumer Side of the Environmental Issue* (Washington, D.C.: Resources for the Future, March 1990).

The **National Audubon Society** boasts a membership nearly as large as that of the Sierra Club. Its concern about reductions in the bird population can be traced back to the discovery that DDT was creating eggs with thin shells that reduced reproduction. Subsequently, the membership of the society mushroomed. Its agenda includes a major emphasis on eliminating pesticides and preserving wetlands and endangered species.

The **National Wildlife Federation** is the largest environmental organization, with about six million members. It is a strong and effective advocate of endangered species, higher grazing fees on public lands, reduced use of pesticides, and the preservation of wetlands. It is clearly a network leader among environmental groups that has broad-based public support.

The **Environmental Working Group** provides a coalition focal point for organizations interested in specific issues to come together for purposes of studying those issues and developing positions on them. Its head, Ken Cook, has done his homework—he understands programs related to agriculture. He helped forge the coalitions that resulted in the development of the conservation reserve program and the requirement that farmers develop conservation plans that control soil erosion in return for farm program benefits. Subsequently, he has threatened increased regulation because of the conclusion that USDA has not effectively enforced the implementation of those plans.

The **National Association of Farm Worker Organizations (NAFO)** is the central lobbyist for more than 50 migrant and seasonal farm worker organizations. It provides services and training to farm workers, but its main agenda includes unionization rights for migrant and seasonal farm labor; improving their living conditions; providing them access to food assistance programs and to education; and studying the effects of working conditions, including pesticides, on the health of farm workers.

The **Humane Society of the United States, Animal Protection Institute of America, People for the Ethical Treatment of Animals, Animal Liberation Fund,** and **Friends of Animals** are a few of the several public interest groups concerned with animal welfare and animal rights. These groups see modern livestock and poultry production methods as promoting cruelty to animals. Caged layers, confinement hog production, and raising veal calves in crates are particular concerns, in addition to traditional issues involving slaughter techniques and the use of animals in research.

INTEREST GROUP STRATEGIES AND THE FUTURE

The days of the agriculture establishment having complete control over the agricultural policy agenda are clearly past. Three major reasons account for this change:

- New public interest groups, recognizing the impact of food and agricultural policy on the people and interests they represent, are continually injecting new ideas into legislative and executive decision processes. More frequently, they are willing to challenge adverse decisions in the courts.

- The members of the agriculture committees of the Congress are no longer responsive only to the agriculture establishment. New interest groups have members on these committees who not only hear them out but also plead their case directly or give them an opportunity to plead it.

- USDA is no longer in full control of the food and agriculture policy decisions in the executive branch, if it ever was. These decisions are viewed as too important to be left to USDA. They may have important foreign policy, budget, balance-of-trade, or food price implications. On specific issues, the secretary of state, director of the office of management and budget, or the secretary of the treasury may be more influential in a policy decision than the secretary of agriculture.

The new public interest groups will not go away. The agriculture committees of the Congress will continue to have members responsive to the appeals of public interest groups, and food and agriculture decisions will probably increase in importance in the executive branch.

Coalitions will be more important in determining the outcome of policy and program decisions. Attempts by the agriculture establishment to organize a broad-based farm bloc coalition have been less than a resounding success.

Agribusiness lobbyists appear to be much more effective than producers at organizing coalitions. The Food Group, composed of agribusiness organizations having an interest in food issues, meets regularly to coordinate positions on farm and food policy. The Agricultural Round Table has been organized as a combination of agribusiness organizations and commodity groups to promote solid opposition to government programs involving mandatory production controls. Numerous other coalitions are formed for brief periods to address issues of interest.

Farm organizations continue to be successful in getting farm bills enacted while economists and politicians have been suggesting that farmers' influence has declined. Bills that provide farmer subsidies of more than $10 billion in one year and more than $60 billion over four- or five-year periods can hardly be considered the result of ineffective or splintered lobbying! Yet the defeat of the wool and mohair program during the 1993 budget debate sounded an alarm that all agriculture organizations should heed. This defeat represented the first crack in the armor of the agriculture establishment's farm program shield. However, a conservative budget-cutting Republican Congress passed a farm bill that will cost $44 billion over seven years from 1996 to 2002 and refused to repeal the permanent 1949 Agricultural Act.

What counts is relative influence at the time the bill is being developed. In the 1980s, farmers proved more powerful than consumers—although both groups received their share of benefits from the 1985 and 1990 farm bills. In the 1990s, the environmentalists lost key court decisions on property takings and pesticide issues. Yet there is still considerable question regarding the balance of power between agriculture and the public interest groups.

The farm lobby is most effective when its position is compatible with agribusiness. *Networking* (mutual support among organizations) has enabled some groups, notably consumers, to become stronger. When farm organizations truly have their backs to the wall, they may be forced to develop such networks as well.

To remain politically effective into the twenty-first century, agriculture may have to practice the politics of the minority:

- Finding allies and building coalitions by issue rather than by philosophy. Agricultural interest groups too often may depend on philosophical niceties and folklore.
- Looking for common ground and compromise. Getting a portion of what you are after is better than getting nothing.
- De-escalating arguments with adversaries. An adversary on one issue may be an ally on another. Be positive by playing down small differences, playing up mutual interests, and working within the system.
- Basing policy positions on facts. Political contests not based on fact and solid analysis will generally be lost.
- Avoiding identification with either political party. To succeed, agriculture needs the support of both political parties.[22]

ADDITIONAL READINGS

1. William P. Browne, *Private Interest, Public Policy, and American Agriculture* (Lawrence, Kans.: University of Kansas Press, 1988) is an excellent analysis of the politics affecting contemporary farm and food policy.
2. Carol S. Kramer, ed., *The Political Economy of U.S. Agriculture* (Washington, D.C.: Resources for the Future, 1989) contains several excellent articles on the role of the rapidly expanding numbers of food and agriculture interest groups.
3. Ardith Maney's *Consumer Stakes in Food and Agricultural Policy* (Washington, D.C.: Resources for the Future, July 1991) is a very good summary of contemporary consumer forces affecting policy.
4. Patrick J. Kigen, "The Grainmaker," *George*, August 1996, is an excellent discussion of the influence of agribusiness. It profiles Dwayne Andreas, of Archer-Daniels-Midland.

---

[22]B. L. Flinchbaugh, "It's Easy to Be Ignored If You Don't Have Your Act Together," speech presented at the National Institute on Cooperative Education, Colorado State University, Fort Collins, Colo., July 28, 1981, p. 1.

INTERNATIONAL TRADE AND MACROECONOMIC POLICY

The world economy is highly interdependent. Agriculture operates in a world economic environment that can be insulated neither from events in the U.S. economy nor in the world economy. Domestic policies cannot be evaluated in terms of either their effects or their effectiveness without a basic understanding of the interrelationships among the U.S. economy, the world economy, and agriculture. It has been argued that domestic policies are less effective because of the stronger world economic forces. The next four chapters are designed to provide an understanding of these international forces and interrelationships.

Chapter 5 describes the world food problem. After an opening discussion of the factors influencing the international demand for and supply of food, the emphasis shifts to the problem of hunger, its causes, and the policy options for dealing with it. The chapter ends with a discussion of U.S. policies regarding economic development.

Chapter 6 discusses the role of trade. It develops the theory of trade and barriers to trade. The competitive position of the United States in international markets is evaluated.

Chapter 7 explains the major policy issues, the strategies, and the institutions for resolving them. The impacts on U.S. agriculture, as well as on other countries, are explained.

Chapter 8 explains the economic interrelationships between the U.S. economy and agriculture. Linkages, thereby, are provided between the international economy, the domestic economy, and the agricultural economy. Emphasis is placed on those general economic variables that most directly affect the agricultural economy. Then the general macroeconomic tools that are used to influence those variables are described.

5 THE WORLD FOOD PROBLEMS

The gap in our economy is between what we have and what we think we ought to have—and that is a moral problem, not an economic one.

—Paul Heyne

In the 1970s, the American farmer was thrust overtly, as a matter of public policy, into the world agricultural economy. Foreign demand became the major determinant of farm prices and incomes. Expanding foreign demand for American farm products had multiple objectives:

- Agricultural exports were seen as a means of reducing the need for government subsidies and controls.
- Agricultural exports had the potential for becoming a major source of export earnings and bolstering the value of the dollar.
- Agricultural exports held the potential for dealing with major world issues including reducing hunger and malnutrition, demonstrating the inadequacies of communism, and bolstering the deteriorating position of the United States as a world power.

In reality, American farmers and U.S. policymakers had little choice in making U.S. agriculture part of the world economy. World food supplies were becoming visibly tighter in the late 1960s. The American economy badly needed new sources of export earnings to offset the rising cost of imported oil. Vivid television portrayals of the plight of the hungry and starving engendered concern among the American people about future food supplies and the need to help. Whereas at one time it may have been possible for U.S. agriculture to ignore the rest of the world, that was no longer possible.

Two major famines occurred during the past two decades.[1] One (1974) was in Bangladesh with one million people dying of hunger. In the second famine

---

[1] Phillips Foster, *The World Food Problem* (Boulder, Colo.: Lynne Rienner Publishers, 1992), p. 2.

(1983–1985), 300,000 people died of hunger-related causes in Ethiopia. The first famine of the 1990s, in Somalia, likewise took a toll of 300,000 lives.[2] Ironically, most hunger-related deaths do not occur in widespread famines. It is conservatively estimated that 10 million people, roughly the population of Illinois, die each year because of hunger.

The purpose of this chapter is to explain the dimensions of the world food problem, its complexity, and the major policy options for addressing it.

THE WORLD SUPPLY–DEMAND BALANCE

Thomas Malthus published his famous essay, *"An Essay on the Principle of Population, As It Affects the Future Improvement of Society,"*[3] in 1798. The essay attacked theories of eternal human progress by arguing that the standard of living cannot be indefinitely improved because the growth of the population will exceed the capacity of the earth to produce. Malthus asserted that population, unchecked by war, disease, or famine, increases geometrically, whereas food production increases only arithmetically. He contended that the implication of his argument was a clear need to restrain global population growth.

At the end of the second century since these predictions were made, it is generally concluded that Malthus has been proven wrong.[4] The reasons generally cited include Malthus's failure to recognize that population growth slows with increases in per capita real income, with increases in population density, with improved health and nutrition, and with the realization that less food is available per capita. Further, Malthus did not foresee the advances in food production that have enabled the food supply to grow faster than arithmetically. Increased food production has come not only from increases in cultivated acres but also from technological innovation, the substitution of capital for labor, and from the very fact that in a market economy, when demand increases relative to supply, it creates a reward for increased productivity.

Although such denials of the Malthusian theory represent the clear majority of contemporary thought, there are those who still predict eventual exhaustion of the world's resources and a decline in the standard of living. Such thinking has been fostered by short crops such as that which developed in the early 1970s when the combination of corn blight, bad weather, and an Organization of Petroleum Exporting Countries (OPEC) supply cutback resulted in high energy prices that led to sharp increases in food prices and export embargos. Almost simultaneously, in 1972 a widely publicized study of the implications of continued worldwide economic and population growth embraced the concept of exponential population

[2]*United Nations Operation in Somalia II* (George Mason University, Institute of Public Policy, 1996).

[3]D. V. Glass, *Introduction to Malthus* (London: C. A. Watts & Co., 1953).

[4]Kenneth Smith, *The Malthusian Controversy* (London: Routledge & Kegan Paul, 1951).

growth.[5] It then asserted the need to plan both population growth and resource use at levels consistent with the earth's carrying capacity. In addition, it was noted that world population has, from time to time, been checked by war, disease, and famine, as Malthus predicted.

As indicated in Chapter 2, the 1970s world food crisis was followed by a 1974 World Food Conference. This conference emphasized the hunger problem, the need to increase food production, the need to provide more food aid, and the need to reduce barriers to trade. It also confirmed the right of every person to an adequate diet.

Subsequently in 1982 a major government study of future economic trends requested by President Carter fortified the prediction. The resulting report, *Global 2000*, painted a bright economic outlook for farmers and a bleak picture of their ability to satisfy world food needs. The report concluded: "After decades of generally falling prices, the real price of food is projected to increase 95 percent over the 1970–2000 period."[6]

In the 1980s this pessimistic scenario for the future ability to satisfy world food needs failed to play out. Analysts such as D. Gale Johnson,[7] a U.S. expert on international economics, indicates five reasons for the failure of long-term prosperity to materialize:

- Overestimation of the growth in income resulting in a less than predicted increase in feed grain demand
- Underestimation of the ability to develop new agricultural lands
- Underestimation of the supply of water available to agriculture
- Underestimation of the impacts of high energy prices
- Overestimation of the impacts of high energy prices

In reviewing past events that resulted in price changes, Johnson observes that one of the major sources of world price instability is government intervention.[8] When governments attempt to stabilize domestic prices or farm income, they contribute to international price instability. Johnson's rank ordering of commodities in terms of the degree of international price instability from least stable to most stable is sugar, wheat, corn, and cotton. This ordering exactly coincides with studies of the degree of government prices and income interference. In other words, efforts to achieve price stability in domestic markets often translate into instability in world markets as supplies are added to the world market at subsidized prices or demand is curtailed by import restrictions.

[5]Donella H. Meadows *et al., The Limits of Growth* (New York: Universe Books, 1972).

[6]Gerald O. Barney, *The Global 2000 Report to the President* (New York: Penguin Books, 1982), p. 17.

[7]D. Gale Johnson, "The World Food Situation: Recent and Prospective Developments," in D. Gale Johnson and G. Edward Schuh, eds., *The Role of Markets in the World Economy* (Boulder, Colo.: Westview Press, 1983), pp. 1–33.

[8]Ibid., p. 18.

ASSESSMENTS OF THE SUPPLY–DEMAND BALANCE

Farmers responded to the world food crisis of the 1970s by increasing production. Yet the 1970s became a turning point from chronic surpluses to periodic surpluses and deficits. The result was greater instability in food prices and increased concern over future supply–demand conditions. In 1995–1996 there was another production shortfall. Coarse grain supplies became sufficiently short that prices rose to record levels. This precipitated another round of studies of the world food supply–demand balance.

Brown's Perspective (1996)

Lester Brown is generally credited with being pessimistic about the ability of world leaders to come to grips with the problems of resource scarcity. Without remedial action in terms of efforts to restrict population growth, expand research to increase food production, and reduce environmental degradation, Brown sees spiraling human demands outstripping the capacity of the earth's natural systems.[9]

As a result of rapid economic growth of the East Asian and Chinese economy, encompassing 1.4 billion people, Brown sees competition among exporters for markets being replaced by competition among importers. The result is the development of what Brown sees as a seller's market with access to food becoming the defining issue.

Brown criticizes projects of food production that are based on past trends because recent years have consistently been below trend. The result has been substantially reduced stocks.

USDA (1995)

USDA's outlook is considerably more optimistic. It projects a relative balance between production and consumption.[10] While indicating the expectation of a decline in the rate of increase in production to 1.5 percent per year down from 2.4 percent during the 1960–1994 period, USDA sees a slowing of the population growth rate to 1.5 percent from more than 2 percent.

USDA does not see income growth as being the driving force that Brown projects. However, it does recognize that disparities in income growth across countries will increase the need for food aid in a relatively tight market. Commodity prices are projected to decline but at a slower rate than has been the case in the past.

---

[9]Lester R. Brown *et al.*, *State of the World, 1996* (New York: W. W. Norton & Co., 1996).
[10]Margaret Missiaen, Shahla Shapouri, and Ron Trostle, *Food Aid Needs and Availabilities*, GFA-6 (Washington D.C.: ERS/USDA, October 1995).

FAPRI Study (1996)

The **Food and Agriculture Policy Research Institute's (FAPRI)** annual 1996 international outlook projection for the Congress reflected a strong demand for coarse grains and concern about stock levels.[11] Coarse grain demand was driven by strong income growth in China and most of Asia, which would lead to increased consumption of poultry and pork. This upgrading of the quality of the diet would require imports of corn and soybeans, which would effectively drive the agricultural economy.

FAPRI expressed concern about the level of commodity stocks that would be maintained by the private sector. The result would be increased price instability. Overall, the outlook presented was fairly positive from a farmer perspective but not nearly as pessimistic as Brown's from a consumer perspective.

20–20 Vision Project (1996)

Anderson and Garrett, working for the 20–20 Vision project, like Brown and FAPRI, saw reduced stocks as major destabilizing factors.[12] Like Brown, they put substantial emphasis on the need to expand investment in agricultural research and technology as a key to production keeping up with demand. China, Eastern Europe and the former Soviet republics were seen as having a major influence on the future supply–demand balance.

Developing countries were advised to take the following courses of action:

- Hold small stocks to provide insurance.
- Use world futures and options markets to hedge against future price increases.
- Invest in transportation, communication, and agricultural research to ensure competitiveness.

OECD Study (1995)

The **Organization for Economic Cooperation and Development (OECD)** saw the major factor affecting the future supply–demand balance as being the high rate of economic growth in Asia and the Latin American countries.[13] This will lead to higher levels of food consumption and a gradual shift toward diets richer in protein. The result will be increased coarse grain and soybean use.

---

[11]Stanley R. Johnson *et al.*, *FAPRI 1996 International Agricultural Outlook*, Staff Rpt. 2-96 (Ames, Iowa: Iowa State University, September 1996).
[12]Per Pinstrup-Anderson and James L. Garrett, *Rising Food Prices and Falling Grain Stocks*, (Washington D.C., IFPRI, January 1996).
[13]*Agricultural Outlook 1995–2000* (Paris, France: OECD, 1995).

Office of Technology Assessment (1986 and 1992)

The purpose of the **Office of Technology Assessment (OTA)** studies[14] was to assist the Congress in projecting the impact of technological change on the supply of food and natural fibers. Although not designed to be studies of the future world supply–demand balance, the OTA studies add an important missing dimension to the other studies by analyzing the rate of technological change. They utilized the expertise of leading agricultural scientists to identify major expected technological breakthroughs, their timing, and their potential impact on the production of particular commodities.

The results of the OTA studies stand in stark contrast to prevailing opinions during the mid-1970s when it was thought that the shelves of available new technologies were becoming bare. In the 1980s, however, OTA identified a multitude of new and evolving forms of biotechnology and information technology. The pervasiveness and potential impacts of these new technologies were projected to increase at an increasing rate through the turn of the century. Rapid technological change was projected to impact animal agriculture sooner than plant agriculture. Despite this lag, the long-range impact on plant agriculture and its crop producers could be profound in terms of types of plants grown, cultural practices, yields, and structural changes in farming.

If the OTA studies have any degree of accuracy, they render obsolete the studies of the world supply–demand balance that assume a constant rate of technological change. This is particularly the case as time progresses toward the year 2000 and beyond. The case for declining real farm prices is, thereby, fortified.

Foster Study (1992)

The Foster study is the most comprehensive study of the world food problem available. It concludes that the problem is not one of food versus population, as believed by the Malthusians. He notes that despite the most rapid population growth in the history of the world during the last half of the 1900s, per capita food production has increased, fueled by a chemical and biological revolution in farming, not by more land in cultivation.[15]

While recognizing the dangers posed by pressures on the available water supply for irrigation and the vagaries of weather, Foster appears to subscribe to the induced innovation theory of Ruttan[16] by suggesting that food crises may be a stimulating factor for increased productivity, thereby embracing the OTA assess-

---

[14]Office of Technology Assessment, *Technology, Public Policy, and the Changing Structure of Agriculture* (Washington, D.C.: U.S. Congress, March 1986).

[15]Foster, *The World Food Problem*, p. 173.

[16]Vernon W. Ruttan, *Agricultural Research Policy* (Minneapolis, Minn.: University of Minnesota Press, 1982), pp. 26–36.

ments of the importance of technological change. Ruttan's theory is related to that of Boserup, who implies that population growth creates a kind of crisis situation that stimulates the invention of new technology.[17] However, Foster cautions against the implication that population growth helps the world hunger problem by initiating increased food production.

As Foster noted, despite the fact that Malthus has been proven wrong, it is still possible that a series of adverse political and weather circumstances could once again thrust the world agricultural economy into a situation of food shortages. Because of that possibility, the governments of most countries prefer to err on the side of abundance and food security. This, of course, contributes to the likelihood of surpluses and protectionist policies.

Conclusions on Supply–Demand Balance

The following conclusions evolve out of the studies of the food supply–demand balance in the world:

- China is the major production, consumption, and economic uncertainty in the world. With 1.2 billion people, it greatly affects the balance.
- Income growth in Asia is an important factor shifting the world market in the direction of coarse grains and soybeans to feed livestock and poultry.
- Investment in research and technology is the key to future supplies. Otherwise, resources are largely fixed in quantity.
- A major commitment is required to cut hunger in half and feed the extra 3 billion people during the next 35 years, as committed to by the 1996 World Food Summit.

THE PROBLEM OF HUNGER

Even in the 1980s, when U.S. farm policymakers were trying to figure out what to do about surpluses and low prices, starving African children appeared regularly on television. Hunger and malnutrition in the midst of world plenty is a complex problem without simple solutions. The complexity is indicated by the magnitude of the problem alluded to at the beginning of this chapter—10 million people dying annually because of hunger. Despite enormous food programs (about 11 million tons of food aid annually), 1 billion people were estimated to be hungry in the mid-1990s.[18] These people, unfortunately or fortunately, depending on one's

---

[17]Ester Boserup, *Population and Technological Change* (Chicago, Ill.: University of Chicago Press, 1981).

[18]*Food and Fiber Letter* (Washington, D.C.: Sparks Companies, November 19, 1993), p. 2.

perspective, are not all located in one place or on one continent. A large portion of those who die are children (six months to two years old) because they are unable to make the transition from nursing (breast-feeding) to other foods.[19] Pregnant women, lactating mothers, and old women are the next most vulnerable.

While 10 million die of hunger annually, many more suffer from hunger with long-term adverse consequences. In 1985, Reutlinger estimated that 730 million people (21 percent of the world's population) have energy-deficient diets, while 340 million are so deficient that stunted growth and serious health risks result.[20] Most of these people are located in South Asia and sub-Saharan Africa. Compared to the current 11 million tons of food aid, 23 million tons would be needed to satisfy minimum nutrition standards.[21]

While the percent of undernourished in the world has declined from more than 20 percent to about 12 percent since 1945, the absolute number has increased. In other words, the population is growing sufficiently fast in countries plagued by hunger that the world economies have been unable to increase income and/or food production fast enough to keep up.

Causes of Hunger

Hunger is frequently characterized as a problem of distribution. This characterization is too simplistic and can be very misleading. Foster summarizes the causes of hunger in the concept of household food security (see box). In essence, household food security is determined by the availability of income and liquid assets to buy food not produced by the household at the prevailing price. Although this concept could once again be characterized as a restatement of the economic concepts of effective demand and the distribution problem, Foster is careful to point out that household production, the price of food, and income are a function of a host of public and private sector variables, many of which are beyond the control of the households. These variables are the causes of hunger. In many respects, the hunger policy problem is one of changing the public policies that influence those causes of hunger.

Income

One of the most basic principles of consumption economics and demand theory is that demand is not just a function of population; it is also a function of income. In a market economy, which is basically characteristic of the world economy and has become increasingly characteristic of the economies of individual countries, income is a driving force behind demand as it is in Foster's concept of food security.

---

[19]Foster, *The World Food Problem*, p. 101.
[20]Schlomo Reutlinger, "Food Security and Poverty in LDCs," *Finance and Development* (December 1985), pp. 7–11.
[21]*Food and Fiber Letter*, p. 2.

The Concept of Household (HH)*Food Security*

$$\text{Value of food production deficit in a HH} \leq \text{Income and liquid assets available to purchase food}$$

$$\text{Food purchase requirement } 3 \times \text{Price of food} \leq \text{Income and liquid assets available to purchase food}$$

$$\left\{ \begin{array}{l} \text{HH food} \\ \text{consumption} - \text{production} \\ \text{requirement} \end{array} \right\} \times \text{food} \leq \text{assets}$$

You are more food-secure as the left-hand side gets smaller relative to the right, or as the right-hand side gets bigger relative to the left. The risk of food insecurity is the probability that the left-hand sides are bigger than the right.
Factors influencing each element in the final equation are listed below:

- HH food consumption requirement
  - Number of people in household
  - Age, sex, working status of individuals
  - Health status of individuals
  - Childbearing status (pregnant, lactating)
- HH food production
  - A complex set including amount of land, technology, capital, education of farmer
  - Government policies (tariffs, price controls, export taxes, input subsidies, research, etc.)
- Price of food
  - Quantity produced
  - Size of population
  - Income of population
  - Government policies (tariffs, price controls, export taxes, input subsidies, research, etc.)
- Income and liquid assets available to purchase food
  - A complex set including education of members of household, capital position of household, land position, employment opportunities, attitudes towards work, transportation cost to and from work, and health

Source: Quoted from Phillips Foster, *The World Food Problem* (Boulder, Colo.: Lynne Rienner Publishers, 1992), p. 111.

Effective demand *exists only when the ability to buy exists.* Such effective demand corresponds most directly with increases in income. Without increases in income, demand becomes increasingly dependent on government food assistance programs as population increases. Without effective demand in terms of either income growth or government assistance, potential demand dictated by population growth goes unsatisfied.

The diet in much of the world today is grain based. Increasing incomes enable people to buy more food and different types of food. As incomes increase, more grain is consumed initially. Further income increases lead to shifts in consumption from grains to fruits, vegetables, meat, milk, and eggs. The **income elasticity** *of demand measures the percentage change in demand associated with a 1 percent change in real income.* The income elasticity for food differs among countries and falls fairly steadily as an economy develops. In other words, higher incomes are associated with a lower income elasticity.

Shifts in consumption patterns as incomes increase suggest considerable variation in the income elasticity among commodities and among countries at different stages of development. Income elasticities indeed vary dramatically by foods, by level of income, and, therefore, by country. Gray provides an excellent illustration of the differences in income elasticity for rural Brazil by income level (Table 5.1). Income elasticities are generally positive, although for staples, it may be negative. This is the case for cassava[22] flour in rural Brazil. For the lowest income people, the income elasticity is large. However, it declines rather consistently and dramatically as incomes increase. This decline is explained by the mean caloric intake, which indicates that 30 percent of the population in rural Brazil with the lowest incomes have lower caloric intake (1,963 calories) than the approximately 2,500

TABLE 5.1 Income Elasticities for Calorie Intake, Selected Foods, by Income Group, Rural Brazil, 1974–1975

	Income Group		
	Lowest 30 Percent	*Middle 50 Percent*	*Highest 20 Percent*
Cassava flour	-3.50	-1.59	-.356
Rice	1.99	.172	.173
Milk	2.27	.147	.172
Eggs	1.93	.630	.114
Mean per capita calorie intake	1,963	2,432	2,771

Source: Cheryl W. Gray, *Food Consumption Parameters for Brazil and Their Application to Food Policy,* IFPRI Research Report No. 32 (Washington, D.C.: International Food Policy Research Institute, 1982), p. 26.

[22]Cassava, or tapioca, is a root plant extensively relied on as a staple source of calories in the tropics.

calories recognized as being required by the World Health Organization (WHO). For these people, increases in income will go largely to food to fill the **calorie gap** *or the deficit between current consumption and requirements.* Thus, these lower income people generally have a higher income elasticity.

Note also from Table 5.1 that the highest income people exceed the caloric requirement. In fact, it is not unusual in developing countries for there to be an extreme mixture of undernourishment and overnourishment (overweight) individuals. Table 5.1 suggests that at least 30 percent of the population in rural Brazil is undernourished and 20 percent is overnourished.

Income Distribution

The combination of undernourished and overnourished people results from the inequality in the distribution of income. Pareto's law states that if the population's income is arrayed from lowest to highest on a graph, with household income being on the vertical axis, the result is suggestive of a letter J (Figure 5.1). The J distribution implies that a relatively small proportion of the people earn most of the income. For example, in the Philippines, 43 percent of the income was concentrated with the highest one-tenth of the population, while 1 percent of the income was concentrated with the lowest one-tenth of the population.[23] Moreover, a Nobel Prize-winning agricultural economist, Kuznets, found that during the early phases of development in a country, the concentration of income will increase but as development progresses, it will subsequently decline.[24]

The result of the unequal distribution of income is unequal purchasing power. Many of those who do not grow enough food for their household consumption do not have enough income to buy sufficient food for their sustenance. The result is the Reutlinger triangle which indicates the size of the calorie gap between current consumption and food nutritional requirements.

Price of Food

How much food can be bought with a given level of income depends on the price of food. The responsiveness of changes in the quantity demanded to changes in price is referred to as the price **elasticity of demand.** The elasticity of demand is almost always negative. For example, if the elasticity of demand is -0.2, it means that a 10 percent increase in price results in only a 2 percent decline in demand. This is referred to as an **inelastic demand** because the percent increase in price is higher than the percent reduction in demand. Inelastic demand for food is quite common. An elastic demand is relatively rare and tends to be limited to foods having very good substitutes.

---

[23]Foster, *The World Food Problem,* p. 144.
[24]Simon Kuznets, "Economic Growth and Income Inequality," *American Economic Review* (January 1955), pp. 1–28.

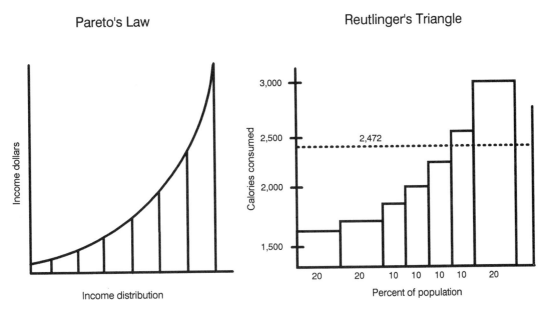

FIGURE 5.1   Stylized representation of Pareto's law indicating the distribution from lowest to highest, and Reutlinger's triangle indicating the distribution of caloric consumption by income group, lowest to highest.

The price elasticity of demand has also been found to vary with the level of income. Specifically, as incomes increase, the elasticity of demand tends to become more inelastic. Thus, while the price elasticity of demand for cassava flour by low-income people may be -0.2, for high-income people, it may be zero (Table 5.2). In other words, high-income people may not change consumption at all in response to a lower price for cassava flour. The result is an important lesson for economic development and solving problems of hunger. *The poor benefit most from increases in the production of those items having the most inelastic demand.* The reason lies in the fact that for these items, the poor do not have to compete as directly with the incomes of the rich, and the price on these items falls the most as production increases.

Low Education Levels

One of the keys to increasing income involves education. Improved education means increased purchasing power. It also means reduced fertility in women. "Educated women are more likely to put off marriage in order to enter the work force, more likely to delay having children in order to remain in the work force, and are more likely to know about and use contraception than are uneducated women."[25]

[25]Foster, *The World Food Problem,* p. 229.

TABLE 5.2   Estimated Direct Price Elasticity of Demand by Income Group, Cali, Colombia, 1969–1970

	Low Income			High Income		
	I	*II*	*III*	*IV*	*V*	*Average*
Cassava	-0.23	-0.28	-0.25	-0.00	-0.00	-0.19
Potatoes	-0.41	-0.42	-0.31	-0.00	-0.00	-0.26
Rice	-0.43	-0.40	-0.40	-0.26	-0.18	-0.35
Maize	-0.63	-0.55	-0.44	-0.00	-0.00	-0.44
Bread/pastry	-0.65	-0.56	-0.32	-0.24	-0.00	-0.31
Beans	-0.82	-0.78	-0.64	-0.45	-0.25	-0.60
Peas	-1.13	-1.13	-0.76	-0.59	-0.52	-0.70
Eggs	-1.34	-1.23	-1.26	-0.75	-0.35	-0.92
Oranges	-1.39	-0.96	-0.79	-0.64	-0.29	-0.69
Milk	-1.79	-1.62	-1.12	-0.64	-0.20	-0.77
Pork	-1.89	-1.61	-1.12	-0.82	-0.70	-1.01
Daily caloric intake as percentage of requirement	89	99	117	132	178	119

Source: Per Pinstrup-Andersen *et al.*, "The Impact of Increasing Food Supply on Human Nutrition: Implications for Commodity Priorities in Agricultural Research and Policy," *American Journal of Agricultural Economics* (May 1976), pp. 137–138.

Unfortunately, only a small share of the population receives an education through high school. In 1985, only 34 percent of the secondary school-age population was in school in 39 countries where per capita income was less than $450 per household. In 18 high-income market economies, 93 percent were in school.[26]

The challenges for achieving higher levels of education in developing countries are exceedingly difficult. The reason lies in the high and increasing proportion of the population that is of school age. As a result, the requirements for schools and teachers increase each year. This contrasts with developed countries in which the school-age population is relatively static or may actually be contracting. Thus, in developed countries, school infrastructure needs are limited largely to maintenance and replacement. In developing countries, the problem is finding resources for growth in the capacity of schools. In the absence of the tax revenue required to expand classrooms and hire new teachers, class size often balloons to more than 60 students. As a result, the opportunities, incentives, and rewards from school attendance decline.[27]

Poor Health

Poor health results from poor nutrition. Infection exacerbates malnutrition and malnutrition exacerbates infection. Therefore, poverty, hunger, and poor health

[26]Ibid., p. 181.
[27]Ibid., p. 181.

foster one another. The interrelationships are too complex to be adequately discussed here, but they stand to reason.

Children in developing countries are most likely to encounter health problems at weaning.[28] The reason lies largely in the lack of a clean water supply. The result, more often than not, is diarrhea. The potential for diarrhea appears to be enhanced by child-feeding programs that encourage the use of infant formula milk substitutes for breast-feeding.

Once the problems of malnutrition and infection are encountered, they tend to endure, although at a declining rate. Children fare worse than adults because, in contrast with what might be expected, adults tend to receive preference in the food line.

Views on Policy Approaches to Hunger

Like most lingering policy issues, the problem of hunger is much more complex than is generally recognized. The complexity is enhanced by differences in philosophy regarding the hunger problem. In other words, people who deal with the problem of hunger differ on how to approach it and its solutions. The philosophical differences in approach to the hunger problem have been addressed by both Foster[29] and Cassidy.[30] The following discussion is an attempt to capture the essence of the views of Foster and Cassidy in three basic philosophical approaches to the hunger problem.

1. The **activist position** is the one seen most often on television with personalities pleading for support of causes such as CARE or UNICEF in the presence of starving children. Activists believe strongly that something should be done to relieve the problems of hunger and malnutrition. Often, they believe that *every human being has a right to food* and that those who are not suffering from hunger and malnutrition have an obligation to supply it even if it means self-sacrifice. The focus is on helping individuals or societal population segments that face severe adversity.

    Activists tend to emphasize the need for food assistance to minimize the economic costs of relieving hunger and the potential long-term adversities that could arise from a food assistance strategy that does not consider the long-run requirements for an effective development program. In other words, the immediate problem of hunger relief takes precedence over longer term concern about development and how people live. However, some activists are more enlightened than others, recognizing longer run economic and social trade-offs.

---

[28]Ibid., p. 195.
[29]Ibid., pp. 205–215.
[30]Claire Cassidy, "World-View Conflict and Toddler Malnutrition: Change Agent Dilemmas," in Nancy Scheper-Hughes, ed., *Child Survival: Anthropological Perspectives on the Treatment and Maltreatment of Children* (Norwell, Mass.: Reidel Company, 1987).

2. The **adaptor position** recognizes that we live in a world of scarcity. Both within and among countries, decisions must be made on how to allocate limited resources between the short-run needs for food assistance and the long-run needs for food production, water systems, family planning, transportation systems, electrification, and communication. It recognizes that unless these long-run needs are satisfied, the child saved by short-run food assistance could live a life of misery with impaired mental, social, and emotional development. Resource scarcity leads to the realization that someone (a policymaker) must make the decision of who lives and who dies—otherwise referred to as the **lifeboat theory.**[31]

   For an economist, the adaptor approach essentially requires cost–benefit analysis of alternative expenditures of limited resources. On the cost side of the equation, both actual monetary costs (aid) and opportunity costs (return from alternative investments such as infrastructure, education, or hospitals) would be considered in food aid decisions. The adaptor approach also recognizes that undernutrition may have some benefits such as shortening the number of years a woman is fertile and reducing the risk of certain diseases (cancer and heart disease). More productive individuals receive greater assistance than less productive ones because the net benefit is greater. In other words, the adaptor responds to the totality of conditions at hand. The adaptor reasons from the perspective of what is good for the group (country) as a whole rather than for the individuals within the group.

3. The **acceptor position** does not think there is a problem, does not see a need for intervention, and believes that intervention has greater potential for making the situation worse than for making it better. Acceptors have lived with hunger long enough to consider it normal. They have come to accept the process of self-selection, or survival of the fittest, as being natural. They recognize the benefits of undernutrition in terms of lower fertility rates and selectivity, favoring males over females, resulting in fewer children to feed. Moreover, they have seen the longer run adversities associated with providing food aid to deal with hunger and starvation situations, only to result in increased fertility, a lack of long-term development assistance, and a population that is made worse off by an inadequately planned and supported development strategy.

In reality, these three positions represent a continuum stretching from the television-oriented activist to the most conservative status quo acceptor. Most of those who work on problems of "hunger," whether in private or public organizations, tend to lean toward the activist's philosophy. They are often referred to as **interveners** or **interventionists.** Cassidy classifies most developing country villagers as being adapters, receptive to intervention but skeptical. In other words,

---

[31]The lifeboat theory has its origin in a sinking ship far at sea with limited lifeboat capacity and the need for the captain to decide who gets in the lifeboat and who does not—effectively, a decision of who lives and who dies.

the acceptor position is a minority. However, these philosophical differences lead to conflict and misunderstanding regarding the appropriate direction for policies designed to deal with hunger and economic development in the Third World.

Policy Options for Dealing with the World Food Problem

Foster identifies four major policy alternatives that might be used by countries to deal with hunger issues:

1. Improve health.
2. Reduce population growth rates.
3. Reduce income and wealth inequalities.
4. Lower the price of food.

A number of different interventionist programs are available for addressing each policy alternative. Some of these programs have substantially different consequences. The following discussion summarizes these differences.

Improving Health

It was noted previously that good health promotes good nutrition and good nutrition promotes good health. Programs designed to improve health are widely understood to include the following:

- Food can be fortified with iodine, iron, and vitamin A to eliminate micronutrient deficiencies.
- Mother and child health centers can be developed to emphasize disease prevention and promote breast-feeding and family planning. Promotion of breast-feeding demonstrably lowers fertility.
- Sewer and water systems can interrupt the transmission of diarrhea.

Reduce Population Growth Rates

Rapid population growth exacerbates all dimensions of the world food problem. The concentration of hunger in the youngest children (particularly female children) of large families serves to emphasize this point. Alternatives for lowering the fertility rate include these:

- Pronatalist policies can be eliminated. Such policies include tax deductions proportional to the number of children, subsidized maternity leaves, and child care subsidies. These policies exist in most countries of the world. Originally designed to encourage population growth following wars and famines, they have more recently become family and women's rights issues. They also foster large families, which are liabilities to developing countries.

- Antinatalist policies including contraceptive research, subsidized family planning, economic disincentives for large families, and raising the legal marriage age can be adopted to reduce fertility.
- The status of women can be upgraded to reduce fertility. Basically, this involves equality for women in education and employment opportunities. Educated, working women have fewer children.

Reduce Income and Wealth Inequalities

Reducing income disparities decreases the tendency for the rich to bid food away from the poor. Interestingly, the nutritional status of both groups is improved—the poor eat more and the rich eat less! The options for reducing income and wealth inequality cover a wide range of policies and related consequences:

- Land reform involves a redistribution of patterns of land holding. In many developing countries (Eastern Europe, the former Soviet republics, and China), land holdings tend to be highly concentrated. Dispersed land ownership patterns are said to have benefited development in Western Europe, the United States, Canada, Japan, and Taiwan. However, land reform is a drastic measure and runs the risk of resulting in uneconomic production units.[32]
- Progressive taxation shifts wealth by requiring that the rich pay a higher proportion of the costs of government, including social programs, education, and development assistance.
- Elimination of tax shelters and subsidies for farm machinery reduces the transition costs associated with the movement of workers from the farm to the nonfarm sector. The benefits of machinery tax shelters and subsidies go to large shelters and subsidies go to large farmers—not to laborers.
- Encouragement of rural financial institutions to serve as intermediaries between savers and borrowers reduces the incentives for credit subsidies, which generally benefit the rich relative to the poor.
- Rural road subsidies improve the access of farmers to markets and reduce transportation costs.
- Elimination of minimum wage laws encourages the substitution of machines for labor and encourages population concentration in cities.

---

[32]Alain de Janvry, "The Role of Land Reform in Economic Development: Policies and Politics," *Am. J. Agric. Econ.* (1981), pp. 384–392.

- Improvement of primary and secondary education reduces the gap in income-earning potential between the rich and the poor. Improved education is particularly important for women since fertility is reduced and more balanced nutrition results within the family.[33]

Lower the Price of Food

The lack of purchasing power was previously identified as one of the major causes of malnutrition. Under these conditions, it should make sense to reduce the price of food. However, the following three policy options have very different consequences:

- Explicit food subsidies include disaster relief, marketwide subsidies, and targeted subsidies such as food stamps. Although such subsidies are popular and frequently advocated by those having an activist philosophy and are sometimes essential, they are costly and often counterproductive. The rich often receive most of the benefits. The poor actually may be worse off in the long run, and farm prices may be lowered if the food aid is supplied by another country.
- Cheap food policies implicitly subsidize consumption by explicitly lowering food prices. The consequence is lower farm prices and reduced farm production. Cheap food policies used by developing countries commonly include government procurement of cheap foreign sources of food that are often subsidized, price controls, export taxes, overvaluing the domestic currency, and limiting the growing of cash crops for which the country has a comparative advantage.
- Encouraging increased production is accomplished by providing incentives for production. Farmers universally respond to production incentives. This can be done without direct farmer subsidies by government support for technology creation, education, and the development of rural infrastructure. On the other hand, input subsidies for fertilizer and chemicals can encourage excess use, damage the environment, and drain away needed capital, although increased production likewise results.

ECONOMIC DEVELOPMENT POLICY

U.S. economic development policy as discussed here emphasizes the agriculture component of development. It is not as comprehensive as that envisioned by Foster in describing the policy options for dealing with hunger and malnutrition.

---

[33]Benjamin Senauer, "Determinants of the Intrahousehold Allocation of Food in Rural Philippines," *Am. J. Agric. Econ.* (1988), pp. 170–180.

However, reality suggests that there is no comprehensive policy to deal with these complex issues.

U.S. policies to foster economic development have a rationale that is closely related to, and compatible with, food aid programs. Specifically, the rationales are as follows:

- Economic development satisfies basic human needs, reduces tension, and fosters political stability. U.S. development policy is thus an integral part of U.S. foreign policy.
- Economic development facilitates and encourages trade. Development aid is a means of market development.
- Economic development has humanitarian motives created by a basic desire to improve the lot of people throughout the world. In terms of food, this humanitarian motive is fostered politically by the realization that solving the world food problem requires that people work together to increase production.

A wide range of development strategies has been pursued by the United States. In the process of pursuing these strategies, the United States has learned that an agriculture in the early stages of development cannot be readily transformed into a dynamic capital-intensive and machinery-intensive agriculture in the image of the U.S. farmer.[34] Instead, the development process involves helping to accelerate the rate of growth of agricultural output and productivity consistent with resources available and the growth of other sectors of the economy. In other words, it is necessary to build on the resources that are available, which often includes an excess supply of labor. This requires the capacity to educate and develop new technical knowledge adapted to country conditions; to increase the capacity of the industrial sector to develop, produce, and market the new technical inputs for farmers that are adapted to local conditions; and to expand the capacity of farmers to acquire new knowledge and effectively use the new inputs.[35]

Although a better knowledge of the development process now exists, that does not mean that uniform rapid progress is being made in moving poorer, developing countries into the "takeoff" stage of growth. Progress remains slow, with political, resource, debt, and social impediments to progress.

U.S. agricultural development assistance is of two basic types: bilateral and multilateral. Bilateral assistance is provided by the United States directly to the recipient country. Multilateral assistance is extended to developing countries in combination with assistance primarily from other developed countries through organizations such as the World Bank, Inter-American Development Bank, and the United Nations. The major tools of development include commodity assistance, institution building and infrastructure development, technical assistance, credit, and research.

---

[34]Vernon W. Ruttan, Thinking About Agricultural Development, Texas A&M University Lecture Series (College Station, Tex.: Texas A&M University, October 30, 1975), p. 4.
[35]Ibid., p. 7.

Commodity Assistance

U.S. bilateral commodity assistance under P. L. 480 has been one of the most extensively utilized, yet hotly debated, forms of development assistance. In addition to its market development role, P. L. 480 has also been used as a development tool in at least two ways:

1. P. L. 480 fills the interim food needs of a country, allowing release of labor from subsistence agriculture.
2. P. L. 480 assistance normally involves agreements that the recipient country will use funds obtained from the sale of commodities for specified development programs.

The P. L. 480 controversy arises from the potential adverse impact of food aid on farm prices in the recipient country. Economic theory suggests that commodity assistance increases the supply of food available to the recipient country.[36] This, in turn, has the effect of reducing price and production. These effects have been confirmed in cases where P. L. 480 assistance accounted for a large proportion of the recipient countries' production.[37] However, recent research suggests that this price- and production-suppressing relationship may not be as clear-cut as theory would suggest.[38] The theoretical price-suppressing effect can be offset by developing country policies that support farm prices as a stimulant to production. Unfortunately, the inclination of some developing countries has been to overtly hold prices down as a consumer policy, thus discouraging production.[39]

Institution and Infrastructure Development

Institution building involves the development of an infrastructure of those facilities, industries, and agencies that are crucial to development and use of productivity-enhancing knowledge and innovation. Examples include the development of irrigation systems, credit agencies, universities, agriculture ministries, statistical collection systems, fertilizer production plants, and marketing systems.

Institution building is costly, requiring substantial credit and, in some instances, direct monetary investments and assistance. The U.S. center for bilateral institution building activities is the Agency for International Development

---

[36]T. W. Schultz, "Value of U.S. Farm Surpluses to Underdeveloped Countries," *J. Farm Econ.* (December 1960), pp. 1019–1030.

[37]Dale W. Adams, *Public Law 480 and Columbia's Economic Development* (East Lansing, Mich.: Medillin, 1964); and Frank D. Barlow and Susan A. Libbin, *Food Aid and Economic Development*, Foreign Agric. Econ. Rep. 51 (Washington, D.C.: ERS, USDA, 1969).

[38]Brady J. Deaton, "Public Law 480: The Critical Choices," *Am. J. Agric. Econ.* (December 1980), pp. 988–992; and Paul Isenman and H. Singer, *Food Aid: Disincentive Effects and Their Policy Implications*, AID Discussion Pap. 31 (Washington, D.C.: Agency for International Development, 1975).

[39]G. E. Schuh, *Improving the Developmental Effectiveness of Food Aid* (Washington, D.C.: Agency for International Development, September 1979).

(AID). In terms of multilateral assistance, the United States supports several development lending institutions, of which the World Bank is the largest. Its activities are coordinated with several regional development banks, such as the Inter-American Development Bank.

Technical Assistance

Technical assistance involves providing developing countries with the service of experts in specific need areas. Those who provide technical assistance may be looked on as troubleshooters who identify problems and recommend solutions. They may be agricultural engineers, food scientists, agronomists, animal scientists, or economists from universities, business, or government. Some are retired volunteers. Most are highly qualified in dealing with current problems. For example, the Food and Agriculture Organization (FAO) of the United Nations offers technical assistance in pasture and crop improvement, conservation, water resources, land reform strategies, establishment of cooperatives, price stabilization programs, nutrition improvement projects, and plant and animal disease protection.

AID has an equally wide range of technical assistance activities that were given a substantial boost by developments in Eastern Europe and the former Soviet republics. The need to assist in converting centrally planned economies to market economies represented a major new challenge and opportunity for economists and management experts.

Research

The development of a strong international agricultural research component to U.S. development activities has occurred only since the 1960s. The U.S. center for international agricultural research is once again in AID. AID-supported agricultural research was given a major boost in 1966 when President Johnson, in his War on Hunger message, emphasized the need to help countries to balance agricultural productivity with population growth.

As in other development activities, international agricultural research has bilateral and multilateral components. The U.S. component has as its major thrusts conducting research in major problem areas, in strengthening research capability in areas where future needs are anticipated, and in training the future scientific manpower from and for developing countries. With leadership and funding from AID, scientists from both USDA and the universities are involved in these activities.

The initial scientific breakthroughs that led to the Green Revolution, which sharply expanded grain yield in developing countries, were the product of a combination of country research in the United States and Japan implemented by the international agricultural research system. This system is composed of a number of international agricultural research centers, such as the International Maize and Wheat Improvement Center in Mexico and the International Rice Research Institute in the Philippines. Financial support for these centers comes

Fighting with Guns and Butter

Drought is a given in her life. It hasn't rained properly here in four years. In that time, her husband hasn't been able to raise a decent crop. Outside her dirt-floor hut, her babies play in a dusty moonscape—rocks, mountains, nothing that is green. But food aid keeps rolling in.

Asked where the food comes from, Kadija replies that it comes from guerrillas, the Eritrean People's Liberation Front. As Kadija testifies, it doesn't matter much who donated the food. Hearts and minds are won by the people who deliver it.

As long as the EPLF holds these hills and the rebels can truck in food donated by the West and funneled through bordering Sudan, Kadija and her family are famine-proof. As one might suspect, she and her neighbors have only good things to say about the EPLF.

The food weapon is wielded not only by the Eritrean rebels, who want to create an independent nation out of Ethiopia's northernmost region, but by guerilla fighters in the bordering Tigray region, who want to overthrow Ethiopia's central government.

It also is used by the Ethiopian government, which refuses to negotiate with either rebel organization and repeatedly has mobilized Africa's largest standing army in attempts to crush them.

It is the famine fight, the battle over rights to fill the stomachs of the peasantry. Part of the battle is operational—actually delivering food by airplane, truck, and camel. The other half is rhetorical—propaganda attacks on the "enemy" for "playing politics" with food.

"He who controls roads controls food. He who controls food controls people. This time around, the rebels in Eritrea and in Tigray are prepared to play much harder ball in terms of controlling the people," says Shun Chetty, deputy representative for the U.N. High Commissioner for Refugees in neighboring Sudan.

In the game to control the hungry, the best-known and most powerful player is the Ethiopian government. In the next eight months, the authoritarian Marxist regime in Addis Ababa will supervise distribution of the lion's share of the 1.3 million tons of donated food that the FAO says is needed for Ethiopia's drought victims.

Source: Blaine Harden, "Fighting Guns With Butter: Food Has Become a Weapon in Eritrea's Civil War," *The Washington Post Weekly Edition*, January 18–24, 1988, pp. 6–7.

from a consultative group of international institutions, such as the World Bank, private foundations, and countries.

The multilaterally funded international research centers are generally considered to be excellent research organizations with fine facilities and highly qualified staffs. Their activities are coordinated with basic and applied research conducted in both developed and developing countries. Emphasizing applied research, their direct impact in increasing productivity has been demonstrated.

U.S. Development Support: A Controversy

The impression could be obtained from this discussion of international development that U.S. support for international development activities has been substantial. In one sense, this is true; the United States is the largest single contributor to development activities. However, as a proportion of gross national product, the United States is one of the lowest contributors of the developed market economies of the world. Yet the role of the United States is controversial from a producer perspective. The charge is frequently made that U.S. development assistance increases production of commodities that are competitive with U.S. exports. The increased production either reduces the recipient country's imports or, worse yet, ends up on the world market in direct competition with U.S. production.

In the short run, there is a degree of merit to the argument. But in the long run and from a broader perspective, denial of development assistance would be counterproductive. As indicated previously, development begins with agriculture, which generates export earnings and releases people from subsistence to work in other jobs. Agricultural development thus becomes crucial to getting developing economies into the takeoff phase of growth. Once in the takeoff phase, the country, more often than not, becomes an importing country, particularly of feed grains and soybeans for which U.S. farmers are most competitive.

In addition, development research frequently leads to short-run benefits. For example, cooperative research on soybeans in Brazil led to greater variety in germplasm for the improvement of U.S. varieties. In fact, the primary origins of most crops were in developing countries where native species have provided genes for dwarf stature, resistance to insects and disease, day-length insensitivity, and high yield potential. For example, semidwarf wheat varieties, grown on two-thirds of the U.S. acreage, contain genes brought from Asia. U.S. agriculture would surely suffer if a decision was made not to participate in international research.

The case of Eastern Europe and the former Soviet republics (FSR) vividly illustrates another often overlooked reason for helping countries develop. Market economies and democracies are hand-in-glove relationships; one fosters the other. Therefore, helping the Eastern European and FSR countries successfully make the conversion to market economies has enormous political significance. Yet the trade-off could be reduced subsidies for U.S. farmers.

Food Diplomacy

Food diplomacy *refers to the use of agricultural exports and development assistance as tools to achieve specific foreign policy goals.* Foreign policy considerations play an important role in many, if not most, international agricultural and food policy decisions.

Food and development assistance may be utilized as a tool of diplomacy with either a positive or a punitive strategy:

- *Positive food diplomacy* refers to efforts to expand exports, give food aid, or provide development assistance based on the theory that a government will not bite the hand that feeds it. Positive food diplomacy is pursued on a humanitarian or need basis without regard to the system of government or the long-term objectives of the government receiving assistance. It is food assistance without strings attached and with no intention to withdraw food supplied or development assistance in the future. Historically, one of the most ardent advocates of a positive food diplomacy strategy was Hubert Humphrey, the former senator from Minnesota and vice president in the Johnson administration. Humphrey was always a strong supporter of unlimited access to exports, expanded food aid, and expanded agricultural development assistance.[40] Few such ardent advocates remain.

- *Punitive food diplomacy* refers to making access to exports, food aid, or development assistance contingent on specific action by the recipient government. The most widely recognized example of an effort to utilize food as a tool of diplomacy was the decision by President Carter to partially embargo grain exports to the USSR in January 1980, when the Soviets invaded Afghanistan. Grain exports to the Soviets were made contingent on withdrawal of troops. This is not an isolated example. During 1965 to 1968, the United States unsuccessfully used P. L. 480 food aid to persuade India to support U.S. policies in Southeast Asia.[41] Food aid was used unsuccessfully in 1974 in support of U.S. efforts to achieve military disengagement in the Middle East.[42] U.S. food aid was parceled out in the mid-1970s in return for putting Third World pressure on OPEC countries to discontinue the oil embargo.[43]

Can food be successfully utilized as a tool of diplomacy? The answer to this question probably lies in how it is used and in how success is defined. America's capacity to produce for export, provide food aid, and assist in agricultural development (positive food diplomacy) builds goodwill. Attaching strings to food aid (punitive food diplomacy) creates resentment. For example, in the mid-1960s, U.S. food aid to India was made contingent on the adoption of population control programs. Rothschild observes: "That policy was resented intensely. There would be nothing, I think, more likely to destroy the influence

---

[40]Emma Rothschild, "Food Politics," *Foreign Affairs* (January 1976), p. 295.

[41]Don Paarlberg, *Farm and Food Policy: Issues of the 1980s* (Lincoln, Nebr.: University of Nebraska, 1980), p. 251.

[42]Ibid.

[43]Rothschild, "Food Policies," pp. 285–307.

the United States might have over food-importing countries than the possibility of future coercion."[44]

There is also the question of how effective punitive food diplomacy can be in any event. The USSR effectively demonstrated the ability to import grain in the face of the 1980 partial export embargo by shifting world grain flows. The Soviet Union was also often in a position to capitalize on U.S. refusals to provide food aid or development assistance to Third World countries. Equally important, certain forms of food diplomacy, such as embargoes, create adverse political reactions within the United States. Presidential candidate Ronald Reagan successfully capitalized on farmer resentment of the Carter-imposed Soviet grain embargo. Once elected, however, Reagan was caught in a dilemma of choosing between fulfilling a campaign promise to lift the embargo and the danger of appearing to take action that conflicted with a hard-line policy toward the Soviets. He eventually opted to lift the embargo. The embargo, however, continued to be blamed for reduced exports of farm products throughout the following crop year.

Despite such dangers and conflicts, the use of food as a tool of diplomacy in both a positive and a punitive sense is likely to continue. Tighter food supplies increase not only the temptation to use food as a tool of diplomacy but also diplomacy's chances of being effective. It is not always easy to distinguish between positive and punitive food diplomacy. For example, arguably, food diplomacy has been used, in a contemporary context, against the warlords in Somalia, the Serbs in Yugoslavia, Iraq's Hussein, and the communist hardliners in the former Soviet republics. Is each of these an example of positive or negative food "diplomacy"?

Can We Be Satisfied?

Some ask why the United States should be concerned about world hunger when there are 30 million hungry Americans. Reality indicates that political stability depends on long-run solutions to the world food problem. These solutions are too complex to be solved by any one country acting alone. The United States has been a leader in working with other nations to find solutions to the problems of world hunger, which have not been solved. The willingness of the United States to commit troops to Somalia indicates a level of resolve that had previously been lacking. Regardless of the outcome of that effort, it indicates that we are not satisfied.

ADDITIONAL READING

1. By far, the best publication on the complexity of the issues discussed in this chapter is Phillips Foster's *The World Food Problem* (Boulder, Colo.: Lynne Rienner Publishers, 1992).

[44]Ibid., p. 294.

6

THE ROLE OF TRADE

Free trade, one of the greatest blessings which a government can confer on a people, is in almost every country unpopular.

Thomas Babington

U.S. agriculture has become dependent on world markets for its products; the crops from two of every five cultivated U.S. acres are exported. Similarly, many countries of the world have become dependent on American farmers as a principal source of their food supply.

But the growth in agricultural trade has not been a one-way street. American consumers depend on substantial imports of agricultural products from other countries, such as orange juice and coffee from Brazil. Brazil, in turn, uses a portion of the revenues from the sale of orange juice and coffee to buy wheat from the United States.

Because of increased trade, agricultural economies of the world have become much more interdependent in recent years. This interdependence in trade is not limited to agriculture. Japan uses a portion of its export earnings from automobiles to buy agricultural products from the United States. The United States uses export earnings from grain to help pay for imported oil from the OPEC countries.

The purpose of this chapter is to explain why trade is beneficial to society and, at the same time, why barriers to trade emerge. The characteristics and economic impacts of various types of barriers to trade are also discussed.

TRADE-OFFS IN TRADE

Trade among nations is one of the keys to expanding food supplies and reducing food costs. However, emotions frequently run high when trade issues are discussed. U.S. wheat producers located in North Dakota and Montana fret about the increased imports of wheat and barley from Canada.[1] Corn Belt farmers want to be able to export more corn, grain sorghum, and soybeans to Canada as well as other countries. Ironically, they join beef producers in favoring restrictions on beef imports, because imports potentially reduce the domestic demand for feed grains and soybeans. The same producers then express concern about restricting imports of Japanese automobiles for fear that Japan will retaliate by curtailing imports of U.S. grain.

Trade restrictions may make sense from the viewpoint of an individual adversely affected by imports, but society as a whole pays the cost of each and every trade restriction in terms of quantity, quality, and price of the goods received—an illustration of the fallacy of composition. The proponents of trade are often neither as visible nor as effective as the opponents of trade. There are a number of reasons for this:

- The cost of trade restrictions is hidden in the price of goods, whereas those who lose jobs as a result of imports are visible and get media attention.
- Exports have a price-increasing effect; thus, the net benefits of trade expansion to the consumer are not as apparent.
- Consumers are only periodically an effective lobbying force. In addition, a strong free-trade stance may conflict with the goals of organized labor, a supporter of the consumer movement.
- Free-trade concerns have sometimes been overridden by foreign policy concerns such as tension over China's civil rights policies.

Despite the often expressed opposition to freer trade, on some occasions, such as the momentous vote confirming the North American Free Trade Agreement (NAFTA), arguments for free trade have prevailed. In fact, since the early 1970s, the freer trade trend has been evident.

WHY TRADE?

There are both economic and practical reasons for trade. From a practical perspective, without trade, the world could not feed itself nearly as well as it does currently. The distribution of food production in the world does not parallel the distribution of population (Figure 6.1). Even though there are vast differences in the quantity and quality of food consumed in different countries, variation in production alone, due to weather, makes trade essential.

From an economic perspective, advantages accrue to countries that produce those products for which they are best qualified in terms of their resource endowment, export their excess production, and buy other products that other countries

---

[1]Interestingly, increased U.S. wheat and barley imports are partially the result of a highly effective U.S. subsidy program referred to as EEP (Export Enhancement Program), which increased demand for U.S. grain (thus raising its price) and lowered the demand for Canadian grain (thus lowering their returns). Canadian farmers can thus get higher returns by selling their wheat and barley in the United States.

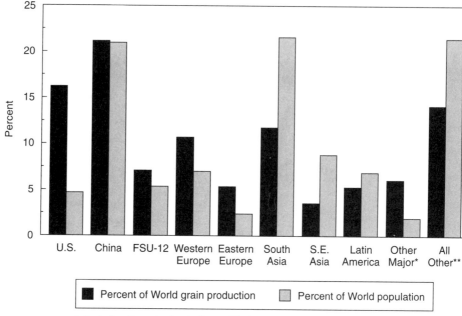

FIGURE 6.1   Share of world grain production and population by region, 1995.

Sources: World Agricultural Production (Washington, D.C.: USDA/ERS, July 1996), p. 14; and World Population and Projections to 2050 electronic database (Washington, D.C.: USDA/ERS, August 1995).

are best qualified to produce. As a result, total world output increases and resources are allocated to their best use.

Despite reduced agricultural world trade volume in the early 1980s, the longer run trend has been toward increased trade (Figure 6.2). As might be expected from previous discussion of the world food situation, population growth, and income elasticities, trade with developing countries has increased more than trade with developed countries (Figure 6.3).

The major products exported by the United States include grains and oilseeds (Figure 6.4). Other significant exports include livestock products, fruits, vegetables, nuts, and cotton. The United States generally exports about half of its wheat, rice, soybean, and cotton production. It exports about 23 percent of its coarse grain production. The U.S. share of the world market is about 63 percent in coarse grains, 71 percent soybeans, 32 percent in wheat, 17 percent in rice, and 30 percent in cotton.[2] Overall, the U.S. share of world trade in grain generally runs around 40 percent (Figure 6.2).

[2]Percentages obtained from USDA-FAS Circular Series, various issues.

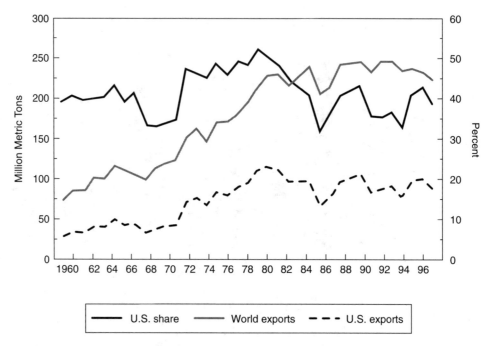

FIGURE 6.2   Volume of world trade in grain and the U.S. share, 1960–1996.

Source: PS&D View electronic database (Washington, D.C.: USDA, 1996).

Specialization

The benefits of trade arise from advantages of specialization. Production conditions vary from country to country. Any country that specializes in producing those commodities in which it has the greatest advantage or the least disadvantage will reap the benefits of the largest production at the lowest cost.

It is, for example, widely believed that the United States has a comparative advantage in the production of feed grains and soybeans. Japan, on the other hand, has a comparative advantage in the production of automobiles. It is logical then that Japan would be the United States' largest customer for grains while the United States is Japan's largest market for automobiles. Both countries are, as a whole, better off because of this trade—even though U.S. automakers and their employees protest Japanese auto imports, and Japanese rice farmers protest U.S. grain imports. Social welfare is improved because, with a given income level, more automobiles and grain can be enjoyed by more Japanese and American people with trade than in its absence.

The advantages of specialization, although intuitively obvious, are encompassed in two economic principles: absolute advantage and comparative advantage.

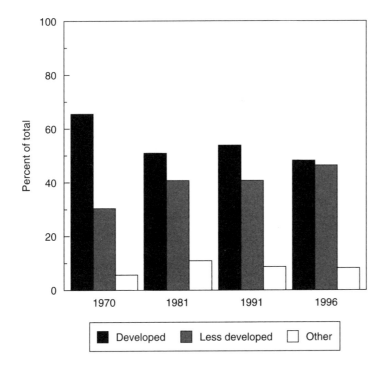

FIGURE 6.3
U.S. agricultural exports by region.

Source: Agricultural Outlook (Washington, D.C.: USDA, various issues).

Absolute advantage.   *A country is said to have an* **absolute advantage** *when it can produce a product at less cost than another country.* Suppose, for example, that the European Union (EU)[3] and the United States have options of growing wheat and corn on 10 million acres of land each. In the United States, the yield is 25 bushels per acre of wheat and 130 bushels per acre of corn. In the EU, the yield is 20 bushels per acre of wheat and 60 bushels per acre of corn (Table 6.1). If the cost per acre is the same in both the United States and the EU, the United States is said to have an absolute advantage in the production of both wheat and corn.

Suppose that the United States requires a minimum 125 million bushels of wheat and 650 million bushels of corn from the 10 million acres of land. The EU requires a minimum of 100 million bushels of wheat and 300 million bushels of corn. This can be accomplished in each country by using half the land for wheat and half the land for corn (Table 6.1). Total U.S. and EU production of wheat and corn under this self-sufficiency strategy is 1,175 million bushels (225 + 950).

[3]The EU, also frequently referred to as the *Common Market*, comprises 15 countries that have developed a Common Agricultural Policy (CAP), which includes both internal price supports and protectionist trade policies. The 15 EU countries are Belgium, Germany, France, Italy, Luxembourg, the Netherlands, Denmark, Ireland, the United Kingdom, Greece, Portugal, Spain, Austria, Finland, and Sweden. Countries being considered for EU membership in the future include: Bulgaria, Cyprus, the Czech Republic, Estonia, Hungary, Latvia, Lithuania, Malta, Poland, Romania, Slovakia, and Slovenia.

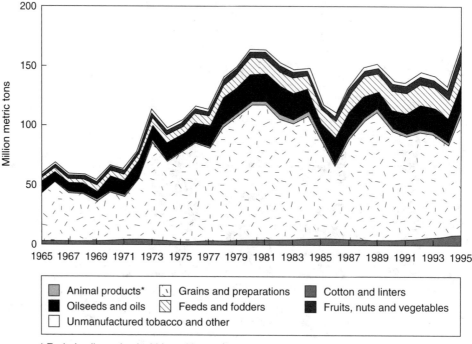

FIGURE 6.4   Quantity of U.S. agricultural exports by commodity, 1965–1995.

Source: Foreign Agricultural Trade of the United States, calendar year 1995 supplement (Washington, D.C.: USDA/ERS, June 1996), p. 36.

Comparative advantage. *The principle of* **comparative advantage** *recognizes that total output can be increased when each country specializes in producing that commodity for which it has the greatest advantage or the least disadvantage.* According to the data presented in Table 6.1, even though the United States has an absolute advantage in the production of both wheat and corn, output can be increased by specialization. This can be seen in Table 6.2. The minimum wheat requirement of 225 million bushels is attained by having the EU devote all of its land to wheat and the United States produce wheat on 1 million acres. The remaining 9 million acres of land in the United States are devoted to corn, with a total corn production of 1,170 million bushels. The resulting total U.S. and EU production of wheat and corn is 1,395 million bushels, 220 million more bushels than without specialization.

Of course, to achieve this higher level of production, it is necessary to specialize and trade. The EU must buy a minimum of 300 million bushels of corn from the United States, and the United States must buy a minimum of 100 mil-

TABLE 6.1   Production in the EU and the United States from 10 Million Acres of Land Assuming That Each Is Self-Sufficient

Country	Wheat			Corn		
	Yield/Acre (bushels)	*Acres (millions)*	*Production (millions of bushels)*	*Yield/Acre (bushels)*	*Acres (millions)*	*Production (millions of bushels)*
United States	25	5	125	130	5	650
EU	20	5	100	60	5	300
Total		10	225		10	950

TABLE 6.2   Production in the EU and the United States from 10 Million Acres of Land Assuming Specialization and Trade

Country	Wheat			Corn		
	Yield/Acre (bushels)	*Acres (millions)*	*Production (millions of bushels)*	*Yield/Acre (bushels)*	*Acres (millions)*	*Production (millions of bushels)*
United States	25	1	25	130	9	1170
EU	20	10	200	60	0	0
Total		11	225		9	1170

lion bushels of wheat from the EU. They may buy even more because with more total supplies, the price will be lower even after transportation costs are considered.

Differences in costs of production exist among countries because the inherent endowment of natural resources (land, water, and minerals) and other productive resources (labor, capital, management research, and technology) varies from country to country and because the production of different commodities requires different resources in varying proportions. The advantage enjoyed by the American farmer in the production of feed grains and soybeans results largely from the favorable soil and climate conditions of the Corn Belt. The EU farmer specializes in producing wheat and barley. Canada produces wheat, barley, and rapeseed. New Zealand is the lowest cost producer of milk in the world because of soil and climatic conditions that are favorable to roughage, the contentment of cows, and milk production. Over time, a country's resources, and for that matter, the world's resources, are more efficiently and fully utilized when allocated among commodities and industries according to the principle of comparative advantage. This involves specialization and trade.

THE THEORY OF TRADE

Trade occurs because governments, businesses, farmers, and consumers, as buyers or sellers, realize benefits from trade. Trade is beneficial when it allows buyers access to goods that would otherwise be either unavailable or more expensive. *Trade will occur to the point where the costs of goods (including transportation and transaction costs) are equal among countries.*

All people do not benefit equally from trade. Consumers in exporting countries generally pay a higher price for a commodity being exported than they would if it were not being exported. Producers, of course, receive higher prices. On the other hand, consumers in importing countries pay a lower price for the commodity than they would if it were not being imported. Similarly, the importing country's producers receive a lower price.

These complexities are best visualized in a two-country trade model for a single product—in this case, wheat.[4] The model is based on conventional economic logic and highlights the interdependence among the agricultural sectors of countries linked by trade. For simplicity, the model consists of two world or "country" segments: exporting countries and importing countries. The model considers only one commodity, wheat. We make the assumption that everything is measured in the same currency and that there are no transportation costs.

Assume initially that there is no trade; a closed economy condition called *autarky.* Prices are determined independently for importing countries and exporting countries utilizing their respective supply and demand curves (Figure 6.5).

FIGURE 6.5   Prices and production under autarky.

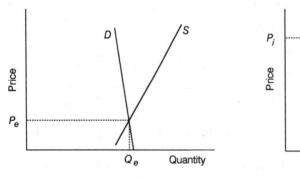
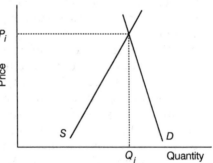

Exporting countries   Importing countries

[4]The theoretical analysis draws on the following excellent works: James P. Houck, *Elements of Agricultural Trade Policies* (Prospect Heights, Ill.: Waveland Press, 1992); Phillip L. Paarlberg *et al., Impacts of Policy on U.S. Agricultural Trade* (Washington, D.C.: ERS, USDA, December 1984); and Bob F. Jones and Robert L. Thompson, "Interrelationships of Domestic Agricultural Policies and Trade Policies," in *Speaking of Trade: Its Effects on Agriculture,* Agric. Ext. Spec. Rep. 72 (St. Paul, Minn.: University of Minnesota, November 1978), pp. 37–57.

The resulting equilibrium is price P_e and quantity Q_e for the "exporting" countries and price P_i and quantity Q_i for the "importing" countries.

Trade is introduced into the two-country trade model by providing for a world market in the middle graph of Figure 6.6. The quantity available for trade by exporting countries, referred to as excess supply (ES), is the horizontal distance between the supply and demand curve above price P_e. At price P_e, supply equals demand in the exporting country and there is nothing available for export. This provides the vertical axis intercept in the world market graph.

At prices below P_i, the importing countries have an excess demand (ED) of the difference between their demand and supply schedules. The world market excess demand curve ED, therefore, begins at P_i and slopes downward and to the right. The quantity imported at each price is the difference between the quantity supplied and the quantity demanded at each price below P_i.

In the world market graph, at price P_w, excess supply and excess demand are identical with the quantity traded being Q_t. Note that in the exporting countries' graph, the quantity exported ($S_e - D_e$) is identical to the quantity imported ($D_i - S_i$) in the importing countries' graph, which is Q_t in the world market graph.

Now assess the benefits of trade. The quantity and price received by the exporting countries are increased from the level it would have been at under autarky. The quantity consumed by the importing countries is increased at the world price. World production and consumption are increased by trade. But not everybody is better off. Exporting country consumers pay a higher price and importing country producers receive a lower price. Therefore, while trade is beneficial to society as a whole, it is also controversial.

Barriers to Trade

Despite the benefits of trade, interest groups in both exporting and importing countries may object to trade taking place. Protectionist policies emerge when any

FIGURE 6.6 Price, production, and quantity traded under free trade.

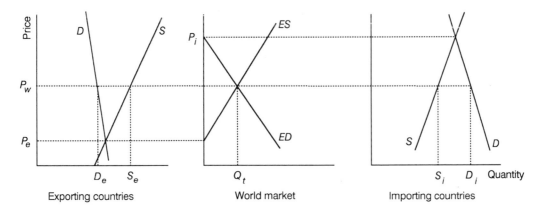

group of producers or consumers attains enough political strength to secure barriers to trade that will insulate the group(s) from international competition.

From a national policy perspective, a variety of reasons is often advanced to rationalize the pursuit of protectionist policies. The following are leading ones[5]:

- *Protection against painful economic adjustment* probably constitutes the most widely used reason for protectionism. Over time, the comparative advantage of industries changes among countries. As this occurs, substantial economic pressure is placed on producers in the less efficient country. When faced with this situation, the adversely affected industry will tend to first seek protection from government rather than make the adjustments required to adapt to the new economic situation.

- *Maintaining domestic government programs* implemented to achieve specific objectives may require protectionist policies. Suppose, for example, that the EU supports the price of wheat above the world price to improve its farmers' income. Unless protectionist policies exist, EU millers would find it less expensive to import their wheat. EU farmers would then be unable to sell their wheat and the objective of the price support program would be thwarted. For many years before 1995, Section 22 of the Agricultural Adjustment Act of 1933 allowed import restrictions to be imposed by the president whenever imports posed a threat to domestic programs that supported farm prices. This legislation was frequently invoked to reduce imports of cotton, sugar, milk products, and peanuts.

- *Protecting national security* has been a frequently advanced justification for protectionism. For example, Japan has been particularly concerned about maintaining domestic rice production as part of a contingency "food security reserve." In addition, the United States imposed a grain export embargo against the Soviet Union as a punishment for actions in Afghanistan that were said to threaten national security. Nations tend to maintain industries that produce the essentials of war—food and weapons—even though the principles of free trade suggest that they would be better off economically without trade barriers. Also, rather than face the political risk involved in high domestic prices and shortages, governments sometimes resort to export embargos, thus depriving the world market of the product.

- *Protecting infant industries* until they can compete in the world markets is another argument advanced for protectionist policies. Frequently, costs of production are higher in the initial stages of industry development. As the size of the industry expands, costs fall, making the industry competitive in world markets. One problem associated with this approach to development is that the protected industry must later be weaned from the protectionist policies under which it grew up.

---

[5]This discussion draws heavily on James P. Houck and Peter K. Pollak, "Basic Concepts of Trade," in *Speaking of Trade: Its Effect on Agriculture*, Agric. Ext. Serv. Spec. Rep. 72 (St. Paul, Minn.: University of Minnesota, November 1978), pp. 21–35.

■ *Protecting national health* leads to outright prohibitions of trade between some countries. Such prohibitions or regulations are often referred to as sanitary and phytosanitary regulations. For example, the United States will not allow beef to be imported from countries whose cattle have a history of hoof-and-mouth disease. Also, restrictions are imposed on imports of vegetables containing residues of pesticides banned in the United States. Yet public health may be used in an arbitrary manner even when there is no health hazard, such as rBST.

■ *Retaliation against policies* of another country deemed to be unfair sometimes results in the adoption of protectionist trade policies. For example, most countries have policies that protect their industries against another country "dumping" its products on the world market by selling its products below the cost of production. Dumping may be done either to gain access to a market or to get rid of burdensome surpluses. Retaliatory responses have an inherent danger of triggering a worldwide round of trade restrictions, frequently referred to as a *trend toward protectionism.*

■ *Improving balance of payments* provides a rationale for protectionist policies if the country has a persistent balance of payments problem and the value of its currency is threatened. Restricting imports has the potential to improve an adverse balance of payments situation as long as the supplying countries do not retaliate against the importing countries' protectionist policies.

■ *Improving international terms of trade* is a monopsonist's argument for imposing a tariff, with the effect of forcing down the world price. Only large importers have this power.

■ *Providing revenue* was once a reason for the British imposition of a tea tax on the original 13 American colonies. Developing countries still frequently impose export taxes for the purpose of generating revenue. U.S. tariffs are not designed primarily to generate revenue. They may be a negotiated substitute for quotas. Or they may be designed to cover the cost of border inspection operations—sort of a "user fee." When NAFTA was negotiated, proposals were made to continue certain tariffs as a means of improving the infrastructure near the Mexican border.

The nine major types of barriers to trade are import tariffs, variable levies, import quotas, quality restrictions, export subsidies, export taxes, export embargoes, state traders, and exchange-rate distortions.[6] Each of these barriers to trade is sufficiently important to warrant an explanation of its effect in a two-country trade model context.

Import tariffs.   The classic method of import protection is the tariff, also called an *import tax* or *customs duty.* A tariff may be either a fixed charge per unit

---

[6]This list does not include a host of domestic programs that have an impact on trade and, therefore, are implicit barriers to trade. These are discussed in subsequent chapters as elements of domestic policy.

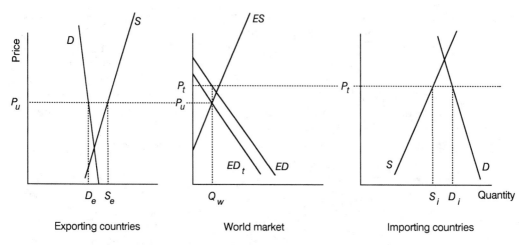

FIGURE 6.7   Price, production, and quantity traded under an import tariff.

of product imported (specific tariff), a variable levy per unit of product imported, or a fixed percentage of the value of the product imported (*ad valorem* tax). In each case, the effect of the tariff is to raise the price of the imported good equal to or above the price of similar domestically produced goods. Tariffs thus make foreign goods economically unattractive to potential importers.

Referring to Figure 6.7, a tariff of $P_t - P_u$ causes the importing country's price to rise. In effect, the excess demand curve is shifted down by the amount of the tariff from ED to ED_t. At the higher price P_t, the importing countries' producers increase the quantity supplied. Consequently, less is imported $(D_i - S_i)$ than would be under free trade. The effect is to drop the world price to P_u. This lower price is transmitted to the exporting countries' consumers and producers whose consumption rises to D_e and production falls to S_e. The tariff results in lower overall production (GNP) than under free trade, due to a misallocation of resources.

Variable levy.   One of the most interesting tariffs is the variable levy, which has been used by the EU to provide price support for imported commodities such as corn. To illustrate the variable levy, Figure 6.8 isolates the EU on the right graph and the rest of the world on the left. Suppose that the level of price support for corn chosen by the EU is P_s. At this price, S_i will be produced and D_i demanded with $D_i - S_i$ being imported. The levy results in a world price of P_w, which leads to D_w being demanded in the rest of the world and S_w supplied.

The important feature of the variable levy is that once the price support level is set at P_s, the levy is constantly adjusted to reflect the difference between P_s and the prevailing world price. Therefore, the EU price is always supported at P_s and the world price adjusts to yield Q_w imports. All price adjustments are transferred to the world market, increasing world price variation. Price variation in the world market is absorbed by adjustments in the variable levy. Changes in supply or

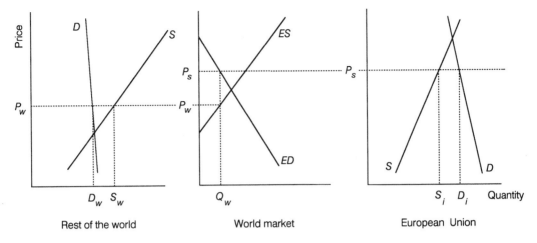

FIGURE 6.8   Price, production, and quantity traded under a variable levy.

demand within the EU are transmitted to the world market through changes in import demand at price P_s. The variable levy effectively acts as an import quota at quantity Q_w in Figure 6.8.

The variable levy was the primary means by which the EU supported its prices until the 1990s when the inevitability of the implications of the Uruguay Round of GATT became clear. EU policies then began to take on a form of direct farmer payments that more closely approximated those of the United States prior to the 1996 farm bill.

Import quotas.   An import quota sets an absolute limit on the quantity of a product that can be imported. A zero quota, of course, prohibits all imports. The United States used quotas to limit imports of beef, dairy products, and sugar until the GATT negotiations forced conversion to tariff rate quotas, which sharply increase the tariff at the quota level.

Quotas are easy to analyze in a two-country trade model. Figure 6.9 represents the case of the U.S. quota on beef imports with the United States shown in the left graph and the rest of the world on the right. If the United States sets the quota at Q_q (or $D_q - S_q$), the excess demand curve is truncated at Q_q, therefore taking the shape of ED_q rather than the straight line ED_w. The result is a lower price P_w in the rest of the world than would exist under free trade and a higher price P_q in the United States. The quota, therefore, limited competition to U.S. beef producers. A quota can be set to have the same effect on prices and quantities as an equivalent tariff.

Voluntary import restraints are a subtle form of import quota with the same economic impacts. *They involve agreements negotiated between an importing country and one or more exporting countries to limit their exports to some specified quantity.* The implied "quota" becomes the product of negotiation.

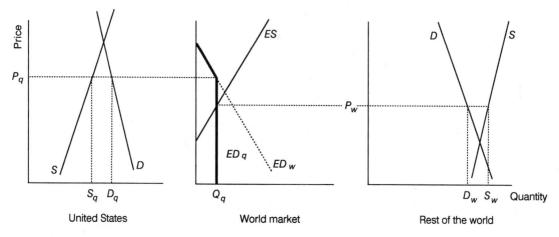

FIGURE 6.9   Price, production, and quantity traded under U.S. beef import quota.

Quotas, whether mandated or voluntarily negotiated restraints, are valuable to both the country that establishes the quota and the quota holder. Quotas are generally divided up among potential suppliers. Since the price received for commodities imported under a quota is higher than the world price, quotas are valuable, and giving a country a quota is a political favor. Economists refer to the process of obtaining the *higher price associated with protectionist government policies as* **rent seeking.** The *extra income* associated with a quota is referred to as quota **rent.** Yet rent-seeking activities are by no means limited to quotas. Producers seeking tariffs or other forms of protection are also rent seekers (see box). For that matter, anyone who seeks competitive protection from the government is a rent seeker.

Import licenses are another substitute for a quota because they are generally given to export countries for a specific quantity of product being imported. The total of import licenses issued is the importing country's quota. Licenses, therefore, also perform the function of allocating the quota among exporting countries. However, once again, the GATT negotiations put a substantial constraint on the ability of countries to utilize import licenses.

Technical restrictions. **Sanitary and phytosanitary regulations** are technical restrictions that limit the quality of products entering the country for reasons of protecting health or human, animal, and plant life.[7] While all countries maintain such regulations on both imports and domestic production, sometimes the regulations on imported products are more strict than on domestic products or they effectively ban all or nearly all production from another country. For

---

[7]Jimmye S. Hillman, *Technical Barriers to Agricultural Trade* (Boulder, Colo.: Westview Press, 1991), pp. 44–45.

Rent Seeking in International Trade: The Great Tomato War

The tomato has been the center of many legal and political international trade battles designed to gain rents through both tariff and nontariff barriers. This rent-seeking activity began when vegetable producers got the Tariff Act of 1883 enacted. This act assessed a 10 percent import duty (tariff) on fresh vegetables but not on fruit. Importers (supermarket buyers of the day) argued that tomatoes were actually a fruit. The Supreme Court ruled that in the common language of the people, tomatoes are vegetables. The Tariff Act of 1938 established duties on vegetables that for tomatoes and cucumbers were higher in harvest periods.

In the early 1950s, Florida tomato producers turned their attention toward creating nontariff barriers to trade. In 1954, Florida Senator Holland got enacted the "golden rule agreement" in marketing orders making imports subject to the same marketing order regulations as domestic fruits and vegetables. The Florida Tomato Committee immediately got a regulation establishing that vine ripe tomatoes marketed from Mexico had to be one-quarter inch smaller than mature, green tomatoes marketed from Mexico. This quarter-inch size difference, which would exclude many tomatoes from Mexico, was justified under the golden rule because vine-ripened tomatoes had to grow longer and become larger than mature green tomatoes. This order regulation stood until 1970 when an extended political battle broke out between Arizona-New Mexico interests in imports and Florida producers. Consumer groups got in the fray forcing the issue to hearings. In the hearings, it was pointed out that only 165 Florida producers benefited from the regulations, one of which was Gulf and Western Industries, which had one-eighth of the production.

Having lost the marketing order battle, in 1978 three Florida producers filed an antidumping petition with the Department of Treasury, charging that the Mexicans were selling tomatoes at less than their fair value with the intent of destroying the Florida fresh winter vegetable industry. Although dumping was never found to exist, extensive negotiations were held at top levels of both governments to obtain voluntary Mexican import restraints.

In 1996, after NAFTA was supposed to have solved all these problems, the war heated up once again. During the heat of the 1996 presidential election, the Clinton administration bowed to the demands of Florida tomato interests by negotiating a floor on the price of tomatoes exported to the United States. Mexico's agriculture minister objected to the pact by indicating that this new barrier to trade would damage Mexico's producers and would cost jobs in a country already plagued by unemployment.

Source: Abstracted in part from Maury E. Bredahl, Andrew Schmitz, and Jimmye S. Hillman, "Rent Seeking in International Trade: The Great Tomato War," *Am. J. Agric. Econ.* (February 1987), pp. 1–10.

example, the United States has strongly objected to an EU ban on hormone-produced beef; most fed beef in U.S. feedlots is produced with hormones that regulate growth and produce muscle more efficiently than if hormones were not used. Likewise in 1996 the EU held up imports of genetically altered Bt corn and Roundup Ready soybeans at the urging of environmental interests such as Greenpeace.[8]

Sometimes it is not clear whether a particular sanitary or phytosanitary regulation resulted from a legitimate health concern or was erected as a barrier to trade. For example, was the EU hormone-produced beef ban a result of a legitimate concern about the health impacts of the hormones, or was it designed to protect the EU's less efficient means of producing beef primarily with dual-purpose dairy cows on smaller size farming units? Should a country have a means of setting higher health standards or protecting the structure of its agriculture even if it imposes higher costs on its consumers and suppresses world prices? This is obviously a political decision involving value judgments, not an economic decision.

In the two-country trade model context, quality restrictions at their extreme can lead to autarky (Figure 6.5). This would happen if the exporting countries' producers were unable to meet the requirements of the quality restriction. Prices would be determined entirely by domestic supply and demand. A more likely case is one in which some countries can comply and others cannot. In any case, quality restrictions can distort trade patterns as effectively, or perhaps more effectively, than tariffs or quotas.

Export subsidies.   An **export subsidy** *is a government payment per unit of product exported.* Export subsidies and related practices are designed to expand exports by placing products on the world market at prices lower than would exist under competitive conditions. The intent is to reduce the competitive pressures on producers or to dispose of surplus production.

In the two-country trade model, a U.S. export subsidy of $P_e - P_w$ is perceived by the rest of the world as a shift in the excess supply function to the right from ES to ES_s (Figure 6.10). The subsidy increases the U.S. price and reduces the export price. U.S. exports are increased relative to free trade, but production in the rest of the world falls. The subsidy produces an economic gain for U.S. producers and foreign consumers. It causes an economic loss to U.S. consumers, U.S. taxpayers (who pay for the subsidy), and foreign producers.

Export subsidies come in many forms, some of which are quite subtle. In the 1950s and 1960s, the U.S. regularly provided subsidies to foreign buyers, totaling as high as $822 million in 1964. A 1972 decision to sell the Soviets grain at a sub-

---

[8]Bt corn is transgenetic. It carries a gene from *Bacillus thuringiensis* (Bt), a soil bacterium that is toxic to the European corn borer (a pest which has been known to reduce yields by as much as 32 bushels per acre). Roundup Ready soybean varieties are tolerant to glyphosate, the active compound in some herbicides. The concern in the case of Bt corn is the potential for the development of resistance in the targeted insects. Roundup Ready soybean varieties have encountered opposition since they are engineered to withstand larger doses of herbicides. Critics are concerned that increased herbicide use benefits producers at the expense of consumer's health (many herbicides are known carcinogens) and the environment.

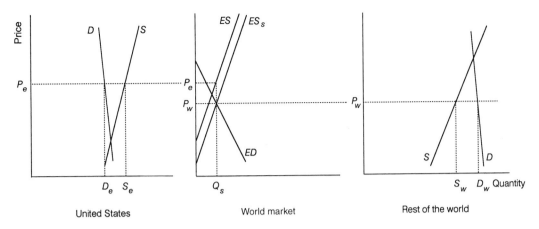

FIGURE 6.10   Prices, production, and quantity traded under U.S. export subsidy.

sidized price was very controversial. Recently, the United States has begun to use targeted export subsidies as opposed to the general subsidies provided in the 1950s and 1960s. The 1972 Soviet subsidy was, in a sense, a targeted subsidy. A targeted subsidy is limited to a particular country or to a sale designed to attract or regain a customer.[9] How effective the targeted subsidy is depends on the longer run dependability of the buyer as a customer and on the reaction of competitors.

Under marketing orders, commodities such as almonds are *sold in the world market at a lower price than in the domestic market, which is, in essence, a subsidy to foreign consumers.* Such a **two-price plan** was initially proposed in the 1920s as the *McNary-Haugen plan,* designed to dispose of surplus commodities on the world market. It has been used for peanuts for a number of years, with strong producer support.

The United States is by no means the only country that subsidizes exports. The EU used export subsidies as a part of its Common Agricultural Policy (CAP). For example, price supports on wheat and sugar were maintained sufficiently high that excess production occurred. The surpluses were often sold in the world market at a lower price. This was accomplished by an export subsidy, referred to as a **restitution.** This two-price plan caused EU consumers to spend a larger share of their income on food. However, through the GATT negotiating process, the EU has been required to substantially modify and reduce these subsidies.

As discussed in Chapter 10, direct income payments from the government to farmers are a form of export subsidy—although both domestic and foreign consumers receive the benefits of the resulting lower price.

---

[9]Targeted export subsidies are also referred to as expanded export assistance, which exists on many different commodities under several different programs. Targeted export assistance by the United States was a primary target of the GATT negotiations.

A much more subtle form of export subsidy involves subsidies on the use of inputs for production. For example, it is common in developing countries to supply farmers with fertilizer and seed at little or no cost. Such subsidies shift the exporting countries' supply curve to the right, as it does in turn to the excess supply curve, resulting in a lower price.

The ultimate export subsidy and two-price plan involves dumping. Technically, **dumping** *means selling on the world market below the cost of production.* Dumping gets its name from an uncontrolled sale on the world market of a commodity at whatever price it can obtain. Such a commodity is said to be "dumped" on the world market.

Export taxes.   An **export tax** *is imposed by the exporting country; it raises the price of goods entering international trade.* Such taxes may serve as a source of revenue for an exporting country. It may also be a means of reducing international demand for a product that is desired for domestic consumption. For example, during the 1995–1996 crop production shortfall, the EU placed a tax on exports of wheat and barley as a means of rationing its short supply in the face of high world market prices. This, of course, raised the world market price even higher.

Recall from Chapter 5 that export taxes penalize producers and are a source of production shortfalls in developing countries. An implicit export tax occurs when the government puts a ceiling on the producer price as a means of controlling food costs. Consumers benefit from the lower price but may not get enough production to satisfy domestic needs.

Export embargoes.   An **embargo** *is a suspension of exports of one or more commodities by a country to one or more countries.* Export embargoes are imposed by government fiat and may be imposed either when supplies of a commodity are very short and prices are extremely high or as part of a foreign policy strategy. The United States, for example, imposed an embargo on the export of soybeans, cottonseed, and related products in June 1973 when the price of soybeans rose above $12 a bushel and concern arose regarding the adequacy of protein feed supplies for poultry and hog production. The domestic price dropped almost immediately after the embargo was imposed to $6 per bushel. The disruption in supplies available to traditional U.S. foreign customers severely strained trading relations with them.[10] The high price of soybeans and the embargo have been correlated with the development of the Brazilian soybean industry, which eroded the dominance of the United States in the world soybean market.

The Soviet grain embargo of 1980 was imposed as a retaliatory reaction to the Soviet invasion of Afghanistan. Virtually every Republican candidate, and some Democrats, made "political hay" of the Soviet grain embargo during the 1980s. The embargo was blamed for all of the ills that have beset agriculture.

---

[10]John T. Dunlop, "Lessons of Food Controls, 1971–74," in John T. Dunlop and Kenneth J. Fedor, eds., *The Lessons of Wage and Price Controls—The Food Sector* (Cambridge, Mass.: Harvard University Press, 1978), p. 244.

Interestingly, a comprehensive economic study of the 1980 Soviet grain embargo mandated by Congress found that it was not effective and had little long-run impact.[11] Although it does not condone embargoes, the study highlights the extent of exaggerated political claims regarding both their effects and effectiveness. This study and related experiences such as the Persian Gulf War against Iraq suggest that foreclosing the supply of food to a country that has money and/or allies is extremely difficult. Effective embargoes in these circumstances require a level of military commitment that typically has not existed.

During the 1995–1996 crop production shortfall, talk arose once again regarding the embargo issue. In this case poultry integrators, beef feedlots, dairy farmers, and hog producers had problems securing a supply of feed grains, so they sought a means, other than price, of rationing the available supply. In retrospect, however, price did effectively ration supplies as export demand declined and high prices attracted early marketings from the newly harvested crop.

State traders.   One of the most frequently ignored barriers to trade is the state trader. **State traders** *are government or quasigovernment monopolistic sellers or buyers of a commodity.* In other words, the government or a government-authorized agency sells in the export market on behalf of producers and buys on behalf of consumers.

State traders constitute a significant share of world trade in a number of commodities. Some of these countries are in a sufficiently strong position to exercise a degree of market power in influencing price and/or quantity traded.[12] Canada, Australia, and New Zealand are state trading exporters. Each of these countries operates with a marketing board structure on major export commodities. For example, the Canadian and Australian wheat boards are both major competitors of the United States. Many developing countries operate with state trading import policies that may involve issuing import licenses or may allow food processors to purchase jointly commodities such as wheat. The Soviet Union was perhaps the strongest of the state trading importers. Its breakup had a material impact on the structure of international trade, although most of the former Soviet republics still operate as state traders in their own right.

Exchange-rate distortions.   The exchange rate functions as a relative price between the currencies of different countries. In market economies, changes in exchange rates affect supply and demand for commodities, which, in turn, alter trade levels and flows. As in the case of price changes, exchange rates must be adjusted for inflation in each country to evaluate effects on trade. Also, since countries export and import different goods, it is necessary to calculate a trade-weighted exchange rate for those countries actually importing a commodity or group of commodities. Figure 6.11 gives the Federal Reserve's trade-weighted average exchange rate for all exports, for corn, and for cotton.

---

[11]Alex F. McCalla, T. Kelley White, and Kenneth Clayton, *Embargoes, Surplus Disposal and U.S. Agriculture* (Washington, D.C.: ERS, USDA, November 1986).
[12]Paarlberg *et al., Impacts of Policy,* pp. 48–49.

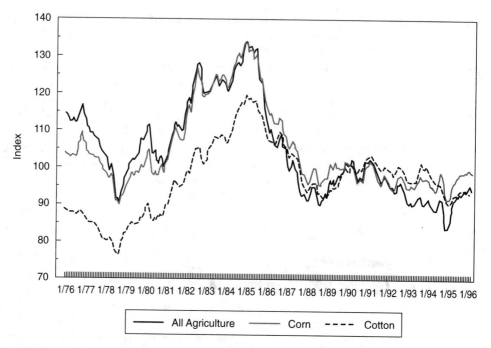

FIGURE 6.11   Index of the monthly trade-weighted exchange rate of the U.S. dollar, 1976–1996.

Source: Data provided by USDA/ERS, Commercial Agriculture Division, November 1996.

In a two-country trade model, the exchange rate has the effect of rotating the excess demand function (Figure 6.12). A real appreciation of the dollar rotates the excess demand function ED downward to ED_a. The downward rotation means that foreign buyers must sell more goods in their currency to purchase a given amount of goods in U.S. dollars. The effect is to restrict export demand to Q_a, as the price to the rest of the world is raised to P_b, and the price in the United States is lowered to P_a. When the value of the dollar falls, just the opposite occurs. That is, the excess demand function is rotated upward, indicating that U.S. goods are more favorably priced. Export demand rises, as do U.S. prices.

We can see that exchange rate changes have much the same effect as export taxes and subsidies. When the value of the dollar declined in the 1970s (see Figure 6.11), it took less foreign currency to buy U.S. commodities. Exports, therefore, expanded rapidly as if there were a large export subsidy. When the dollar appreciated in the early 1980s (see Figure 6.11), exports plummeted as if they were being taxed. It took more foreign currency to buy a dollar's worth of U.S. wheat, corn, or cotton. Beginning in 1985, a decline in the value of the dollar once again increased the competitiveness of U.S. commodities. Instability in exchange rates is translated into instability in commodity markets.

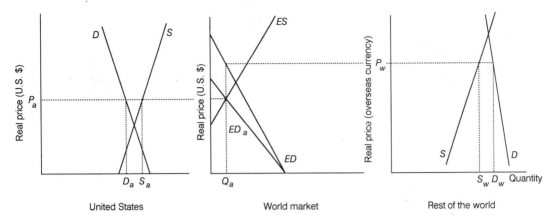

FIGURE 6.12   **Prices, production, and quantity traded under appreciated dollar.**

To prevent such adverse consequences, countries will sometimes manage their own exchange rates. Overtly lowering the exchange rate operates like an export subsidy, expands trade, and tends to fuel inflation. Raising the exchange rate operates like an export tax and holds down domestic prices. For example, in the face of high inflation rates, several South American countries overvalued their currencies. This forced their farm prices below world prices. In the early 1980s, Uruguay overvalued its currency to the extent that its cattle prices were less than half the world price.[13]

TRADE WARS AND PROTECTIONISM

This analysis of the impacts of trade barriers on price, production, and quantity traded assumed that protectionist policies can be pursued without eliciting retaliatory response from exporters. That is seldom the way the real world operates. Countries do not stand idly by when their producers are experiencing lower prices due to an export subsidy. They retaliate with either their own subsidy or with a countervailing duty. *A* **countervailing duty** *is a tariff imposed to offset the competitive advantage that the export subsidy gives the foreign product.*

In other words, subsidies are matched with subsidies or tariffs. The result is commonly referred to as a *trade war* or an *increase in protectionism*. In such periods, it is not unusual for discussions to go on in the offices of policymakers as to what it would take in subsidies to break the bank of the EU, for example. At some point, there is the realization that the trade war has gone too far. Policies to reduce trade barriers are discussed in Chapter 7.

---

[13]Lucio G. Reca, "Price Policies in Developing Countries," in D. Gale Johnson and G. Edward Schuh, eds., *The Role of Markets in the World Food Economy* (Boulder, Colo.: Westview Press, 1983), p. 124.

LEVELS OF BARRIERS TO TRADE

There are no free traders among the world's agricultural trading countries. Without exception and with varying degrees of comprehensiveness and success, all governments intervene in agriculture.[14]

In a nutshell, these were the findings of a major USDA study designed to quantify the extent of barriers to agricultural trade. Producer subsidy equivalents were calculated based on the combination of government expenditures on subsidies and the difference between the domestic price and the world price. In essence, a **producer subsidy equivalent (PSE)** *measures the proportion of producers' gross income (including government payments) from government intervention.*

A comparative ranking of the level of producer subsidies is presented in Figure 6.13. Japan has by far the highest level of subsidies with 77 percent of producer receipts being a result of government intervention. The EU ranked second, with 49 percent of its producer receipts attributable to government. Canada at 27

FIGURE 6.13   Levels of government intervention in agriculture for latest available year, by country, 1995.

Source: Agricultural Policies, Markets and Trade in OECD Countries: Monitoring and Evaluation 1996 (Paris, France: OECD/OCDE, 1996), p. 187.

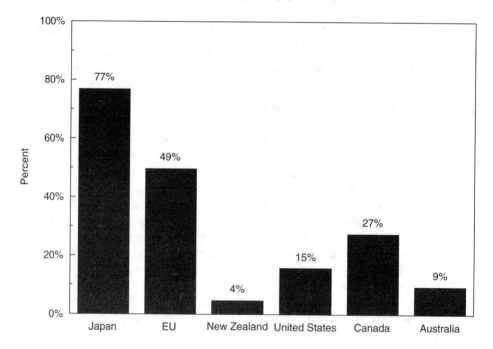

[14]Economic Research Service, *Government Intervention in Agriculture: Measurement, Evaluation and Implications for Trade Negotiation*, FAER-229 (Washington, D.C.: USDA, April 1987), p. v.

percent ranked third. The United States had 15 percent of their receipts from the government. Australia provides 9 percent producer benefits while New Zealand provides only 4 percent.

As a general rule, importing countries provide higher levels of producer assistance than do exporting countries, reflecting the drive for food security and self-sufficiency. Food grains, dairy products, and sugar were found to be the most highly subsidized. In less developed countries, producers of some commodities were taxed while others were subsidized.

TRADE POLICY: A WORLD ISSUE

The benefits of free trade are obvious. Yet protectionism is pervasive and has a tendency to become increasingly so. It is driven by a desire to protect an important industry, by the desire to expand economic activity, and by the need for food security. Yet for the world to feed itself effectively and efficiently, we must engage in trade. If world markets degenerate into a dumping ground for surplus production, the world's economic growth potential is stifled. Particularly adversely affected are those developing countries that must depend on trade in agricultural products for their revenue to grow and improve their living conditions. Trade policy, therefore, is a world issue, not just a U.S. issue.

ADDITIONAL READINGS

1. James P. Houck's *Elements of Agricultural Trade Policy* (Prospect Heights, Ill.: Waveland Press, 1992) is an excellent reference on trade theory written at about the same level of economic expertise as this text. Of course, it goes into each topic in greater detail.

2. A detailed reference written in preparation for the Uruguay Round of Trade Negotiations is the book by Jimmye S. Hillman, *Technical Barriers to Agricultural Trade* (Boulder, Colo.: Westview Press, 1991). It provides considerable information on the extent of the use of individual barriers to trade with emphasis on the United States, the EU, and Japan.

3. The Economic Research Service staff paper by Philip L. Paarlberg, Alan J. Webb, Arthur Morey, and Jerry A. Sharples, titled *Impacts of Policy on U.S. Agricultural Trade* (Washington, D.C.: USDA, December 1984) provides an excellent review of the theory of trade. It includes an analysis of farm programs and macroeconomic policy using the two-country trade model approach.

4. The original reference on trade policy is the publication *Speaking of Trade,* Spec. Rep. 72 (St. Paul, Minn.: Minnesota Agricultural Extension Service, University of Minnesota, November 1978), edited by Martin K. Christiansen. This publication contains a series of outstanding articles on trade policy. While *Speaking of Trade* is getting a bit old, it is still very relevant and well worth reading.

5. The most comprehensive treatment of agricultural trade policy issues is by Alex F. McCalla and Timothy E. Josling, *Agricultural Policies and World Markets* (New York: Macmillan Publishing Company, 1985). Although parts of this book are a bit heavy for the novice in economics, other sections are very readable.

6. The book *Farm Incomes, Wealth and Agricultural Policy* (Aldershot, England: Ashgate Publishing Limited, 2nd ed., 1996) by Berkeley Hill is an excellent analysis of the impacts of farm policies on EU farmers.

7 INTERNATIONAL TRADE AND MARKET DEVELOPMENT POLICY

Political internationalism without economic internationalism is a house built upon sand. For no nation can reach its fullest development alone.

Wendell L. Wilkie

In the 1970s, the United States decided to become an active participant in the world market for agricultural products. It expanded production to take advantage of the high prices and profits that were available to crop producers during this period of booming exports. Once in the world market, it had little choice but to stay.

The purpose of this chapter is to provide the setting for the development of trade policy. The emphasis is on the United States, although the positions of other countries and groups of countries are developed when it is appropriate. Of particular significance is the development of the World Trade Organization and a series of country trading blocs made possible through free-trade agreements such as the North American Free Trade Agreement (NAFTA).

IMPORTANCE OF TRADE TO THE UNITED STATES

Beginning in the 1970s, agricultural exports soared from less than $6 billion to nearly $44 billion in 1982. They then declined to about $27 billion in 1987 and then resumed a steady growth trend, reaching an all-time record of $60 billion in 1996. Exports account for about 30 percent of overall farm production, about $50 billion in added gross national product, and about 1.4 million jobs. In 1996, about 40 percent of the wheat and roughly one-third of the rice, cotton, and soybeans were exported, while about 20 percent of the feed grains were exported. Agriculture exports are about twice imports.

Farm groups sometimes assert that if foreign buyers are not willing to pay the cost of production for a commodity, it should not be sold. But if it is not sold, it should not be produced. The consequence of abandoning the export market would be a reduction in production of almost half of the wheat acreage, one-third of the rice, soybean, and cotton acreage, and 20 percent of the feed grain acreage. Production adjustments of this magnitude would send economic reverberations throughout America. In addition to putting a large number of farmers out of business, the hardest hit sectors would be the agricultural input industries, which would lose as many as 400,000 jobs, and export-related financing, transportation, storage, merchandising, and port operations, which could lose as many as 1 million jobs.[1]

COMPETITIVENESS IN TRADE

If U.S. farmers are to be a part of the world market, they must be competitive. Competitiveness has two basic economic requirements:

- The U.S. price must be at a level that allows it to meet or beat the competition to maintain market share.
- The U.S. cost of production must be sufficiently low that its farmers can remain in business at the prevailing world price.

The price requirement must be met in the short run to make a sale. The cost requirement must be met in the long run to stay in business. The relevant costs include not only farm costs but also those costs involved in moving commodities from the farm gate to the end users. This concept of costs recognizes the importance of inland transportation, storage, and labor costs, as well as production costs.

The United States is generally recognized to be the most competitive country in the world for corn, soybeans, and grain sorghum, reflecting a share of world trade that is more than 70 percent. In wheat, barley, rice, and cotton, other countries appear to hold an advantage over the United States.

While global import demand for food has experienced persistent growth, the U.S. market share declined from a high of 22 percent in 1980 to a low of 17 percent in 1986, with a rebound to about 23 percent in the mid-1990s.[2] The decline in market share is the result of a combination of adverse U.S. macroeconomic policies, a change in geopolitical relationships (particularly related to the European Union and the former Soviet republics), and more aggressive competi-

---

[1]David Harrington, Gerald Schluter, and Patrick O'Brien, *Agriculture's Links to the National Economy: Income and Employment*, Agric. Inf. Bull. 504 (Washington, D.C.: ERS, USDA, October 1986).
[2]Foreign Agriculture Service, *Long-Term Agricultural Trade Strategy* (Washington, D.C.: FAS/USDA, 1996).

Competitiveness Depends on Technology

How can U.S. farmers be competitive when labor costs are so much lower in other countries? This is a question frequently asked by U.S. farmers. The answer to the labor cost advantage lies in superior technology and management as well as in larger scale farm operations. U.S. agriculture has dealt with the labor cost issue by substituting other inputs, such as machinery and chemicals, for labor. In many respects, U.S. agriculture competes on the basis of technology. Continued technology investments and accelerated investments in human capital will be the keys to the future competitiveness of U.S. farmers.

---

tion from other major exporting countries. The growth in market share in the 1990s was due to the use of aggressive export promotion strategies (such as the Export Enhancement Program), favorable macroeconomic policies, and tighter global supplies.

Public policy plays an important role in both short-run and long-run competitiveness. In the short run, the United States cannot be competitive if it is priced out of the market by any combination of high domestic farm price supports, an overvalued dollar, or the aggressive policies of other countries to expand trade.

In the long run, competitiveness is a function of the basic endowment of resources, investment in production and marketing infrastructure, technology, and the growth of demand. Investment, technology, and demand are greatly affected by public policy. Available evidence clearly shows that many countries are committing substantial resources to the research and development of agricultural technological breakthroughs.[3] While technologies are transferable, the country that is the first to discover, develop, and apply a new technology can gain a critical competitive edge. More important, since technological development and change are long-term processes, the country that has the long-run public and private commitment to invest in research is more likely to be competitive in the long run (see box).

In the past decade, the United States has made some significant public policy changes affecting private sector agricultural research incentives. The most important of these was the extension of intellectual property rights to discoveries of new life forms and computer software.[4] This change has resulted in greatly

---

[3]U.S. Congress, Office of Technology Assessment, *A New Technological Era for American Agriculture,* OTA-F-474 (Washington, DC.: U.S. Government Printing Office, August 1992).

[4]Robert E. Evenson, "Intellectual Property Rights and Agribusiness Research and Development: Implications for the Public Agricultural Research System," *Am. J. Agric. Econ.* (December 1983), pp. 967–975.

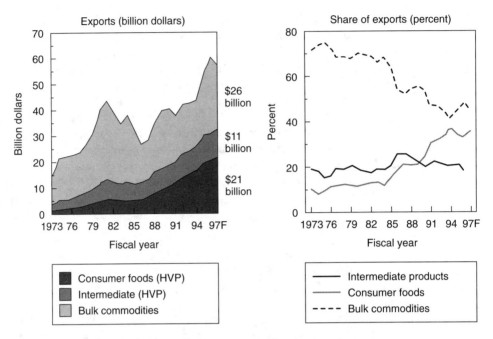

FIGURE 7.1   Growth of bulk commodities versus high valued products, 1973–1997.

Source: Michael J. Dwyer, "Situation and Outlook for U.S. Agricultural Exports to the 21st Century," *Southern Regional Outlook Conference* (Washington, D.C.: USDA/Foreign Agricultural Service, November 12, 1996), p. 2.

increased private sector interest and investment in basic agricultural research. Continued public sector support for agricultural research, combined with new private investment in research, holds the potential for improving the competitive position of U.S. agriculture.[5]

To be competitive, the emphasis in U.S. exports must change as the market changes. Traditionally, U.S. market emphasis has been on bulk commodities— wheat, corn, soybeans, cotton, and rice. As incomes expand in importing countries, the opportunities for exports of consumer foods (such as ready-to-cook chicken pieces) or intermediate products (such as soybean meal) grow. Initially, the United States was slow to take advantage of that potential. The failure to be perceptive and responsive to market change reflected a lack of sophistication in export strategy on the part of both the public and private sectors. The emphasis on consumer food and intermediate exports changed in the late 1980s (Figure 7.1). Since then, consumer foods have experienced consistent growth.

[5]James T. Bonnen, "A Century of Science in Agriculture: Lessons to Science Policy," *Am. J. Agric. Econ.* (December 1986), pp. 1065–1080.

U.S. Trade Policy Goals

Since 1970, the major thrust of U.S. trade policy has been in support of freer trade. This commitment is seen on a bipartisan basis in both policy statements and actions of government including the following:

- Policy statements strongly supporting freer trade in virtually every *Economic Report of the President* since 1970
- A commitment to an international monetary system based on market-determined currency values
- A commitment to pursue bilateral and multilateral trading agreements, such as NAFTA, to ensure that market access grows and that competition takes place under internationally agreed-on rules

Goals of Agricultural Trade Policy

The USDA reexamined its trade policy goals and strategy in 1996 and stated as its mission: " ... to open, expand, and maintain global market opportunities through international trade, cooperation, and sustainable development activities which secure the long-term economic vitality and global competitiveness of America's rural communities and related food and agricultural enterprises."[6]

USDA has adopted a very specific goal of increasing the value of farm, food, fish, and forestry exports 50 percent over 1994 levels by 2000. To carry out its mission and achieve this goal USDA has identified five strategies to guide its policies:

- Improve the level of market intelligence regarding export opportunities and competitor export programs, practices, and products.
- Identify, develop, and promote new market channels for consumer food products.
- Monitor, negotiate, and assist in securing full implementation of free-trade agreements, intensify pressure for science-based sanitary and phytosanitary standards, and monitor potentially exclusionary practices of state traders or trading blocs.
- Provide credit and export subsidy assistance to markets having the greatest growth potential with the framework of agreements reached in trade negotiations.
- Build sustainable export markets through technical assistance and food security to less developed nations and emerging democracies such as the former Soviet republics.

---

[6]Foreign Agriculture Service, Long-Term Agriculture Trade Strategy, p. ii.

Such goals and strategies espouse lofty ideals but belie the many conflicts that exist between domestic policy and trade policy as well as the strong political forces favoring greater protectionism. Before discussing these conflicts, it is necessary to explain the World Trade Organization (WTO) and its relationship to its predecessor the General Agreement on Tariffs and Trade (GATT).

WTO and GATT

WTO is a permanent formal international organization that oversees the conduct of trade as agreed to by its member nations. It is a multilateral treaty among some 130 governments, including the United States. Its purpose is to liberalize and expand trade through negotiated reductions in trade barriers. WTO was formed in 1995 by the Uruguay Round trade negotiations as the successor organization to GATT. Under GATT member countries were required to adhere to three basic principles which still apply to WTO:[7]

- Once a country has agreed to make a tariff concession, it cannot unilaterally raise that tariff at a later time.
- Any advantage given any product to any country must be given to all countries. This is the so-called most favored nation provision (MFN) that is designed to ensure fair and equitable treatment of all GATT/WTO countries. For example, when the United States grants MFN status to a country it ensures that country that it will have access to the U.S. market at the lowest tariff offered to any other country.
- Imported goods must be treated the same as domestic goods in terms of regulations and taxes.

From its inception through 1981, seven major GATT negotiating conferences or "rounds" have been held. Agreements on reduction of agricultural trade barriers have not come easily. More progress was made in the liberalization of industrial product trade because agricultural trade barriers tend to be rooted in domestic subsidies to agriculture for most of the major exporting and importing countries. The major achievements in GATT negotiations that have been of particular significance to U.S. agriculture follow.

In the 1960–1961 Dillon Round, the EU provided assurance of duty-free entry for soybeans, soybean meal, other oilseeds, and cotton. Variable levies on feed and food grains were continued.

The 1963–1967 Kennedy Round resulted in tariff reductions on a wide range of farm products. These reductions were less than the reductions on industrial products but were still considered to be significant.

[7]Stephen D. Cohen, Joel R. Paul, and Robert A. Becker, Fundamentals of U.S. Foreign Trade Policy (Boulder, Colo.: Westview Press, 1996), p. 163–164.

The 1973–1979 Tokyo Round emphasized reduction of nontariff barriers. Codes of conduct were established to discourage the use of export subsidies and unique product standards, and to remove procedural obstacles to obtaining import licenses. Concessions were obtained from Japan for increased imports of beef and citrus fruits. The most substantial U.S. concession involved increased dairy imports and an agreement not to increase the palm oil import tariff.

In the 1987–1994 Uruguay Round, the United States renewed its commitment not to separate an agreement on agriculture from one on industrial goods, services, and intellectual property rights, the latter two of which had no authority under GATT. After seven arduous years of negotiation covering the Bush and Clinton administrations, an agreement was reached that is now generally recognized as being the most successful trade liberalization round.

From an agricultural policy perspective, the principal agreements included:

- Reducing the level of budget outlays for export subsidies by 36 percent and in the quantity of products subsidized by 21 percent.
- Reducing the level of domestic subsidies to agriculture by approximately 20 percent relative to the base period 1986–1990.
- Converting quotas to tariffs and tariff rate quotas and then phasing them out over a specified number of years.
- Providing minimum levels of market access for products currently barred from entering countries, such as rice into Japan, Korea, and Taiwan, and then to gradually increase the access levels.
- Establishing rules and disciplines of sanitary and phytosanitary regulations to make them more science based.

From an overall perspective, the most significant contribution of the Uruguay Round was the creation of the WTO and the broadening of its jurisdiction to cover services and intellectual property rights. The latter includes the products of biotechnology, which are very important to agriculture.

The functions of the WTO, as a successor to the GATT include:

- Acting as a forum for future trade negotiations
- Administering and implementing the results of trade negotiations
- Seeking to resolve trade disputes
- Overseeing national trade policies that may conflict with freer trade objectives
- Cooperating with other international institutions involved in global economic policy such as the international monetary fund

As of this writing, plans for the next round of trade negotiations beginning in 1999 were being developed. Further reductions in export subsidies, domestic subsidies, and increased market access will be on the agenda.

U.S. Trade Policy Conflicts with WTO Objectives

The existence of WTO implies that many countries' trade and agricultural policies conflict with WTO objectives, rules, and guidelines. The key conflicts are among exporting countries, including the domestic farm subsidies, the variable levy and export subsidies of the European Union, the Japanese protection of its rice and beef markets, and the Canadian protection of its dairy and poultry markets.

The United States has several programs that directly affect access to U.S. markets and/or subsidize exports of farm products. Some of these programs are agriculture specific, while others apply to all U.S. businesses and industries:

- **Section 201** *of the Trade Act of 1974 provides temporary protection for import-sensitive industries to facilitate adjustment to the loss of competitiveness internationally.* This statute, and its counterpart in WTO, is known as the "escape clause" because no demonstration of unfair trade practices is necessary to justify temporary protection. Only serious injury or the threat of serious injury must be demonstrated. Devices such as countervailing duties are used in Section 201 cases, but disputes are settled more often by a voluntary export restraint.

- **Section 301** *of the Trade Act of 1974 prohibits unfair trade practices such as dumping with demonstration of injury to a U.S. industry.* Section 301 might legitimately be referred to as a retaliatory provision, allowing the president to impose restrictions on imports with a finding of injury.

- **Section 8e** *of the Agricultural Marketing Agreements Act provides that grade, size, quality, or maturity regulations established for fruits, vegetables, or nuts under a marketing order shall apply equally to imported products.* This regulation, frequently referred to as the golden rule of marketing orders, was added to the act in 1954. Prior to its existence, imported products that did not meet order standards could be imported to compete with the U.S. products that were required to meet the order standards. The fear was that the imported products would capture a substantial market share in the face of more severe marketing restrictions.

- *Targeted export assistance and export enhancement provisions of domestic farm policies are designed to recover lost export markets and to counter subsidized competition.* The basic export enhancement tools included export payment-in-kind (PIK) when the Commodity Credit Corporation (CCC) had surplus stocks and expanded export assistance (such as EEP) in the form of cash subsidies when surpluses do not exist. (See the following market development discussion.) Such programs are legal, under WTO provisions, when a country is recovering lost market shares. However, the Uruguay Round specifically limited both the amount of such subsidies and the quantities that could be subsidized affecting primarily the United States and the European Union.

Uprooting the Root of Protectionism

The root of protectionism lies in domestic farm programs. Every country of the world has decided for one reason or another to protect its farmers from international competition. Trade policy is sometimes an integral part of farm policy. This is the case in Japan where, prior to the Uruguay Round, no rice imports were allowed. The Uruguay Round agreement was designed to begin rooting-out the root of protectionism by mandating a 20 percent reduction in the level of domestic subsidies. In the United States, the 1996 farm bill goes a step further by decoupling subsidies from production. The challenge for the European Union is to make adjustments in its policies to eliminate effectively the tie between farm support and international trade, which requires the elimination of its variable levy and restitution export subsidy framework of farm support

---

Each of these programs affecting trade conflicts with WTO objectives and thus is the continuing subject of discussion in the multilateral trade negotiations. It can readily be seen that trade negotiations cannot be successful unless the domestic farm programs employed by each country are a part of the negotiation. For farm organizations, negotiating the future of farm programs under GATT was a contentious point because they have spent their time and resources seeking to maintain and enhance these programs in their respective countries.

ECONOMIC INTEGRATION AND FREE-TRADE AGREEMENTS

The major contemporary economic movement in the world is economic integration to take advantage of the benefits of freer trade. Countries have come to realize that their economic future is dependent on the products, services, and progress of other countries. They also have learned that they cannot individually raise their prices without closing themselves off to the benefits of trade. Thus increased world economic interdependence has led to economic integration.

The forms of economic integration include free-trade agreements, customs unions, common markets, and economic unions. These are listed in order of the degree of integration. In a sense, these forms of economic integration may be looked on as trading preferences like MFN status under GATT/WTO. However, they may also be looked on as trading blocs, which are multilateral alliances of countries that rationalize their comparative economic advantages through freer trade. Individually and as a trading bloc, these countries become more powerful economic forces internationally.

Free-Trade Agreements

Members of a free-trade agreement lower or eliminate barriers to trade on broad categories of products while maintaining their own domestic policies and trade policies toward nonmember nations. The best contemporary example of a free-trade agreement is NAFTA, negotiated by Mexico, Canada, and the United States, which sets a timetable for the elimination of barriers to trade over as much as a 15-year period (see box). Yet NAFTA contains no provisions regarding the conflicting domestic marketing board policies of Canada or the *ejidos* policies of Mexico designed to maintain a small farm structure. These policies may need to be rationalized for NAFTA to work as intended.

Free-trade agreements are a first step toward economic integration. They achieve efficiency gains by allowing the principles of comparative advantage to operate more fully, thus benefiting employment and income. Yet invariably, they prove controversial because during the transition some farmers, businesses, and employees are adversely affected.

Customs Union

The second step toward economic integration is a customs union. In addition to an internal lowering or elimination of trade barriers, a customs union involves the establishment of a common trade policy toward nonmembers. Therefore, for NAFTA to become a customs union, the United States, Canada, and Mexico would have to agree on a common trade policy toward other countries. The Canadian Wheat Board and U.S. sugar policy, for example, could be major stumbling blocks to NAFTA becoming the North American customs union agreement.

Common Market

A customs union plus free movement of factors of production (labor, capital, and businesses) among the participating countries is a common market. This requires at least partial alignment of macroeconomic policies as well as sectoral policies such as agriculture. This implies the delegation of some country policy decisions to the common market governing body.

The European Union, prior to 1994, was a common market. Since 1957, it has been moving toward more complete forms of economic integration, beginning with the development of its Common Agricultural Policy (CAP). Several South American countries have aligned themselves in blocs with a goal of becoming common markets. Examples include MERCOSUR, a common market alignment of Paraguay, Argentina, Brazil, and Uruguay, and the Andean Group, made up of Venezuela, Colombia, Equador, Peru, and Bolivia.[8]

---

[8]Kenneth Forsythe and Liana Neff, *The U.S. Enterprise for the American Initiative*, Agric. Inf. Bull. 660 (Washington, D.C.: ERS/USDA, April 1993).

North American Free Trade Agreement (NAFTA)

NAFTA created the largest free-trade zone in the world, encompassing Canada, Mexico, and the United States. As such, it is designed to phase out tariff and nontariff barriers to trade, establish rules for investment, strengthen intellectual property rights, and provide a mechanism for dispute settlement.

Canada and Mexico, respectively, have been the United States' second and third largest export country markets (after Japan). Likewise, the United States is the largest export market for both Mexico and Canada.

NAFTA is actually composed of three bilateral agreements: one between the United States and Mexico, a second between Mexico and Canada, and the Canada-U.S. Free Trade Agreement (CUSTA). These are the major areas of agreement among the three countries:

- Import barriers converted to tariffs and phased out over up to 15 years.
- Tariff increases provided for as safeguards for sensitive commodities if imports exceed specified volumes.
- Domestic farm support programs for the respective countries allowed to continue.
- Export subsidies discouraged except when meeting competition of outside countries.
- Rules of origin established to minimize products imported into any of the three countries from being reexported to the other countries.
- Sanitary and phytosanitary restrictions used only if scientifically justified.
- Investment restrictions eliminated.
- Procedures and time limits for dispute resolution established.

Source: General Accounting Office, *North American Free Trade Agreement: Assessment of Major Issues*, GAO/GGD-93–137B (Washington, D.C.: U.S. Congress, September 1993).

Economic Union

In an economic union, the involved nations agree to unify all social, agricultural, tax, monetary, and fiscal policies, including the acceptance of a common currency. This implies a substantial degree of political integration, with national boundaries becoming increasingly transparent. In 1994, the European Community made a sufficient number of these changes in policy to be officially referred to as the European Union.

FOREIGN MARKET DEVELOPMENT POLICY[9]

Expansion of agricultural exports has been a major goal of USDA programs since the 1970s. The lead agency in USDA for foreign market development activities is the Foreign Agriculture Service (FAS). The activities of FAS include assisting in the development of trade policy, supervising agricultural counselors located in the U.S. embassies in major U.S. trading partner and competitor countries, analyzing foreign market developments, referring export opportunities to U.S. exporters, and assisting in and directing foreign market development activities. In addition, FAS's responsibilities include coordinating USDA's development activities with other agencies of government. The lead agency for economic development is the Agency for International Development (AID) in the Department of State.

Because elements of foreign policy, foreign relations, and diplomacy are involved in nearly all FAS activities, many of them are conducted jointly with the Department of State, the Office of the U.S. Trade Representative (USTR), and other cabinet departments. One of the prime responsibilities of the agricultural counselors or attaches is to identify markets for U.S. farm products and to assist firms selling U.S. farm products.

The precipitous drop in exports during the 1980s led to new programs designed to expand exports. The following is a summary of the major old and new market enhancement initiatives.

Market Development and Promotion Programs

Market development and promotion programs are aimed primarily at countries with potential for increased cash purchases. An outgrowth of the producer-oriented cooperator program, such programs combine private and government support to increase the quantity and effectiveness of U.S. market development activities. FAS works jointly with farm organizations and agribusiness firms to expand markets. The cooperators are producer-oriented farm groups such as U.S. Wheat Associates and the American Soybean Association; export associations such as the U.S. Feed Grains Council; and agribusiness firms such as ConAgra, Cargill, McDonald's, and PepsiCo. They combine the export promotion efforts of producers, processors, handlers, and exporters. Market development activities are planned, implemented, evaluated, and financed jointly by FAS and the cooperator organizations.

Market development and promotion programs emphasize market information and technical assistance in servicing the needs of importing countries to utilize products effectively, enhancing buyer awareness and appreciation of U.S. farm products, and assisting in conducting consumer promotion activities in importing countries. These programs have come under attack as corporate welfare for using government monies to support the marketing programs of large

---

[9]This section draws heavily on the description of market development programs contained in Mark Smith, *Increased Role for U.S. Farm Export Programs*, Agric. Inf. Bull. 515 (Washington, D.C.: ERS, USDA, April 1987).

agribusiness firms such as McDonald's, Tyson Foods, and ConAgra. However, expanding higher valued consumer food markets has become a major agricultural trade policy goal for which recent success has been demonstrated (see Figure 7.1). As agriculture becomes more industrialized, it is anticipated that the issue of corporate welfare will become increasingly contentious—possibly extending to domestic farm subsidies.

Food Aid

For countries in the early stages of economic development, the first step in establishing a healthy export market may be either through direct food aid or concessional sales. This is accomplished primarily under P. L. 480, the Agricultural Trade Development Act of 1954.

This program has an interesting history and an important contemporary application. P. L. 480 was established as the Food for Peace program following World War II. Its success in aiding Europe's recovery following the war led to its use as a tool of economic development, market development, and foreign policy. In light of this history, it should not be surprising that P. L. 480 became an important policy tool aiding in the transition of Eastern Europe and the former Soviet Union to market economies.

Therefore, P. L. 480 now plays the dual role of assisting the poorest of the poor Third World countries and helping build market economies in emerging democracies—tasks that are really quite different because of the substantial resource base that exists in some countries. These dual functions are performed under three P. L. 480 titles.

1. **Title I** is a concessional program that provides for the sale of agricultural products under long-term dollar credit sales at low interest rates for up to 40 years. Also sales for local currencies were reauthorized in 1985. Actual sales are made by private U.S. suppliers to foreign government agencies or private trade entities, which, in turn, resell in the recipient countries. Title I country agreements generally contain provisions with regard to how the recipient country uses the proceeds from sales of commodities purchased under Title I.

2. **Title II** is the principal U.S. program for responding to emergency food relief needs. Under Title II, food commodities and associated costs are donated to the recipient countries. Such donations are designed to provide commodities directly to specific nutritionally vulnerable groups through child health clinics, school lunch programs, and food-for-work programs. Title II distribution programs are carried out partially by voluntary groups such as CARE and Catholic Relief Services.

3. **Title III** is designed to provide the poorest countries with additional encouragement for economic development. Multiyear commodity commitments and loan forgiveness are made in return for agreements to undertake specific development initiatives designed to increase production and improve the lives of the poor.

In the 1950s and 1960s, it was not unusual for concessionary sales under government programs, including P. L. 480, to account for as much as one-third of U.S. exports. However, when commercial exports mushroomed in the 1970s, the proportion of P. L. 480 sales declined. In the 1980s, P. L. 480 sales once again became more important, making up about 20 percent of rice, wheat, and vegetable oil exports. However, as the level of Commodity Credit Corporation (CCC) stocks dwindled in the 1990s and commercial markets strengthened, P. L. 480's relative importance again declined.

As these trends indicate, P. L. 480 has been more than just a food aid and market development program. It has also been used to dispose of surplus farm products with P. L. 480 sales declining when commercial sales rose. Ironically, this typically is the time when the need for food aid for the poor is the greatest. On the other hand, when farm prices are low, political pressures increase for more money to be appropriated for P. L. 480.

Despite these multiple objectives, P. L. 480 appears to have contributed substantially to export market growth. Japan, South Korea, Taiwan, Brazil, Peru, Chile, Colombia, and Spain all have graduated from being P. L. 480 recipients to being primarily commercial customers. This success is one of the factors that has contributed to its broad-based agricultural support.

Export Credit

A country enhances its position to compete for export sales if it has an ample supply of competitive credit available to potential buyers. The extension of credit terms tailored to the needs of particular customers has been important to progressively developing markets from concessional sales into full-cash customers that rely on commercial credit sources.

USDA's Commodity Credit Corporation plays a major role in seeing that credit is available for eligible foreign customers. Each year, 25 to 30 overseas markets are financed by $1 billion to $2 billion of CCC export credit. This has been accomplished under two major programs:

- **Direct short-term credit** programs extend CCC loans to foreign buyers for six months to three years at interest rates that approach commercial levels. In the early 1980s, the policy of direct lending changed to that of placing greater reliance on guarantees.
- **Guaranteed or assured credit** involves CCC guarantees of repayment in the event of a default on the loan under so-called GSM (General Sales Manager) provisions. Loan repayment terms and interest rates are the product of negotiation between the U.S. banks and the foreign buyers.

Both of these programs depend on annual appropriations from the Congress. As a substitute for annual appropriations, the 1981 farm bill provided for the establishment of a revolving fund to support USDA export credit programs. Revolving fund credit would be established from a large lump-sum appropriation by the Congress to the CCC for either making or guaranteeing loans to export customers. Such a fund would allow the CCC to manage export loans on a longer term basis as opposed to unpredictable year-to-year congressional funding. The fund would be self-perpetuating from loans repaid. To the extent that direct loans are made from the fund, accumulated interest would provide a basis for growth of the fund. Appropriations for the revolving fund have not, however, been provided.

Export Enhancement Programs

In times past, high domestic price supports or the subsidies of other countries have made U.S. producers noncompetitive in particular foreign markets. Alternatively, markets have been lost by the United States because of macroeconomic policies that were particularly adverse to trade—as was the case in the early 1980s. These conditions spawned a series of export enhancement programs initially referred to as export PIK (payment-in-kind) or targeted export assistance (TEA), both aimed at recovering particular markets. *Under export PIK, USDA provided an in-kind commodity bonus for each regular commercial purchase—a "buy one, get one free" sale.* For example, a country purchasing 1 million metric tons of wheat might receive 100,000 tons of wheat or other commodities from government stocks as a bonus.

When surplus CCC commodity stocks ran low, cash export subsidies were substituted for commodities. Moreover, the cash export enhancement program (EEP) became complemented by a series of individual commodity export subsidy programs such as the dairy export incentive program (DEIP), the sunflowerseed oil assistance program (SOAP), and the cottonseed oil assistance program (COAP), the latter two of which are no longer authorized. These programs provided USDA with the flexibility to promote expanded trade utilizing either surplus commodities (export PIK) or a cash subsidy to meet or beat prices offered by competitors. These programs have been used primarily for bulk commodities that are highly price sensitive rather than value-added products. One of the provisions of the WTO agreement is to scale down and phase out these programs over several years.

Export enhancement programs have had some very interesting and unexpected consequences. For example, a Canadian study found that EEP had the effect of raising the price of U.S. wheat and barley while depressing the price of Canadian wheat. The lower Canadian price resulted from reduced and lower priced sales by the Canadian Wheat Board. Canadian producers then increased their sales directly across the U.S. border into North Dakota and Montana.[10]

---

[10]A. Schmitz, R. Gray, and A. Ulrich, *A Continental Barley Market: Where Are the Gains?* (Saskatoon, Saskatchewan, Canada: Department of Agricultural Economics, University of Saskatchewan, April 1993).

POLICIES DESIGNED TO ENHANCE MARKET CONTROL

Some of the most controversial trade policies are those designed to enhance the level of country control over markets. This may be accomplished, with varying degrees of success, unilaterally with marketing boards, bilaterally with trade agreements, or multilaterally with cartels and/or international commodity programs.

Marketing Boards

A **marketing board** *is a compulsory marketing organization set up under government legislation to perform specific marketing functions.* The board can perform a wide range of functions, including collection and dissemination of market information; product promotion, research, grading, operation, and supervision of selling facilities; collective bargaining; and the purchase, storage, and sale of products.[11] Sometimes boards become a conduit for, or a participant in, country subsidy programs.

The functions associated with marketing boards in this discussion include direct participation in the handling, pricing, and marketing of commodities such as grain for export. Such boards have the primary objectives of (1) increasing producer prices and incomes, (2) reducing fluctuation in prices and incomes, and (3) equalizing returns among producers.[12]

Marketing boards are distinguished from competitive marketing systems, such as the U.S. grain marketing system, by the following:

- There is only one agency responsible for purchasing and/or selling the commodity.
- The board is directly involved in market operations.
- All producers are required to participate.
- All producers receive the same price before adjustment for quality and location.

Board Operations

The best known marketing boards are the Canadian and Australian Wheat Boards. Although these boards differ in detail, the following operational features are common:

- Upon harvest, grain is delivered to a public or privately owned elevator.
- Upon delivery, the producer is paid an advance that is a guaranteed price.

---

[11]Martin E. Abel and Michele M. Veeman, "Marketing Boards," in *Marketing Alternatives for Farmers* (Washington, D.C.: Committee on Agriculture and Forestry, U.S. Senate, April 7, 1976), pp. 73–81.

[12]C. E. Bray, P. L. Paarlberg, and F. D. Holland, *The Implications of Establishing a U.S. Wheat Board*, Foreign Agric. Rep. 163 (Washington, D.C.: ERS, USDA, April 1981).

- The board sells grain either through direct negotiation with a foreign government or to private grain-trading companies that buy grain from the board for resale. There are some indications that boards prefer to sell direct to foreign governments and that foreign governments prefer to buy from marketing boards. Private companies obviously would prefer not to have to deal with marketing boards.

- The board controls movement of grain from local elevators to port elevators.

- Proceeds from the grain sales are pooled. Costs, including administrative, interest, insurance, and storage, are subtracted from the proceeds.

- Producers are paid a uniform average price based on proceeds from sale, less cost, divided by the total volume of grain sold. Adjustments are made for transportation and grain quality.

- Producers are paid the difference between the pool price and the advance. If the pool price is less than the advance, the government makes up the difference. A portion of the difference between the pool price and the advance may be retained for a "stabilization" fund, which is drawn on when the pool price is below the advance.

Board Impacts

Marketing boards have been advocated as a means of raising producer returns, stabilizing prices, offsetting the market power of parastatal buyers, preventing other exporting countries from undercutting the U.S. price in the world market, facilitating the establishment of a cartel commodity agreement, and rationing available grain supplies. Nearly all of these reasons imply that a marketing board could do a better job of marketing grain at higher and more stable prices than the present U.S. open-market system. The evidence, however, as indicated previously, is inconclusive when the U.S. and Canadian systems provide the basis for comparison.

Boards stabilize prices to producers within the year because the producer is paid the pool price. However, the cost of that stability is the opportunity for a farmer to make a higher (or lower) price than the pool price. Some Canadian farmers prefer to sell their wheat and barley in U.S. markets themselves; this happens even though the Canadian Wheat Board has exclusive export jurisdiction. Most U.S. farmers appear to make the sales decision. Whether boards stabilize price from year to year is questionable. A USDA study found that there was no significant difference in year-to-year variation in prices received by producers in the United States, Canada, and Australia.[13]

A marketing board would substantially alter U.S. methods of grain merchandising. Open-market methods of producer price determination would no longer exist. The role of the futures market would be radically changed—it might

---

[13]Ibid., p. 8.

Do Blocs Facilitate Free Trade?

Trading blocs are designed to facilitate trade among countries within the bloc but what effect do they have on countries outside the bloc? While the European Union has facilitated trade among its member countries, its agricultural policies (CAP) have been a thorn in the side of other exporting countries.

A major challenge facing WTO is to see that blocs do not become barriers to trade. A positive development involves the formation of interregional trading blocs. One such bloc is the Asia-Pacific Economic Cooperation (APEC) forum which includes 17 countries, 13 of which are on the Pacific Rim, the 3 NAFTA countries, and Chile. Although these 17 countries have not signed a free-trade agreement, they have set a goal of free trade among its developed country members by 2010 and among its less developed economies by 2020.

---

be changed to a predominantly international device for hedging sales, or it could disappear entirely.[14]

Cooperative and private elevators and exporters would continue to operate but would perform predominantly market intermediary functions such as internal grain storage, handling, port facility operation, and arranging for ocean transportation. They would perform sales functions on only that portion of the grain that the board did not sell directly to another government or purchasing agent.

Because of the ability of a marketing board to undercut U.S. sales of commodities in export markets by simply making an offer at a lower price with no apparent direct export subsidy, boards have become a major target of U.S. foreign policy in the WTO and in NAFTA. Thus board policies are increasingly being attacked both from within (see box) and through the trade negotiation process.

Trade Agreements

One policy instrument used by two exporting countries to secure a share of the market from competitors is a trade agreement. A **trade agreement** *is a contract between two countries regarding the quantity of a commodity to be traded within a specified time period.* The principal objectives of trade agreements are supply assurance for the importing country and market assurance for the exporting country. Related objectives may include the normalization of trading agreements and development

---

[14]A. F. McCalla and A. Schmitz, "Grain Marketing Systems: The Case of the United States versus Canada," *Am. J. Agric. Econ.* (May 1979), pp. 200–212.

Running the Border

Not all Canadian farmers are happy with their government's marketing board policies. When grain prices are high in the United States they look across the border with envy. Some Canadian farmers have taken the situation into their own hands, loaded their grain trucks, and proceeded to market their grain to U.S. merchants who gladly purchase the Canadian grain at the U.S. market price, which is much higher than these farmers can ever hope to get from the Canadian Wheat Board.

When this first happened in the early 1990s, the board failed to enforce its exclusive marketing authority, which encouraged more Canadian farmers to run the border. U.S. farmers reacted negatively to the new competition and to waiting in line at border elevators for Canadian farmers to unload their grain.

The unhappiness of some Canadian farmers with their board has led to hearings, court challenges, and referenda on whether the board should be continued as the exclusive marketing agency.

---

of markets. In addition to quantity, trade agreements may contain provisions with respect to quality, price determination, or the exchange of information on production conditions, available supplies, and expected needs.

After the emergence of increased concern about the availability of global food supplies in the early 1970s, trade agreements became more prevalent but then declined in relative importance beginning in the late 1980s. They are most common between major developed state trading exporting and importing countries such as Canada, Australia, China, and Japan. Prior to the breakup of the Soviet bloc, the United States had a long-term agreement requiring Soviet purchase of six million tons of grain annually.

The advantages of trade agreements are limited largely to the two trading partners. The importing country is assured a source of supply and the exporting country is assured a market. In addition, trade agreements have, from time to time, been considered market development tools in that commitment to trade is increased. They have foreign policy consequences in the sense that they increase cooperation and communication between the parties to the agreement.

The parties outside of a trade agreement potentially are disadvantaged. As agreements are extended to additional countries, the potential adverse consequences become apparent: Free-market supplies decline, potentially denying countries without agreements access to a commodity and leading to increased price instability. Those countries most likely to be denied commodity supplies are developing countries, typically not strong cash customers and not regularly active buyers.

Private traders are highly skeptical of trade agreements because they represent departures from traditional open-market concepts and increased government "meddling" in grain marketing. This government interference has been defended on the grounds that trade agreements "preserve intact the American system of private grain trade contracts but, at the same time, attempt to provide, through a government-to-government agreement, more regularity in future grain exports."[15]

World Cartels

World cartels of grain-exporting countries are sometimes suggested as a potential solution to the problems of producers in exporting countries. Those who see the need for a cartel not only see the world price as being too low but also contend that importers have a sufficiently strong market position to suppress world price. In other words, they believe that the world is a buyers' market with monopsony (single buyer) pricing. This market power originates from state trader activities of large importing countries, such as Japan's Food Agency and China. A world cartel, in theory, would offset this market power.

In its pure form, a cartel is designed to increase producer returns through the exercise of market power. That market power would offset the power of monopsonistic buyers; serve as a lever against the use of tariffs, variable levies, and quotas by importing countries; and potentially extract pure monopoly rents by raising the market price above the competitive equilibrium.

Cartels may have less lofty and more benevolent objectives such as providing international price stability and/or holding commodity reserves. (These more modest objectives are discussed in a later section on International Commodity Agreements).

The pure cartel with price enhancement objectives must have a sufficient market share to raise price. At the higher price, less will be demanded. Therefore, the cartel faces a downward-sloping excess demand function. Figure 7.2 provides a two-country trade model explanation of the operation of a cartel and provides substantial insight into its potential pitfalls. Since the exporting country cartel faces the excess demand curve *ED*, its marginal revenue curve is MR_{ed}. To maximize profit, it equates marginal revenue with marginal cost as represented by *ES* and charges the demand price P_m for export quantity Q_m or D_m minus S_m. Note that at price P_m, the desire of the exporting cartel countries is to produce S_c, which is larger than the quantity Q_m that can be sold at P_m. Therefore, in a cartel, there is always an incentive to sell more than the market will take at the cartel price. Excess production, which has been the principal weakness of OPEC, is inherent in a cartel.

To be effective at raising price over a period of years, a cartel must:

- Be dealing with a commodity that has an inelastic demand. Only with an inelastic demand over the relevant price range does an increase in price result in an increase in revenue.

---

[15]Leo J. Mayer, "The Russian Grain Agreement of 1975 and the Future of United States Food Policy," *Univ. Toledo Law Rev.* (Spring 1976), p. 1032.

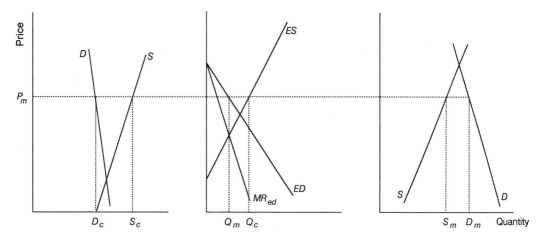

FIGURE 7.2   Exporting country cartel creates incentive for excess production.

- Be able to control supplies and entry into production. Without the power to control production, a cartel's policy to raise price will be self-defeating. Production control in a multilateral-country context involves allocation of a quota to each country. Cartel agreements, therefore, also have been referred to as quota arrangements.
- Be able to prevent substitute products from penetrating the market. For products with close substitutes, an effective cartel also must control the substitutes.

OPEC was not successful in holding its cartel together. It appears that an agricultural product would have even greater difficulty for the following reasons:

- Production control programs in agriculture have had few successes. Establishing coordinated production control programs across countries would have an even lower probability of success than OPEC had.
- With price incentives, most agricultural products can be produced in a large number of countries. Price increases on the cartel commodity will tend to attract productive resources from other commodities. This will be particularly likely in countries that are marginal exporters.
- Consumers will, over time, tend to substitute a lower-priced commodity for the higher priced cartel commodity.
- Agricultural commodities are renewable resources that are grown each year. Once oil is used, it is gone. More wheat or corn will be available each year.
- A cartel in a commodity such as wheat may draw strong adverse political reactions from Third World countries. This is particularly true in the case of wheat since the major exporting countries, except Argentina, are higher income, developed countries.

International Commodity Agreements

An **international commodity agreement** *is a multilateral agreement among countries that affects the terms of trade.* The terms of trade directly affected by commodity agreements may include the price, quantity sold, quantity produced, or the quantity held in reserve stocks. Commodity agreements are cartels with less lofty goals. Legally, commodity agreements are treaties among nations.[16] They have been used and/or proposed to accomplish two primary goals:

- Stabilize the world price within a specified corridor.
- Maintain an international reserve of commodity stocks.

Corridor commodity agreements *establish a range within which prices are allowed to move.* Provisions thus require participating exporting countries to restrict sales and build stocks when prices are low and sell stocks when prices are high. The result is an attempt to stabilize price within an acceptable range. This may be visualized in Figure 7.3 where ES_1 and ED are the initial supply and demand schedules in the export market. The price corridor is established within the range of P_1 and P_2. That is, the agreement specifies that sales will not be made below P_1 or above P_2.

FIGURE 7.3   Price corridor established by a commodity agreement.

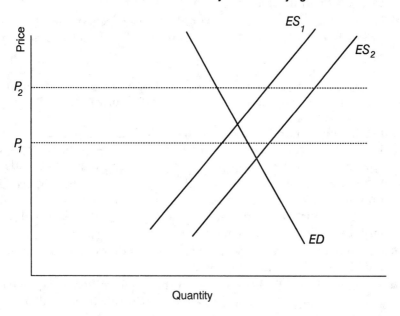

[16]Mayer, "Russian Grain Agreement," pp. 1047–1051, contains a very good and interesting discussion of the legal status of international commodity agreements, including the controversy over the authority of the president to enter into and enforce trade commodity agreements without the advice and consent of the Senate.

Corridor agreements encounter problems when supplies increase to ES_2 due to favorable weather and price thus falls below P_1—the commodity agreement minimum. Even with an agreement to accumulate stocks, a clear incentive exists to cut price and undermine the agreement. Even if the minimum price (P_1) were adhered to, at that price surplus production would result. Surplus stocks would undermine the price structure in the next production season.

Corridor agreements tend to break down when the price range specified is sufficiently narrow that the market price would frequently fall outside the range. This conclusion is supported by the following observation by Paarlberg on price stabilization attempts in wheat:

> An International Wheat Agreement worked fairly well for a decade or so after its establishment in 1949. This success was due in large measure to the wide latitude for competitive price discovery and to the fact that economic conditions were fairly stable during the period. A subsequent International Wheat Agreement failed when it prescribed a price range less compatible with market conditions.[17]

The agreements to which Paarlberg refers included provisions that guaranteed prices and purchases, recorded transactions, and spelled out enforcement rights and stocks to be established—policies that have cartel characteristics.

Most of the commodity agreements that have been entered into under the United Nations' Conferences on Trade and Development (UNCTAD) establish a price corridor. The United States has participated in several of these agreements, more as a gesture of goodwill and a desire to achieve other foreign policy objectives than as a commitment to the concept. The United States generally has been in a position of advocating a relatively wide corridor within which prices can move to allow relatively free interplay of market forces.

Reserve commodity agreements *are designed primarily to ensure supplies in times of production shortfalls.* Proposals for reserve agreements evolved largely out of the 1974 World Food Conference where resolutions called for stocks of grain to be maintained in amounts sufficient to provide food security. It recommended a study of the feasibility of establishing an international system of reserves at strategic locations.[18] This issue arose once again at the World Food Summit in 1996.

International reserves are to be distinguished from domestic reserves held by an individual country. The objectives of domestic reserves are country specific. The primary objective of an international reserve would be to provide food security for countries that could not afford to purchase commercially.

The primary interest in international reserve agreements comes from developing countries. History demonstrates that when global conflicts occur, widespread hunger and starvation are a reality. Moreover, the world supply-demand balance now periodically becomes sufficiently tight that a production

---

[17]Don Paarlberg, *Farm and Food Policy: Issues of the 1980s* (Lincoln, Neb.: University of Nebraska Press, 1980), p. 248.
[18]Joseph W. Willett *et al., The World Food Situation and Prospects to 1985,* Foreign Agric. Rep. 98 (Washington, D.C.: ERS, USDA, December 1974), p. 90.

shortfall in a major exporting country raises significant public concern about the adequacy of stocks. Yet some very difficult issues arise in establishing an international reserve agreement including the size of the reserve, control over purchase and release decisions, who should supply the grain, and who should pay the costs. These issues proved difficult enough in the 1970s that they were not resolved and no international reserve agreement has been signed.

HOW U.S. PRODUCERS FARE IN THE EVOLVING GLOBAL MARKETPLACE

The world market is composed of three main components: sellers (exporters), buyers (importers), and market intermediaries (private trading companies). Each is affected by the government policies of the countries in which it is trading. The sellers and buyers may be either state traders or private traders. As noted previously, state traders are government or parastatal agencies that buy or sell all of a particular commodity for a country. State trading sellers may be either marketing boards, such as the Canadian Wheat Board, which sells all wheat and barley exported from Canada on behalf of its producers,[19] or governments, such as China's, in cotton.

State trading buyers may be either a quasi-government-authorized associations of firms, such as a flour millers' association, which acts as a buyers' monopoly, or a government agency, such as the Japan Food Agency. Many developing countries and all centrally planned economies are state traders. The former Soviet republics are in transition from state trading to private trading, but the change is occurring slowly.

Market intermediaries are the agents of the grain trade. They are private companies that buy, sell, hedge, store, ship, and insure grain; handle currency transactions related to sales; and make arrangements for any of these functions. Private trading companies are generally headquartered in the United States, Western Europe, or Japan. It is generally accepted that a few multinational grain firms account for a large portion of the U.S. exports, and at some point in the marketing process, handle well over three-fourths of the world grain trade.[20] Two privately held companies may account for 50 percent of the U.S. trade.

Private trading companies are involved to some degree in most international commodity trading. However, the degree of involvement varies greatly. For example, for grain originating in the United States, private traders may engage in all of the functions, from country buying through delivery to a foreign port. On the other hand, for transactions between state traders, private trading companies may arrange for only shipping and insurance.

---

[19]This is not precisely correct. Under a recent change in policy, Canadian producers can sell their wheat across the border to private U.S. merchants. These sales have become the source of considerable conflict with U.S. wheat producers.

[20]Andrew Schmitz *et al.*, *Grain Export Cartels* (Cambridge, Mass.: Ballinger Publishing Co., 1981), p. 279.

However, during the current period of dramatic global economic and political change, the trend involves a large increase in the relative importance of state versus private traders. Prior to the 1990s, state rather than private traders were in the dominant position—particularly on the buyer side of the market. For example, in wheat trades during 1973–1977, sales from private exporters to private importers were less than 5 percent. State trading importers were dominant. In the future, state trading buyers are expected to be less important due to economic transformations under way in the former Soviet Union, Eastern Europe, and China.

The importance of state trading by exporters depends on the relative importance of exports by the United States and the European Union. In the 1970s, the private trader share in wheat increased to about 40 percent. It probably held at about that same percentage throughout the 1980s and into the 1990s, with reduced sales by the United States and increased sales by the European Union.

State trading is less important in the feed grains. Australia is not a feed grain state trading exporter; Japan is not a state trading importer of feed grains; the United States sells a larger share of the world's production and trade; and several importers perform state trades in food grains but not in feed grains.

With this background on the structure of international markets, how do U.S. producers fare under current trading agreements in which many barriers to trade continue to exist? The research on this issue is indecisive. Most research compares the U.S. system with the Canadian Wheat Board. McCalla and Schmitz find strengths in both systems, making it extremely difficult to determine which is "best."[21] Peltier and Anderson indicate that U.S. producer returns have exceeded Canadian returns.[22] Yet the considerations of a conversion from state trading in barley to private trading raises questions about how Canadian producers feel about the choice.[23]

It seems fair to conclude that under the current structure of agriculture policies in the United States, Europe, Japan, Canada, Australia, and New Zealand, the following general observations can be made:

- The total volume of trade continues to be reduced relative to free trade, but not by a large amount. Producer subsidies continue to keep production of a smaller number of commodities relatively high with less excess than in the past. Excess production is placed on the world market at low prices. The volume of trade in beef and dairy is probably substantially reduced by current limitations.

[21]McCalla and Schmitz, "Grain Marketing Systems," p. 206.

[22]K. Peltier and D. E. Anderson, *The Canadian Grain Marketing System*, No. 130 (Fargo, N.D.: Department of Agricultural Economics, North Dakota State University, 1978).

[23]Colin A. Carter, *An Analysis of a Single North American Barley Market* (Ottawa, Ontario, Canada: Agriculture Canada, March 31, 1993).

U.S. as a Residual Supplier: Policy Choice or Inevitable?

A residual supplier tends to be the seller of last resort—the last country from which commodity supplies are purchased in international trade. A residual supplier holds more than its share of the world stocks of a commodity. The United States has frequently been referred to as the world's residual supplier. This characterization had roots in policy but may also be endemic to a private trading system operating in a world dominated by state traders.

No doubt U.S. policies that supported farm prices held commodities off the world market, in the loan program, or in government storage. State traders could readily sell their supplies just under the U.S. support price. Private traders have not been competitive dealing in U.S. grains, so they concentrate their merchandising activities on the commodities of other exporters.

The 1996 farm bill reduces the U.S. role as a residual supplier. The decoupling of subsidies from market prices and production makes farmer decisions on what and how much to produce a function of market prices. The secretary of agriculture is directed not to allow the accumulation of anything but emergency reserves in the hands of the government.

Even in a freer trade environment with fewer state trading importers, the United States still might be a residual supplier. The reason lies in the ability of state trading exporters to undercut the U.S. price even in a free market. Private traders set their bid price or tender offer in a potential sale at a price consistent with the cost of acquiring the commodity in the United States plus handling charges. Acquisition costs for the commodity are generally determined from the futures market. A competitive state trading exporter can always underbid the U.S. price because the commodity is not priced to them; they determine its price.

The conclusion is that a state trader can always underbid a sale from the United States regardless of the policy. If their objective is to move the grain, state traders have the advantage. Note that the export enhancement programs can be used by the United States to offset some of this advantage. But export enhancement subsidies give the United States many of the characteristics of a state trader—the government becomes directly involved in the sale!

Source: The residual supplier issue is discussed further in Maury E. Bredahl and Leonardo Green, "Residual Supplier Model of Coarse Grain Trade," *Am. J. Agric. Econ.* (November 1983), pp. 786–790.

- World market prices are lowered by producer subsidies, but not as much as in the past. Consumers benefit from those lower prices except when access is restricted.

- Taxpayers bear a major burden of current trade barriers—particularly in the United States and the European Union.

- The role of the United States as a residual supplier and a holder of a disproportionate share of world stocks is changing with the adoption of the 1996 farm bill. Yet, the United States is likely to continue to be a residual supplier of those commodities where state traders predominate.

Freer Trade

One of the major issues is how the United States would fare under a system of substantially freer trade. Note that the emphasis is on *freer trade, not free trade. With state traders as prominent as they are in the world, it is unrealistic to anticipate a world structure that approximates free trade in the classical economic sense.* On a more realistic level, assume that the United States and the EU mutually agreed to abolish all producer subsidies and related barriers to trade—considerably freer than that comprised in the Uruguay Round of GATT. What would be the impact?

Economic Research Service analyses of removing subsidies on food grains (wheat and rice) and feed grains (corn and barley) estimate reductions in U.S. producer returns by about 19 percent and EU producer returns by 10 percent.[24] World prices for food grains would rise by 10 percent while feed grain prices would rise by 7 percent. Since 43 percent of U.S food grain producer returns are a result of government subsidies and 19 percent of U.S. feed grain producer returns come from the government, the rise in price is not sufficient to offset the elimination of subsidies.

The Economic Research Service (Leuck) analysis is not the final answer to the impact of free trade on prices and producer returns. Previous Australian research by Tyers and Anderson indicated that world prices would rise by about 30 percent and U.S. farmers would have been about as well off as under the U.S. farm program.[25] At the other extreme, Josling has reported results that indicate increases in world grain prices of as little as 3 percent.[26]

If the Leuck or Josling studies come anywhere close to reflecting the magnitude of the market price effects of freer trade, U.S. producers would be worse off with free trade *as long as they were able to retain subsidies without free trade.* Political

[24]Dale J. Leuck, *The Effects of Decoupling Agricultural Subsidies on the United States and the European Union on Budget Expenditures and Producer Surplus* (Washington, D.C.: ERS, USDA, 1988).

[25]Rod Tyers and Kym Anderson, *Liberalizing OECD Agricultural Policies in the Uruguay Round: Effects on Trade and Welfare*, Working Papers in Trade and Development 87/10 (Canberra, Australia: Australian National University, July 1987).

[26]Timothy E. Josling, "Bilateral Movements toward Harmonization," American Agricultural Economics Association Annual Meeting, East Lansing, Mich., August 3, 1987.

reality indicates that U.S. farmers have not been able to retain the level of subsidies. The 1990 farm bill and the related budget compromise resulted in a 15 percent reduction in acreage on which payments are made under the flexibility provisions. The 1996 farm bill terminated acreage control programs for nearly all crops and gave farmers almost complete flexibility on what to produce.

Questions have arisen as to whether unilaterally reducing its subsidies undermines the U.S. negotiating position in WTO and related bilateral negotiating arrangements. Ironically, as the level of U.S. subsidies declines, the relative magnitude of free trade benefits increases! In other words, high U.S. subsidies discourage farmer support for freer trade; low subsidies encourage their support. In any event the United States has assumed a position of leadership in reducing subsidies as an advocate of freer trade.

U.S. as a Price Leader

Farmers previously asserted that the United States set the world price and everybody else simply followed. An extension of this assertion was that the United States could raise the world price by increasing its farm program price support level.

A U.S. **price leader strategy** *takes a naive perspective of the world grain market, the factors influencing it, and the U.S. position in the world market.* With the U.S. setting its price at a higher level, grain producers in the rest of the world would increase production. The United States would clearly be in a residual supplier status with a potentially large surplus production. It is very important to keep in perspective the U.S. position in world agricultural trade and production (Table 7.1). The share of production is important because it indicates the potential for supply response from the rest of the world. U.S. strength is in coarse grains and soybeans. Even in soybeans, the share of world production and trade has been eroded by expanded acreage, particularly in South America.

Future Trading Blocs

The formation of the European Union, NAFTA, MERCOSUR, and related forms of economic integration suggests that trading blocs in the future could become the

TABLE 7.1   U.S. SHARE OF WORLD PRODUCTION AND TRADE, 1996

	Percent of World Production	Percent of World Exports
Wheat	11	28
Rice	2	18
Coarse grain	31	71
Corn	41	80
Soybeans	48	71
Cotton	21	22

Source: World Outlook and Situation Board, *World Supply-Demand Estimates* (Washington, D.C.: USDA 1996).

predominant structure influencing trade flows among countries of the world. Three such blocs likely could predominate in that structure:

- The European Union is expected to expand beyond its 15 members to include several Eastern and Central European countries around the turn of the century.
- The Asian Pacific Economic Cooperation Forum (APEC) with Japan as a focal point includes the rapidly growing developing countries of the Pacific Rim, as well as Australia, New Zealand, United States, Canada, Mexico, and Chile.
- NAFTA can be expected to expand to include much of South America. Chile is in line as the next entrant, then MERCOSUR with the result being the Free Trade Agreement of the Americas (FTAA).

With economic integration comes political integration, which creates increased uncertainty regarding the nature of the evolution. Among the most important political uncertainties are the future of the former Soviet Union, the political and economic orientation of China, the stability of the oil-rich Middle East countries, and the alignment of Australia and New Zealand.

Almost forgotten in this future design are the poor developing countries of Africa and the remainder of the so-called Third World. Where these countries fit in the ultimate design of economic integration remains to be seen. Their importance, from a population perspective, means that they obviously cannot be eliminated from the economic integration and development process. The success of NAFTA in integrating Mexico may foretell future patterns of integration, even extending to Africa.

There should be no doubt as to the importance of food and agricultural policy in the future design of world political and economic events. In many respects, it makes the remainder of this book, which concentrates on domestic policy issues, seem trivial. Yet, domestic policy has a major impact on the course of events, as was seen in the outcome of the 1994 elections. In addition, domestic policy will continue to be increasingly affected by strides and setbacks in world political and economic integration.

ADDITIONAL READINGS

1. While recommended at the end of Chapter 6 as an excellent, albeit heavy, reference, Alex F. McCalla and Timothy E. Josling, *Agricultural Policies and World Markets* (New York: Macmillan Publishing Company, 1985) do an excellent job of integrating the economics and politics of trade issues. Particular attention should be paid to pages 193–278, which discuss international policy decisions and goals.

2. The book *Grain Export Cartels* by Andrew Schmitz *et al.* (Cambridge, Mass.: Ballinger Publishing Co., 1981) is a must for anyone interested in understanding world grain markets and policy. It is not necessary to believe in

cartels to find this book enjoyable and useful. It is important to recognize that both McCalla and Schmitz have a background in Canada.

3. An excellent book, *Fundamentals of U.S. Foreign Trade Policy*, by Stephen D. Cohen, Joel R. Paul, and Robert A. Blecker (Boulder, Colo.: Westview Press, 1996) provides an insightful discussion of the integration of GATT policies into WTO and their implications for U.S. trade policy.

4. Vernon W. Ruttan provides an in-depth discussion of assistance policy in *United States Development Assistance Policy* (Baltimore, MD.: The Johns Hopkins University Press, 1996). This book reflects a life of dedication to helping the peoples of developing countries help themselves.

8 THE MACROECONOMICS OF AGRICULTURE

Practical men, who believe themselves to be quite exempt from any intellectual influences, are usually the slaves of some defunct economist . . . It is ideas, not vested interests, which are dangerous for good or evil.

—*John Maynard Keynes*

Macroeconomics *is concerned with the economy as a whole and how it functions.* The **macroeconomics of agriculture** *is concerned with the relationship between the general U.S. economy and the agricultural economy.* Some economists would prefer to define it even more broadly to include the relationship between the world economy and agriculture. There is merit in that broader perspective. One of the main effects of macroeconomic policy is on interest rates and, in turn, on the value of the dollar, which, as was seen in Chapter 6, has international trade implications that affect other countries and deflect back on the United States.

The impact of the macroeconomy on agriculture has been very apparent since the 1970s. In 1976, Schuh effectively argued that changes in macroeconomic and international economic policy had thrust agriculture into a new era.[1] Subsequently, he asserted that these changes made agricultural policies that support prices and control production outdated and counterproductive.[2] Although these arguments may have been overdrawn to make the point, it is clear that macroeconomic policy has had, and continues to have, a major impact on agriculture. If domestic farm programs (subsidies) are further reduced or eliminated, agriculture becomes more directly exposed to macroeconomic forces. The purpose of this chapter is to (1) describe the place of agriculture in the macroeconomy, (2) explain the impacts of macroeconomic variables on agriculture, and

---

[1] G. Edward Schuh, "The New Macroeconomics of Agriculture," *Am. J. Agric. Econ.* (December 1976), pp. 802–811.
[2] G. Edward Schuh, "U.S. Agricultural Policy in an Open World Economy," Testimony to the Joint Economic Committee (Washington, D.C.: U.S. Congress, May 26, 1983).

(3) provide a summary of the major macroeconomic policy tools that affect agriculture.

AGRICULTURE IN THE MACROECONOMY

The farm level is by no means the limit of the influence of agriculture on the U.S. economy. Putting food on the table and cotton or wool on people's backs requires many more nonfarmers than farmers. The importance of agriculture in the economy may be viewed either from the perspective of the farm sector or the overall food and fiber system. For every farmer and hired farm worker, 12 additional people are employed in the food and fiber system. About 17 percent of the civilian labor force is employed in agriculture.

Farm Sector

The farm sector represents a substantial but generally declining share of the U.S. economy. Measured in terms of share of the gross domestic product (GDP), the farm sector declined from nearly 7 percent in 1950 to 1.2 percent in 1994 (Figure 8.1).[3] The rest of the economy has grown more rapidly than the farm sector. That should not be surprising. From a domestic perspective, **Engel's law** *indicates that as income increases, consumers spend a smaller* **share** *of their disposable income on food.* The U.S. population is expanding at about 1 percent annually. Shoemaker explains the consequences of Engel's law in the following terms.[4] The inevitable results of a decline in consumption of agricultural products relative to other goods were

- The fall of real agricultural commodity prices
- The decline of the percentage of national resources devoted to agriculture
- The decline of farm income as a percentage of national income

If farmers' share of GDP is to increase, the declining share of income spent on food would have to be made up for by rapidly expanding exports. While exports experienced a boom in the 1970s, they declined in the 1980s and surged again in the 1990s. The decline in the 1980s was at least partially related to macroeconomic policy changes. The 1990 increase in exports largely reflects increases in income in the world but particularly in the Pacific Rim.

The substitution of technology for labor has made it possible to reduce dramatically the agricultural workforce. As a result, total farm employment has

---

[3]The spike in 1973–1975 occurred during the world food and energy crises resulting in double-digit inflation.

[4]Robbin Shoemaker, *How Technological Progress and Government Programs Influence Agricultural Land Values*, Agric. Inf. Bul. 582 (Washington, D.C.: ERS/USDA, January 1990), p. 2.

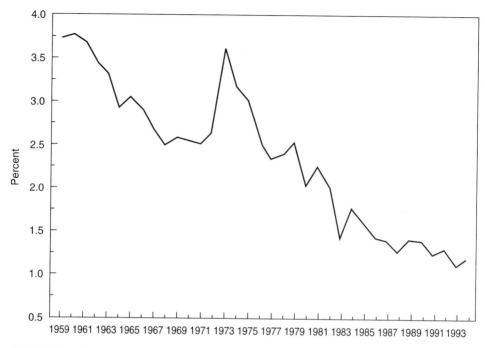

FIGURE 8.1   Gross farm product share of gross domestic product, 1959–1994.

Source: Council of Economic Advisers, *Economic Report of the President* (Washington, D.C.: Executive Office of the President, February 1996), p. 290.

declined considerably (Figure 8.2). Hence, output per farm worker increased more than sixfold from 1950 through the early 1990s. The reduced number of farmers resulted in the release of people from agriculture to produce other goods and services in the nonagricultural economy. Politicians, farmers, and the media bemoan the decline in the number of farmers, but the movement of labor from agriculture has facilitated economic progress in the rest of the economy.

A reduced farm workforce has been made possible by technological change, as reflected in greater farmer reliance on purchased inputs (Figure 8.3). Before mechanical power was developed, purchased inputs accounted for less than 50 percent of cash receipts. Since 1980, about 75 percent of cash receipts went to purchased inputs, which made agriculture more subject to inflation and to financial risk. In other words, in earlier years it was easier for farmers to "tighten their belts" and cut down on farm expenditures when times got tough. Now there is a smaller margin for belt tightening. That is, back when the farm family provided a large share of the inputs, farmers could operate less like a business. Today, farming is a business. Yet in periods when farm incomes decline, purchased inputs likewise decline as a percent of receipts. This occurred in the late 1960s and again in the mid-1980s.

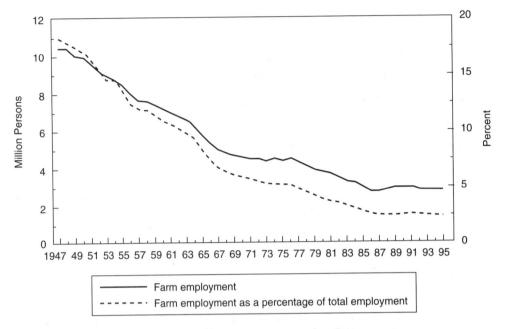

FIGURE 8.2   Total farm employment and total farm employment as a percentage of total civilian employment, 1947–1995.

Sources: USDA/NASS survey data, Council of Economic Advisers, *Economic Report of the President* (Washington, D.C.: Executive Office of the President, February 1996), p. 316.

Food and Fiber System

In 1995, the food and fiber system accounted for 13.5 percent of the U.S. gross domestic product. Farming is linked to the rest of the food and fiber system, and ultimately to the general economy, through its input purchases and product sales.

Figure 8.4 indicates the linkages between agriculture and the general economy through the gross domestic product and employment accounts. The inputs used in agriculture amounted to 4.4 percent of GDP and 3.3 percent of the jobs in the United States. The largest single category of inputs includes services such as interest, insurance, and rent.

Farming itself accounted for only 0.9 percent of GDP and 1.2 percent of the jobs. Note that this is the value added to the inputs by farmers. The production as it comes off the farm (the sum of the inputs and farming) is 5.3 percent of GDP and 4.5 percent of the jobs.

The food manufacturing and distribution sector is about 55 percent larger than the combination of the input and farming sector. Food manufacturing and distribution accounted for 8.3 percent of the GDP and 12.7 percent of the jobs. The largest component is food wholesaling and retailing with 4.5 percent of GDP and

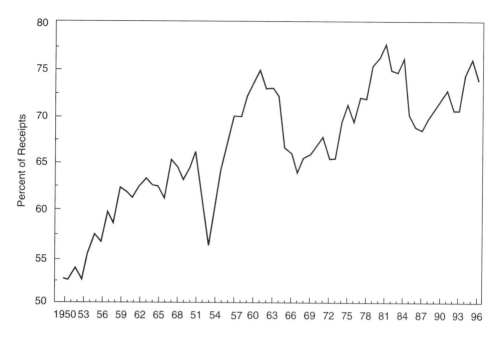

FIGURE 8.3 Cash expenses account for an ever-increasing share of cash receipts including government payments, 1950–1996.

Sources: Economic Research Service, *Economic Indicators of the Farm Sector: National Financial Summary, 1992* (Washington, D.C.: USDA, January 1994), p. 23 and revised and updated data provided by ERS/Rural Economy Division.

4.9 percent of the jobs. It may be surprising that food service accounts for only 1.8 percent of GDP and 5.3 percent of the jobs. Yet, by the mid-1990s, 40 percent of the consumer's food dollar will go to food service—food consumed outside the home.

Impact of Agricultural Prosperity on the Rest of the Economy

Agricultural fundamentalists in the 1970s and the 1980s asserted that if agricultural prices were raised by administrative fiat to 100 percent of parity, the whole economy would boom. Their contention was that agriculture is the primary determinant of the level of economic activity in the entire nation. This agricultural fundamentalist perspective asserts that the solution to all economic problems is raising agricultural prices and incomes. Farmers will, in turn, spend this income, increasing demand for farm equipment, fertilizer, energy, and other farm inputs, which has a multiplier effect throughout the economy (see box).

In the early days of the American economy, there was some merit in this point of view. In fact, throughout the nineteenth century, agriculture directly employed and accounted for a large proportion of the economy's total production.

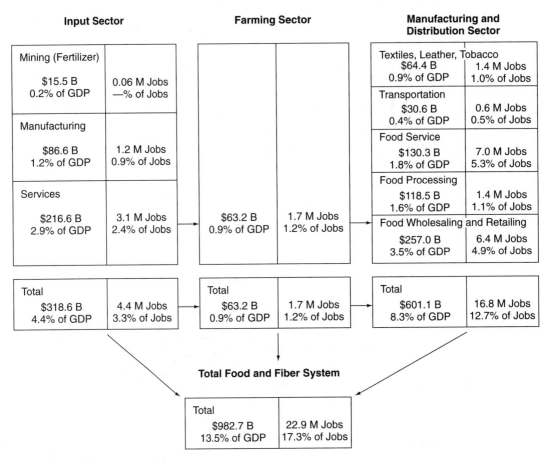

FIGURE 8.4   Food and fiber linkage and importance to the macroeconomy, 1995.

Source: William Edmondson *et al., Measuring the Economywide Effect of the Farm Sector: Two Methods* (Washington, D.C.: USDA/ERS, July 1995), p. 35.

In the late 1920s, the farm population accounted for 25 percent of the total U. S. population.[5] Even today, for many rural communities and states, the economic health of agriculture is an important factor influencing the level of business activity, employment, and income in a number of states, particularly in the Great Plains and Upper Midwest (Figure 8.5). Yet agriculture is only one link in the total economy. It neither drives nor pulls the whole system. The farm sector cannot unilaterally increase either its output or its prices and expect a proportional increase in economic activity from the rest of the economy. Agriculture can purchase inputs or sell its products only if the demand is there for the final products. The economic

[5]*Economic Report of the President* (Washington, D.C.: Council of Economic Advisers, 1981), pp. 263, 339.

The Raw Material Theory of Value

Farm protest movements in the 1970s and the early 1980s embraced the raw material theory of value as their justification for higher farm prices . . . [This theory] constitutes the economic basis for favored treatment of agriculture, and creates frustration when such treatment is not forthcoming. . . .

The central principle of the [raw material] theory [of value] is that raw materials are the sole source of the nation's wealth. Corollaries are that (1) pricing of raw materials at "full" or "honest" parity (never clearly defined but interpreted by many farmers to mean the same ratio of prices received for raw materials to prices paid by producers as prevailed in the 1910–14 period) will restore and sustain the nation's economic health, and (2) that national income is a simple multiple of income from raw materials. . . .

In its simplistic logic, the raw material theory of value is akin to Karl Marx's labor theory of value which gives sole credit to labor rather than to raw materials for creating value. A theory which attributes value solely to labor is as absurd as one which attributes value only to raw materials.

Advocates of the raw material theory of value maintain that 100 percent of parity will make not only agriculture but their entire economy boom. Carl Wilken's leverage principle of $1 of farm income creating $7 of national income grew out of the 1930s when farm receipts happened to be one-seventh of national income. As consumers' income rises, they choose to spend a declining proportion of it for food and fiber ingredients supplied by farmers. The pattern may appear invidious to farmers but is a normal expression of consumers' preferences. It follows that national economic growth reduces the ratio of farm sector income to national income. In 1985, national income was 20 times gross farm income. Are we to believe that raising farm income by $1 will raise national income $20?

Source: Luther Tweeten, "Sector as Personality: The Case of Farm Protest Movements," paper presented at E. T. York Distinguished Lecture Series (Gainesville, Fla.: University of Florida, March 1986).

system is driven by the combination of output volume and value. Farmers may benefit from increased prices, but many agribusiness segments are hurt by the reduced demand and/or production that goes with it. Thus, simply raising farm price does not generate income and employment in the economy as a whole.

Assuming no reduction in output, every $1 billion increase in demand-driven gross farm income generates about $3 billion additional GDP and 30,000 to 35,000 jobs. In agriculture, many of those extra jobs would be created through increased purchases of farm machinery and exports of farm products. Farm income and machinery purchases are closely correlated. The large losses experienced by farm machinery manufacturer International Harvester during the 1980s

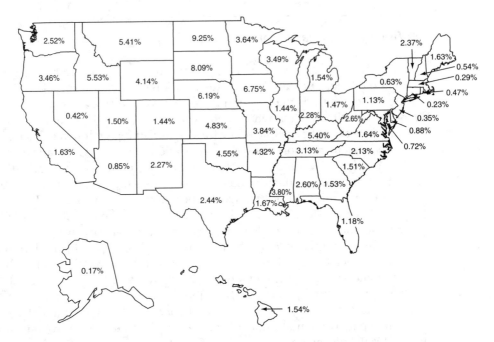

FIGURE 8.5   Farm employment as a percentage of total employment, 1995.

Source: Compiled from data supplied by the U.S. Department of Commerce, Bureau of Economic Analysis, November 1996.

and its eventual consolidation into Case were consequences of depressed farm income conditions and the resulting farm consolidations. The prosperity experienced in 1996 as a result of high exports, high prices, and farm program transition payments resulted in enhanced equipment purchases.

Because exports are a barometer of agricultural prosperity, reduced agricultural exports cause impacts that extend throughout the U.S. economy. Each dollar earned from agricultural exports results in another $1.38 of output in the U.S. economy.[6] Thus, exports of $60 billion reached in 1996 generated more than $82 billion in supporting U.S. economic activity, 85 percent of which is in the nonfarm sector. Each $1 billion in added exports would be expected to mean 18,000 new jobs. The largest factors affecting exports are macroeconomic policy and domestic farm policy. Those sectors most affected by exports, in addition to farmers, include the manufacturers of farm and export-related inputs, merchant and credit services, and transportation. Exports are also important. Agriculture is one of the few sectors of the economy that generates a trade surplus (Figure 8.6). This surplus

[6]Michael J. Dwyer, "Situation and Outlook for U.S. Agricultural Exports to the 21st Century," *Southern Regional Outlook Conference* (Washington, D.C.: Foreign Agriculture Service, November 12, 1996), p. 8.

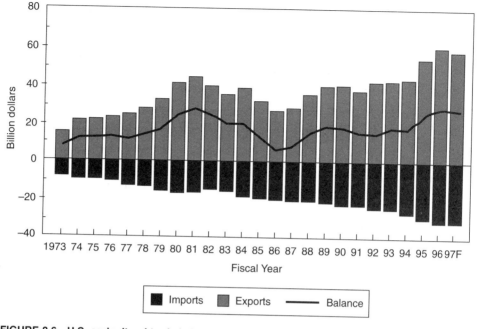

FIGURE 8.6   U.S. agricultural trade balance, 1973–1997.

Sources: Michael J. Dwyer, "Situation and Outlook for U.S. Agricultural Exports to the 21st Century," *Southern Regional Outlook Conference* (Washington, D.C.: USDA/Foreign Agricultural Service, November 12, 1996), p. 3.

strengthens the value of the dollar and makes it easier for the United States to finance its debt.

MACROECONOMIC VARIABLES AFFECTING AGRICULTURE

While the impacts of agriculture on the macroeconomy are sometimes exaggerated, until the mid-1970s, the effects of the general economy on agriculture tended to be ignored or at least understated. Four macroeconomic variables have a particularly pronounced impact on agriculture: (1) income growth, (2) inflation rate, (3) interest rate, and (4) value of the dollar. Each is sufficiently important to warrant separate discussion and comment.

Income Growth

One of the major objectives of macroeconomic policy is to create stable economic growth that fosters low levels of unemployment and persistent increases in real income. The domestic income elasticity of demand for food and beverages is about 0.35. Thus, a 10 percent increase in real per capita income will generate about a 3.5 percent increase in aggregate farm-level demand.[7] If real income trends upward by 2 percent annually, food demand trends upward by only 0.7 percent due to income growth and an additional 1.0 percent due to population growth for a total annual growth trend of 1.7 percent annually.[8]

The income elasticities are different in three primary respects:

- As income increases, the food income elasticity declines—becomes more inelastic. Low-income people may spend more than 50 percent of additional income on food while high-income people spend less than 20 percent.

- As income increases, more of the added income is spent on food consumed away from home than at home. The income elasticity of demand for food consumed away from home is about 0.57 while at home, it is 0.2.[9]

- Incomes for individual foods cover a wide range. For example, cheese has an income elasticity of 0.38 while shortening has an income elasticity of −0.5. Foods with a negative income elasticity are referred to as inferior goods. That is, as income increases, their consumption declines. Other inferior goods include sugars, flour, and eggs.[10]

The income elasticity of demand for food in developing countries is more elastic (in the 0.7 to 0.9 range).[11] Not surprisingly, therefore, in times of an expanding world economy, developing countries have been a major growth area for exports of agricultural products. The difference in income elasticity between the domestic demand and the income elasticity of developing countries has important implications for U.S. macroeconomic policy. The United States is a major world economic force. It accounts for about 22 percent of the world's economic output.[12] If it conducts its macroeconomic policy in a manner that provides the stimulus for domestic as well as world economic growth, U.S. exports of agricultural products stand to benefit. Coordinated macroeconomic policies by developed countries such as the United States, Japan, and European Union would be even more effective.[13] In addition to favorable macroeconomic policies, open markets for devel-

[7] Ben Senauer, Elaine Asp, and Jean Kinsey, *Food Trends and the Changing Consumer* (St. Paul, Minn.: Eagan Press, 1991), p. 139; J. R. Blaylock and D. M. Smallwood, *U.S. Demand for Food: Household Expenditures, Demographics, and Projections*, Tech. Bul. 1713 (Washington, D.C.: ERS/USDA, 1986).

[8] Paul T. Prentice and David A. Torgerson, "U.S. Agriculture and the Macroeconomy," *Agricultural-Food Policy Review: Commodity Program Perspectives* (Washington, D.C.: ERS/USDA, July 1985), pp. 9–24.

[9] Senauer, *et al.*, *Food Trends*, p. 139.

[10] Ibid., p. 140.

[11] Timothy Josling, "World Food Production, Consumption, and Trade," in *Food and Agricultural Policy in the 1980s* (Washington, D.C.: American Enterprise Institute, 1981), p. 95; Michael K. Wohlgenant, "Conceptual and Functional Form Issues in Estimating Demand Elasticities for Food," *Am. J. Agric. Econ.* (May 1984), pp. 211–215.

[12] *The World Factbook 1995*, Internet version (Washington, D.C.: Central Intelligence Agency, 1996).

[13] Alex F. McCalla, "Impact of Macroeconomic Policies upon Agricultural Trade and International Agricultural Development," *Am. J. Agric. Econ.* (December 1982), pp. 861–868.

oping country products would need to be maintained, and development assistance increased.

These observations have special meaning for U.S. policy toward the nations of the former Soviet republics (FSRs)where higher incomes could be expected to increase the effective demand for food, for which the United States could continue to be a major supplier. The implication is the existence of a high-income elasticity. In other words, U.S. agriculture has a substantial stake in seeing the FSRs restructure their economies rapidly to get back on a growth path.

Inflation Rate

Inflation refers to a general rise in the price of goods and services. Inflation occurs by two primary means:

- Demand-pull when the economy is running near full capacity and the demand for goods and services exceeds the supply.
- Cost-push when costs are rising rapidly due to the market power of labor unions, big business, or commodity monopolies such as OPEC.

Macroeconomic policy is the main determinant of the rate of inflation.

In the mid-1980s, it was sometimes said that what agriculture needed was a dose of inflation. Such statements may reflect, more than anything else, the need to stimulate the demand for agricultural products. The impact of inflation on agriculture is the subject of considerable controversy. One argument is that because farm prices are more responsive to changes in supply and demand than prices in the nonfarm sector, farm prices respond more rapidly to unanticipated growth in demand (demand-pull inflation) than prices in the nonfarm sector. Therefore, Starleaf, Myers, and Womack found that farmers have been net beneficiaries from inflation.[14] Tweeten, however, has concluded that as inflation rises, prices paid by farmers increase faster than prices received (cost-push inflation).[15] Inflation does allow those in debt to pay it off with lower valued dollars, but that is the case only if farmers receive higher prices. Inflation does not necessarily mean higher real land prices. The price of land is most closely related to the realization and expectation of higher income to land.[16] More specifically, Shoemaker finds that the following factors are the most influential in determining farm land prices:

- Income currently earned from land that depends on commodity prices, costs, yields, and government programs.

[14]Dennis R. Starleaf, William H. Myers, and Abner W. Womack, "The Impact of Inflation on the Real Income of U.S. Farmers," *Am. J. Agric. Econ.* (May 1985), pp. 384–389.

[15]Luther Tweeten, "An Economic Investigation of Inflation Passthrough to the Farm Sector," *West. J. Agric. Econ.* (1980), Vol. 5, pp. 89–106.

[16]J. M. Alston, "Growth of U.S. Land Prices," *Am. J. Agric. Econ.* (February 1986), pp. 1–9; and Oscar R. Burt, "Econometric Modeling of the Capitalization Formula for Land Prices," *Am. J. Agric. Econ.* (February 1986), pp. 10–26.

- Expectation of future earnings inside or outside agriculture.
- Discounting earnings from the future back to the present.

Summing up, farmers quite surely benefit from inflation during periods of strong demand. But farmers probably benefit most from stable growth in the U.S. and world economies.

Interest Rate

The interest rate is the price of money—the cost of money to a borrower and the return on money to the investor. As such, it plays a very important role in the macroeconomy by influencing the level of saving and the value of the dollar. It is the allocator of capital currently and over time.

The interest rate affects agriculture in many different ways. Some of these impacts, such as its effect on interest expenditures, are obvious. A 1 percent change in the real rate of interest leads to about a $2 billion change in farm interest expenses. The impacts on investment decisions, the price of land, the storage of commodities, or the value of the dollar are considerably more obscure.

In the case of investment decisions and land prices, the real rate of interest is more important than the nominal rate. The real rate of interest is the nominal rate minus the rate of inflation. A low (or negative) real interest rate is a spur to investment because the real cost of money is less and, as a result, there would be an inclination to use a lower discount rate on the flow of future earnings.

The real rate of interest has varied widely during the past two decades (Figure 8.7). During the period 1973 through 1979, real interest rates were negative for five of seven years. The clear signal during this time was to borrow, to invest, and to expand. Buoyed by high farm prices and incomes, farmers borrowed in a big way. Farm debt rose threefold from $59 billion in 1972 to $182 billion in 1981.[17] Among other things, they bought farm machinery and land. The average value of farmland rose from $219 per acre in 1972 to $819 per acre in 1981.[18] A change in macroeconomic policy in 1981 shifted real interest rates to the opposite extreme. The result was a sharp rise to a peak real interest rate of more than 6 percent. Investment in land and farm machinery tumbled with a corresponding decline in asset values. The value of farmland declined to a low of $599 per acre in 1987. The subsequent relative stabilization of real interest rates in the late 1980s and early 1990s, combined with relatively stable farm policy, led to recovery in land prices to surpass the 1981 level in 1995.

Interest rates also affect the willingness of farmers to store commodities. When interest rates rise, the cost of holding commodities in storage likewise rises.

---

[17]*Agricultural Statistics* (Washington, D.C.: USDA, 1985); *Agricultural Finance* (Washington D.C.: ERS, USDA, March 1986).

[18]John Jones and Charles H. Barnard, *Farm Real Estate: Historical Series Data, 1950–85* (Washington, D.C.: ERS, USDA, December 1985).

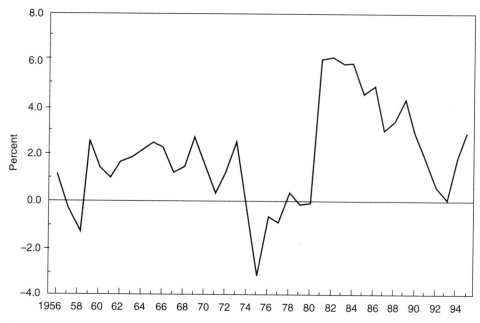

FIGURE 8.7   The real rate of interest, 1956–1995.

Source: Council of Economic Advisors, *Economic Report of the President*
(Washington, D.C.: Executive Office of the President, February 1994), pp. 346
and 360.

The incentive is to sell farm products from the field and pay off money borrowed
to buy inputs as opposed to storing commodities in hopes of receiving higher
prices later. Alternatively, if available, farmers will move commodities from private storage into government-financed storage and take advantage of lower interest costs.

Value of the Dollar

In Chapter 6, the impact of exchange rates on the level of exports was discussed.
It was noted that while U.S. farm exports benefited from declining dollar values
in the 1970s, exports suffered from the high-valued dollar in the early to mid-
1980s. It was also noted that macroeconomic policy is one of the prime determinants of the value of the dollar. In a free market, the linkage between the value of
the dollar and macroeconomic policy is through the real rate of interest.

As the real U.S. interest rate rises relative to other countries, it becomes more
attractive for investors to hold dollars as opposed to other currencies. The effect is
to bid the price of the dollar up relative to other currencies. In the pre-1972 days
of fixed exchange rates, macroeconomic policy received little attention as a factor
affecting exports. However, since the advent of floating exchange rates in 1973

and a more export-oriented farm policy, macroeconomic policy and related real interest rate and value of the dollar changes have been the center of attention for all segments of agriculture, including policymakers. Exports heighten the importance of low and stable interest and inflation rates.

MACROECONOMIC POLICY TOOLS

The objective of macroeconomic policy is to achieve full employment and stable economic growth without inflation. The tools of macroeconomic policy fall into two general categories: fiscal policy and monetary policy. Fiscal policy deals with the power of government to tax and spend. Spending more than tax revenues leads to deficits that must be financed either by issuing more money or by borrowing. Deficit financing falls in the realm of monetary policy, which deals with the power of government to borrow and create money. Eliminating the deficit, on the other hand, involves fiscal policy in that it affects government spending and/or taxes.

Fiscal Policy

Fiscal policy influences the level of aggregate demand for goods and services by changing the level of government spending and taxes.

Government Spending

Government spending has a substantial impact on both the overall level of economic activity and the level of activity in specific sectors, including agriculture. While the need to control government spending has been front and center in macroeconomic policy debate during the past three decades, outlays as a percent of GDP have declined. Contrary to popular belief, the federal government has been more effective at controlling spending than state and local governments (Figure 8.8). Part of the reason is that state and local governments have been mandated to pick up spending responsibilities previously handled by the federal government.

Government spending constitutes a direct source of demand for specific goods and services, thus affecting the overall level of production and employment. When government spending lags, total economic growth tends to lag. Surges in spending lead to rapid economic growth and, eventually, inflation.

Government spending is politically difficult to control. This is because much spending is based on entitlements, which are indexed to the rate of inflation. Thus, when unemployment rises, more people receive government unemployment benefits. The food stamp program falls under the category of entitlement programs. Farm programs, while having some of the characteristics of entitlements, are not

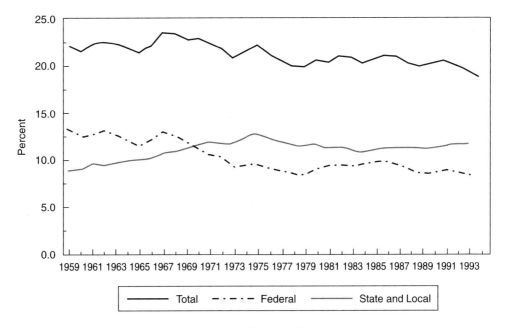

FIGURE 8.8   Government consumption expenditures and investment as a percentage of gross domestic product, 1959–1994.

Source: Council of Economic Advisors, *Economic Report of the President* (Washington, D.C.: Executive Office of the President, February 1994), pp. 282 and 298.

officially considered so because the payments are made for productive activities in the eyes of the government.

Efforts to control federal spending have had substantial impacts on agriculture through the budget process. Historically, government payments have been a significant share of net farm income. In the early to mid-1980s, they were about equal. By the mid-1990s, government expenditures had been reduced to 20 percent or less of net income—still not insignificant. Much of this reduction has been the result of the combination of budget pressures, increased export subsidies that supported farm prices, and changes in farm policies. The policy changes arguably resulted from budget pressures. This most certainly was the case for the flexibility provisions of the 1990 farm bill, which reduced deficiency payments by 15 percent to bring spending in line with the budget resolution. In the 1996 farm bill, payments were fixed, in part, to make spending more predictable and within budget limits.

An important caveat needs to be added at this point. Despite pressures to reduce federal spending, history demonstrates that if a clear need arises, the government has the ability to come up with the cash for farmers and farm programs. This was true in the farm financial crisis of the early 1980s and whenever a natural disaster has wiped out crops over a substantial area.

Taxes

Taxes are among the most politically sensitive of the macroeconomic policy instruments. Reductions in the level of taxes lead to increased money being available for private sector spending. As a result, reduced taxes tend to be politically popular. Increased taxes, on the other hand, are difficult for the political process to deal with and normally occur the year after an election.

For farmers, the composition of taxes may be as important as the level. Tax reform, therefore, has received substantial attention, as it did in the mid-1980s. Under the federal income tax laws, many tax shelters were agriculturally related. Land development, cattle feeding, and purebred breeding stock provided excellent examples. One of the major targets of the 1985 tax reform law was to reduce the incidence of tax shelters while raising corporate taxes, reducing the maximum tax rates for individual income taxes, and eliminating tax averaging. The elimination of tax shelters had a selective impact on agriculture. The elimination of tax averaging was more adverse to farmers because of the greater price and income variability that exists in agriculture. At the state and local levels, farmers in most states have obtained the benefits of property taxes being based on the income earning potential from agriculture (use value assessment) as opposed to the market value of real estate.

Monetary Policy

Monetary policy influences the money supply with the objective of stabilizing output, investment, and the price level. Working through the banking system, the tools of monetary policy include changes in reserve requirements, changes in discount rates, and open-market operations.

Reserve Requirement

Reserve requirements *specify the proportion of a bank's deposits that must be kept as reserves and, therefore, not loaned.* The money supply expands or contracts by a multiple of the reserve requirement. Thus, a change in the reserve requirement has a direct and pronounced impact on the money supply. As a result of this sharp impact, reserve requirement changes are seldom made.

Discount Rate

The **discount rate** *is the interest rate that Federal Reserve Banks (FED) charge commercial banks for lending them reserves.* Banks borrow from the FED when they do not have sufficient reserves to expand their loans. Raising the discount rate, of course, has a direct effect on interest rates that banks charge their customers. Therefore, the discount rate is the most direct means available to the FED for influencing interest rates.

Open-Market Operations

Open-market operations *involve the purchase and sale of government securities by the FED.* A sale of government securities reduces the supply of money available for banks to lend. With fewer free funds available, banks will tend to charge higher interest rates. Because they are the least harsh monetary policy tool, open-market operations are used the most frequently to influence the money supply.

MONETARY POLICY LINKAGES TO AGRICULTURE

Monetary policy has important linkages to agriculture because interest rates, inflation, and the value of the dollar all directly affect farm income, expenses, and asset values. For purposes of discussing monetary policy linkages to agriculture, two general monetary policy strategies can be distinguished: easy money and tight money. The results presented here are based on a USDA study by Kitchen and Mack of the impacts of macroeconomic shocks.[19]

An **easy money or expansionary policy** *involves a more rapid rate of money supply growth than is indicated by the growth of the overall economy.* The scenario studied by Kitchen and Mack involves an increase in the money supply growth from 5.5 to 7 percent. A larger money supply has the most immediate effect of lowering the interest rate. A lower interest rate makes the dollar less attractive internationally (thus lowering the value of the dollar and increasing export demand), raises GDP growth, and lowers the unemployment rate. The result of a tighter macroeconomic supply-demand balance in the macroeconomy is increased inflation.

Farmers generally are not any better off as a result of an easy money policy because the prices paid for inputs rise more rapidly (1.4 percent) than farm prices (1 percent).[20] Yet producers of some export-oriented commodities, such as soybeans and corn, are substantially better off. Overall, however, the temporary benefits of an expansionary policy occur at the expense of higher inflation.

A **tight money policy** *restricts money supply growth relative to GDP.* Tight money may be accomplished by increasing the discount rate, making it less attractive for banks to borrow from the FED, and selling government securities, thus reducing the money supply. These actions increase the real interest rate and, thereby, discourage business investment and consumer purchases, particularly of durables. In addition, with higher real interest rates, the dollar is a more attractive investment that increases its value and reduces export demand. The end result is reduced economic growth and low levels of inflation.

This tight money policy scenario was typical of the early to mid-1980s. Its effect on agriculture was to reduce exports of farm products, increase interest

---

[19]John Kitchen and Elizabeth Mack, *Macroeconomic Shocks: Effects on the General Economy, Agricultural Policy and Rural Unemployment* (Washington, D.C.: ERS, USDA, August, 1991).
[20]Ibid., p. 14.

expenditures, encourage government farm commodity stock holdings, reduce farm income, and reduce investment in farm machinery and land. But the tight money policy was applied following the robust expansionary (easy money) policies of the late 1970s. The brakes were applied very firmly, precipitating the farm credit crisis.

Since the mid-1980s, the FED has attempted to chart a middle-of-the-road course on monetary policy—that is, neither easy money nor tight money. This policy has led to relatively stable interest and inflation rates that have benefited agriculture and the general economy substantially.

Restrictive Fiscal Policy

One of the major problems facing the United States is how to deal with the federal deficit. It should be apparent that this issue has major significance to agriculture. Kitchen and Mack studied the impacts of a restrictive fiscal policy that reduced federal spending by $20 billion each quarter and increased the effective tax rate by 0.5 percent—generating about $20 billion in revenue. In other words, a $40 billion dent was placed in the deficit.

The increase in the tax rate and decline in spending leads initially to a decline in GDP. The resulting slack in the economy reduces the rate of inflation and the interest rate. The lower interest rate decreases the value of the dollar, which increases net exports and farm prices. In this case, crop prices rise more than input prices due to the slack economy, whereby unemployment rates surge 3 percent higher.

Kitchen and Mack suggest that one way to deal with the adversities to the general economy resulting from a combination of spending reduction and tax increase would be to utilize a combination of an expansionary monetary policy and a restrictive fiscal policy. They caution, however, that such a policy would need to be carefully designed to avoid excessive inflation.

Financing the Deficit

Financing the federal deficit provides two interesting options for agriculture:

- **Increased borrowing** is essentially the tight money policy pursued by the Reagan administration through most of the early to mid-1980s. That is, the deficit was financed by selling government securities, which has increased real interest rates, kept inflation low by discouraging economic growth, and kept the value of the dollar relatively high. Agriculture suffered under this policy with high interest expenses, low exports, and declining asset values. When interest rates declined, agriculture was relatively better off with higher exports and moderate increases in land values.

- **Monetizing the deficit** is essentially an easy money policy involving lowering the discount rate and purchasing government securities. This policy lowers the interest rate and, thereby, promotes economic growth. Agriculture benefits from this policy as long as inflation does not become excessive. Lower real interest rates are accompanied by a lower valued dollar, which tends to stimulate farm exports. However, monetizing the deficit is a dangerous policy because of the potential for inflation getting out of hand.

Balancing the Budget

During and after the 1996 election, both parties put a substantial emphasis on the need to balance the federal budget. Running continuous deficits places a great burden on interest spending for both present and future generations. As a practical matter, budget balancing is a worthy objective as long as it is pursued on a pragmatic basis. This means that consideration must be given to the effects of spending reductions on overall economic activity. In other words, the pursuit of a balanced budget requires careful management of all of the tools of monetary and fiscal policy to maintain a stable growth path. That is not an easy task.

ADDITIONAL READINGS

1. One of the most complete explanations of the linkages between macroeconomic policies and agriculture is contained in Gordon C. Rausser, "Macroeconomics and U.S. Agricultural Policy," in Bruce L. Gardner, ed., *U.S. Agricultural Policy: The 1985 Farm Legislation* (Washington, D.C.: American Enterprise Institute, 1985).

2. Paul T. Prentice and David A. Torgerson, "U.S. Agriculture and the Macroeconomy," in *Agricultural-Food Policy Review: Commodity Program Perspectives* (Washington, D.C.: ERS/USDA, July 1985), pp. 9–24, do a good job of explaining the state of knowledge regarding the magnitude of the effects of macroeconomic policy.

3. An elementary theoretical explanation of the effect of macroeconomic policy on the general economy and, in turn, on agriculture is contained in John Penson, Rulon Pope, and Michael Cook, *Introduction to Agricultural Economics* (Englewood Cliffs, N.J.: Prentice-Hall, 1986), pp. 319–421.

4. Luther Tweeten, "The Economic Degradation Process," *Am. J. Agric. Econ.* (December 1989), pp. 1102–1111, provides an excellent analysis of the potential worst case scenario.

P A R T III

DOMESTIC FARM AND RESOURCE POLICY

Domestic farm policy has been the bread and butter of agricultural and food policy. In 1985, the Congress began to reduce the level of support for farm prices and incomes. In 1996, it decoupled support for farm income from the level of farm prices. Although the cuts in spending were fostered by the budget process, there are those who argue that decoupling has put U.S. agriculture on an irreversible course of consistently less government involvement. Even long-standing subsidy programs such as dairy, sugar, and tobacco subsidies are under siege. In recent farm bills, farm policy has become increasingly interwoven with resource policy through conservation compliance, the conservation reserve program, and green payments.

The next four chapters are devoted to analyses of the domestic farm and resource policy issues.

Chapter 9 is a diagnosis of the farm problem. It distinguishes between the symptoms of the problem, which are frequently the center of discussion, and the root causes. Although the symptoms change over time, the causes are of a more enduring nature.

Chapter 10 explains how the programs have evolved over time from an emphasis on supporting the level of farm prices to providing direct payments that support farm income.

Chapter 11 looks to future policy options and their consequences, while always recognizing the potential for return to policies of the past.

Chapter 12 involves an analysis of the structure of agriculture issues. It focuses both on the economic forces affecting structure and on the alternative programs for influencing structure. It concludes that while highly emotional issues, such as family farm survival, receive much political ballyhoo, very little has been done explicitly to preserve family farms.

Chapter 13 explains the economic principles underlying policy developments in the increasingly important resource arena. It recognizes that since 1985 farm bills have become increasingly green.

9   THE FARM PROBLEM

A chicken farmer noticed that one of his hens was not up to par and was afraid there might be some disease in the chicken yard. He decided to have the chicken diagnosed, wrung the neck and sent it to the county agent's office. He received a report some days later which said, "cause of death was a broken neck."

                 —Maynard Speece

The most crucial step in prescribing policy is correct diagnosis of the problem and its causes. It is easy to fall into the trap of concluding that current conditions are typical of the agricultural economy over the long run. During the past three decades, American agriculture has gone from chronic surpluses to a world food crisis, back to surpluses, through a farm financial crisis, and to a decade of what would appear to be a relatively balanced supply–demand situation. In the process, farmers have gone through several periods of booms and busts. With this historical perspective, the farm problem could probably best be characterized as involving instability and unpredictability.

Although these are important dimensions of the contemporary farm problem, agriculture has become much more diverse from a structural perspective. Different segments of agriculture have very different problems. Arguably, the problems of the large farm sector are little different than firms in the nonfarm economy. Small and moderate-size farms have efficiency, price, and risk management problems. Larger farms are much better at dealing with these problems. Designing policy to ameliorate the problems of these very different segments is exceedingly difficult.

The purpose of this chapter is to provide a diagnosis of the farm problem as U.S. agriculture approaches the next millenium. The hazards of predicting the future are apparent in a review of agricultural economic history. First, post-Depression agriculture through the mid-1990s will briefly be reviewed. Out of this review arise some conclusions concerning long-term trends. The remainder of the chapter is devoted to specific analyses of the symptoms and the causes of the farm problem.

HISTORICAL PERSPECTIVE ON THE FARM PROBLEM

The economic history of depression and post-Depression agriculture may be divided into four major periods:

1. The 1920–1970 period was the mechanization era, characterized by a rapid transition of labor out of agriculture.
2. The 1970s were characterized by booming economic conditions comparable to the golden years of agriculture (1910–1914).
3. The 1980s were characterized by deflation of land prices.
4. The 1990s appear to be taking on the characteristics of being transition years to an industrialized agriculture.

The Mechanization Era: 1920–1970

The overall problem confronting agriculture during the mechanization era was excess capacity fostered by technological change, fixed resources, and government policy. Agriculture was in a period of constant adjustment to new technology dominated by the trend toward mechanization but compounded by yield-increasing hybrids, expanded use of commercial fertilizer, and the development of chemicals to control weeds and pests. Except for interludes during World War I and World War II, this period was characterized by chronic excess capacity.

The capacity of agriculture to produce increased continuously throughout the mechanization era. From 1920 to 1970, the index of farm output doubled while U.S. population increased by 66 percent.[1] This increase in output was accomplished with virtually no change in total inputs used in agriculture—although the mix and quality of inputs changed dramatically (Figure 9.1). While total cropland remained amazingly static, the use of purchased inputs nearly doubled, and the quantity of labor declined by nearly three-fourths. The overall productivity of agriculture more than doubled, with labor productivity increasing eightfold.

Farm numbers declined from 6.5 million in 1920 to about 3 million in 1970, while the farm population dropped from 32 million to less than 10 million (Figure 9.2). Agricultural fundamentalists and politicians bemoaned the reduction in farm numbers much as they do today. Yet the release of labor from agriculture contributed materially to the overall development of the economy and the transformation of the United States from an agricultural to an industrial economy.

With the exception of the war periods, real farm prices during the 1920–1970 period were generally declining.[2] Even in nominal terms, these were long periods

---

[1]Economic Research Service, *Economic Indicators of the Farm Sector: Production and Efficiency Statistics, 1980,* Stat. Bull. 679 (Washington, D.C.: USDA, January 1982).

[2]Gary Lucier, Agnes Chesley, and Mary Ahearn, *Farm Income Data: Historical Perspective,* Stat. Bull. 740 (Washington, D.C.: ERS/USDA, May 1986), p. 31.

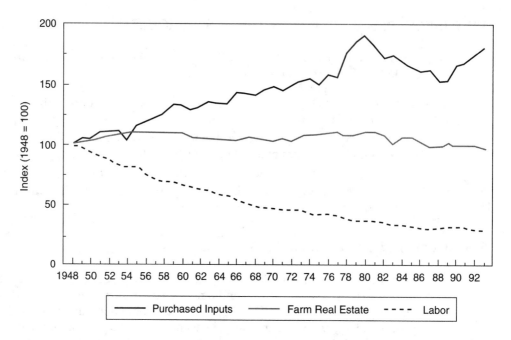

FIGURE 9.1   Changes in the mix of agricultural inputs measured in index numbers, 1948 base, 1948–1993.

Source: Data supplied by USDA/ERS, Natural Resources and Environment Division, Environmental Indicators and Resource Accounting Branch.

of declining and depressed farm prices. For example, from 1950 to 1970, the price of corn declined from $1.52 per bushel to $1.33, and the price of wheat fell from $2.00 per bushel to $1.33.[3]

During this period, when there were few direct payments from the government, low farm prices resulted in low farm incomes. Per capita incomes of persons on farms were frequently less than half of the incomes of persons not on farms.[4] These conditions fostered considerable unrest in rural America. Out of this unrest arose high levels of government intervention in agriculture, which are discussed in Chapter 10.

From the late 1930s through the 1960s, there was one bright spot. Farmland prices were on a general upward trend (Figure 9.3). Land was purchased primarily to expand the size of operation. Land was by far the farmer's best and safest investment. The combination of relatively low incomes with the upward progression of land prices led to the frequently used expression that farmers would live poor but die rich.

---

[3]Douglas E. Bowens, Wayne D. Rasmussen, and Gladys L. Baker, *History of Agricultural Price Support and Adjustment Programs: 1933–84,* Agric. Inf. Bull. 485 (Washington, D.C.: ERS/USDA, December 1984), p. 45.
[4]*Agricultural Statistics: 1962* (Washington, D.C.: USDA, 1962), p. 570.

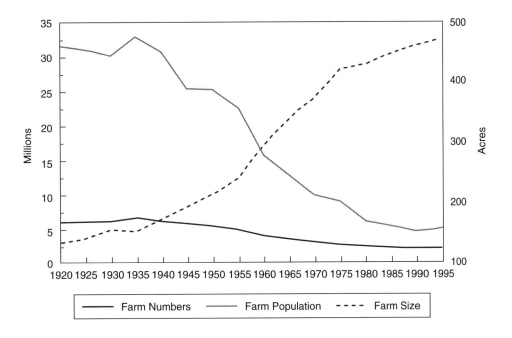

FIGURE 9.2   Farm population, farm numbers, and farm size, selected years, 1920–1995.

Source: Agricultural Statistics (Washington, D.C.: USDA, various issues).

Golden Years: The 1970s

In the late 1960s, the progress of constantly increasing output appeared to stall. For three years, output remained at the same level. Then in the early 1970s, a confluence of political and national forces combined to change agricultural economic conditions radically. Early frosts and corn blight markedly reduced feed grain production and, in turn, total farm output.[5] In 1972, the Soviet grain sales were made, signaling a major change in the USSR's policy. Imports as opposed to rationing were used by the Soviet government to compensate for production shortfalls. Bad weather in the United States was replicated in several production areas around the world. A floating exchange rate was adopted by the United States while OPEC made oil-producing countries flush with income. Adverse weather, the Soviet policy change, and expansionary economic policies, combined with a falling value of the dollar, caused exports to surge from $7.3 billion in 1970 to $34.7 billion in 1979 (Figure 9.4).

[5]Economic Research Service, *Economic Indicators of the Farm Sector: Production and Efficiency Statistics, 1985*, ECIFS 5-5 (Washington, D.C.: USDA, April 1987).

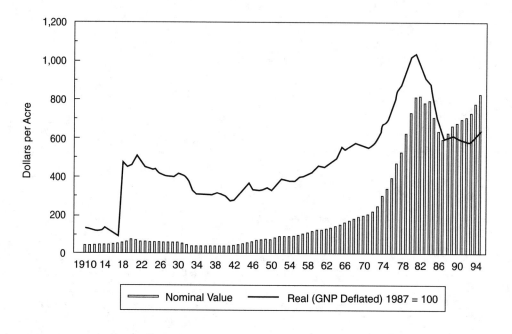

FIGURE 9.3   Nominal and real values of U.S. farmland per acre, 1910–1995.
Source: Data supplied by USDA/ERS, Natural Resources and Environment
Division, Environmental Indicators and Resource Accounting Branch.

Crop prices responded to the increased export demand by rocketing 224 percent from 1970 to 1974 while production input prices escalated by 146 percent,
thus leaving farmers a healthy profit margin.[6] The result was a doubling of real
net cash income per farm from 1970 to 1973. Higher income meant more money to
buy farm machinery and farmland. Farm machinery capital expenditures
increased 290 percent from 1970 to 1979.[7] Land values took a sharp upward swing
from $196 per acre in 1970 to $628 in 1979—320 percent nominally and 72 percent
in real terms (Figure 9.3). The price increase was more than could be justified on
the basis of the long-run agricultural income earning potential of land, since by
1979 real income had returned to the 1970 level. Most of the increase in nominal
values was eaten away by inflation. Even in the short run, farmers frequently
found that an adequate cash flow was not possible.

Back on Trend: The 1980s

As fast as the economics of agriculture improved in the 1970s, it reversed in the
1980s. Actually, the decline in the agricultural economy began in the mid-1970s,

---

[6]Lucier *et al., Farm Income Data,* p. 29.
[7]Ibid., p. 29.

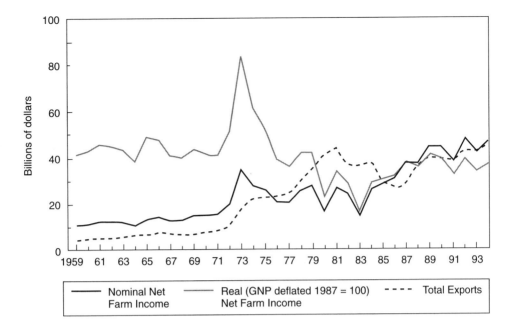

FIGURE 9.4   Boom in farm exports and net farm income, 1959–1994.
Source: Council of Economic Advisors, *Economic Report of the President*
(Washington, D.C.: Executive Office of the President, February 1996), pp. 386
and 391.

even though exports did not peak until 1981. The significance of the agricultural decline was not realized until land values began to drop in 1983.

Crop prices, adjusted for inflation, declined 43 percent from their 1973 peak to 1985. Most of this drop (34 percent) occurred from 1980 to 1985.[8] Net farm income correspondingly tumbled from a high of $34 billion in 1973 ($83 billion 1987 dollars) to a low of $16 billion in 1981 and then recovered to $45 billion by the end of the decade. In 1984, the government subsidies to farmers cost $7.4 billion and subsequently rose to $26 billion in 1986. In real 1987 dollars the income swing was from $83 billion in 1973 to $16 billion in 1983 to $41 billion in 1989.

Low farm incomes were bad enough in the mid-1980s, but the decline in land values was worse. In 1983, continuous inflation in land values came to an abrupt end, falling 27 percent in five years (see Figure 9.3). For decades, farmers and their lenders had banked on increases in land values. The land price decline eroded farmers' equity. The pain was greatest for those who bought land after 1979 or who borrowed on land to buy farm machinery. Many of these farmers soon faced bankruptcy. For example, in 1985, 3 percent of U.S. farms went out of business—a marked increase from the 1.6 percent average annual rate of decline in the 1970s.

[8]Council of Economic Advisers, *Economic Report of the President* (Washington, D.C.: Executive Office of the President, January 1987), p. 363.

At the same time, agriculture exhibited a surprising amount of resilience during the 1980s. This resilience was primarily a result of substantial off-farm income. Estimates of off-farm income did not begin until 1960 at which time off-farm income of the farm operator and family was below farm income (Figure 9.5). By 1980, off-farm income was more than twice that of farm income. Farmers sought off-farm employment to provide financial stability to their farm operation. Moonlighting for farmers and farm wives working off-farm is becoming as common in agriculture as in the nonfarm sector. Most small and moderate-size farm families must either work off the farm or sell it.

Increasing Diversity: The 1990s

Off-farm income grew from 1989 to 1994 at an amazing rate. Farm income had increased progressively from 1983 to 1989. By the beginning of the 1990s, the averages suggested that agriculture was relatively well off. Or was it? The answer to this question very much depends on the commodity, the size of the farm being evaluated, and the expectations regarding farm versus off-farm income dependence.

FIGURE 9.5   Level of net farm and off-farm income, 1960–1996.

Sources: Economic Indicators of the Farm Sector: National Financial Summary, 1990, ECIFS 10-1 (Washington, D.C.: ERS/USDA, November 1991), p. 11, and revised and updated data provided by ERS/Rural Economy Division.

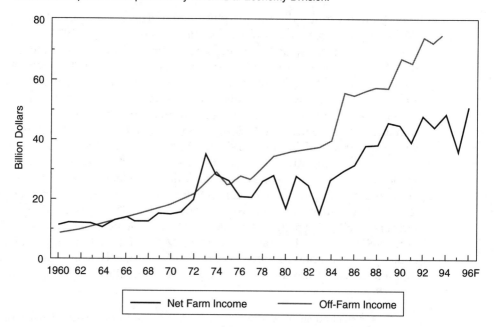

Producers of commodities that had become farm program dependent such as rice, cotton, and wheat began to suffer by the mid-1990s as government withdrawal from agricultural subsidies progressed. These farmers, likewise, became more directly exposed to the instability of agriculture. Industrialization added another dimension to the diversity of agriculture. Moderate-size farms encounter ever-increasing competitive problems. The 1990s may be recognized as the decade in which industrialization became widely recognized as the force of the future in agriculture, although industrialization began much earlier.

DIAGNOSIS OF THE FARM PROBLEM

In the past, the farm problem has been characterized as being an income problem, an excess capacity problem, a problem of declining farm numbers, a debt (financial stress) problem, an instability problem, and a structural problem. This section evaluates the extent to which each of these appears to be a problem in a contemporary context and the origin of each of these problems to the extent that they still exist.

The section ends by exploring two additional potential dimensions of the farm problem—diversity and technological change. These are problems that perhaps merit further discussion and attention because they are at the root of the structural change occurring within agriculture and the mixed signals regarding the nature of the farm problem.

Low Farm Incomes

In 1995, the average farm household earned $44,392. Of this, $4,720 was farm income and $39,190 was off-farm income.[9] Nearly 92 percent of the 1.7 million farm households reported off-farm income. Farm household income in 1995 was approximately equal to nonfarm income. In 1993, farm household income was 97 percent of the average nonfarm household income. Such comparisons of farm and nonfarm household income are viewed by some as being useful in making a case *for or against* the existence of a farm income problem.

Those arguing that a problem does exist might compare the $4,720 in farm income with the $39,190 nonfarm income. They will note that comparable statistics going back to the 1930s indicate that farmers have consistently earned less from their farm operation than was earned by nonfarm households. Their basic belief is that a farm family should be able to earn *enough income from the farm alone* to make a living. In 1993, Secretary Espy sided with those who argue for the average income for a farm operator household from farming alone—excluding off-farm income. According to Espy, the average income from farming more nearly reflects "... the true nature of farming, making it easier for us to continue to make changes in farm policy that will benefit production agriculture."[10]

---

[9] *Agricultural Outlook,* A.O. 235 (Washington, D.C.: ERS/USDA, November 1996), p. 49.
[10] *Food and Fiber Letter* (McLean, Va.: Sparks Companies, May 24, 1993), p. 3.

Those arguing *against* the existence of a farm income problem will point out that the relevant comparison is between the total incomes of each group whether the income is from farm or off-farm sources. That total household income in 1993 for farm operator households was 97 percent of that for U.S. nonfarm households is persuasive to those arguing for fewer government subsidies. They will point out that these income measures do not consider the fact that farmers have far more net worth—$396,570 for farm households versus $220,000 for U.S. households in 1993. In addition, they indicate that past statistical series indicating lower farm household income were unreliable and understated the income of farm households—a charge that appears to be true.

Regardless of the merits of such charges and countercharges, aggregate average estimates are not revealing in terms of the extent and nature of the farm income problem. At a minimum, it is essential to break agriculture down into farm size groups to begin to provide a diagnosis of the problem in terms of where and if it exists.

Table 9.1 provides a breakdown of household income by size of farm. It indicates that for farms with up to $100,000 in sales, off-farm income vastly exceeds farm income. It also indicates that the vast majority of people who live on farms do not depend on farming for their incomes. If a household earns more money from an off-farm job than its net farm income, it can hardly be classified as a farmer household. Rather, these workers are mechanics, carpenters, teachers, truck drivers, assembly line workers, lawyers, or professors. They can logically be referred to as *noncommercial farmers* because they do not depend primarily on agriculture for their living. They have also been referred to as *nonfarm-farmers*. This is true of farms with less than $100,000 in sales.

TABLE 9.1   Farm Household Numbers and Income Sources by Sales Class, 1993

Item	Noncommercial Farms with Gross Farm Sales		Commercial Farms with Gross Farm Sales			All Farms
	Less than $50,000	$50,000–$99,999	$100,000–$249,999	250,000–$499,999	$500,000–or more	
Number of farm operator households	1,498,460	206,402	221,184	68,278	41,368	2,035,692
Percentage of farm operator households	73.61%	10.14%	10.87%	3.35%	2.03%	100.00%
Net farm income/farm	($2,815)	$7,728	$21,118	$40,551	$120,487	$4,827
Off-farm income/farm	$38,413	$31,934	$21,850	$25,457	$32,840	$35,396
Off-farm income share of total household income	107.91%	80.52%	50.85%	38.57%	21.42%	88.00%
Total household income/farm	$35,597	$39,662	$42,968	$66,008	$153,328	$40,223

Source: Robert H. Hoppe *et al., Structural and Financial Characteristics of U.S. Farms, 1993* (Washington, D.C.: USDA/ERS, October 1996), pp. 31–32.

More than 1.7 million (84 percent) of the 2.0 million farm households are noncommercial farms (Table 9.1). These farms account for about one-fourth of the production. In 1993, the average net income of these farms in total was negative. Noncommercial farms with gross sales of $40,000 to $90,000 earned a profit of $7,728. But even for these farms, their farm income was only 19 percent of household income.

One can legitimately question whether this noncommercial farm segment is part of the farm problem. No doubt the farm lobby would prefer to claim the numbers as part of its farm constituency. However, it is difficult to justify government farm subsidies to households that do not depend on farming for the majority of their income. Persons in this group may contend that they should be able to earn enough income from farming so that they are not required to work off-farm. But that same argument could be applied to many part-time jobs. School teachers would like to earn enough income so they do not have to moonlight.

In 1993, only 109,646 farm households (commercial farms with gross farm sales greater than $250,000) clearly depended on agriculture for the majority of their income (see Table 9.1). Representing only 5 percent of the total number of farms, these operations accounted for 57 percent of production. Their farm income was 71 percent of total household income. Therefore, even this more narrowly defined commercial sector is fairly dependent on off-farm income, with more than 29 percent having off-farm wages or salaries. The broader commercial segment is highly diverse, with net farm income ranging from an average of $21,118 per household for the smaller commercial farms to $120,487 for the large farms. Farms having more than $500,000 in sales tend to be concentrated in cotton, vegetables, fruit and nuts, nursery products, poultry, beef, hogs, and dairy cattle.[11] Geographically, a larger proportion of the large farms tended to be located in the South and West, although each region had large farms.

The averages do *not* indicate that commercial farms have an income problem. Yet at least 20 percent of the households have income levels that are below the poverty threshold! More will be said about this diversity issue later in the chapter. Bjornson and Innes found also that farmer returns on assets are generally lower than what they would earn on the investments having comparable risks in the nonfarm sector.[12]

Farms in Financial Stress

In the 1980s, agriculture experienced a cycle of substantial financial stress. The cycle is very apparent in the debt-to-asset ratio, which increased from 16.6 in 1978

---

[11]Robert A. Hoppe *et al., Structural and Financial Characteristics of U.S. Farms, 1993,* Agric. Inf. Bull. 728 (Washington, D.C.: ERS/USDA, October 1996).

[12]Bruce Bjornson and Robert Innes, "Another Look at Returns to Agricultural and Nonagricultural Assets," *Am. J. Agric. Econ.* (February 1992), pp. 109–119.

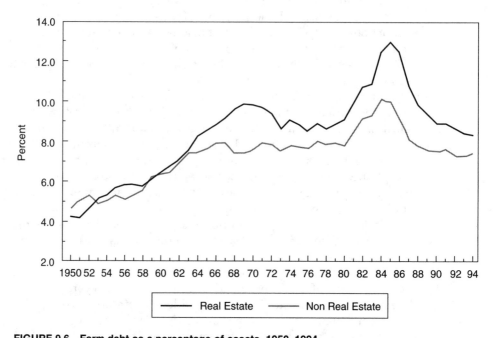

FIGURE 9.6   Farm debt as a percentage of assets, 1950–1994.

Source: Council of Economic Advisors, *Economic Report of the President* (Washington, D.C.: Executive Office of the President, February 1996), p. 387.

to a high of 23.0 in 1985 and then declined to 15.7 in 1994 (Figure 9.6). The decline in asset values, resulting primarily from lower land prices (see Figure 9.3), was primarily responsible for the decline in the value of assets (as reflected in the sum of debt and net worth). The result was a rapid erosion of equity from $816 billion in 1980 to $567 billion in 1986 (Figure 9.7). The combination of reduced debt and an increase in asset values resulted in a sufficiently sharp improvement in net worth to finally exceed the 1980 level in 1995.

The financial crisis of the mid-1980s took its toll. As many as 300,000 farms became bankrupt and were foreclosed or financially restructured between 1980 and 1988.[13] Agricultural bankers estimated that in 1986, as many as 4.2 percent of the farmers in their lending areas had filed for bankruptcy and as much as 6 percent of the loans were delinquent, compared with 1.5 percent in 1989.[14] In 1985–1986, $5.6 billion in agricultural loans were written off as uncollectible by commercial banks, the Farm Credit System, and Farmers Home Administration.[15]

---

[13]Jerome M. Stam *et al., Farm Financial Stress, Farm Exits, and Public Sector Assistance to the Farm Sector in the 1980s,* Agric. Econ. Rpt. 645 (Washington, D.C.: ERS, USDA, April 1991), p. 14.
[14]Ibid., p. 19.
[15]Ibid., p. 18.

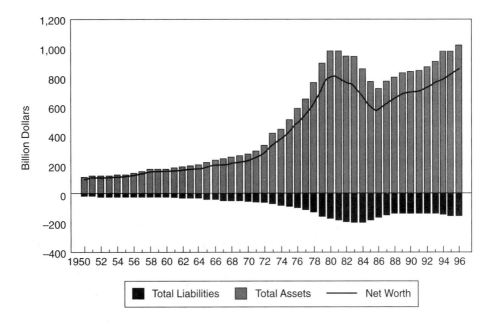

FIGURE 9.7 U.S. farm balance sheet, 1950–1996.

Sources: Economic Indicators of the Farm Sector: National Financial Summary, 1992. (Washington, D.C.: ERS/USDA, January, 1994), p. 21, and *Agricultural Income and Finance: Situation and Outlook Report* (Washington, D.C.: USDA/ERS, September 1996), p. 47.

Many farm financial institutions failed with the number of agricultural banks declining: 19 percent from 5,156 in 1982 to 4,077 in 1991. To prop up farm credit institutions, subsidies totaling $7.2 billion were paid during 1986–1988.[16]

Decline in Farm Numbers

Since 1930, farm numbers have consistently declined (Figure 9.8). Since 1960, the decline was generally at a decreasing rate, although in the 1969–1974 and 1982–1987 periods, the rate of decline accelerated in the face of economic adversity.[17]

Declining farm numbers are viewed as a problem by many, perhaps because they are viewed as a symptom of the demise of family farm agriculture. As indicated previously, during the 1980s, there was a marked acceleration in the number of such farms going out of business. Recall from Chapter 1 that in the late 1980s, 82 percent of the public felt that the family farm should be preserved. Two-thirds felt that special policies should be established to see that it is preserved, and only 22 percent felt that greater efficiency is more important than preserving the

[16]Ibid., p. 36.
[17]Stam *et al., Farm Financial Stress,* p. 25.

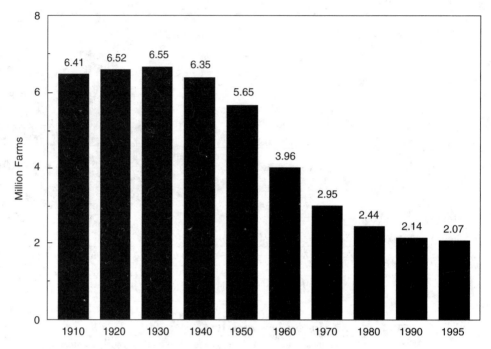

FIGURE 9.8   Number of farms, selected years, 1910–1995.
Source: Agricultural Statistics (Washington, D.C.: USDA, various issues).

family farm.[18] However, it may be argued that the decline in farm numbers is justified by these factors:

■ Fewer farms mean an increase in the overall efficiency of agriculture. Generally, large farms are demonstrably more efficient than smaller farms. The only farm size groups consistently increasing in numbers since 1980 are those having more than $500,000 in sales.[19]

■ Fewer farms leave a larger amount of net income to divide because those who leave, more often than not, are losing money.

A principal cause of the decline in farm numbers involves the advantages of large farms in terms of their ability to achieve lower costs and higher net returns, referred to as **economies of size.** These economies are of three types:

[18]Brenda Jordan and Luther Tweeten, *Public Perceptions of Farm Problems,* Res. Rep. 894 (Stillwater, Okla.: Oklahoma State University, June 1987), p. 3.
[19]Economic Research Service, *National Financial Summary, 1990: Economic Indicators of the Farm Sector,* ECIFS 10-1 (Washington, D.C.: USDA, November 1991), p. 53; Economic Research Service, *Farm Business Economics Report, 1994,* ECI 1995, (Washington, D.C.: ERS/USDA, August 1996).

■ **Technical economies** *exist when an increase in farm size results in a reduction in the average cost of production.* Until the 1980s, the perception was that farms having $100,000 to $250,000 in sales were the most efficient. This position had been espoused the most strongly by USDA.[20] More recently, a host of studies has confirmed substantial economies of size extending beyond $250,000 in sales for both crop and livestock production.

One of the first studies contradicting the notion of limited economies of size was by that of Matulich. It demonstrated economies of size in California dairies extending to 750 cows, with no evidence of diseconomies to 3,600 cows.[21] Subsequently, Smith *et al.* found that larger Texas High Plains cotton farms of more than 4,000 acres have lower costs.[22] Based in part on questions raised by the Smith study concerning the magnitude of economies of size in crop agriculture, the Office of Technology Assessment (OTA) commissioned a major investigation of economies of size in the production of major crops. The resulting study by Cooke found economies of size extending to 5,920 acres in cotton, 1,113 acres in corn, and 3,909 acres in wheat. Diseconomies were found in a limited number of production areas for cotton, corn, and rice.[23] In 1984, Tweeten reported the results of analyses of agricultural census data that indicated economies of size extending far beyond the level indicated in previous analyses of the same type.[24]

Building on the technique developed by Tweeten, Knutson *et al.* utilized census data to evaluate economies of size in each of the three major states producing wheat, cotton, corn, sorghum, and rice.[25] Their study indicated economies of size extending throughout the full range of farm sizes reported by the agricultural census, with no conclusive evidence of diseconomies. Economies of size were found to extend to at least 7,981 acres in California cotton, 4,097 acres in Kansas and Texas wheat, 3,769 acres in

[20]J. P. Madden and E. J. Partenheimer, "Evidence of Economies and Diseconomies of Farm Size," in *Size, Structure, and Future of Farms* (Ames, Iowa: Iowa State University Press, 1972), pp. 91–107; T. A. Miller, "Economies of Size and Other Growth Incentives," in *Structure Issues of American Agriculture*, ERS Agric. Econ. Rep. 438 (Washington, D.C.: USDA, 1979); and T. A. Miller, E. Rodewald, and R. G. McElroy, *Economies of Size in United States Field Crop Farming*, ERS Agric. Econ. Rep. 472 (Washington, D.C.: USDA, 1981).

[21]S. C. Matulich, "Efficiencies in Large Scale Dairying: Incentives for Future Structural Change," *Am. J. Agric. Econ.* (November 1978), pp. 642–647.

[22]E. G. Smith, R. D. Knutson, and J. W. Richardson, "Input and Marketing Economies: Impact on Structural Change in Cotton Farming on the Texas High Plains," *Am. J. Agric. Econ.* (November 1986), pp. 716–720.

[23]S. Cooke, "Size Economies and Comparative Advantage in the Production of Corn, Soybeans, Wheat, Rice, and Cotton in Various Areas in the United States," in *Technology, Public Policy, and the Changing Structure of American Agriculture, Vol. II: Background Papers* (Washington, D.C.: OTA, May 1986).

[24]L. Tweeten, "Diagnosing and Treating Farm Problems," in *Farm Policy Perspectives: Setting the Stage for 1985 Agricultural Legislation,* S. Prt. 98–174 (Washington, D.C.: Committee on Agriculture, Nutrition, and Forestry, U.S. Senate, April 1984), pp. 75–118

[25]R. D. Knutson *et al., Economic Impact of Payment Limits* (College Station, Tex.: Department of Agricultural Economics, Texas A&M University, Texas Agricultural Experiment Station, 1987).

Nebraska corn, 3,992 acres in Kansas sorghum, and 6,225 acres in California rice. The only evidence of diseconomies was in Iowa corn where the minimum cost farm was at 1,374 acres.

■ **Pecuniary economies** *of size are the advantages gained by larger firms, resulting in either a lower cost per unit of inputs purchased or a higher price per unit for the products marketed.* Although there has been considerable speculation on the extent of pecuniary economies for larger farms, the empirical evidence remains scarce. Krause and Kyle found that corn farmers having more than 2,500 acres could purchase inputs, such as fertilizer, for as much as 20 percent less than could smaller farmers.[26] However, they provide little detail on either the prevalence or methods of determining such lower cost inputs. The Smith study suggests that pecuniary economies may be available only to those farmers who are large enough or have sufficient resources to be integrated (backward) into the farm supply (input) business.[27] Eighty percent of cotton farmers operating more than 4,400 acres of land were found to be integrated into a source of supply. Such integration resulted in reduced costs of individual inputs by as much as 28 percent, an average cost reduction across all inputs of about 16 percent, and an average reduction in the cost per unit of cotton produced of 5 percent.

Pecuniary marketing economies may also exist for large farms. The Texas study also found pecuniary economies of marketing of as much as 10 percent for individual farmers; they averaged 5 percent for farms having more than 1,600 acres of cotton. In contrast with input economies, pecuniary marketing economies did not appear to be directly related to integration. Instead, it appeared that farms of more than 1,600 acres simply had more management and marketing expertise, thus obtaining a higher price.

Pecuniary economies may substantially alter the shape of the traditional long-run average cost curve for very large farmers. Average costs may perceptibly fall at the size when such pecuniary economies become possible. The average revenue also increases at that point.

■ **Technological change economies** *occur when large farms are more likely to adopt or gain the benefits of technologies.* Until the 1980s, it was generally assumed that technology was size neutral. That is, all farm sizes were in an equal position to adopt and gain the benefits of a new technology.[28] Relatively few studies of the process of technology adoption are available. The studies that have been done suggest that midsize family farms may be

[26]Kenneth R. Krause and Leonard R. Kyle, *Midwestern Corn Farmers: Economic Status and the Potential for Large- and Family-Sized Units,* Agric. Econ. Rep. 216 (Washington, D.C.: ERS/USDA, November 1971), p. 53.

[27]Smith *et al.,* "Input and Marketing Economies."

[28]F. C. White, "Economic Impact of Agricultural Research and Extension," in *Technology, Public Policy, and the Changing Structure of American Agriculture, Vol. II: Background Papers* (Washington, D.C.: OTA, May 1986).

slower than larger scale farms in adopting new technologies. Two recent examples from dairy can be cited.

In a survey involving 2,712 southern dairy farms, Carley and Fletcher found that size of herd was one of the factors related to higher output per cow.[29] Size of herd was also positively related to the use of management practices such as dairy herd improvement testing, artificial insemination, forage quality testing, and ration balancing. Each of these factors has been found to have a substantial, statistically significant impact on output per cow. The study concluded by stating that the typical dairy farm operator who would use several management factors associated with high output per cow "would tend to have a large herd of cows producing above average, be younger than average, be in a partnership operation, and have a college education."[30]

In a study of milk production costs, Stanton found a positive relationship between herd size and milk output per cow.[31] Higher milk yields were particularly pronounced for herds of more than 200 cows, which had average milk yields of 2,120 pounds more than herds of fewer than 40 cows.

Keeping pace technologically may be one of the major challenges facing moderate-size farmers throughout the 1990s and into the next decade. OTA has devoted extensive resources to studying future agricultural technology and its potential impact on the future structure of agriculture. It sees agriculture entering a new technological era with the potential benefits of increased productivity, competitiveness, environmental friendliness, safeness, and quality of the food supply.[32] But the challenges for adjustment will be substantial for farmers lacking managerial skills and progressiveness. While OTA finds the new technologies to be relatively size neutral in their application, it wonders whether moderate-size farms will have the managerial and integrative skills to effectively adopt and compete (see box). This is particularly the case for technology involving a combination of large investments and management skill such as precision farming or the utilization of rBST in milk production.

The conclusion drawn from a review of literature regarding economies of size is that there has been progressive expansion in the optimum size of farms over time and, as a result, midsize family farms have been losing, or have lost, their comparative advantage. This trend is an integral part of the industrialization process that agriculture is going through. Considerable pressure is being placed on farmers to keep pace with technological change and to grow. From this perspective, unpopular as it is, one dimension of the farm problem may be too many

[29]D. H. Carley and S. M. Fletcher, "An Evaluation of Management Practices Used by Southern Dairy Farmers," *Dairy Sci.* (1986), pp. 2458–2464.

[30]Ibid., p. 2464.

[31]B. F. Stanton, *Complexities of Northeast Milk Producers in the National Market*, Cornell Univ. Agric. Exp. Sta. 87-5 (Ithaca, N.Y.: Cornell University, Department of Agricultural Economics, March 1987).

[32]Office of Technology Assessment, *A New Technology Era for American Agriculture*, OTA-F-474 (Washington, D.C.: U.S. Congress, August 1992).

Technology and Management Skill Drive Farm Structure

The post–World War II era of farm mechanization made it virtually impossible for small, unmechanized production units to compete and survive with farming as the sole source of family income. Some past chemical and biological technologies such as insecticides and hybrid seed, on the other hand, have been scale neutral except for price discounts afforded producers who were able to purchase them in large volume. The emerging biotechnology and information industries appear to have the potential for being relatively scale neutral in their application on those farms already large enough to support mechanization technology. But two qualifying considerations are important. First, the implementation of these emerging technologies will generally require increased management skills and, for some, computer literacy. Second, at least some of these technologies will be effective and profitable only if they are integrated into rather technically complex production systems at the farm level. Some of these systems in animal agriculture may involve environmentally controlled housing and scientifically based feeding and management procedures. Thus, increased managerial skills, and, in some cases, additional capital in the form of specialized buildings and equipment will be important components of successful farming in the future. This will most likely mean increased concentration of farm production among larger units with more sophisticated technology and management capabilities.

Source: Office of Technology Assessment, *A New Technological Era for American Agriculture,* OTA-F-474 (Washington, D.C.: U.S. Congress, August 1992), p. 149.

farmers—not too few. The decline in farm numbers may be a blessing in disguise. It releases farmers from an industry in which they are unable to compete. Many of them prefer to continue to live in the country and become part-time farmers (nonfarm-farmers). Others move completely out of agriculture.

Structural change in agriculture is more complex than indicated by studies of the extent of technical, pecuniary, and technological economies of size. OTA and related studies have identified three additional forces impacting structure, which could have as much impact on the survival of family farms as these economies:

- **Managerial skill** is becoming more important. Although managerial quality may involve innate ability, it is also fostered by education. Ahearn finds that large farm operators (over $500,000 in sales) are likely to have more than a high school education, with 28 percent now having college degrees and beyond.[33] Attracting this level of skill and training requires a scale of operation that can earn a substantial net income—perhaps more than $50,000. This level of income only exists on the average for farms having more than $500,000 in sales.

- **Integration** of input supply, technology development, production, processing, and marketing functions are often fostered by technological change. As production systems become more complex, consumers become more sophisticated buyers, food handlers demand product uniformity, and more integrated systems are a likely consequence. This has happened in poultry, fruit, and vegetables; is happening in hogs; and may be pending in fed beef.

- **Size begets size** because large farms have the capacity to earn a higher net income. OTA points out this reality in studying the impact of the rBST technology.[34] Large farms tend to have substantially higher net farm income, which provides the basis for perpetual expansion or duplication of their operations. For example, while a 50-cow Wisconsin dairy had a 1992 net income of about $60,000, the 190-cow Wisconsin dairy income is $172,000.[35] The large dairy clearly is in a stronger position to expand. Of course, a 2,150-cow California dairy, with a 1992 net of $1.8 million annually, is in an even stronger position to grow.

   Because early adopters of technology earn the highest level of profit and because most technology increases the optimum size of farms, these are the farms that are in the best position to grow. They grow by buying out those farmers who failed to adopt the latest technology. **Economic cannibalism** *is the process of the large farmers continuously getting larger by buying out the smaller farmers.*[36] Economic cannibalism is one of the major reasons for the continuous decline in the number of farms as well as for the concern about the longterm ability of family farms to survive.

The implications of size economies and the related trend toward large farms are not clear for consumers:

- In the short run, benefits to consumers from increases in farm size are limited to the extent of technical and pecuniary economics. These appear to vary considerably among crops geographically, and between crop and animal agriculture.

---

[33]Economic Research Service, *Economic Well-Being of Farm Operator Households*, Electronic Database (Washington, D.C.: ERS/USDA, May 1996).

[34]Office of Technology Assessment, p. 144.

[35]Ronald D. Knutson *et al.*, *Status and Prospects for Dairying, 1992–1997*, Working Paper 93-2 (College Station, Tex.: Agricultural and Food Policy Center, Texas A&M University, March 1993).

[36]Phillip M. Raup, "Some Questions of Value and Scale in American Agriculture," *Am. J. Agric. Econ.* (May 1978), pp. 301–308. A very good description of the relationship between the treadmill theory and economic cannibalism is contained in Willard W. Cochrane, "The Need to Rethink Agricultural Policy and Perform Some Radical Surgery on Commodity Programs in Particular," *Am. J. Agric. Econ.* (December 1985), pp. 1002–1009.

- In the long run, technological innovations may change the whole shape of the long-run average cost curve. Generally, the effect is to shift the long-run average cost curve downward and to the right, thus benefiting consumers.
- High levels of concentration resulting from increases in farm size could lead to firms gaining sufficient market power so that the efficiencies would not be reflected in product prices paid by consumers. However, acquisition of such market power will occur at a much higher level of concentration. This point is discussed further in Chapter 12.

Instability

In a free-market environment, farm prices are inherently unstable because of the combination of production and demand changes with the inelastic nature of the supply and demand for farm products. The inelasticity of supply and demand is discussed subsequently as a cause of the farm problem.

From the 1930s through the 1960s, agricultural instability was masked by large surpluses and government price supports that held domestic prices above world market levels. During the 1970s, price variability increased markedly. An analysis of price and income variability for four time periods is shown in Table 9.2. These data indicate the following:

- From 1955 to 1963, farm prices and income were relatively stable. The intervening years, 1964 to 1971, were a transition; crop prices were still relatively stable while livestock prices were becoming considerably more variable.
- The variability in prices for all products increased 4 to 10 times during the period from 1972 to 1979 relative to 1955 to 1963. The variability in crop prices tended to increase more than livestock prices due to the more inelastic and unstable demand for crops. The variability in farm income was more than eight times as great in the 1970s as during 1955–1963. Income variability was reduced marginally by government payments and reduced further when income from relatively stable nonfarm sources was included.
- During the 1980–1995 period, price variability declined but not to the level of 1955–1963. Government price supports in the 1980s were generally maintained low relative to the 1955–1963 period. But farm income was more unstable, reflecting the rapid inflation of the early 1980s and the conservative macroeconomic policies of the mid-1980s. The high government payments in the mid-1980s and subsequent moves to withdraw government support resulted in an acceleration in the variability of net farm income including government payments. This increased variability has been interpreted by Cochrane and Runge[37] as an indication that, in a world economy, government can no longer influence the farm economy. Although there is

[37]Willard W. Cochrane and C. Ford Runge, *Reforming Farm Policy: Toward a National Agenda* (Ames, Iowa: Iowa State University Press, 1992).

TABLE 9.2 Mean and Standard Deviation in Farm Income and Product Prices, Selected Periods, 1955–1995

Item	Units	1955–1963 mean	1955–1963 s.d.*	1964–1971 mean	1964–1971 s.d.	1972–1979 mean	1972–1979 s.d.	1980–1995 mean	1980–1995 s.d.
Prices received, marketing-year average									
All wheat	($/bu.)	1.872	0.112	1.363	0.121	3.148	0.838	3.320	0.538
Corn	($/bu.)	1.139	0.111	1.156	0.096	2.324	0.432	2.452	0.453
Rice	($/cwt)	4.864	0.224	5.026	0.151	9.406	2.368	7.623	1.889
Upland cotton	(cts./lb.)	0.316	0.012	0.254	0.038	0.501	0.119	0.618	0.076
Beef cattle	($/cwt)	19.333	2.772	23.513	3.720	40.850	11.644	63.356	7.323
Hogs	($/cwt)	16.000	1.786	19.738	2.936	39.363	7.073	45.463	4.687
Fluid grade milk	($/cwt)	4.626	0.084	5.450	0.616	9.296	1.853	13.164	0.553
Manufacturing grade milk	($/cwt)	3.363	0.200	4.319	0.323	8.000	1.908	11.789	0.469
Eggs	(cts./doz.)	0.361	0.027	0.353	0.036	0.517	0.088	0.620	0.057
Cash receipts									
Crops	(million dollars)	14,809	1,524	19,156	1,710	47,080	10,648	77,076	9,986
Livestock	(million dollars)	18,505	1,555	25,664	3,696	48,516	10,718	78,573	8,441
Net farm income less government payments including government payments	(million dollars)	10,591	653	10,071	1,163	23,054	5,302	26,109	9,785
	(million dollars)	11,614	728	13,211	1,481	24,912	5,057	34,496	11,123
Off-farm income	(million dollars)	9.65	1.07	15.19	2.52	26.79	3.89	54.57	14.81
Gross farm income	(million dollars)	38,213	3,573	52,379	6,497	107,468	23,477	179,634	21,663
Farmland value	($/acre)	108.56	15.73	172.25	23.51	392.13	143.34	729.31	73.21

*s.d. is the standard deviation which measures the amount of unexplained price variability about the mean.

Computed by the Agricultural and Food Policy Center, Texas A&M University.
Major data sources: USDA/NASS, *Prices Received by Commodity: Historic Data Series*, electronic database (Washington, D.C.: USDA/NASS, June 1994),
USDA/NASS, *Agricultural Prices* (Washington, D.C.: USDA/NASS, various),
USDA/ERS, *Farm Sector Statistics*, electronic database (Washington, D.C.:
USDA/ERS, June 1993), USDA/ERS, *U.S. Farm Income: Historical Perspective*, electronic database (Washington, D.C.: USDA/ERS, October 1996) and
USDA/ERS, *Agricultural Outlook* (Washington, D.C.: USDA/ERS, November 1996).

truth in this perspective, it is also true that government has lost its will to influence the farm economy. This is another case of government-induced variability due to shifts in farm policy. It is indeed important to note that net farm income has become much more unstable by all measures for all four periods. Risk and uncertainty are clearly increasing.

Overall, these estimates confirm that farm income variability has increased for the entire sector in recent years. This variability in income has both favorable and unfavorable aspects. From a favorable perspective, the movement in income reflects changes in supply and demand conditions and is a signal for producers regarding the needs in the marketplace. Yet the price signal was masked by government income payments that were tied to market prices. This effect was corrected in the 1996 farm bill by the elimination of the target price and the implementation of decoupled transition payments. When prices become highly unstable, the signals may be misinterpreted and mistakes may be made in production and marketing decisions. The result frequently is misallocation of resources. In addition, variability in prices and income increases the risk and uncertainty to the farm business. The result is a higher incidence of failure of business firms.

The implications of economic instability in the farm sector are perhaps more significant today than in previous times when farm families were thought to be very resilient. In the past, during periods of adverse economic conditions, family farms would tighten their belts, reduce personal consumption expenditures, and weather the period until economic conditions improved. They were much less dependent on purchased inputs from the nonfarm sector, and their fixed annual cash obligations were relatively small. Today, however, farmers purchase a high proportion of production inputs and may have substantial debt repayment obligations for their fixed assets (see Figure 9.1).

The changed situation is evidenced by the ratio of cash production expenses to gross farm income; this ratio has trended upward since World War II (see Figure 9.3). The implications of the ratio of cash production expenses to gross receipts are illustrated in Table 9.3 by the effects of increased production expenses on net

TABLE 9.3   Sensitivity of Annual Net Income to Changes in Production Expenses

| Item | Production Expenses (Dollars) as Percentage of Cash Receipts | | | |
	60%	*70%*	*80%*	*90%*
Gross cash receipts	100	100	100	100
Cash production expenses	60	70	80	90
Net cash income	40	30	15	10
10 percent increase in production expenses	66	77	94	99
Net cash income	34	23	6	1
Decrease in net cash income (%)	15	23	60	90

income. The greater the dependence on purchased inputs, the more adverse impact the 10 percent increase in production has on net income. Similarly, the higher the proportion of cash production expenses, the more vulnerable farms are to changes in product prices.

The role of government in reducing price variability is an important policy issue. Yet it is difficult to design a set of government programs that will reduce the unnecessary and disruptive effect of price variability while allowing the market to provide clear signals to both producers and consumers. How much intervention is optimal is, in fact, the principal point of divergence in the farm policy views of the major political parties and among economists.

The inelastic nature of both the supply and demand curves for farm products is the major cause of instability of farm prices and incomes. As a result of an inelastic supply and demand, a small increase in supply can result in a multiple reduction in price. Thus, a 5 percent increase in supply or reduction in demand can easily lead to a 10 to 15 percent drop in price. Moreover, these price changes result in proportionately larger changes in net farm income. Demand or supply changes of this magnitude are not at all unusual, with variation in weather being a primary short-run cause, but pests, politics, farm policy, and macroeconomic policy are also contributing factors.

Price variability increases tremendously under free-market conditions because without government supporting the price, stocks tend to be lower. Lower stocks mean more price variability because the stocks remaining at the end of one year are a part of the next year's supply. Stocks are a cushion against adversity. Thus, in the 1970s, price variation was at record levels—a direct reflection of a highly inelastic supply and demand (see Table 9.2).

Cost of Farm Programs

During the 1980s, government expenditures on farm programs reached record levels. For years, government subsidies to farmers had totaled less than $6 billion nominally and $13 billion in real terms. Then, in 1982, subsidies increased to nearly $12 billion ($14 billion real); this was followed by a rise to nearly $26 billion in 1986 (Figure 9.9). In the late 1980s, farm program spending declined to about $10 billion and remained at approximately this level through 1994. In addition to the level of spending, questions have arisen regarding the level of payments to large farms and the lack of targeting of expenditures to farmers with the greatest need. The cost dimensions of the farm program and how they arise are discussed in greater detail in Chapter 10. Suffice it to say that program costs attracted increasing attention as a dimension of the farm problem.

Excess Capacity

Excess capacity *is defined as the difference between potential supply and commercial demand at prevailing politically acceptable prices, higher than market-clearing prices.*

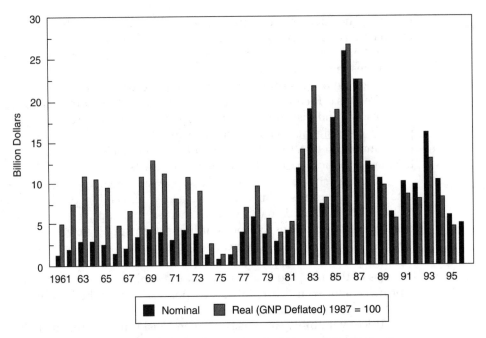

**FIGURE 9.9   Nominal and real costs of government programs for agricul-
ture, 1961–1996.**

Source: Economic Research Service, *Agricultural Outlook* (Washington, D.C.:
USDA, various issues).

Since price is maintained by the government above market-clearing levels, excess
capacity results. This excess capacity may be reflected in:

■ Increased commodity stocks generally held under government programs
■ Commodities given away by the government under both domestic and for-
  eign food-aid programs
■ Land removed from production by the government.

We noted earlier that excess capacity was a root cause of the farm problem
throughout the mechanization era of the 1920s to 1970s. Since then, capacity problems have been more periodic than chronic. The extent of excess capacity has been
the subject of considerable study. Tweeten estimated that in the early 1980s, excess
capacity totaled about 11 percent. A substantial proportion of this excess capacity
was in the form of land retirement programs that began with the payment-in-kind
program in 1983.[38]

[38]Tweeten, "Diagnosing and Treating Farm Problems," p. 30.

The most detailed estimates of excess capacity have been made by Dvoskin in an excellent comprehensive study of the issue.[39] This study found that the long-run excess capacity for the four major crops (wheat, feed grains, cotton, and soybeans) was *above 20 percent* in 1985 and 1986. In the long run, when averaged over all crops, excess capacity was *8 to 9 percent* in the mid-1980s. Dvoskin found that in the mid-1980s, excess capacity was the highest on record. Measured in acreage equivalents, excess capacity averaged more than 60 million acres in 1985 and 1986, or 22 percent of harvested acres.

The Economic Research Service of USDA has not kept up the capacity series developed by Dvoskin. The procedures utilized by Dvoskin relied heavily on acres retired and commodities in Commodity Credit Corporation storage. By these standards, the amount of excess capacity declined at least 50 percent from the mid-1980s to 1992 when the amount of land in retirement was about 35 million acres of highly erodible farmland in the conservation reserve program (CRP), government stocks were low, and other forms of land retirement were generally in the 5 to 10 percent range. Such numbers, however, need to be used with care. The quality of land in both CRP and other forms of land retirement is low, implying low productivity—certainly over the long run. Moreover, during the 1973–1980 period, excess capacity spread over all crops averaged only about 2 percent. The 1996 farm bill further reduced government-induced excess capacity by eliminating annual forms of land retirement such as set-aside.

Three explanations can be given for the excess capacity that, except during wars and the 1970s, has been a chronic problem in agriculture and more recently appears to have become periodic:

1. **The agricultural treadmill theory** notes that new technology constantly flowing into agriculture creates the potential for higher-than-normal profits by the first adopters.[40] As an increased number of farmers adopt new technologies, the supply schedule shifts right and forces down the price. Later adopters, having higher costs, run the risk of falling hopelessly behind. Therefore, farmers are said to be on a treadmill with supply shifting to the right faster than demand; they either adopt or go out of business. The consequence of supply shifting faster than demand is chronic excess capacity.

2. **The fixed-asset theory** suggests that once resources enter farming, they tend to be locked in by the fact that they are worth little outside agriculture compared with their cost.[41] In other words, high-cost farm machinery or buildings

[39]Dan Dvoskin, *Excess Capacity in Agriculture: An Economic Approach to Measurement,* Agric. Econ. Rep. 580 (Washington, D.C.: ERS, USDA, February 1988).

[40]Willard W. Cochrane, *Farm Prices: Myth and Reality* (Minneapolis, Minn.: University of Minnesota Press, 1958), pp. 85–107.

[41]Glenn L. Johnson, "Supply Functions—Some Facts and Notions," in Earl O. Heady *et al.,* eds., *Agricultural Adjustment Problems in a Growing Economy* (Ames, Iowa: Iowa State University Press, 1958), pp. 74–93. Perhaps the best explanation of Johnson's theory is contained in Dale Hathaway, Government and Agriculture (New York: Macmillan Publishing Company, 1963), pp. 110–130.

have a very low salvage value outside agriculture. The fixed-asset theory has been applied to human resources as well as to farm assets. However, with an increasing proportion of farmers being college educated, human capital should be less subject to the fixed-asset theory in the future. Yet even though human capital leaves, it does not mean that the assets themselves leave. Land tends to be farmed as long as its returns cover the variable costs of production and make some contribution to fixed costs of farming. Of course, in the long run, agricultural resources must earn a normal profit.

3. The **government theory** holds that excess capacity is caused by farm programs that prevent markets from clearing. Were it not for high price supports, direct farm subsidies, and barriers to trade, Paarlberg contends that there would be no excess capacity and no surpluses. "Given the price objectives specified in the law, there is undeniably excess production capacity in agriculture. . . . The result of holding farm prices and farm incomes continuously and substantially above equilibrium levels is to create excess capacity."[42]

An issue not often addressed is the desired level of excess capacity. Clearly, with a commodity as important as food, some excess capacity is needed. That may be a long-run justification for farm programs. Equally as clear, 20 percent excess capacity is too much. The issue is how much excess capacity is desired to deal with environmental issues such as erosion and water quality, and with risk factors such as weather variability and pests. Private sector storage decisions are based solely on economic criteria. The public sector may prefer a higher level of food security. Some of this publicly justified excess capacity may be held in the form of stocks to deal with short-run risks while land may be held in reserve to deal with long-run risks. Despite its importance, the nature and type of excess capacity desired is seldom explicitly addressed as public policy.

Market Position

Perhaps the most neglected potential reason for the farm problem is that farmers are caught in a position without market power between large input suppliers and large marketing firms, except in integrated industries such as poultry. Farmers buy inputs from oligopoly sellers and sell to oligopsony buyers. Lanzillothi suggested this structural difference as one of the causes of the farm problem.[43] Market signals to farmers are distorted by market power.

Although it is generally recognized that in most of agriculture, farmers are price takers on both the input and output side of market, there has been little research on the effects of this market position. The best recent discussion of the

[42]Don Paarlberg, *Farm and Food Policy* (Lincoln, Neb.: University of Nebraska Press, 1980), pp. 28–30.

[43]Robert F. Lanzillothi, "Market Power and the Farm Problem," *J. Farm Econ.* (December 1960), pp. 1228–1257.

market position of the farmer is by Schmitz *et al.* in their analysis of the farmers' position in international grain markets.[44] They postulate a situation in which multinational grain companies have not only monopsony (buyer) power with regard to producers but also monopoly (seller) power with regard to importing countries. In this environment, it is in the grain companies' best interest, Schmitz *et al.* contend, to manufacture price instability to create "buy low, sell high" opportunities.

There are only a few instances in which farmers have been able to organize sufficiently through cooperatives to offset the market power of buyers and sellers. Generally, cooperative realization of power is accomplished with the aid of marketing orders such as in milk, fruits, and vegetables. However, in major commodities for which traditional forms of price and income support exist, cooperative strength has been nil.[45]

The most recent treatment of agribusiness structure as a dimension of the farm problem is by Tweeten, who concludes that the agribusiness impact is more favorable than unfavorable.[46] The favorable impact is in terms of increased productivity. According to Tweeten, "There is no evidence that farm problems of annual and cyclical instability and squeezing out of commercial family farms would be any different today if the agribusiness sector were perfectly competitive. Finding scapegoats for farm problems in bankers, the multinational grain trade, the futures market, the Trilateral Commission, or in some ethnic group at best detracts from the real causes of farm problems and at worst establishes a climate of hate, violence, and even murder." This issue is revisited in Chapter 12.

Macroeconomic Policy

The impact of macroeconomic policy on agriculture was discussed in Chapter 8. The only purpose here is to recall that the highly expansionary monetary and fiscal policies of the 1970s fueled inflation and undervalued the dollar. The resulting increases in export demand were contributing factors to the overexpansion of output. These conditions also fostered land values that exceeded its income-generating capacity. The macroeconomic policies of the mid-1980s were at least equally disastrous—particularly following the policies of the 1970s. Tight money growth fostered high real interest rates, led to an overvalued dollar that choked off export demand, and put debt-burdened farmers in an impossible cash-flow squeeze. In other words, government policy was an important contributor to the problems of the 1980s but not the only contributor.

---

[44]Andrew Schmitz *et al., Grain Export Cartels* (Cambridge, Mass.: Ballinger Publishing Co., 1981), pp. 38–48.

[45]R. D. Knutson and W. E. Black, *Cooperative Involvement in Issues of Domestic Farm Policy*, Dep. Agric. Econ. DIR 86-4 (College Station, Tex.: Texas A&M University, 1986).

[46]Luther Tweeten, "Is the Family Farm Being Squeezed out of Business by Monopolies?" in *Research in Domestic and International Agribusiness Research* (Greenwich, Conn.: JAI Press, 1988), pp. 213–243.

SO WHAT IS THE FARM PROBLEM?

Several symptoms and potential problems have been discussed in this chapter. Several are from the past and no longer appear to be particularly relevant. These include the following:

- **Chronically low farm prices and incomes** can no longer be considered a dimension of the farm problem. This is the case not only because (1) it is a symptom of the problem of chronic excess capacity but also because (2) off-farm income has become a significant component of total household income. Since the late 1960s, excess capacity has been more periodic than chronic. Off-farm income has added an element of stability to agriculture that previously was not sufficiently large to add stability.
- **Reduced farm numbers** cannot be considered a problem. Despite political rhetoric to the contrary, it may be a positive sign of adjustment to a more efficient and competitive agricultural system. A broad base of economies of production, marketing, technology, integration, and management favors a continuing trend toward fewer but larger farms among which to divide the available net farm income base.
- **Financial stress** cannot be considered to be a general agricultural problem even though it became a major problem in the mid-1980s. It had not previously been a major problem since the 1930s. Overexpansion in the 1970s and inflation in the 1980s, followed by highly restrictive macroeconomic policies, were the causes of financial problems encountered in the 1980s. These problems have not reappeared since.
- **Market position** advocates, more often than not, believe in the "villain theory." Agribusiness has contributed much to the advancements made in U.S. agriculture. Agribusiness is an integral part of the industrialization process that is occurring in agriculture. While there may be cases in which agribusiness has become a dominant, if not a monopolistic force, the clear record is that of a highly competitive structure.

Agriculture has economic problems. Some of these problems, as indicated earlier, have become periodic rather than chronic. Periodic problems are characteristic of all industries. The bottom line is that U.S. agriculture has four major problems that may merit the consideration of policymakers:

- **Instability** is inherent in agriculture. It results from the interaction of a highly inelastic supply and demand. We will see in the next chapter that one of the alternatives for government is to structure policy in a manner that tends to place bounds on that instability in terms of prices and incomes.
- **Diversity** is another problem in U.S. agriculture that has been mentioned only in passing. Farms cover a wide range of sizes operating under substan-

tially different conditions with various levels of dependence on agriculture. Some of these farms have existed for many years and have developed a strong equity base. Others are operated by beginning farmers with relatively high debt. Some farmers have less than a high school education; others are highly skilled managers with one or more college degrees. It is difficult to design government policy to deal with this high degree of diversity.

- **Food security** was discussed in Chapter 5 as an important dimension of the world food problem. There is little danger that U.S. consumers will experience a food shortage, but the world perspective on this issue is significantly different because the United States is the only country in the world that has consistently maintained substantial commodity stocks. Policy analysts at the University of Missouri and Iowa State University concluded in 1992, prior to the record corn crop, that U.S. grain stocks had dwindled to the point where serious policy concerns were developing.[47] This low stock situation was further aggravated in 1996 when stocks of corn became sufficiently low that a second world food conference was assembled to revisit food security issues.

- **Resource scarcity and agricultural externalities** are identified as a dimension of the farm problem, broadly defined, in Chapter 13. Farm programs may be viewed both as a contributor and as a policy tool for dealing with resource problems. Resource policy would need to be substantially restructured were it not for the incentives for conservation provided by current farm programs.

A WORD OF CAUTION

Any casual or periodic observer of agriculture needs to be cautioned against getting swept up in current events. Agriculture is sufficiently unstable that the minute one gets swept up in conditions of surplus, a production shortfall creates deficits or a military conflict creates uncertainty over food supplies and needs.

The income problem of the past is much less apparent today in the commercial farm sector on which government policies can be expected to impact most directly. Increased off-farm incomes have become significant stabilizing factors. Yet in any given size category, low or negative farm profits exist for up to one-fourth of the farms.

The food supply–demand balance has tended to lie on the side of surpluses most of the time since 1930. By the 1990s, there appeared to be some tightening of the balance. Yet short-run aberrations from that trend are inevitable. Policies must be sufficiently flexible to deal with both the long-run trends and the short-run aberrations. This requires a careful matching of policy solutions with problem causes, which is the subject of Chapters 10 and 11.

[47] Abner Womack *et al. U.S. Agricultural Outlook, 1992* (Ames, Iowa: FAPRI, May 1992).

ADDITIONAL READINGS

1. Willard W. Cochrane's *Farm Prices, Myth and Reality* (St. Paul, Minn.: University of Minnesota Press, 1958) does an excellent job of explaining why agricultural prices are so unstable and how technology, combined with agriculture's competitive structure, generates surpluses. Although this book was written four decades ago, it is still an excellent reference. Yet in reading *Myth and Reality,* be aware that Cochrane's policy prescription has changed. The reason for this change and his new policy prescription are presented in Willard W. Cochrane, "The Need to Rethink Agricultural Policy in General and to Perform Some Radical Surgery on Commodity Programs in Particular," *Am. J. Agric. Econ.* (December 1985), pp. 1002–1009; and Willard W. Cochrane "Focusing on the Specific Problems of Agriculture: A Fresh Look at an Old Policy Approach," *Am. J. Agric. Econ.* (December 1986), pp. 1102–1108. Cochrane's most recent contribution is coauthored with C. Ford Runge, *Reforming Farm Policy: Toward a National Agenda* (Ames, Iowa: Iowa State University Press, 1992).

2. The major forces influencing U.S. agriculture through the end of the century are summarized in Roy Frederick and Dennis Henderson, eds., *Forces of Change: Policy Alternatives for the 1980s* (Columbus, Ohio: Department of Agricultural Economics, Ohio State University, 1987).

3. A different perspective on the farm problem can be obtained from James Bovard, *The Farm Fiasco* (San Francisco, Calif.: ICS Press, 1989).

4. During the 1990 farm bill debate when criticisms of farm policies became heated, the House Committee on Agriculture asked the Agricultural and Food Policy Center to develop a summary statement providing a rationale for farm programs. The resulting publication was Ronald D. Knutson and Edward G. Smith, *Rationale for Farm Programs,* AFPC Policy Issue Paper 90-1 (College Station, Tex.: Texas A&M University, July 1990). The publication views the justification for farm policy from a considerably broader perspective than just farmers.

DOMESTIC FARM POLICY: A HISTORICAL PERSPECTIVE

Ezra Taft Benson [Secretary of Agriculture, 1953–1961] . . . announced that he was going to work night and day on farm problems. The [congressional] protesters requested that he refrain from night work, since he was doing enough damage to farmers in a normal day's effort.

—*Eugene J. McCarthy*

The focus of this chapter is on **farm policy**—*the set of government programs directly influencing agricultural production and marketing decisions.* The economic problems of farmers have occupied public policy attention for more than 60 years. During this time, farm policy has evolved slowly. In the process, experience has been gained on what works and what does not work. Yet, as indicated early in the text, there is a cyclical nature to policy proposals and even to the policies themselves. History, therefore, is important in understanding the evolution of policy and its consequences.

This chapter treats farm policy primarily from an historical perspective. In the process each of the major program dimensions is discussed including their economic underpinnings and consequences. Chapter 11 then picks up with the Freedom to Farm provisions of the 1996 farm bill with a look toward future policy options and their consequences. Farm policy development is segmented into six reasonably distinct time periods. Throughout the discussion, an effort is made to provide perspective on how policies were implemented in terms of the specific tools utilized and their consequences.

FARM POLICY GOALS

The changing nature of the farm problem explained in Chapter 9 suggests that the goals of policy have also tended to change over time. Although this is the case in a marginal sense, the goals of policy have been amazingly stable over time. The following goals have played an important continuing role in policy development.

These goals have, however, changed in relative importance as well as in the mechanics of implementation.

- **Expanding farm production** to utilize America's bountiful agricultural resources has been an important goal, extending over time from the initial land settlement programs to continuous support for the creation of technology and its adoption. Programs have been implemented to expand irrigated land area and to drain wetlands with the effect of expanding the productive capacity of agriculture. The beneficiaries of this goal have extended beyond farmers to agribusiness firms and consumers. Programs to expand food production have contributed to the goal of providing an adequate and secure supply of food at reasonable prices. The existence of this goal may be the origin of frequent charges by farm groups of a cheap food policy. At the same time, it can readily be argued that without an adequate supply of food at reasonable prices, the whole complexion of government programs with respect to agriculture would change dramatically—probably involving much higher levels of government involvement and controls.

- **Supporting and stabilizing farm prices and incomes** began in earnest as a policy goal with the depression conditions in the 1930s and has continued in varying degrees through the present. Calls for parity prices permeated legislative debate into the 1960s. Stabilizing farm prices became reemphasized as a policy goal in the 1970s when world economic and political events had a sharp destabilizing effect on American agriculture. However, the 1980s and the early 1990s once again signaled a decline in the relative importance of the support goal; yet the government's role in stabilizing prices remains a controversial issue. Family farm preservation appears to be a driving force behind the price and income policy goal. Interestingly, we will see that although nearly all farm bills espouse preservation of the family farm, little has been accomplished. The political attractiveness of appealing to farm audiences with the goal of preserving family farm institutions remains.

- **The adjustment of agricultural production to market needs** became a major goal of farm policy after World War II. Attempts to achieve this goal have covered a myriad of programs, ranging from voluntary acreage reduction programs to mandatory production controls. An integral part of this goal has involved encouraging the transition of excess resources out of agriculture. Reductions in price and income support have, at times, been utilized to encourage this adjustment process.

- **Expanding agricultural exports** became a goal of domestic farm policy in its own right in the 1970s. The pursuit of this goal had a major impact on the composition of farm programs in the 1985 farm bill and subsequent efforts to modify farm price and income supports policy. To a degree, direct export subsidies increased in importance as farm subsidies were reduced in the late 1980s and the 1990s. However, as indicated previously, the Uruguay Round GATT agreement in 1994 put substantial curbs on export subsidies. Arguably, strong export demand has made them less necessary.

■ **Resource conservation and preservation** have periodically been emphasized as a goal of policy with farm programs having their origin in the depression and Dust Bowl days of the 1930s. The importance of this goal, which is the focus of attention in Chapter 11, accelerated markedly beginning with the 1985 farm bill.

SIX POLICY PERIODS

Farm policy changes are evolutionary, not revolutionary. Watershed changes in policy seldom occur; yet at certain points, major changes in the direction of farm policy, or at least in the tools utilized, took place. Therefore, the history of American farm policy is divided into six periods.

The Settlement Period: 1776–1929

Government intervention in the marketplace on behalf of farmers began with the depression era of the 1930s. Yet government involvement occurred much earlier in colonial times. Government programs played a major role in the immigration and settlement of the vast land area of this country. The federal government at one time owned as much as 1.4 billion of the approximate 1.5 billion acre land area of the country.

The disposal of public lands in relatively small, widely dispersed parcels gave rise to the family farm structure of agriculture. This structure became an end in itself. Its perpetuation remains prominent in the rhetoric of public policy to this day.

The scope of government influence extended far beyond land settlement, as is illustrated by a list of some of the major pieces of legislation enacted by the Congress in the nineteenth and early twentieth centuries.

■ The Homestead Act of 1862 made vast acreages of federal lands widely accessible in small parcels at little or no cost to would-be farmers.

■ The creation of the USDA in 1862 provided a focal point for institutionalization of its farm constituency and influence in the executive branch.

■ The Morrill Act of 1862 created the land-grant college complex by giving federal lands to the states to endow colleges in the agricultural and mechanical arts.

■ The Hatch Act of 1887 provided annual grants to each state for agricultural research, leading to the system of state agricultural experiment stations.

■ The Land Reclamation Law of 1902 provided subsidized irrigation water from federally financed projects to family farms up to 160 acres in size, a limit that was seldom, if ever, imposed.

■ The Smith–Lever Act of 1914 created the cooperative federal–state Agricultural Extension Service, completing the system of teaching, research, and extension whereby the benefits of teaching and results of research were extended to farmers.

- The Federal Farm Loan Act of 1916 created the 12 cooperative Federal Land Banks and the beginnings of today's Farm Credit System, the largest lender of short-, intermediate-, and long-term farm credit.
- The Smith–Hughes Act of 1917 provided federal support for teaching vocational agriculture in high schools.

Although the programs authorized by these statutes aided agriculture and farmers, the government presence did not extend to influencing directly farmers' economic decisions, such as the selection of which crops to plant, how many acres of each to cultivate, and how to market the products. This did not come until the farm depression of the 1920s, which subsequently engulfed the entire economy early in the next decade.

The New Deal Era: 1929–1954

During the 1920s and the 1930s, farmers endured the longest period of financial stress in the twentieth century. It began with a precipitous break in farm prices during the 1920–1921 crop year and continued almost uninterrupted throughout the next two decades.[1] During the Great Depression, economic conditions went from bad to worse. From 1929 to 1932, the index of farm prices fell by 56 percent, and net farm income fell 70 percent.[2] Radical new answers were sought to the nation's problems. Several ideas were proposed, debated, and rejected.[3]

Some of these ideas have since surfaced and been debated again. One of the more interesting was the McNary–Haugen bill, which, if President Coolidge had not vetoed it, would have put into place a two-price plan. A "fair price" was to be set in the United States, and a government corporation would sell the rest on the world market—a policy that resembled the policy used by the European Union (EU) (until it was substantially modified in the late 1980s and early 1990s), and in the United States, for peanuts, and has been advocated in dairy.

An idea that was adopted in 1929, but that failed shortly thereafter, was the Federal Farm Board. A $500 million loan was made by the Treasury to USDA for the purpose of storing commodities during times of surpluses and disposing of them during times of crop failure. The problem was that no shortages developed. This idea was subsequently tried time and again and was generally unsuccessful.

Drastic depression conditions called for major reform and a more comprehensive approach that became known as the *New Deal farm policy*—based on President Roosevelt's campaign promise for a "new deal." The Agricultural Adjustment Act of 1933 and subsequent farm bills enacted during the 1930s set up

---

[1]Harold F. Breimyer, "Conceptualism and Climate for New Deal Farm Laws of the 1930s," *Am. J. Agric. Econ.* (December 1983), pp. 1152–1157.

[2]Don Paarlberg, "Effects of the New Deal Farm Programs on the Agricultural Agenda a Half Century Later and Prospects for the Future," *Am. J. Agric. Econ.* (December 1983), pp. 1162–1167.

[3]Douglas E. Bowers, Wayne D. Rasmussen, and Gladys L. Baker, *History of Agricultural Price Support and Adjustment Programs*, Agric. Inf. Bull. 485 (Washington, D.C.: ERS/USDA, December 1984).

several of the major farm program institutions that exist today.[4] Parts of the initial 1933 bill were declared unconstitutional, but the basic farm policy tools remained intact, including the following:

- The establishment of price support loans through the Commodity Credit Corporation
- Provisions for controlling production through diversion payments
- Provisions for commodity storage
- Provisions for crop insurance and/or disaster assistance

Price Support Loans and Commodity Purchases

The most important program in the eyes of many farmers is the nonrecourse loan. It was developed in the early 1930s and has survived as a policy foundation ever since. Price support loans and purchase programs set a floor on the market price. Loans and purchase programs provide both price and income support by two means:

- Under a **purchase program,** the government sets a price floor purchase on any products offered to it at the support price. This is how the milk price support program operates. The government stands ready to buy cheese, butter, and nonfat dry milk from anyone offering them at the support price.
- The **price support loan** is more complex than the government purchase program. At harvest, the government offers eligible farmers a nonrecourse loan from which the commodity is the collateral. The source of funds is the Commodity Credit Corporation (CCC), a government corporation that finances farm programs. Administration of farm programs is by the Farm Service Agency (FSA), which has federal, state, and county offices.

The farmer receives the loan at the support price for each unit of commodity placed under loan. The farmer pays the cost of storage and is free to sell the commodity at any time but, of course, must immediately pay off the loan plus interest costs. The interest rate normally is set at a level that reflects the costs of borrowing by the federal government, which are somewhat lower than the commercial interest rate.

The loan is a **nonrecourse loan,** *meaning that if the farmer does not sell the commodity by the due date, the commodity becomes the property of CCC in full payment of the loan.* There is no incentive for the farmer to sell unless the market price rises above the loan rate plus accumulated interest costs. The **loan rate** *becomes a floor on the market price because if the farmer cannot receive a higher price from the market, normal-*

---

[4]Wayne D. Rasmussen, "The New Deal Farm Programs: What They Were and Why They Survived," *Am. J. Agric. Econ.* (December 1983), pp. 1157–1162.

ly it is forfeited to the government. Therefore, no market sales occur at less than the market price if producers are economically rational.

The original purpose of the loan was to provide farmers a source of credit to facilitate storage and prevent all of the commodity from being marketed at harvest. With farmers being able to take out nonrecourse loans, they could pay their production bills and market when the price became more favorable later in the year.

In theory, that sounds as if it should work. However, the "gut" reaction of any Congress or administration to farmer agitation over low farm prices and income has been to raise the loan rates. This reaction has characterized farm policy since the 1930s. In 1994, one of the first major actions by Secretary Espy was to seek approval from the Office of Management and Budget (OMB) and the president to increase loan rates in reaction to farmer pressure. If the loan rate is set too high, the market does not clear and surpluses inevitably develop. *The basic dilemma of supporting farm prices and incomes above market-clearing levels is that it inherently leads to increased production, thereby creating the incentives for further government involvement in agriculture through programs that control production.*

Figures 10.1 and 10.2 present two contrasting loan rate situations. Instead of the two-country trade model utilized in Chapter 6, the graph is drawn to reflect only U.S. supply (*S*) and demand conditions. Both domestic demand (*DD*) and export demand (*ED*) are reflected in the graph to obtain the total demand (*TD*). Export demand is obtained from the excess demand function in the two-country trade model (see Figure 6.10, for example). In Figure 10.1, export demand is the area between *DD* and *TD*. Domestic and export demands, therefore, are added horizontally to obtain total demand.

In Figure 10.1, the loan rate P_ℓ is below the competitive equilibrium. The market, therefore, clears at price P_e and quantity Q_t. Quantity Q_d is sold in the domestic market while $Q_t - Q_d$ is exported.

In Figure 10.2, the loan rate is set above the market-clearing equilibrium at P_ℓ, which becomes the floor price and, therefore, the market price. At this higher price, supply is increased to Q_s while Q_t is the total demand. Quantity $Q_s - Q_t$ is the resulting surplus, which is forfeited to CCC. Note that the export demand falls sharply at the higher price, reflecting its greater elasticity. That is, when the United States raises its loan rate (price) above the market-clearing equilibrium, export demand suffers inasmuch as the U.S. price is less competitive in the world market.

The options for handling the surplus include the following:

- It could be stored. This was frequently done; however, there are limits to the willingness and ability of government to store commodities. Costs of storage are high. In addition, once the government purchases commodities, their eventual resale has a depressing effect on market prices. Thus, decisions to sell commodities from government storage are highly political.

- The surplus could be purchased and disposed of through nonmarket outlets such as school lunch or other nutrition programs and foreign food-aid pro-

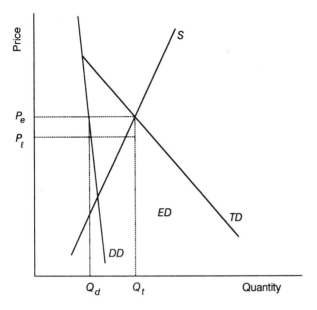

FIGURE 10.1
Low loan rate has no effect on market price.

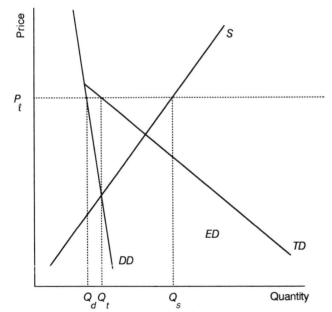

FIGURE 10.2
High loan rate creates surplus, reduces exports, and illustrates the basic dilemma of farm policy.

grams. However, such outlets, to a degree, displace domestic and foreign commercial sales. They, therefore, also tend to have a price-suppressing effect.

■ The surplus could be utilized as an export assistance program to capture or recapture foreign markets. Recall from Chapter 7 the discussion of the EEP,

TEA, and DEIP programs, all of which have used CCC stocks as a two-for-one bonus or a cash subsidy to make U.S. commodities more competitive in international markets. Of course, the two-for-one bonus can be pursued only if CCC has commodity stocks. If this is not the case, export promotion through these programs depends on cash subsidies.

Setting the Price Support Too High

A dilemma becomes apparent from this discussion. *Attempts to support farm prices above market-clearing levels lead to even higher levels of government involvement in farmers' production decisions in terms of encouraging government involvement in commodity storage, domestic commodity disposal programs, foreign food aid, or export subsidies (see box). If bad goes to worse and the prices supported are too high, storage stocks and disposal*

Monument to the American Cow

In 1981, deep beneath the ground in Independence, Missouri, in more bags, barrels, and boxes than the mind can imagine, the awesome triumphs of the prodigious American milk cow rested enshrined in dark, cool, and costly comfort.

What they kept here was government-owned milk, butter, and cheese. It kept piling up, costing the treasury millions upon millions of dollars, and nobody knew what to do with it. . . .

In its subterranean freezers and cooling rooms, CDC (Commercial Distribution Center, Inc.) alone stored more than 47 million pounds of dairy products that the government had to buy from U.S. farmers through the controversial dairy price-support program.

But that's only a drop in the bucket. The Department of Agriculture held about 1.9 billion pounds of surplus, with only a smidgen of that committed to sales. . . .

About 44 million pounds of cheese (of a 625 million pound inventory) were identified as moldy. . . .

Dairy product storage and handling came to about $42.5 million a year, although interest charges on a $2 billion inventory could lift the total daily cost to around $100 million. . . .

These inordinately high stocks led to cheese disposal programs and eventually to dairy herd buyout programs where the government purchased whole milking herds and sent them to slaughter. Subsequently, price support levels were reduced until the 1996 farm bill provided for elimination of the milk price support program in 1999.

Source: Ward Sinclair, "Under Missouri: A Monument to the Output of the American Cow," *The Washington Post,* December 21, 1981, p. A2.

programs can become overburdened, leading to production controls. At one time or another, all of these have happened.

As noted previously, one of the early guidelines for setting the loan rate was the parity price, which was developed for the original 1933 farm bill. *The* **parity price** *is defined as that price that today gives a unit of the commodity the same purchasing power as it had in 1910–1914.* Paarlberg uses the following illustration: "If a bushel of wheat would buy a pair of overalls in 1910–1914, then, to be at parity, a bushel of wheat should be priced so as to buy a pair of overalls today."[5]

The parity concept became the political standard by which to judge the economic position of farmers in the early 1930s and continued to play a major role in setting price support levels through much of the 1960s. The situation of 100 percent of parity has been achieved only twice since the turn of the century—in the base period 1910–1914 and during World War II. For the remainder of most of the twentieth century's farm policy history, farmers were forced to accept a price support level specified as some percentage of the parity price. Any higher price was considered politically unacceptable because of its potential for stimulating even higher levels of surplus production.

Yet parity was worshiped by farmers, their organizations, and by politicians attempting to capture their votes. According to Thomsen and Foote: "The concept of parity price has become part of the economic faith of farmers, accepted without question as an objective measure of a fair price. Actually, it is merely a mathematical expression of somebody's idea of what is fair. There is no possible objective measure of fairness or equity, which is entirely a subjective concept."[6]

Aside from its inherently subjective nature, the parity concept has several more basic flaws:[7]

- Parity assumes that farm commodity price and input costs were in their proper relationship in the 1910–1914 base period. In reality, 1910–1914 was exceedingly favorable to agriculture. This period is frequently referred to as the *golden age of agriculture.* Farmers were not only exceedingly well off related to costs, but also were well off relative to income in the nonfarm sector.

- Inputs used in production have changed radically since the manpower–horsepower era from which the parity concept arose. New inputs, such as tractors, combines, chemical fertilizers, and pesticides, had to be integrated into the index of prices paid by farmers. This process of determining prices of nonexistent inputs in the base period 1910–1914, of course, was arbitrary and, in fact, is very difficult statistically.

[5]Don Paarlberg, *Farm and Food Policy: Issues of the 1980s* (Lincoln, Neb.: University of Nebraska Press, 1980), p. 25.

[6]F. L. Thomsen and R. J. Foote, *Agricultural Prices* (New York: McGraw-Hill Book Company, 1952), p. 268.

[7]An excellent analysis of the history of the parity concept, the flaws associated with it, and some suggested modifications or alternatives are contained in Lloyd D. Teigen, *Agricultural Parity: Historical Review and Alternative Calculations,* Agric. Econ. Rep. 571 (Washington, D.C.: ERS/USDA, June 1987).

- Parity fails to take into account the increased productivity of agriculture over time. Increases in efficiency of production have no recognition in the parity concept.
- Parity for an individual commodity such as milk is determined in relation to the index of prices paid for all agricultural inputs, rather than just those inputs used in milk production. Differences in input mix among commodities, therefore, are not considered in the parity price.

Parity pricing has received sufficient criticism that its use in legislation as a standard for setting price supports has almost vanished. Beginning with the 1985 farm bill, parity has not even been mentioned. However, permanent legislation, which provides the basis for agricultural programs if a current farm bill expires, relies on parity measures. Marketing orders still specify parity as a pricing goal.

Two related concepts of parity merit mention. **Parity income** *suggests that net farm income per household should be the same in the agricultural sector as in the nonfarm sector.* **Parity returns** *would give agricultural resources returns equivalent to what they could earn in the nonfarm sector.* Neither of these concepts has been applied to setting price supports, although they are sometimes confused with parity prices as a farm policy goal.[8] Parity income and returns are considerably more modest policy goals than parity prices.

CCC Storage

When a farmer forfeits a commodity from the price support loan program, it becomes the property of the CCC and goes into government storage. This happens primarily when the price support loan rate is set above the market-clearing price (Figure 10.2). From the initial supply schedule S and with the loan rate set at P_ℓ only Q_t is demanded, leaving a surplus of $Q_s - Q_t$ forfeited to the CCC. In the next year, this surplus becomes a part of the supply. It can readily be seen that as long as the loan rate is kept at P_ℓ, stocks will continue to build.

During the 1950s and the 1960s, the government owned a large amount of storage space. During the world food crisis in the 1970s, Secretary Earl Butz sold the then-empty bins in the hope that government would never again become involved in the storage business. Such a fate was not to be. The government has not gotten back into the business of owning and managing grain bins, but it does contract space with private warehouses to hold forfeited CCC stocks. When commodities are in surplus, these warehouses are full and profitable. Private warehouse owners constitute a group that has a special interest in government storage.

Production Controls

Early in the history of the domestic price support programs, problems of overproduction and surpluses developed as a direct consequence of supporting prices

[8]Ibid., pp. 8–9.

and incomes above competitive equilibrium (see Figure 10.2). Many different production control methods have been tried with an overall general lack of success. Farmers and policymakers have not wanted to regiment themselves to the level of control needed to manage prices and supplies effectively.

A great deal of debate has periodically arisen among economists with regard to production control policies.[9] On the one hand, there are those who assert that farmers are one of the few economic segments that do not manage their markets.[10] They contrast agriculture's persistent tendency to overproduce with other economic segments, such as automobile manufacturers, chemical dealers, and farm equipment manufacturers, that tailor production to market needs. These advocates note that the inelasticity of demand for farm products makes agriculture ideally suited for supply management programs. On the other hand, advocates against supply management point to their adverse impacts on exports, their ineffectiveness in controlling production, and their tendency to increase costs of production (see box). An attempt is made below to capture the economic arguments on both sides of the issue without taking a position for or against production controls.

Acreage allotments and marketing quotas, also referred to as mandatory production controls, trace their origin to the Soil Conservation and Domestic Allotment Act of 1936. These provisions were supplemented by the 1938 farm bill and have been carried forward in the 1949 permanent legislation.

Initially, acreage allotments restricted farmers to planting only a specific number of acres of a specific crop (see box).[11] A national acreage allotment for a crop has been set at a level that would meet anticipated domestic consumption and trade needs. The national allotment was apportioned to individual farms based on their historical plantings of the crop. For many years, planting within the allotment acreage was mandatory with severe civil penalties for violations.

Farmers responded to being restricted to planting on fewer allotted acres by farming the allotment acres more intensely, that is, by applying more inputs such as fertilizer and, perhaps, by closer management. The result was reduced effectiveness of the allotment, requiring further reductions in the allotment acreage and, eventually, leading to marketing quotas.

Marketing quotas are simply restrictions on the quantity of commodities a farmer is allowed to sell. Quotas are usually used in conjunction with allotments. Marketing quotas have traditionally been implemented only if two-thirds of the producers vote for them in a referendum.

---

[9]The main arguments made in a debate on the production control issue are contained in Ronald D. Knutson, "The Case for Mandatory Production Controls," and Bruce Gardner, "The Case Against Mandatory Production Controls," both in *Farm Policy Perspectives: Setting the Stage for 1985 Agricultural Regulation* (Washington, D.C.: Committee on Agriculture, Nutrition, and Forestry, U.S. Senate, April 1984), pp. 217–224.

[10]These arguments were initially made most forcefully by Willard W. Cochrane, *Farm Prices: Myth and Reality* (Minneapolis, Minn.: University of Minnesota Press, 1958).

[11]The peanut and tobacco allotments continue to the present, although peanut farmers can plant over their allotment to produce for a contracted export market. The peanut program, therefore, supports a two-price plan with a high domestic price and a world export market price.

Ten Truths About Supply Control

- Agricultural supply controls transfer purchasing power from non-farm households to those controlling the means of agricultural production (quota holders or landholders).
- The single most important factor determining the impact of supply reduction is the farm level demand elasticity.
- Supply controls are most effective when applied to multiple substitute commodities.
- Substitution in production and consumption is the foremost enemy of supply control advocates.
- Supply control is like a drug habit: the longer it is used, the larger the dose needed to get high prices. Therefore, long-standing price controls are the most difficult to remove.
- Acreage restrictions, as opposed to output controls, have dramatically different effects on input use and innovation in agriculture.
- Acreage controls encourage the use of larger amounts of purchased inputs, which worsens environmental pollution.
- From a societal perspective and compared with free market policies, idling acreage represents a notably costly form of supply control.
- The tendency of those who defend supply control is to compare individual decision making at its worst with centralized decision making at its best.
- By permitting surplus production over a quota, we effectively transfer dollars out of our pockets into foreign households.

Source: Adapted from Thomas H. Hertel, *Ten Truths About Supply Control* (Washington, D.C.: Resources for the Future, July 1989).

Allotments and/or quotas are necessary if the goal of a government program is to raise the price of a commodity above competitive equilibrium. They may be used either as part of a strategy to raise price over the long run, to aid in the adjustment transition when technological change sharply increases yields, or to compensate for a sharp decline in export demand. For example, it has been suggested that production controls may be needed to control production resulting from new milk output-enhancing bovine somatotrophin (a growth hormone).[12]

---

[12]See Ronald D. Knutson, "The Case for Mandatory Production Controls in Milk"; Barnard F. Stanton, "The Case Against Mandatory Production Controls"; and Ronald D. Knutson, "Points of Agreement," in *Balanced Dairying* (College Station, Tex.: Texas Agricultural Extension Service, April 1987).

Warning: The Surgeon General Has Determined the Tobacco Program to Be Dangerous to Your Health

Tobacco is one of the most widely criticized farm commodity programs because of its production control features, coupled with the inconsistency of government support of the prices of a product harmful to health.

One million acres of tobacco are grown by 250,000 farmers, an average of 4 acres per farm. Acreage allotments and quotas give farmers the right to grow and market tobacco. The allotments and quotas, being limited in supply, have a capitalized value of $7,500 to $8,500 per acre. They can be sold or rented but not to foreigners. High allotment and quota rental rates (25 to 35 percent of production costs) also reflect the production control policies. Price supports are based on parity. If tobacco sold at auction fails to receive a bid of 1 cent per pound over the support price, it is bought under a CCC loan by a grower cooperative.

Tobacco interests argue that the program incurs no government cost on $3 billion in producer income, supports many small rural farmers and related businesses, and garners large public revenues from taxes on cigarettes and tobacco products.

Opponents argue that there is no logical basis for the government to subsidize production of a harmful product, the program has outlived its original purpose of increasing farm prices and incomes, and the program is far from costless to society as a whole.

The effect of imposing allotments is to shift the supply curve to the left. Without a price support program, this has the effect of raising the market price. With a price support program, a surplus can be effectively removed by shifting the supply curve from S_1 to S_2, thereby reducing production from quantity Q_1 to Q_2 (Figure 10.3).

As noted previously, the main problem with allotments is that farmers respond by farming their land more intensely. The effect is to distort the optimum combination of inputs such as fertilizer.[13] Production controls can be made more effective by placing a quota on the quantity that can be marketed.

The theoretical effect of a quota is very direct in that the supply schedule can be treated as being vertical at the level of the quota, Q_q (Figure 10.4). Because farmers cannot market any more than Q_q, there is no incentive to change the input mix. Distortion in the input mix, therefore, is avoided.

Whether production controls are effective at raising producer income depends on the elasticity of demand. For commodities that are consumed in the domestic market,

---

[13]See Chapter 17 for an explanation of this effect and its impacts on agribusiness.

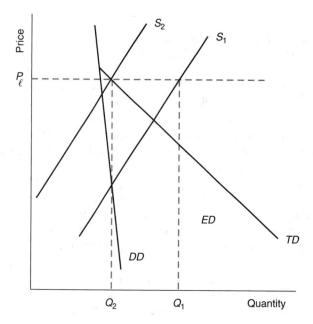

FIGURE 10.3
**Supply reduction required to eliminate sur-
plus with high loan rate.**

demand is generally inelastic.[14] Therefore, production controls on commodities produced for the domestic market, such as milk, would be expected to be quite effective at raising farmers' net income. This helps to explain why proposals to control milk production are debated regularly in the Congress.

As indicated previously, export demand is considerably more elastic than domestic demand. Because of proposals to both raise loan rates and control production, the precise magnitude of export demand elasticities for the major grains and cotton have been of considerable interest and have received extensive study.[15] The findings of these studies generally indicate the following:

- Considering the barriers to trade that exist in the world, the elasticity of export demand for most major grains and cotton is inelastic in the short run (one to two years), but elastic in the long run (four years or more).
- Export demand is more inelastic for groups of commodities such as coarse grain compared with corn, or all grain compared with individual grains.
- Removal of barriers to trade would increase the elasticity of demand materially—perhaps twice or more.

[14]Kuo S. Huang, *U.S. Demand for Food: A Complete System of Price and Income Effects,* Tech. Bull. 1714 (Washington, D.C.: ERS/USDA, December 1985).
[15]S. Devadoss, William H. Meyers, and Michael Helmar, *Export Demand Elasticity: Measurement and Implications for U.S. Exports* (Ames, Iowa: CARD, Iowa State University, April 1988); M. E. Bredahl, W. H. Meyers, and K. J. Collins, "The Elasticity of Foreign Demand for U.S. Agricultural Products: The Importance of Price Transmission Elasticity," *Am. J. Agric. Econ.* (February 1979), pp. 58–63.

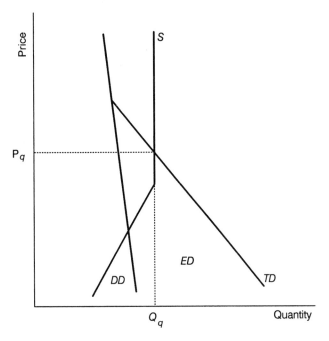

FIGURE 10.4
A marketing quota makes the supply schedule vertical.

Production Controls Revisited

The argument that mandatory controls destroy the export market has been effective in discouraging policymakers from applying allotments and quotas to major grains and cotton. In reaction, the Harkin Bill was introduced in 1986 to set a high domestic price (70 to 80 percent of parity), subsidize exports to maintain market share, and establish marketing quotas to control production. Referendum approval required a vote of only half of the producers to implement. This highly controversial legislation was the subject of much analysis and controversy. The results of three university studies indicated the following[16]:

- With export subsidies, producer returns would rise.
- The increase in producer returns would substantially improve the chances of farm survival.
- The government costs of subsidies to maintain export demand would be at least as large as the current farm program.[17]

[16]Ronald D. Knutson *et al.*, *Policy Alternatives for Modifying the 1985 Farm Bill*, Texas Agr. Exp. Sta. Bull. B-1561 (College Station, Tex.: Texas A&M University, January 1987); Daryll E. Ray and Darrel D. Kletke, *Analysis of Selected Farm Program Alternatives for Wheat*, A. E. 8746 (Stillwater, Okla.: Department of Agric. Econ., Oklahoma State University, June 1987); Food and Agricultural Policy Research Institute (FAPRI), *The Commodity Supply Management Program*, FAPRI 2-87 (Columbia, Mo.: University of Missouri, February 1987).

[17]The FAPRI study assumed an international export cartel with market agreements. As a result of the cartel assumption, government costs associated with this version of the Harkin bill were less than under the 1985 farm bill.

USDA Republican political appointees, free-market advocates, and the agribusiness commodity organizations were irate at the results of these three university studies. Much of their criticism treated only the mandatory controls portion of the bill, ignored the export subsidies, and contended that the domestic market could not be protected against reimport. The latter criticism was a legitimate concern in commodities such as cotton, where reimport would be in the form of textiles whose origin could not be identified.

In 1989, long after the dust had settled on the Harkin bill debate, USDA released its study[18] of the impacts of mandatory production controls covering essentially the same time period and support levels as contained in the Harkin bill and the university studies. Interestingly, the USDA study results were essentially the same as the university studies. Of course, the study was released long after it had been completed and long after the controversy was over.

Capitalization *is the process by which farm program benefits are bid into the price of the control instrument.* The instrument may be land or a chicken on which an allotment is held, or a quota that gives the holder a right to market a given quantity product. The value of the quota or allotment is most visible and easily determined if it is freely transferable from one producer to another. When production control programs are pursued with the objective of raising the price above the market-clearing level over a substantial time period, farmers often use the resulting profits to expand the size of their operation.

The more profitable farming is, the more farmers are willing to pay for land or livestock on which production is controlled. For example, marketing quota rights to sell milk from an average cow in Ontario, Canada, cost a potential producer about $3,500—considerably more than the cost of an average milk cow.[19] For a new dairy operation in Canada, quota costs represent 30 percent of the total initial investment. Under the current allotment program for tobacco, the right to produce and market tobacco sells for as much as $3,000 per acre.[20]

The degree of capitalization would be expected to be directly proportional to the vigor with which production control and price enhancement is pursued. One way to estimate the value of an asset that is expected to generate a given level of return to perpetuity is to divide the extra profit earned by the required rate of return.[21] For example, if a quota adds $10 profit per acre at a 10 percent rate of return, the value of the land would increase by $100 per acre ($10/0.10).

The capitalized value of an asset on which there are production controls also depends on the certainty that those returns will be maintained in the future. Recent doubt about the future of the tobacco program materially reduced the

---

[18]C. Edwin Young *et al., Economic Effects of Mandatory Production Controls*, Agric. Econ. Res. Rpt. 595 (Washington, D.C.: ERS/USDA, March 1989).

[19]Dan Dvoskin, "Some International Experiences with Mandatory Supply Controls," *Agricultural Outlook*, AO-130 (Washington, D.C.: ERS/USDA, May 1987), p. 32.

[20]Economic Research Service, *Mandatory Production Controls*, Agric. Inf. Bull. 520 (Washington, D.C.: USDA, July 1987).

[21]John B. Penson and David A. Lins, *Agricultural Finance* (Englewood Cliffs, N.J.: Prentice Hall, 1980), p. 102.

value of tobacco allotments.[22] Potential depreciation in the value of allotments or quotas creates strong producer resistance to modifying the program. Declines in the values of recently acquired quotas or allotments are just as devastating to producers as are falling land values.

Quotas and allotments are not the only source of capitalization. When a farm program generates higher returns, capitalization occurs. In the case of target prices and loan rates, capitalized values are more difficult to detect because they are hidden in land values and changes in the general price level.

Even if all the obstacles associated with political acceptance of mandatory controls could be overcome, questions arise as to whether they would be supported by farmers voting in a referendum. The referendum on whether mandatory controls should be implemented occurred in 1963. The proposal was for a system of penalties for planting wheat in excess of acreage allotments and for marketing quotas based on domestic and export needs. With over a million farmers voting, the referendum was soundly defeated. The 1985 farm bill mandated a poll on whether wheat producers would prefer a system of mandatory production controls. A slim majority voted yes.

Throughout this period, the principal objective of the New Deal programs was to raise farm prices. *These initial public policy initiatives were hampered by erroneous diagnosis of the problem.* Instead of correctly diagnosing the problem as being one of too many resources committed to agriculture, low farm prices and incomes were initially incorrectly attributed to reduced demand resulting from the depression conditions of the 1930s.[23] *The public policy tendency was, therefore, to treat the symptoms rather than the problem.* That is, initial policies and programs emphasized inadequate grain storage and price support programs as opposed to efforts to move resources out of agriculture and curb production. Policies evolved in response to a growing understanding of the fundamental nature of the farm problem and to the changing economic circumstances; yet lapses in memory periodically occurred.

Permanent Legislation: 1949–?

After World War II there was great concern that after gearing up for war-time production there was substantial danger of accumulating large surpluses. The 1949 farm bill recognized the contribution farmers had made to the war effort by setting support levels near parity but also by providing the tools for controlling production.

---

[22]Daniel A. Sumner and Julian M. Alston, *Effects of the Tobacco Program: An Analysis of Deregulation* (Washington, D.C.: American Enterprise Institute, 1986).

[23]In a review of the post–World War II agricultural economic policy literature, George Brandow ["Policy for Commercial Agriculture," in *A Survey of Agricultural Economics Literature* (Minneapolis, Minn.: University of Minnesota Press, 1977)] notes that the first comprehensive diagnosis and description of the farm problem did not occur until 1945 by T. W. Schultz in his book, *Agriculture in an Unstable Economy* (New York: McGraw-Hill Book Company, 1945).

The 1949 farm bill has special significance because it was the last farm bill enacted without an expiration date. Therefore, all subsequent farm bills have been amendments to the 1949 law. If a new bill is not enacted before the expiration of the previous one, the provisions of the 1949 law go into effect. This would mean high price supports for farm products and an automatic vote of producers on whether to implement mandatory production controls.

The threat of implementing the provisions of the 1949 law provides the impetus for Congress to vote on a new farm bill before the expiration of the current one. Since 1954 that threat has been sufficient to get congressional passage and presidential signature before the expiration deadline. In 1994, there was serious thought given to making the 1990 farm bill permanent as part of the vote confirming the Uruguay Round GATT agreement. This, it was thought, might give the farm bill greater stability.

The threat of the 1949 Act going into effect came up again, in 1995, when the 1990 farm bill was due to expire. With a Republican-controlled Congress, federal budget issues divisively front and center, and the nature of the decoupling proposals themselves, agreement on a new farm bill was not reached until March 1996. (The previous four farm bills were signed in November or December of the year they expired.) In the interim between October 1995 and March 1996, a continuing resolution was passed preserving the provisions of the 1990 farm bill. The 1996 farm bill explicitly removed the 1949 act as permanent legislation right up to its final vote. It was in the Senate bill but not in the House bill, which left it up to the conference committee, that voted to retain the 1949 permanent provisions.

Flexible Price Supports: 1954–1970

Until 1954 Congress set the price support level sufficiently high (90 percent or more of parity) that the minimum mandated price became the floor price. Under the urging of Secretary Ezra Taft Benson, Congress passed the 1954 bill that gave him discretion in setting the price support level between 75 and 90 percent of parity (see box).

Three other important distinguishing attributes of the 1954 legislation follow:

- It was the first general farm program in that all major grains and cotton were treated in a group under the same general program context. Previously, each commodity had its own program; raising price supports or controlling production on one commodity would have adverse spillover effects on other commodities. This arguably began the move away from commodity programs, but commodity titles remained in the farm bill with price and income support levels established explicitly for each commodity.

- The first comprehensive land retirement program (soil bank) was established to openly address the excess capacity issue.

- The export market was recognized as a potential commercial market for farm products through the enactment of P.L. 480.

Secretarial Discretion

The issue of the degree of discretion authorized to the secretary of agriculture has been an item of contention in virtually every farm bill since Secretary Benson was given the authority to lower supports. For example, in framing the 1981 bill, Secretary John Block asked for a bill that would give him maximum flexibility in implementation. The Congress, fearing a swift move toward the free-market policies espoused by President Reagan during the 1980 campaign, passed a bill with little flexibility. Interestingly, in 1985 the administration got what it asked for in 1981: a wide margin of flexibility and discretion. However, since 1981, the farm bills have given the secretary relatively less flexibility. The 1996 farm bill could be a reversal of that trend as indicated by Chapter 11.

Land retirement programs pay producers to take land out of production. Payments are based on the productivity of the land being removed from production. From a farmer's perspective, participation is warranted if the payment exceeds the potential earnings from the land to be retired plus the cost of maintaining the retired acres. Discontent with the regiment and the consequences of mandatory control programs have led to extensive experimentation with voluntary programs to control production. Since the 1930s, three major land retirement programs have been undertaken: the soil bank, payment-in-kind, and conservation reserve programs.

The soil bank program paid farmers in cash to remove land from production. The soil bank, implemented in 1956, was really two different programs. The first, and least effective, involved paying farmers to convert allotment land to conservation uses on an annual basis. This program was discontinued after two years of operation because of high costs. The long-term retirement portion of the soil bank put land into a conservation reserve for a 10-year period. Farmers found this program was sufficiently attractive to put nearly 30 million acres in the soil bank by 1960.

Rural communities located in high-participation areas objected to the whole-farm retirement provisions of the soil bank program.[24] It reduced input purchases and product marketing, undermining agribusiness. In some instances, families would simply retire from farming and move. One of the higher levels of participation in the soil bank program was in areas of the Southeast where farmers planted trees on the retired land. In the short run, these farmers got soil bank payments and, in the late 1980s, they began to receive receipts from harvesting lumber.

In the early 1960s, the Kennedy administration attempted to establish mandatory production controls, which were rejected in a producer referendum. In

---

[24]Bowers *et al., History of Agricultural Price Support,* p. 22.

the late 1960s, increased emphasis was placed on direct payments to farmers in return for not producing on certain acreages. This was the first move toward more market-oriented farm programs.

Market Orientation: The 1970s

Prior to the 1970 farm bill, supporting farm prices and incomes were one and the same. However, in 1973, farm price and income support were overtly separated. Price support was provided by conventional CCC loans, whereas income support was provided by direct farmer payments. Through much of the 1970s and the 1980s, the size of direct payments increased, giving rise to the government cost problem.

The motivation for increased direct payments was to lower price supports to restore competitiveness in the world market. Buoyed by strong foreign demand, exports boomed. With market prices relatively high, income support payments were low. The target price policy, officially established in the 1973 farm bill, appeared to work better than might have been anticipated.

Target Prices

A **target price** *is the level of returns per unit of commodity on certain acreage guaranteed to farmers who participate in farm programs. Target prices provide for direct payments to producers of the difference between the target price and the average market price whenever the average market price for a specified time period falls below the target price. The difference between the target price and the average market price is referred to as a* **deficiency payment.** Target prices have been established for all major food grains, feed grains, and cotton as a means of supporting farm income. *Target prices separate price support from income support. Loan rates support price and income. Target prices support only income.*

Congress has generally set the target price. When first authorized in the 1973 farm bill, the target price was set on the basis of what was thought to be the national average cost of production. USDA was mandated to make annual average cost of production estimates, presumably as a basis for setting target prices and evaluating the performance of farm programs.

The target price is used to calculate the amount of deficiency payment due producers of a commodity. If the market price is less than the target price, the difference (deficiency payment) is paid to farmers. The maximum deficiency payment is the difference between the target price and the loan rate.[25]

---

[25]Since the loan rate should be the market floor price, the deficiency payment should never be any larger than the difference between the target price and the market price. However, we will see that market prices can fall below the loan rate when the CCC is releasing commodities onto the market or when all producers are not eligible for the loan because they do not participate in the program.

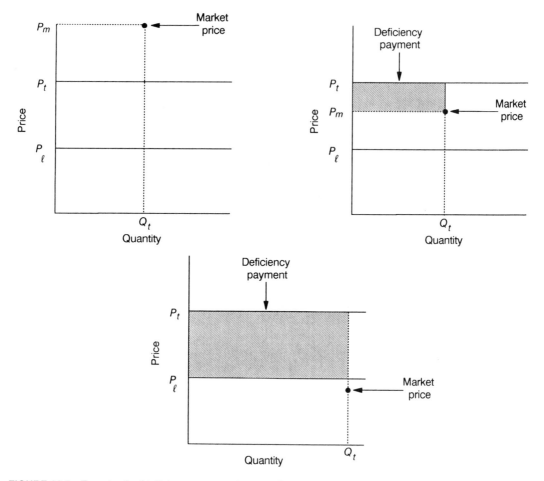

FIGURE 10.5   Target price/deficiency payment concept.

Operation of the target price program is illustrated considering three situations (Figure 10.5):

1. The market price P_m is above the target price P_t. This is illustrated in the top left panel of Figure 10.5, which indicates that no deficiency payments would be made.

2. The market price P_m for the specified period is below the target price but substantially above the loan rate (P_l in the top right panel). Deficiency payments would be based on the differences between the target price and the market price times the quantity of eligible production ($P_t - P_m)Q_t$.

3. The market price for the specified period is below both the target price and the loan rate (bottom panel). Deficiency payments of ($P_t - P_l)Q_t$ would be made—the maximum payment rate is the differential between the target price and the

loan rate. Producers would probably elect to place commodities in the nonrecourse loan program. If unredeemed, the marginal revenue received for the product would be the loan rate plus the additional deficiency payment, $P_\ell + (P_t - P_\ell)$.

Note that the target price is paid only on producers' normal production. Normal production is determined by a farmer's planted acres up to his or her base acreage and farm program yield. The farmer's production history is used to determine both the base acreage and yield. In the following discussion, unless noted otherwise, it is assumed that the base acreage is fixed and that the yield is regularly updated.[26] Base acreages and yields are used for ease of administration in issuing deficiency payments. Later, it will also be seen that they play a role in production control programs.

Because the target price is a guaranteed return to producers having sufficient base and yield, it becomes the marginal revenue on which production decisions are based. As in the case of the loan rate, if the target price is set below the market-clearing price, there is no deficiency payment and no effect on production.[27]

If the target price P_t is set above the competitive equilibrium price P_c and all production is eligible, as in Figure 10.6, production Q_t is encouraged. At this higher level of production, the market price has to fall to P_m before the market is cleared. Note that the quantity supplied Q_t is equal to the quantity demanded at price P_m. Therefore, the market is in equilibrium, even though it is not in competitive equilibrium.[28] Note also that the target price does not depress the quantity demanded, as did a loan rate at the same level (Figure 10.2). Target prices expand the quantity demanded because the market price is depressed. The greatest demand expansion occurs in the export market where the demand is more elastic. The target price can, therefore, also be interpreted as being a form of export subsidy due to these factors:

- The quantity of production is expanded, thus artificially adding to the total world supply.
- The market price is depressed below the competitive equilibrium.

The difference between the target price and the market price $(P_t - P_m)$ is the deficiency payment. If the target price is paid on all units of production,[29] total government costs would be $(P_t - P_m)Q_t$, which is the shaded area of Figure 10.6.

---

[26]In reality, the base yields are based on the Olympic average of the farmers' proven yields in the base period of the early 1980s. Yields have been frozen since as a means to control program costs. Of course, technological change has raised the current average yield above this farm program yield. This tends to constrain farmers' production in that the market price rather than the target price becomes the farmers' marginal revenue.

[27]In an unstable real world, production is affected by target prices set below the market price. The target price reduces risk even though it is below the market price.

[28]Recall from Chapter 1 that the reader was cautioned that with government programs, equilibriums exist even though they are not competitive equilibriums. The competitive equilibrium (P_c, Q_c) becomes a norm against which other prices and quantities are compared.

[29]This assumes that the acreage base and farm program yield are sufficient to cover all units of production.

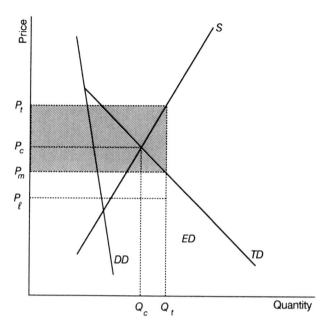

FIGURE 10.6
Target price depresses market price.

In Figure 10.6, the loan rate is below the equilibrium market price. Suppose that the loan rate is set higher, at P_ℓ in Figure 10.7. Because the loan rate is the price floor, P_ℓ is also the market price P_m. At P_m, total demand is Q_t, with $Q_t - Q_d$ being exported. The market does not clear as a result of the high loan rate, and $Q_s - Q_t$ is forfeited to the CCC.

Payment Limits

Large deficiency payments resulted in large total payments to large farms. This reality led to limits on payments per farmer. Payment limits were enacted as early as the 1938 farm bill after the first direct payments were made in the 1936 farm bill. After direct farmer payments were established in the late 1960s, limits were reinstated in 1970. Although payment limits could curb some of the production-enhancing tendencies of the target price, farmers have divided their operations among "persons" sufficiently so that the limit has been less than fully effective (see box).[30] In 1988 an effort was made to tighten the payment limit by preventing the division of farms among more than three persons—referred to as the *three entity rule*. However, this change was also ineffective. Farmers simply reorganized their operations with the help of an attorney to avoid the payment limit. A consequence of these reorganizations was increased inefficiency. The ultimate dodge of the payment limit is a cash rental arrangement.

---

[30]General Accounting Office, *Farm Payments: Farm Reorganizations and Their Impact on USDA Program Costs* (Washington, D.C.: U.S. Congress, April 1987).

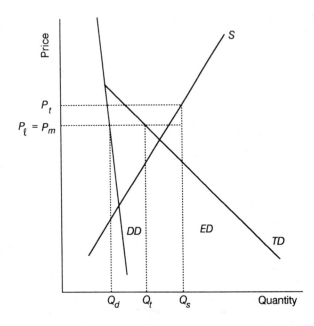

FIGURE 10.7
High loan rate creates surplus with target price.

When the Clinton administration sought ways to reduce spending shortly after its election, a number of proposals were made to further curb payments, including elimination of the three entity rule. However, what ultimately evolved was a proposal to discontinue payments to farmers that have an off-farm income of $100,000 or more. Even then, questions arose as to how such a law would be administered.

The conclusion is that it is virtually impossible to administer a payment limit effectively. Moreover, effectively administered payment limits would generally discriminate against more efficient larger farms (see Chapter 12).

Set-Aside

Set-aside *requires that a certain percentage of a farmer's cropland be removed from production as a condition for receiving farm program benefits.* Those benefits include deficiency payments and the price support loan. Set-aside, therefore, is a conditional land retirement program.

Analytically the impact of a set-aside program in the presence of a target price can be seen in Figure 10.8. The target price P_t and loan rate P_ℓ result in a surplus of $Q_1 - Q_2$. This surplus can be removed by imposing a set-aside requirement sufficient to shift the supply curve from S_1 to S_2.

Set-aside was first introduced in 1970. The 1981 farm bill changed the name of the program to *acreage reduction* although the basic program concept did not

The Mississippi Christmas Tree: Farming the Farm Program

In 1985, a six-member joint venture operated 5,481 acres of farmland that the participants in the venture either owned or cash leased from others. The joint venture comprised a father, his four adult sons, and an adult daughter. USDA officials determined that each individual member in the venture qualified as a person for payment limitation purposes under USDA regulations and could receive up to $50,000 in direct support payment subject to the limit.

The joint venture's 1985 farm operations qualified for about $595,000 in payments subject to the limit. The father, since he was operator of much of the land, exceeded the $50,000 on his payment and, as a result, he did not receive about $315,000 that was earned and attributable to his interest in the operation. Each of his five children received about $46,000 as the result of their interests in the 1985 operation.

For 1986, the father and his children reorganized the operation into a new joint venture that comprised the same six persons as in 1985 plus 15 new corporations they had formed. Each corporation was owned on a 50/50 basis by two individuals, each of whom was a member of the 1985 joint venture. The new joint venture operated 6,870 acres of farmland that were either owned or cash leased by its members.

USDA officials determined that each of the six individuals and 15 corporations composing the 1986 joint venture qualified as a person and could receive up to $50,000 in direct support payments subject to the limit. The 1986 farm operation earned about $1,050,000 in direct support payments. This resulted in $50,000 to each of the 21 persons composing the joint venture for payment limitation purposes for a total payment of $1,050,000. The Mississippi Christmas tree organizational structure became subject to the payment limit under the 1988 revised regulations.

Source: Brian P. Crowley, *Farm Reorganizations and Payments to Foreign Owners of U.S. Cropland,* General Accounting Officer testimony before the Committee on Agriculture (Washington, D.C.: U.S. House of Representatives, April 1, 1987), pp. 19–20.

change. Aside from its voluntary nature, set-aside had the main advantage of reducing government costs. Not only were payments involved in retiring land avoided by set-aside, but also income support payments to those farmers who chose not to participate in the program were avoided. Set-asides also reduced erosion on the least productive land that was taken out of production. On the other hand, the land remaining in production was farmed more intensely, potentially increasing erosion.

The Problem of Slippage

Slippage *measures the difference between the percent of land (or cows) removed from production and the percent reduction in production.* Set-aside slippage is high for the following reasons:

- Producers who do participate set aside their poorest land.
- The remaining acres are often farmed more intensely.

Experience with the set-aside program suggests that a 15 percent set-aside has resulted in only about a 3 percent decline in production. In other words, there is 80 percent slippage. Higher levels of set-aside are more effective because each additional acre of land set aside tends to be more productive. Thus, the PIK program had less slippage because of the large quantity of land removed from production. CRP would be expected to have slippage because the highly erosive land removed from production would be expected to have lower than average yields. The dairy termination program had a high level of slippage because of the movement of heifers and the trading of cows among farmers, particularly during the early stages of the program. This program, enacted as part of the 1985 Farm bill, required the secretary of agriculture to buy whole dairy farms out of production. Enough farms were to be bought out to equal 7 percent of the nation's milking herd in addition to normal culling. All animals from herds bought out were to be slaughtered over a 17-month period. To help pay for the program, the secretary deducted $0.40 per hundredweight from the price received by all dairyfarmers during 1986 and $0.25 per hundredweight during 1987.

The buyout program substantially reduced price support purchases of manufactured dairy products during 1986 and 1987. However, farmers not participating in the program continued to increase production. Some of this increase in production resulted from trading high-producing cows from buyout producers from low-producing cows from nonparticipating producers. Also, some heifers were probably not slaughtered with the buyout herds, thereby reducing program effectiveness. In addition, some of the producers who were bought out probably would have gone out of production. After the five years were up, some of the more modern production facilities went back into production.

The Farmer-Owned Reserve

In a time of volatile world demand during the 1970s, government attention turned to storage programs. A farmer-owned reserve was established as an alternative to government stocks. The reserve was designed to stabilize prices, but policymakers used it as a means of supporting prices.

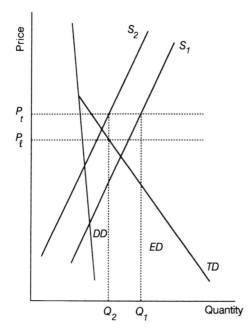

FIGURE 10.8
Supply reduction required to eliminate surplus with target price and high loan rate.

The **farmer-owned reserve (FOR)** *is, in essence, an extended loan program covering a period of up to three years.* FOR was designed in the late 1970s to stabilize prices and provide increased supply assurance to domestic and foreign customers. President Jimmy Carter and Secretary Bob Bergland justified the need for increased management of reserves stock on the grounds of greatly increased price instability, higher priorities placed on exports, and adverse producer and foreign customer reaction to export embargoes imposed by the Nixon/Ford administrations. The Reagan/Bush and Clinton administrations substantially deemphasized FOR stocks.

In return for placing commodities in the FOR, farmers often receive a higher loan rate (the entry price) than the regular price support loan. During the first year in the FOR, this loan could be interest free. In addition, a payment approximating the average cost of storage has been provided by USDA. Interest costs have also, at times, been waived. In return for the higher reserve entry price, interest subsidy, and storage payment, the FOR farmers agree not to market the grain until the market price reaches a specified level referred to as the release price. At the release price, the farmers may, but are not required to, sell their FOR grain. Incentives for sale are provided by an ending of interest subsidies and storage payments. In the event that farmers do not remove their grain from the FOR, the secretary has the authority to call or require payment of the loan.[31]

---

[31]From its inception until the enactment of the 1981 farm bill, the FOR also contained a call price. When the market price reached the call price, farmers were required to repay their loans. This action did not necessarily require actual sale but provided an added incentive for sale. The call provision of the FOR was very unpopular with producers, and USDA, in fact, encountered difficulty getting farmers to pay off their loans when the call price was reached.

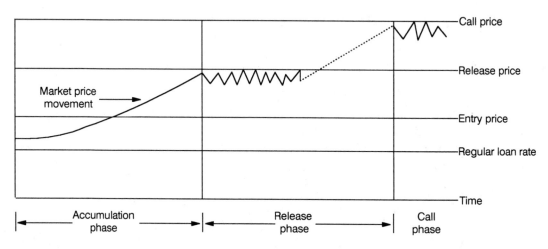

FIGURE 10.9   Illustration of the operation of the farmer-owned grain reserve.

The farmer-owned reserve has important price impacts (Figure 10.9):

- When supplies are abundant, commodities are accepted into the reserve at the higher entry price. As grain is placed in the reserve, the free stocks are reduced and prices rise.
- Once the reserve is established, and if annual production and consumption are at normal levels, the market price fluctuates within the price range bounded by the loan rate and the reserve release price. With sufficient free stocks, the market price would not rise off the loan rate.
- When demand is higher than supply, prices rise to the release price, at which time the reserve stocks may be sold without penalty. Sales of released grain suppress the price somewhat, as indicated by the jagged market price movement line near the release level.
- If available supplies are relatively small, the price soon rises above the release level. Further price stabilization then becomes dependent on action by the secretary of agriculture to encourage marketing, including the potential to call the loans.

It should be clear from this discussion that FOR has a substantial impact on commodity price movements. Therefore, a knowledge of the program and its probable impacts is very important to anyone who deals in grain markets.

As indicated previously, substantial differences of opinion have been voiced over the role of, and need for, government accumulation and holding of stocks. Democratic administrations have been more inclined to hold stocks than have Republican administrations. Private sector grain traders have tended to oppose government-held stocks. Country elevators find themselves caught philosophically between interest in utilizing storage facilities for government stocks and grain traders' views.

When the U.S. government supports the holding of stocks, foreign governments are discouraged from holding stocks on their own account. In fact, there appears to be some tendency for foreign governments to place explicit emphasis on "cash reserves" for the purpose of purchasing commodities when they are needed. Such a policy would have the effect of destabilizing the market in that the bidding power of buyers is increased in the event of commodity shortages.

By the end of the 1970s, stocks once again became burdensome. But surpluses that existed in the 1980s were not the chronic problem that existed in the 1960s; rather, they became periodic with the export market being the safety valve outlet for high levels of U.S. production.

A Hiatus: The 1980s

The robust economic conditions of the 1970s led to double-digit inflation conditions when President Reagan took office in 1981. His economic policy for wringing out this inflation was the tight money policy discussed in Chapter 8. The effect was to stifle exports. As a result, prices once again rested on loan rates for several commodities. As a result, commodities were forfeited under the nonrecourse loan program. CCC stocks accumulated and private storage facilities bulged in the FOR. Questions began to arise as to whether the 1970s had been an aberration.

Payment-in-Kind

Payment-in-kind (PIK) pays farmers in generic certificates to retire land from production. After the 1982 harvest, grain bins were bulging. Another record harvest in 1983 would have meant the forfeiture of large amounts of grain to the CCC. To prevent this occurrence, USDA announced the PIK program in January 1983.

PIK paid farmers 80 to 95 percent of their farm program yield in return for retiring acreage from production for one year. Payment was in the form of generic certificates. Farmers' response to the PIK program was stronger than anyone had anticipated. A record 82 million acres of land were removed from production, 25 percent more than any previous production control effort.[32] This meant that more than one-third of the cropland was retired from production.

Exactly how effective PIK was at reducing production is subject to debate. Its effect was confounded with the worst drought since the 1930s.[33] The combination of these events temporarily lowered stocks and raised farm prices. In contrast to initial expectations, exports increased in spite of the higher market prices. In the

---

[32]Economic Research Service, *An Initial Assessment of the Payment-in-Kind Program* (Washington, D.C.: USDA, April 1983).
    [33]Bowers et al., *History of Agricultural Price Support*, p. 41.

short run, increased exports resulted from PIK-released commodities being made readily available to the export market at competitive prices.

As the farm financial crisis deepened, neither the Congress nor the administration could come to grips politically with the option of reducing price and income supports. Stocks once again accumulated in the hands of the CCC as production returned to normal and PIK land came back into production. PIK was then used in a different way. *Generic PIK pays farm program participants in the form of negotiable certificates that can be redeemed for commodities held by the CCC.*[34] The certificates have a specified face value. A farmer who has commodities under loan may use generic certificates to pay off the loan. Buyers of certificates, other than farmers having commodities under loan, desire to gain control of a commodity in a specific location without having to pay transportation costs. In this case, the commodities purchased with generic certificates are forfeited commodities in CCC warehouses. The certificates are generic in the sense that any commodity held by CCC can be purchased with them. Thus, a farmer who receives generic PIK certificates in lieu of wheat deficiency payments may use them to buy CCC corn or may sell the certificates to a merchant or anyone who wants to buy them.

Generic certificate programs have interesting economic effects that often are not recognized. CCC commodities held in storage or under a price support loan normally are not readily available to the market. Generic certificates release commodities to the market, which has a price depressing impact. As a result, U.S. commodities become more competitive in world markets. The use of generic certificates provides another explanation for the fall of market prices below the loan rate. Release of commodities any time from CCC stocks can cause the market price to fall below the loan rate.

Generic certificates can be used only when CCC has large stocks. From a short-run accounting perspective, generic certificates save the government money because payments do not have to be made in the form of cash, which would be part of current budget expenditures.[35] The commodities have already been purchased by CCC and the government saves storage costs. However, when the government gives commodities away, it is reducing the assets of the CCC, which presumably have to be replenished. Government costs burst through the ceiling as a result of this combination of events.

Marketing Loans

The problem with the 1985 farm bill became one of getting control of production and lowering loan rates to prevent continued CCC takeover of commodities. Target

---

[34]Farmers have received generic certificates in lieu of cash under many government programs other than deficiency payments, including acreage reduction programs (set-asides), paid land diversion, conservation reserve program (CRP), marketing loans, disaster payments, and emergency feed programs. Grain merchants have been issued generic certificates through EEP (export enhancement program) and TEAP (targeted export assistance program).

[35]One of the peculiar aspects of government accounting systems involves treating a price support loan as a cash outlay, as opposed to the purchase of an asset. When the loan is repaid, it is treated as a cash receipt. If the commodity is forfeited to the CCC, it is treated as an asset.

prices were modestly lowered while loan rates were dropped substantially. A new marketing loan policy initiative was added. A **marketing loan** *is a nonrecourse loan that can be paid off by the farmer at the world market price.* The marketing loan was designed to remove the floor price set by a loan rate that was "too high." As explained here, the marketing loan has not generally worked as intended. It was mandated by the 1985 farm bill for rice and cotton and wheat and feedgrains. The 1990 farm bill mandated the use of a marketing loan for all target price negotiated commodities if the Uruguay Round (1989–1992) of the GATT trade negotiations were unsuccessful within specific time constraints.

As in the case of the regular nonrecourse loan, under the marketing loan, a farmer receives a loan from the county Farm Service Agency (FSA) office at the loan rate at harvest. When the commodity is sold, the loan is repaid but *at the world price.* USDA publishes the world price on a weekly basis. The farmer pockets the difference between the loan rate and the world price, which is referred to as a *marketing loan payment.* The marketing loan payment differs from the deficiency payment in that the marketing loan payment is not subject to the payment limit.[36]

The marketing loan was designed to make U.S. commodities competitive in export markets by allowing the market price to fall past the loan rate P_ℓ to P_m (Figure 10.10). At price P_m, the market has cleared with no surplus. Government costs total $(P_t - P_m)Q_t$ and can be divided into a deficiency payment component and a marketing loan payment.

One of the issues involved in the marketing loan is the determination of the world market price. If the U.S. price is the world market price, the marketing loan is ineffective. This arguably is the case for wheat, feedgrains, and soybeans. In rice the world market price is determined in Asia. In cotton the world market price is determined in Europe. In reality, exactly how the world market price is set by USDA to determine the marketing loan payment is often referred to as the "black box." This USDA secret is considered to be necessary to protect the integrity of the program.

Conservation Reserve Program

By 1985 environmentalists had obtained substantial political clout. Their price for support of the bill was the addition of the conservation reserve program (CRP), which removed highly erosive farmland from production. It was designed to accomplish the dual objectives of controlling erosion while reducing overcapacity. Only USDA-designated highly erodible land is eligible to be bid into the program. The 1985 farm bill goal was to retire 40 to 45 million acres of land for a 10- to 15-year period by 1991. However, only 38 million acres of land were retired under CRP.

---

[36]The 1985 farm bill gave the secretary of agriculture the option to lower the loan rate by up to 20 percent. This reduction in the loan rate, which was designed to make the prices of these commodities more competitive in the world market, results in a larger direct payment. The new, lower loan rate is often referred to as the Findley loan rate, resulting in a Findley payment, which also is not subject to the payment limit.

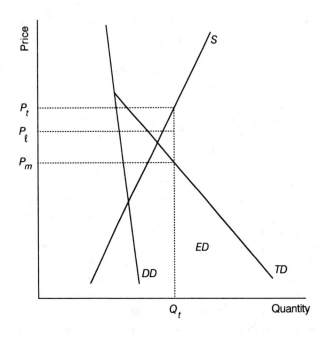

FIGURE 10.10
Marketing loan allows market to clear
despite high loan rate.

To reduce the consequences for rural communities, the total payments per person could not exceed $50,000 annually. In addition, the amount of land retired in any county could not exceed 25 percent of total cropland unless it was determined that there would be no adverse effects on the local economy, in which case 30 percent of the land could be retired. There was a strong tendency for CRP retired lands to be concentrated in certain geographic areas such as the Southern High Plains, the Southeast pine forest areas, and the Southern Iowa/Northern Missouri areas, the northern border of North Dakota and Montana, and parts of the Pacific Northwest.

Budget-Driven Policy: The 1990s

High farm program spending combined with high budget deficits and frustration over an apparent lack of solutions to the farm problem led to the contemporary era of budget-driven policy. Efforts to bring farm program spending under control began with a gradual lowering of the level of income supports in the 1985 farm bill although price supports were reduced even more, resulting in record expenditures.

In the 1990s, however, the budget process took over as the driving force in farm policy. The 1990 farm bill was enacted in two parts:

- The Food, Agriculture, Conservation and Trade Act of 1990, which contained all of the normal provisions of a farm bill.
- The Omnibus Budget Reconciliation Act of 1990, which mandated that deficiency payments not be paid on 15 percent of a farmer's base acreage.

Don't Believe Everything a Professor Tells You

Farm policy has changed sufficiently since the 1930s that the economic impacts are often misstated—even by economists. Professor Milton Friedman won the Nobel Prize in Economics. In a March 23, 1993, article extolling the virtues of the free market, he stated:

> These programs cost taxpayers a multiple of any gains to the intended beneficiaries. In agriculture, for example, most of the money pays for growing, storing, and disposing of food rendered "surplus" by high government fixed support prices and for administering the program. Little money trickles down to the individual farmer. Would consumers regard the lower price that would follow the ending of these subsidies and price supports as a sacrifice?

Professor Friedman obviously has not kept up with the changes that occurred in farm policy during the two decades prior to the publication of this article. Completely ignoring the effects of the target price implemented in 1973 and the substantially reduced loan rate since 1985, Professor Friedman concludes that current policies raise consumer food prices. To achieve this end result, either the loan rate would have to be above the competitive market clearing equilibrium or the government would have to sufficiently control production that the market price rises above the competitive equilibrium. Aside from emphasizing the need to evaluate what you read, there is perhaps a more important lesson: Keep current and voice your opinion only on issues on which you have current expertise.

Source: "Nobelists Rate Clintonomics," *The Wall Street Journal,* March 23, 1993, p. A14.

Farm policy became an issue in every budget debate, and some people began to speculate that the budget process could result in the unraveling of farm policy. Adding fuel to the fire was an appropriations battle over funding for the wool and mohair program and the honey program in the 1993 budget process—the first crack in commodity programs.

In the mid-1990s, the first serious debate over the continuing farm price and income supports since the 1930s occurred. The substance of this debate and how it potentially changes the future policy options is the subject of Chapter 11.

MACROECONOMICS: A FORCE PRECIPITATING POLICY CHANGE

Looking back on 50 years of policy, economics has been an important force that both restrained and led farm policy development. The reaction of farmers to higher levels of price and income support has consistently been increased production. The propensity of farmers to produce eventually outstripped both the available demand and the willingness of Congress to spend. Federal budget deficits increasingly became a restraining force in farm policy development.

Economics also led policy change as export demand provided an opportunity for policymakers to move from price to income support in the early 1970s. This policy change served to further stimulate export demand, thereby making a reversal of policy directions more difficult. That is, as exports grew, the magnitude of adjustment required if agriculture was to revert to production controls became increasingly large.

The big changes that occurred in agriculture were not so much those of farm economics, but rather those of macroeconomics. Changes in the macroeconomy persistently pushed agriculture in the direction of greater market orientation.

ADDITIONAL READINGS

1. An outstanding historical chronology of the development of domestic farm policies and programs is contained in Douglas E. Bowers, Wayne D. Rasmussen, and Gladys L. Baker, *History of Agricultural Price Support and Adjustment Programs,* Agric. Inf. Bull. 485 (Washington, D.C.: ERS/USDA, December 1984). This publication is one of a series issued by the Economic Research Service before the debate began on the 1985 farm bill. In the series are separate publications for each of the major farm program commodities. Each contains a wealth of policy information.

2. After each farm bill is enacted, the Economic Research Service compiles a summary of its provisions. The 1990 farm bill edition was *Provisions of the Food, Agriculture, Conservation, and Trade Act of 1990,* Susan L. Pollack and Lori Lynch, eds., Agric. Inf. Bull. 624 (Washington, D.C.: ERS/USDA, June 1991).

3. An excellent and readable discussion of supply control impacts is contained in the discussion paper by Thomas W. Hertel, *Ten Truths About Supply Control* (Washington, D.C.: Resources for the Future, July 1989). Any graduate student is well advised to know the underlying economic theory behind each truth.

4. For the more economically literate reader, an excellent reference is Bruce Gardner's *The Economics of Agricultural Policies* (New York: Macmillan Publishing Company, 1987).

5. Before each farm bill is enacted, extension economists put together a set of leaflets outlining the options and consequences of policy changes for each title of the farm bill. The set prepared for the 1995 farm bill, edited by Ronald D. Knutson, was *1995 Farm Bill Policy Options and Consequences* (College Station, TX: Texas A&M University System,) October 1994. The provisions of the 1996 farm bill may be found in Frederick J. Nelson and Lyle P. Schertz, eds., *Provisions of the Federal Agriculture and Improvement and Reform Act of 1996,* Agriculture Information Bulletin No. 729, (Washington, D.C.: Economic Research Service/USDA), September 1996.

11

CONTEMPORARY OPTIONS FOR DOMESTIC FARM POLICY

Free at last, free at last, freedom to farm.
Pat Roberts

Chapter 10 explained the history of farm policy from the Homestead Act through the 1990 farm bill. This chapter explains the directional change brought on by the 1996 farm bill and then develops the likely future policy options and their consequences.

We noted previously that policy changes generally occur in relatively small increments. The 1996 farm bill is a contradiction to this principle. It represents a directional change in policy that previously had not occurred since 1973 when the target price program was created. The 1996 farm bill did away with the target price and much more. It truly represents a watershed change in farm policy and likely charts a new policy course, which includes a new set of policy options, that will be followed for years to come.

Since 1973 and, in some instances, back even further, domestic farm policy has included the following major provisions:

- A target price has existed for wheat, feedgrains, cotton, and rice with deficiency payments being made to eligible farmers of the difference between the target price and the market price.
- Eligibility has been determined by a history of consistent production resulting in acreage base. Base drove production decisions because deficiency payments were made on land having base, which had to be planted unless required to be a set-aside.
- Acreage reduction/set-aside requirements were utilized to tailor production to market needs (control CCC stocks) and to manage government costs.
- The nonrecourse loan was available to producers as an alternative to marketing at harvest and as a market of last resort if the commodity's price failed to rise above the loan rate.

- Provision has been made for government storage of commodities taken over by the CCC and/or assistance in on-farm storage through programs such as the farmer-owned reserve (FOR).

The magnitude of change in policy brought on by the 1996 farm bill is indicated by the reality that out of these five key components, only one—the nonrecourse loan—remains in place.

KEY 1996 FARM BILL POLICY CHANGES

The 1996 farm bill made four key policy changes:

- Eliminated the target price program.
- Decoupled transition payments with virtual flexibility.
- Eliminated the set-aside program.
- Eliminated the FOR.

These policy changes individually have important economic impacts on farm management decisions and on the market for farm products.

Target Price Elimination

The target price had the effect of increasing supplies, which, in order for the market to clear, results in a lower market price (Figure 11.1). In recent years, the non-recourse loan rate has been set sufficiently low so as not to interfere with exports and not lead to substantial CCC stock accumulation. This results when the loan rate is set below the market price.

Eliminating the target price has the effect of raising the market price to the free-market equilibrium P_e. However, because P_e is below the target price, the quantity supplied falls to Q_e. With the higher market price, the quantity exported declines more than domestic demand. This is consistent with the notion that deficiency payments have been an implicit export subsidy and a domestic consumer subsidy.

Decoupled Transition Payments with Virtual Flexibility   *gradually wean farmers from gov. sub*

Decoupling *involves the separation of income payments from market prices and from production decisions.* The 1996 farm bill decouples by providing farmers with **fixed transition** payments over the period 1996–2002 with no ties to either production (acreage or yield) or prices. The magnitude of the annual transition payments farmers will receive is known for the full period from 1996–2002 (Figure 11.2). These payments are divided among farmers on the basis of historical program base and yield but are

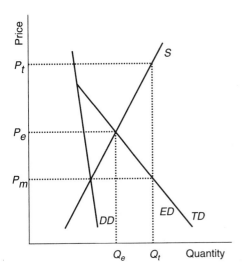

FIGURE 11.1
Decoupling payments from production and market prices results in lower production and a higher market price.

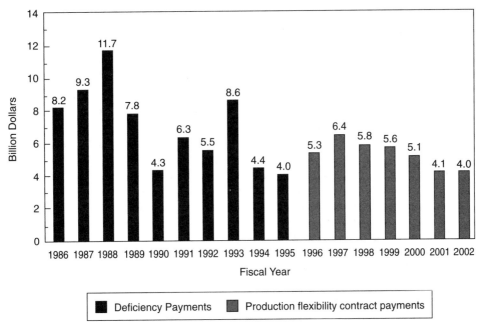

FIGURE 11.2   U.S. Farm Program Deficiency and Contract Payments by Fiscal Year, 1986–2002.

Source: C. Edwin Young and Paul C. Westcott, *The 1996 U.S. Farm Act Increases Market Orientation,* Agriculture Information Bulletin No. 726 (Washington, D.C.: ERS/USDA, August 1996).

The Politics of Freedom to Farm

Historically, farm bills were bipartisan and were passed on the time-honored political principle—"you scratch mine and I'll scratch yours." Not so with the 1996 farm bill. Four factors came together, perhaps accidentally, to result in the greatest change in policy since 1933: (1) The Republicans gained control of the Congress. (2) Budget issues were front and center. (3) Prices of program commodities were at near-record levels. (4) Major players were from the same state (Kansas), and had a long history of working together across party lines on farm bills: majority leader, Senator Bob Dole; chairman of the House Committee on Agriculture, Pat Roberts; and secretary of agriculture, Dan Glickman.

The majority leader of the House, Mr. Armey, had long been an opponent of commodity programs and Speaker Gingrich made it clear that the 1996 farm bill was not business as usual. Chairman Roberts cut a deal immediately after the election with the House leadership (including saving the House agriculture committee from elimination) that agriculture would contribute to budget cuts. Senate Agriculture Chair Lugar fired the first shot by asking if it was time to eliminate commodity programs.

A record number of hearings were held across the country by Chairman Roberts. The committee heard from agribusiness that old archaic programs were stifling our ability to compete in foreign markets. Farm organizations called for more flexibility and less government control. Weather problems produced a unique situation—high prices, low yields, and no deficiency payments under the old farm bill.

Out of this, Chairman Roberts produced what many called a new bold revolutionary approach—declining fixed payments that would save approximately $10 billion over a seven-year period from the budget baseline. Senator Lugar immediately supported Freedom to Farm, which in reality was a 1990s twist on the Boschwitz–Boren decoupling proposal of the early 1980s.

Several southern commodity organizations, in particular the National Cotton Council, immediately opposed the Roberts proposal. Republican Congressmen Combest and Emerson, from cotton country, voted against Freedom to Farm on the first roll call. Secretary Glickman, though supportive of more flexibility, voiced concern about a safety net that was decoupled from commodity prices foreseeing future years of lower payments with lower prices.

Over in the Senate, with then Majority Leader Dole and Chairman Lugar campaigning for the White House against President Clinton, the farm bill became part of presidential politics. Republican Senators and Freedom to Farm advocates Dole, Lugar, and Cochran were opposed by Senate Minority Leader Daschle (representing the Farmer's Union philosophy) and Senator Leahy (representing Vermont dairy farmers).

In the end, farmers and agribusiness agreed. Farm organization changed positions. Conservation enthusiasts signed on because crop rotational practices were enhanced and the CRP was preserved. Urban support was preserved with food stamps remaining part of the farm bill. Eventually, bipartisan support was forthcoming and the president signed it. The right economic circumstances (high prices) and the right political circumstances (budget cuts and agricultural policy leadership) produced a new era in agricultural policy.

independent of what is planted. Thus, each individual participating farmer knows the approximate amount of fixed payments that will be received each year through 2002. As land comes out of CRP with base it goes into a Freedom to Farm pool, and individual bushel payments decline because the total amount per commodity is fixed. Note also that the payments indicated in Figure 11.2 for 1996-2002 are authorized by law, but each year Congress will decide whether to appropriate the funds.

Virtual flexibility is provided in commodities produced regardless of a farmer's acreage base. This means, for example, *that a farmer with wheat base can plant virtually any other crop or not produce at all and still get the fixed payments.* However, the land cannot be commercially developed for nonagricultural uses. The only restriction on planting is that fruits and vegetables cannot be planted unless the farmer has a history of planting them. This restriction was a concession to fruit and vegetable producers who felt threatened by the potential that flexibility could mean increased fruit and vegetable production and lower market prices.

Flexibility was not an entirely new farm policy. Under the 1990 farm bill, farmers could flex or change cropping patterns on up to 25 percent of their base acres to a more restricted set of program crops (15 percent required). Deficiency payments were not received on these flex acres. In contrast, under the 1996 farm bill, virtual full flexibility exists with transition payments on all base acres whether flexed or not.

Since farmers' profit-maximizing production decisions are made on the basis of marginal costs and marginal revenue, the fixed transition payments do not enter into these production decisions. That is, this fixed source of revenue is treated just like a fixed cost in that neither affects cropping decisions. Accordingly, farmers produce based on the market price P_e, quantity Q_e in Figure 11.1.

It would be wrong to assume from this discussion that transition payments are devoid of economic impacts. The following economic effects were experienced in the first two years of the program:

■ Farmers switched cropping patterns in response to high feedgrain prices relative to other crops and because of weather adversities. The flexibility to adjust would not have existed under previous programs.

■ Rental rates increased as landlords demanded a share of the benefits of the fixed transition payments.

■ Land prices increased as the fixed transition payment benefits began to be capitalized into the value of this fixed asset.

■ Some landlords decided to take the transition payment for the 1997 crop (after the 1996 crop lease had terminated) and became farm managers hiring custom services. Issues of potential future structural change associated with the 1996 farm bill and other economic forces are discussed in Chapter 12.

Other important consequences of decoupling include the following:

- Decoupling has the important longer run impact of strengthening the U.S. bargaining position in the next round of trade negotiations, which is expected to begin in 1999. Other countries can no longer charge that U.S. target prices implicitly subsidize exports. Transition payments affect neither prices nor production and, therefore, cannot be export subsidies. Therefore, this change in policy shifts the balance of negotiating position under GATT decidedly in favor of the United States. The European Union is left attempting to defend its explicit export subsidies (restitutions) and its implicit direct payments. Even Canada, New Zealand, and Australia are caught defending their state trading status without the ability to criticize U.S. deficiency payment subsidies. They can, however, criticize the U.S. export enhancement program (EEP), although its expenditure levels were constrained by the Uruguay Round agreement.

- Payments are made regardless of the level of market prices and producer returns. In 1996, this meant that large payments were made to farmers at a time when prices were at near-record levels. For farmers who did not get a crop, these payments were a substitute for crop insurance. For those who did get a crop, transition payments provided the basis for news stories that were critical of farm programs because they gave subsidies to farmers who, it was argued, did not really need them.

- Production will be more responsive to market forces. Under the target price combined with the acreage base, producer response to demand increases was limited unless the market price rose above the target price. Even then, switches in cropping patterns resulted in the loss of base. With transition payments, producers are considerably more sensitive to price changes because the government is no longer there with deficiency payments that make up the difference between the target price and the market price. Likewise, flexibility means that base is no longer a constraining factor on production decisions. If, as has been predicted, feedgrain and soybean demand continues to be buoyant due to rising incomes in developing countries, agribusiness interests associated with these industries will be prime beneficiaries of the changes in policy embodied in the 1996 farm bill. This includes the feedgrain-related poultry, pork, and fed beef industries.

- The instability of agriculture is likely to be magnified as farmers switch production to the crop, livestock, or poultry products that they think will be the most profitable. This expected increase in instability results from the inelastic nature of the supply and demand for farm products interacting with both production and export demand shifts. In addition, the reduction in government support for holding stocks, either in CCC or FOR, eliminates this supply cushion.

- Risk is shifted from the government to the farmer, agricultural banker, agribusiness, and rural community in that if prices go down, payments remain fixed and on a declining scale. Under the old program if price went down, deficiency payment went up. Farmers will need to substitute other risk management tools in place of government programs.

There is a dissenting perspective on the instability issue. It results from the Chicago School perspective that government programs are destabilizing. A reduction in government should then lead to greater stability. In addition, it is argued that decoupling combined with the elimination of set-asides will lead to farmers making adjustments more rapidly and in an incremental manner that is consistent with market signals.

FUTURE POLICY OPTIONS AND CONSEQUENCES

There are three general options for the future of domestic farm policy:

- Return to the policies of the past. Past programs served an important function of facilitating adjustment when the farm problem was one of chronic excess capacity. They did not serve nearly as well when the era of periodic surpluses and deficits was reached. Under changed conditions, past programs were costly, rigid, and did not take full advantage of market opportunities. Yet, options from the past to deal with particular problem situations could surface once again. If this should occur, the consequences of past lessons from employing these policies should be taken into consideration.
- The policies of the 1996 farm bill could be continued. When the 2002 farm bill is debated, the transition payment subsidy budget will be $4 billion. The issue will be how this $4 billion can be most effectively utilized in solving the problems that exist at that time. Those problems could be quite different than those being experienced today.
- Farm programs could be terminated. This however would require repeal of the 1949 Agricultural Act, which remains the permanent legislation.

The remainder of this chapter sets forth additional policy options and their consequences for the future. In some cases, programs utilized in the past are given new life in a contemporary context. The emphasis, however, is on options that are currently being researched and even pilot-tested.

Dairy: Hanging by Itself

Even before federal farm programs, dairy had its own policy. The dairy program that has existed since the 1920s is milk marketing orders, which price milk on the basis of the use processors make of it. Milk used for fluid consumption commands a higher price (Class I) because its demand is more inelastic. Milk used for manufactured products (butter, nonfat dry milk, and cheese), with a more elastic demand, receives a lower class price. Orders set the class prices and compute a blend or average price that farmers receive for milk. State marketing orders were the first to be developed in the 1920s, followed by federal orders in the 1930s.

With the 1949 farm bill, dairy farmers obtained a price support program whereby the government bought butter, nonfat dry milk, and cheese whenever offered to them at the support price. Resulting surpluses from setting the support level too high periodically resulted in huge surpluses.

For a time in the 1970s, dairy farmers were recognized as the most politically potent agricultural lobby. But power often breeds greed and greed breeds fragmentation. Regional differences developed over the appropriate policy direction. In 1996, dairy farmers could not agree on how the revenues from milk ought to be shared geographically. Specifically, dairy farmers in the Upper Midwest wanted to share in the higher prices enjoyed by farmers located further south. Southern farmers had no interest in sharing. In the absence of a clear policy signal, the dairy program was ripe for the picking. It lost its support program, in favor of a recourse loan, and received a mandate for reform of the milk marketing order system.

Dairy had never really been part of the farm coalition. It hung out there on its own. This worked to the dairy farmers' advantage when they were unified and could ride the coattails of the farm bill. But when dairy interests were not unified, their political problems multiplied. They lost the price floor provided by the support program and could, in the future, lose the orders. Previously, the honey, wool, and mohair industries were hung by themselves.

Insurance Policy[1]

The history of crop insurance could have been covered in Chapter 10. It is not a particularly glamorous history in that the insurance policy has been strewn with more failures than successes. Despite this history, policy choices involving insurance could become a centerpiece of farm policy debates early in the twenty-first century.

---

[1]This section draws heavily on the work of G. A. Barnaby, B. L. Flinchbaugh, and Roy Black, "Natural Disaster Protection Policy," *1995 Farm Bill Policy Options and Consequences* (College Station, Tex.: Texas A&M University System, October 1994).

Under **private insurance**, farmers pay premiums which over time are expected to cover any indemnity (damage) payments, plus a profit commensurate with the risk assumed by the commercial provider. When private insurance premiums are established on the basis of the expected loss from a wider range of weather adversities that are typical of agriculture, the resulting actuarially sound premiums become so expensive in risky production areas that farmers cannot afford the insurance. If an attempt is made to spread the risk across broader production areas, farmers in low-risk areas choose not to buy insurance.

Private insurance policies have been successfully written as crop production protection against hail damage. Success in writing private hail insurance policy results from its relatively random occurrence and the fact that it normally does not cover a wide area. As a result, losses are relatively independent among those that are insured. In addition, loss due to hail is not a function of management—a farmer cannot precipitate an indemnity payment. Hail insurance has the advantage of being simple to write and covers spot losses. That is, growers with only a small part of a single field would receive indemnity payments from their private hail insurance policy.

Any farmer will rapidly explain that there are many types of losses due to weather other than hail. The potential liability on this broader range of losses is sufficiently large that it is questionable whether private insurance providers could profitably write policies. In the early part of the twentieth century, several private companies offered policies covering a wide range of natural disasters. However, those companies dropped this coverage within a very short time because widespread drought wiped out their reserves.

Private insurance companies have not been successful in their attempts to insure crops for perils other than hail because of the potential catastrophic losses, and losses are not independent among the insured. One alternative would be for government to be the re-insurer of last resort. The government would re-insure private multiple peril crop insurance contracts for that part of the risk that could not be sold in the world re-insurance market.

There is also the possibility that the catastrophic crop insurance risk could be reduced by using a new futures contract. The Chicago Board of Trade currently trades a catastrophic property–casualty insurance contract. It may be possible to apply those same principles to a catastrophic crop insurance contract.

Disaster protection made direct payments to farmers who lost their crop or were prevented from planting during the 1973–1981 period. Generally, disaster payments were made when yields fell below two-thirds of the normal level and were designed to cover producers' variable costs. Because payments were concentrated in the high-risk states of Texas, Oklahoma, North Dakota, and South Dakota, concern arose that the program fostered production in regions where land was the most fragile and production the least efficient. This, in turn, fostered more disaster payments, leading one congressperson to quip that the disaster program had become a disaster.

Subsequent to its replacement by crop insurance in 1982, disaster payments continued to be used on a periodic basis. For example, in the drought of 1988 and the flood of 1993, disaster payments were made to farmers. This, of course, undermined other public and private crop insurance programs. Why would farmers buy crop insurance if there was a reasonably good chance that the Congress would provide disaster insurance? They would not and did not.

Recognizing this problem, in the 1994 budget debate, Congress made providing disaster payments to farmers considerably more difficult. It provided that any farm disaster program payments would have to come from another farm subsidy program. However, in the great floods of 1996/97, the West and Upper Midwest formed a coalition to reinstate disaster payments as part of the balanced budget initiative. The lesson is that if enough farmers are affected, it is difficult for the Congress to resist disaster relief.

Federal multiple peril crop insurance (MPCI) was proposed in the Congress as early as 1922. Later, the Agricultural Adjustment Act of 1938 established the Federal Crop Insurance Corporation (FCIC) to write multiple peril policies. The program was expanded until 1947, when Congress reduced it to an "experimental program."

After the disaster protection hiatus, the Crop Insurance Act of 1980 was an attempt to make crop insurance the federal government's means for dealing with natural disasters at the individual farm level. Objectives included making MPCI more broadly available and giving farmers the opportunity to make yield risk reduction a cost of doing business. This action was expected to reduce the political and economic incentives to provide "free" disaster assistance programs to farmers.

As indicated previously, MPCI rates often are too high for high-yield producers but too low for low-yield producers. Because participation has been low among producers with high yields, the FCIC began using actual production history (APH) in 1985, which bases yield guarantees on each individual farm's past production history.

The standard MPCI policy involves a combination of minimal catastrophic coverage (CAT) with an option to "buy up" the level of coverage. CAT coverage at 50 percent of the average yield for an individual farm and 60 percent of the top price election was designed to reduce further political demands for ad hoc disaster payments. Farmers can "buy up" additional MPIC insurance from the CAT level to provide guaranteed yields from 55 to 75 percent (in 5 percent increments) of the average yield for an individual farm unit. It is common for large farmers to have more than 10 units and, as a result, more than 10 APHs. Since each unit is insured separately, a yield that is below the guarantee on an individual unit will trigger an indemnity payment. Farmers' indemnity payments are based on level of yield and the price at which the farmer decides to insure. Of course, the higher the percent yield and price insured, the higher the premium. Each lost bushel or pound is then paid based on a forecasted price assuming farmers select the highest price election.

Although the standard MPCI contract is often discussed in terms of bushels guaranteed, in fact, it guarantees bushels only if the market price equals the price election. For example, for each 6,500 bushels of guaranteed corn production, the grower will receive enough indemnity dollars to purchase 6,500 bushels only if the market price equals $2.40, which was the 1994 FCIC forecasted corn price. If market prices were to increase to $3, this same grower would only receive enough indemnity dollars to replace 5,200 bushels of corn per 6,500 bushels guaranteed under the MPCI contract (Figure 11.3). However, growers who forward contracted those 6,500 bushels must either produce the 6,500 bushels or buy those 6,500 bushels at current market value to fill the contract. *Therefore, the remaining 1,300 bushels must be purchased from growers' equity or from borrowed funds.*

Even the most ardent supporters of crop insurance will not argue that the program has worked as expected. The initial goal of the 1980 act for crop insurance was to have 50 percent participation by 1989. Recent voluntary national participation rates are estimated in the 31 to 45 percent range. These figures are low if you count CAT participation. The greatest participation rates are in the Great Plains and the western corn belt. In the Northern Great Plains, current participation exceeds 80 percent. However, GAO estimates indicate that participation in excess of 50 percent nationwide is needed before the pressure on Congress to provide disaster payments will subside. For the period 1980 to 1990, the cost of underwriting losses and subsidies on crop insurance averaged $396 million per year.

Replacement coverage *guarantees all 6,500 bushels at their market value* (Figure 11.4). For example, during the 1993 floods corn prices increased. Those growers who forward priced their corn had to replace any lost production at current market prices. Therefore, replacement coverage is likely to be a niche market that appeals to farmers who use forward pricing tools to cover their normal level of production. The standard MPCI contract offered in 1993 replaced each lost bushel at $2.30. Replacement coverage MPCI paid $2.60 per lost bushel, which was the

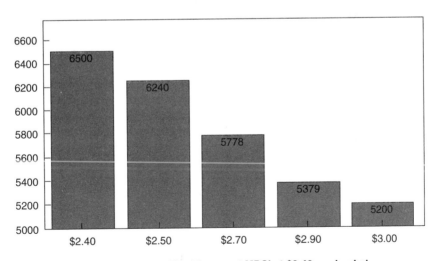

FIGURE 11.3 **Bushels covered by 65 percent MPCI at $2.40 per bushel with a corn price increase.**

Source: G. A. Barnaby *et al.*, "Natural Disaster Protection Policy," in *1995 Farm Bill Policy Options and Consequences* (College Station, Tex.: Texas A & M University System), p. 93.

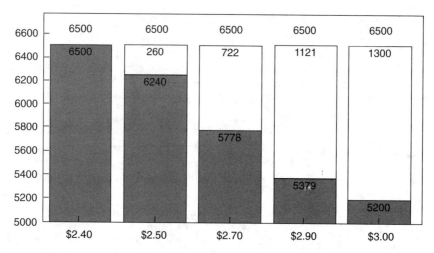

FIGURE 11.4   Replacement MPCI provides full coverage in spite of a corn price increase.

Source: G. A. Barnaby *et al.,* "Natural Disaster Protection Policy," in *1995 Farm Bill Policy Options and Consequences,* (College Station Tex.: Texas A & M University System), p. 93.

cost of replacing bushels at harvest time when farmers were required to make delivery on their forward-contracted bushels.

Private replacement coverage was recently approved by FCIC for catastrophic re-insurance. This policy change should encourage more private companies to provide this coverage and in more states.

Area-wide crop insurance based on county yields only has been pilot-tested by the Risk Management Agency (RMA) of USDA under the name **Group Risk Plan** (GRP). GRP is based on county losses. The group is the county, and everyone in that group will pay the same premium rates for a selected coverage level.

Farmers are able to select dollars of protection (liability) and the percent county trigger. The trigger ranges from 65 up to 90 percent of the expected county yield. Maximum dollars of liability that farmers may select are set by county and crop. RMA uses the established price multiplied by the expected county yield multiplied by 150 percent in order to set the maximum liability. Higher coverages are permitted under GRP than under MPCI because an individual farmer through management cannot trigger an indemnity payment.

Farmers who are considering the purchase of a GRP contract should be aware of the following items:

- The premium rates and the indemnity payments are based only on the county. It is possible for a farmer to have a complete crop failure and receive no crop insurance payment under GRP. It is also possible for a farmer to produce a normal yield and receive a crop insurance payment. GRP provides no individual farm level yield guarantees.

- If the county has a 50 percent yield loss below the expected county yield, then every GRP insured farmer will be paid from 23.1 up to 44.4 percent of his or her liability depending on the percent coverage level selected by the farmer.

- GRP provides the best protection against widespread drought and county-wide freezes. GRP will provide almost no coverage for any peril that tends to be site specific. Therefore, hail, flood, excessive spot rain, quality losses, local winds, or tornados could generate large amounts of damage on an individual farm, but the farmer probably would receive a smaller than expected indemnity payment and, in some situations, no payment.

- The final indemnity payment occurs four to six months after harvest, while the premium is due at harvest.

Farmers who expect their yields to track with county yields are the ones who would probably purchase GRP. Farmers with higher average yields than the county average yield are not penalized because losses are measured in percentages. For example, if the expected county yield is 30 bushels and the actual bushels produced are 15, then the county suffered a 50 percent loss. If the high yielding farmer has an expected yield of 40 bushels but only produced 20 bushels, he or she has also suffered a 50 percent loss, and GRP would cover that loss.

The GRP approach probably would reduce the cost of providing disaster protection to farmers. For that reason, many Washington policymakers have suggested the use of GRP. However, some farmers have opposed the concept because they fear a local loss that does not trigger a county payment.

It may be possible to reduce the area from a county down to a township or a zone. This would reduce the chance of a farmer suffering a loss when the GRP is not triggered. However, the cost and presumably the premium would increase.

Revenue insurance *guarantees farmers a certain level of gross revenue from a crop.* That is, while crop insurance provides yield protection alone, revenue insurance covers the multiple of price and yield (price $\times$ yield). The 1996 farm bill authorized pilot-testing of a revenue insurance option referred to as **crop revenue coverage** (CRC). In 1997, this coverage was being piloted in a limited number of states on corn, soybeans, winter wheat, spring wheat, grain sorghum, and cotton.

Under CRC, agricultural producers are insuring a given percentage of revenue that is based on the higher of a planting-time or harvest-time futures price and the producer's APH yield. *The producer is paid an indemnity if actual revenue (measured by the product of a harvest-time futures price and the producer's actual yield) falls below revenue guaranteed.* The futures prices are taken from prices on the futures contract nearest to the harvest period for the crop. For example, CRC corn prices are based on the December corn futures contract on the Chicago Board of Trade; the February average daily settlement price on this contract provides the planting-time futures price and the November average daily settlement price represents the harvest-time futures price.

If the grower were to forward contract or use other forward pricing tools on 10,000 bushels, he or she must either produce 10,000 bushels or receive enough indemnity dollars to replace the inventory at current market value. *CRC provides this full replacement protection.* If the yield is guaranteed based on a season average price, full income protection is not guaranteed. This pricing method differs from the fixed price election mechanism of traditional MPCI. Consequently, CRC can be viewed as encompassing both yield and price variability.

USDA is also piloting an alternative to CRC referred to as **income protection (IP).** *Under IP, less protection is provided than CRC in that income is insured as a percent of the farm's proven yield at a fixed price based on the harvest futures market contract at planting time.* If the price rises, the level of protection declines. This form of protection does not provide full protection for the farmer who hedges on the futures market during a period of rising prices. However, IP provides protection in the event of lower prices.

Stocks Policy

U.S. stocks policy is likely to continue to be a contentious issue. The tendency in recent years has been to materially deemphasize government-held stocks. The 1996 farm bill took the following steps regarding stocks policy:

- The food security commodity reserve was expanded from two million metric tons of wheat to four million tons of food grains and coarse (feed) grains. The secretary of agriculture makes the decision when to release stocks from this emergency stock reserve.
- The farmer-owned grain reserve (FOR) was terminated. As discussed in the previous chapter, FOR did not work as intended. That is, FOR had a price destabilizing effect when the release price triggers were reached and tended to accumulate excess stocks under surplus conditions.
- The secretary was directed to minimize CCC stock accumulation. The farm bill is not specific in how this is to be accomplished. However, the secretary has several tools that have been used in the past to lower CCC stocks, including reducing the loan rate to curb forfeitures, offering payment-in-kind certificates in lieu of transition payments, using acquired stocks as an export subsidy (2 for 1 sale) or providing stocks at a reduced price for the production of ethanol. Any or all of these options would increase the movement of commodities in the event of temporary surpluses. Except under the ethanol option, most of this increased movement would be into export markets. If protracted, multiyear surpluses develop, the secretary would likely encounter problems expanding export demand in the face of objections and countervailing subsidies by foreign competitors.
- Land retirement programs (reserves) were reduced. Set-aside/average reduction programs were eliminated. The conservation reserve program (CRP) will likely be reduced in scope, although the magnitude of reduction is uncertain. These two programs have typically held 10 to 15 percent of the cropland out of production and, arguably, in reserve for future production.

Future U.S. stocks policies will be influenced by the balancing of the following considerations:

- Food security has become the primary reason for public stock holding. Food security is generally viewed from an international perspective, as it was in Chapter 5. Internationally, the emphasis is on ensuring that there are sufficient reserve supplies to fill production shortfalls where they occur. From a U.S. perspective, problems of international security tend to be viewed both from a humanitarian and political perspective. The humanitarian perspective involves having sufficient reserve supplies to satisfy emergency food needs of developing countries who might not have the financial resources to compete in world grain markets. The political perspective involves contributing to world peace and the stability of governments—particularly toward those that are strategically important to and friendly toward the United States.

- Trade security involves the ability of the United States to satisfy commercial export demand without jeopardizing supplies available for domestic consumption for either food or feed purposes. Although interrelated with food security, trade security, arguably, has more impact on U.S. stock policy decisions. The priority given to maintaining U.S. exports drives the policy objective of having sufficient supplies to satisfy export needs. Equally important, or perhaps more so, however, is the concern that there be sufficient stocks to fill domestic consumption for food and feed needs. For example, sufficiently tight supplies developed in 1996 to raise a specter of questions regarding the availability of sufficient supplies of corn to satisfy the needs of the U.S. livestock and poultry industries. Pending the harvest of the 1996 crop, talk of potential export embargoes once again surfaced. Questions arose as to whether the Congress had correctly assessed government's role in holding stocks.

- Other countries' stock policies influence U.S. stock policies. Both import- and export-dependent countries need to reassess the role of publicly held stocks in a world where the food supply–demand balance appears to have tightened and the United States is no longer inclined to be the stockpile for other countries.

- Freer trade reduces the need for publicly held stocks because total world production is increased when principles of comparative advantage become the allocators of production and consumption. Year-to-year variation in commodity production in virtually every country is higher than for the world as a whole.[2] Countries that rely on imports and exports to balance their supplies are less dependent on stocks. Freer trade is also a world market price stabilizer.

---

[2]Luther Tweeten, "World Trade and Food Security for the Twenty-First Century," paper presented at Bilkent University—FAO Conference, (Colombus: Ohio State University, September 13, 1996).

Tyers and Anderson estimate that the coefficient of variation in the world price of wheat would fall to 30 percent from 45 percent with free trade.[3]

Future U.S. stock policy appears to involve a choice in the relative balance of the following options:

- Private stock holdings motivated by price incentives set by supply and demand in markets have become primary factors in determining carryover supplies. The degree to which reliance is to be placed on private stocks depends on the level of food security, trade security, and price stability that is desired. A desire for greater food and trade security, combined with more price stability, implies less reliance on private stock holdings—and more reliance on public stocks.
- Emergency reserves have become the primary U.S. public policy insuring food and trade security. Reserves also aid in stabilizing prices. Whether the four million metric ton food and coarse grain reserve provided for in the 1996 farm bill is either necessary or sufficient will be debated in the years ahead.
- Multilateral reserves could become more important to the U.S. and world food policy agenda. This likely would involve agreements among developed importing and exporting countries to hold a certain level of stocks. Such an agreement could be made under the auspices of the United Nations, GATT/WTO, or a trading block such as APEC. While not a domestic farm policy issue, agreements among countries to hold reserves could significantly reduce U.S. requirements for emergency reserve stocks. In the process, more reliance could be placed on private holdings of U.S. stocks.

Eliminate Subsidies

Early in the debate over the Freedom to Farm provisions of the 1996 farm bill, the use of the term *transition payments* meant the elimination of farm programs in year 2003. While this interpretation was downplayed by farmer interests, eliminating subsidies could resurface as an option in the 2002 farm bill debate. The political viability of this option depends on the economic status of agriculture at the time, the progress of the new round of GATT negotiations, the evolution of options such as revenue insurance, and the perceived impacts of eliminating subsidies.

Early in the 1996 farm bill debate, there was considerable discussion and study of the consequences of eliminating subsidies. One of these studies[4] led to the following conclusions:

---

[3]Rodney Tyers and Kym Anderson, "Liberalizing OECD Agricultural Policies in the Uruguay Round: Effects on Trade and Welfare," *J. Agric. Econ.* (May 1988), pp. 197–215.

[4]Edward G. Smith *et al.*, *Farm Level Impacts of Continuing the 1990 Farm Bill, Increasing Planting Flexibility, Adopting Freedom to Farm or Eliminating Farm Programs*, AFPC Working Paper 95-16 (College Station, Tex.: Texas A&M University, October 1995); Ronald D. Knutson, statement before the Committee on Agriculture, U.S. House of Representatives, AFPC Briefing Series 95-3 (College Station, Tex.: Texas A&M University, April 25, 1995).

Farm Organizations Without Farm Programs

Most commercial farmers belong to at least two farm organizations—a commodity group for their largest sales and a general farm organization. The major task of these organizations has been to lobby for farm subsidies. It is interesting to speculate what these organizations will do without farm programs.

This issue is particularly relevant to the commodity groups that have devoted almost all of their resources to domestic farm policy. What do they lobby for without programs? Even with decoupled programs, the justification for their existence becomes less relevant. One can visualize farm organization presidents making $150,000 to $400,000 (based on a recent survey) wringing their hands over the future of their organization and their jobs! Even board members, who gain substantially in terms of prestige and influence from their position, have a right to worry about their organization's future.

General farm organizations may be in a somewhat better position to justify their existence in the absence of farm programs. As explained in Chapter 4, their agenda has typically been broader than that of the commodity groups. That is, they have been concerned about the broader issues of environmental policy, trade policy, welfare reform, nutrition programs, and macroeconomic policy. Within these organizations, however, one can visualize staff securing their positions by shifting out of domestic farm policy into these other broader issue areas.

Could commodity groups justify their existence by broadening their scope? The challenge is more difficult for them. Trade policy and environmental policy tend not to be established on a commodity basis. Welfare reform, nutrition policy, and macroeconomic policy are foreign to most commodity groups.

The struggle for farm organization survival could contribute to the decline of farm programs. That is, as the staffs turn their attention to justifying their existence in other areas, domestic farm policy loses its glamour and receives little attention. Distraction frequently causes empires to fall, as does irrelevance.

- For program crops, deficiency payments accounted for 13 to 28 percent of farm level receipts. These would be eliminated.
- For program crops, farm prices would decline by 5 to 15 percent with the largest declines being in wheat, cotton, and rice. The source of this price decline was not the elimination of the target price, which by itself would be expected to raise the market price. Rather, the lower market price resulted from putting land back into production that was retired by the set-aside and

CRP programs. The resulting increase in the quantity produced more than offset the effect of eliminating the target price.

- With increased production, exports and domestic use increased.
- In a U.S. representative farm analysis, ideal net worth was reduced on all wheat and cotton farms, half of the feedgrain farms, and the majority of the rice farms.

It would be incorrect to attribute these same effects to elimination of subsidies in 2003. Substantial adjustments in the direction of freer market policies will have occurred as a result of the provisions of the 1996 farm bill. For example, decoupling and flexibility will result in shifts in acreage in response to market prices. However, with more land being put into production by the elimination of CRP, market prices would decline—although not as severely as indicated by the 1995 study. In addition, the elimination of transition payments would reduce farmers' real net worth as income levels declined. In other words, the direction of effects would be the same although the magnitude would be reduced.

WHAT TO DO WITH $4 BILLION IN 2002

Assuming no change in the budget process, there will be $4 billion on the table for farm subsidies in 2002. That is, according to current procedures, the baseline for next year's budget is the current year's budget. Therefore, the $4 billion in transition payments in 2002 will be the subsidy baseline for the 2002 farm bill.

While $4 billion is a lot of money, questions will arise over how these monies can be most effectively spent. Options such as revenue insurance subsidies could be weighed against food security stocks or payments for the achievement of specific environmental objectives. The latter option is discussed in Chapter 13.

ADDITIONAL READINGS

1. The Agricultural and Food Policy Center (AFPC) at Texas A&M conducts analyses of the impacts of government policy proposals and/or implementation procedures on farmers, agribusinesses, taxpayers, and consumers. Its primary constituency is the U.S. Congress, particularly the Agriculture Committees. For those with Internet access, a comphrehensive database is available online. Direct your browser to **http://afpc1.tamu.edu** and click on **publications.** Accessing the **policy working papers** section, under **1995** and **1996** can be found electronic versions of AFPC reports to Congress concerning alternative policy proposals and the final policy proposals which led to the 1996 farm bill. Also, particular attention should be paid to the **other publications** section, which contains online versions of *Policy Tools for U.S. Agriculture* (an excellent summary and comparison of policy tools, both his-

torical and current) and *1995 Farm Bill: Policy Options and Consequences* (a collection of 32 policy fact sheets).

2. A description of the provisions of the 1996 farm bill is contained in Frederick J. Nelson and Lyle P. Schertz, eds., *Provisions of the Federal Agriculture and Improvement and Reform Act of 1996,* Agriculture Information Bulletin No. 729, (Washington, D.C.:Economic Research Service/USDA), September 1996.

12

THE STRUCTURE OF AGRICULTURE

A staunch supporter went up to a political candidate after a speech and said,
"I admire the straightforward way you dodged those issues."

Maynard Speece

The **structure of agriculture** *refers to the number and size of farms; ownership and control of resources; and the managerial, technological, and capital organization of farming.* The issues of the structure of agriculture are illustrated by such questions as these:

- Who controls production and marketing decisions at the farm level?
- What does the industrialization process mean for future agriculture?
- Will the family farm survive?
- Do farm programs help or injure the chances of family farm survival?
- What is the balance of market power among input suppliers, farmers, and marketing firms?
- What policy options exist for maintaining a balance of market power?

This is not the first time the structure of agriculture issue has been discussed. In the first chapter, basic agricultural values were associated with family farm institutions. In Chapter 4, the family farm coalition and its related network organizations were identified as important forces affecting agricultural policy, particularly in issues such as payment limits and sustainable agriculture. In Chapters 9 and 10, questions arose as to whether the benefits of domestic farm policy are too highly concentrated in the hands of large farmers. The changing structure of agriculture has been highlighted as one of the factors that has led to a questioning of the need to continue the more traditional farm programs involving price and income support, which allocate benefits on the basis of volume of production.

Environmentalists assert that large industrialized farms use more chemicals and cause more soil erosion. Advocates of sustainable agriculture and organic farming believe that establishing requirements for reduced chemical use and targeting farm program benefits are keys to family farm survival.

This chapter is devoted to the issue of structural change in agriculture. It addresses the issues of family farm survival and the industrialization of agriculture. This is, essentially, a chapter on market policy as it relates to the organization of agriculture and the position of farmers within markets.

FAMILY FARM DEFINITION

In policy discussions, the term *family farm* is frequently used but seldom defined. In this book, a family farm must meet four requirements:

- A majority of the management and work must be done by the operator and family.
- A close association must exist between the household and the business.
- Managerial control must be exercised by the operator.
- Family farms must obtain the majority of their income from farming.

None of these requirements should be considered to be hard and fast rules. Yet such a definition excludes certain farming structures and structural trends as being outside the family farm concept:

- Part-time farms are not family farms. If they do not earn a majority of their income from agriculture, they cannot be either farmers or family farmers.
- Contract growers for agribusiness integrators are not family farms. Broiler growers do not own the birds nor the feed and make very few management decisions. Therefore, nearly the whole poultry industry is void of family farmers.
- Hired farm managers are not family farmers.
- Farm operations with ownership extending beyond the immediate family or into businesses other than farming (not including farmer cooperatives) are not family farms. Cargill, ConAgra, and Tysons obviously are not family farmers.

Between the four requirements for meeting the family farm definition and these exclusions is a substantial gray area of large farms that employ substantial labor but are still centered around a family. The dairy industry, particularly that in California, provides an excellent example. Anyone who has met with a group of leading farmers to discuss the family farm survival issue knows the definitional dilemma. Many large farm operations, perhaps the majority, are family owned and controlled with the farm operation being central to the family. Try telling these business operators that they are not family farms even though their net income may approach or exceed $1 million. Most certainly, they are distinctly different from Cargill, ConAgra, or Tysons!

THE INDUSTRIALIZATION PROCESS

The industrialization of agriculture is the process by which agriculture is transformed from a way of life to a business. In its early stages, industrialization involves the formation of large-scale business entities that involve the utilization of substantial off-farm labor and assembly line production techniques—such as the California dairy or the large farrow-finish hog operation. As the industrialization process continues, management becomes more market oriented, combining assembly line production techniques with an integrated systems context, often including the marketing of a finished consumer product: cut-up broilers. In this more advanced stage, industrialization involves the infusion of outside investor capital into agriculture, normally through large agribusiness input suppliers or marketing firms and, increasingly, through a combination of the two.

The concept of industrialization is not new. It was the theme of a 1968 Farm Foundation[1] conference addressing the future structure of agriculture.[2] It subsequently became the subject of a major publication and education program titled *Who Will Control U.S. Agriculture?*, conducted in the 1970s by public policy extension specialists. This educational program pulled no punches regarding the adverse consequences of the industrialization process for family farm agriculture. In the late 1970s, structure became the theme of a major USDA initiative under the leadership of Secretary Bergland. History will record the fact that nothing substantive came out of this initiative in terms of policy. Accordingly, the 1970s may represent the end of the family farm era in agriculture and the beginning of the industrial era.

The industrialization process keys production to market needs. This requires a high level of management control over production, processing, and marketing processes. Control begins with management of the genetic stock for production. As a result, plant and/or animal breeding programs are generally a central part of an integrated system. Maintaining control of genetics is an important component of designing the product being produced. That is, chicks grown in an integrated operation are owned and controlled by the integrator from the egg to the retail grocery chain. The same principle is being applied to integrated hog operations. In both cases, the integrator supplies all inputs except housing and labor. Therefore, feed and all related inputs are supplied by the integrator. The grower is paid a piece

[1]Farm Foundation was initially endowed to address contemporary issues in agriculture by International Harvester, once a leading farm implement company.
[2]Harold Guither, *Who Will Control U.S. Agriculture?*, North Central Reg. Ext. Pub. 32 (Urbana, Ill.: University of Illinois, March 1973).

wage, which represents a return on labor and capital supplied. The same integration principles apply to most fruits and vegetables for processing although the tightness of the contractual arrangement in terms of input supply varies.

Industrialization and integration give rise to issues such as unionization of workers and bargaining over contract terms. The existence of such structures increases the similarities between industrialized agriculture and industry.

Considerable disagreement has arisen among agricultural economists over how far and how fast the integration trend will go. The reality of an integrated poultry industry has existed since the 1960s; debate over future integration of the hog industry began in the late 1960s and continues. Not until the mid-1990s was there widespread admission that integration of hog production and marketing was likely to be the wave of the future. Questions remain over whether integrated hog production is to be concentrated in the traditional Corn Belt producing areas, in the South, and/or in the cattle feeding areas of the High Plains. Within industrialized agriculture, the different forms of integration merit further definition and discussion because their policy implications may be quite different.

Firms are **vertically integrated** *when they control two or more levels of the production-marketing system for a product.* Of particular interest, from a policy perspective, is integration that extends to control of the production and/or marketing decisions of farmers. Such control may be exercised either by contract or by ownership.

1.   **Contract integration** *exists when a firm establishes a legal commitment that binds the producer to certain production and marketing practices.* Contract integration begins before production and requires that the producer sell the product to the buyer (also referred to as the *integrator*). Additional commitments bind the producer to specified production practices and sources of inputs. Contracts containing production and marketing commitments are also referred to as *production contracts.* Controversy exists with production contracts because the farmer loses control; the integrator controls both the production and marketing decisions of producers. In addition, under a production contract, from a legal perspective, the producer may not own the product being grown.

Contract integration can be undertaken by either a cooperative or a proprietary (corporate) agribusiness firm. More questions have been raised by farmers and public interest groups about contract integration by corporations than by cooperatives.[3] However, when a cooperative begins contracting with producers in a traditional family farming region, it can become the talk of the town. Moreover, when the cooperative begins growing its own hogs in competition with traditional farmers, things can get really interesting.

Contract integration was estimated to represent 13 percent of farm sales in 1993–1994 (Table 12.1). Overall, proprietary contract integration has a much larg-

---

[3]When cooperatives themselves become producers of hogs, however, many questions have been asked by family farms as to whether the cooperative is acting in their interest.

TABLE 12.1 **Contract and Ownership Integration in Agriculture
(in percentages)**

	Production Contracts Integration (a)				Ownership Integration (b)			
	1960	1970	1980	1993–94	1960	1970	1980	1993–94
Crops								
Feedgrains	0.1	0.1	1.2	1.2	0.4	0.5	0.5	0.5
Hay	0.3	0.3	0.5	0.5	0.0	0.0	0.0	0.0
Food grains	1.0	2.0	1.0	0.1	0.3	0.5	0.5	0.5
Vegetables for fresh market	20.0	21.0	18.0	25.0	25.0	30.0	35.0	40.0
Vegetables for processing	67.0	85.0	88.1	87.9	8.0	10.0	10.0	6.0
Dry beans and peas	1.5	1.0	2.0	2.0	1.0	1.0	1.0	1.0
Potatoes	40.0	45.0	60.0	55.0	30.0	25.0	35.0	40.0
Citrus fruits	0.0	0.0	0.0	0.0	20.0	30.0	11.2	6.9
Other fruits and nuts	0.0	0.0	0.0	0.0	15.0	20.0	25.0	25.0
Sugar beets	99.0	99.0	99.0	99.0	1.0	1.0	1.0	1.0
Sugarcane	24.4	31.5	29.3	27.3	75.6	68.5	70.7	72.7
Cotton	5.0	5.0	1.0	0.1	3.0	1.0	1.0	1.0
Tobacco	2.0	2.0	1.4	9.3	2.0	2.0	2.0	1.5
Soybeans	1.0	1.0	1.0	0.0	0.4	0.5	0.5	0.4
Feed crops	80.0	80.0	80.0	80.0	0.3	0.5	10.0	10.0
Livestock								
Fed cattle (c)	—	—	—	—	6.7	6.7	3.6	4.5
Calves, slaughter (c)	—	—	—	—	1.5	1.7	1.8	10.0
Other cattle and calves	1.0	2.0	2.8	1.7	—	—	—	—
Sheep and lambs (c)	—	—	—	—	5.1	11.7	9.2	29.0
Hogs	0.7	1.0	1.5	10.4	0.7	1.0	1.5	11.4
Fluid grade milk	0.1	0.1	0.3	0.1	0.0	0.0	0.0	0.0
Manufacturing grade milk	0.0	0.0	0.0	0.0	2.0	1.0	1.0	1.0
Market eggs	7.0	20.0	45.0	35.0	5.5	20.0	43.0	60.0
Hatching eggs	65.0	70.0	70.0	70.0	30.0	30.0	30.0	30.0
Broilers	90.0	90.0	87.0	85.0	5.4	7.0	12.0	14.0
Market turkeys	30.0	42.0	52.0	56.0	4.0	12.0	28.0	32.0
Total farm output (d)	8.3	9.3	10.1	12.9	4.6	5.6	6.3	8.0

(a) Production contracts. Contracts entered into before production begins, excludes marketing contracts.

(b) Ownership integration. The same firm owns farms and other vertically related operations such as a hatchery, feed mill, processing plant, or packer-shipper. Excludes direct marketing to consumers such as producer-dealers of milk, roadside stands, or pick-your-own operations.

(c) Feeding of livestock by the meatpacker, some of which is under contract in feedlots owned by others.

(d) The percentage of total farm output under production contracts and ownership integration includes only the products listed in the tables and calculated using the same weights in each year so that changes in the share of, say, broilers do not affect the figure. The weights are the average share of cash receipts of each product in 1960, 1970, 1980, and 1993–94.

Sources: Alden Manchester, *Transition in the Farm and Food System* (Washington, D.C.: National Planning Association, March 1992), p. 37 and updated and revised data provided by USDA/ERS, Commercial Agriculture Division, Animal Products Branch, November 1996.

er share of this 13 percent than cooperative contract integration. The following observations appear to be valid as to the extent of contract integration:

- Contract integration is the most prevalent in vegetables (fresh and processed), potatoes, sugar beets, sugarcane, seed crops, and poultry products.
- The trend is clearly toward increased contract integration.
- Hogs and tobacco are in the takeoff phase in terms of an increasing percent of contract integration.
- Production contracting appears to be associated with commodities for which breeding and control of genetic factors play an important role in either productivity determination or quality control.

(2) **Ownership integration** *exists when single ownership interests extend to two or more levels of the production-marketing system.* The primary interest in this chapter is in vertical ownership interests that involve production of agricultural commodities. Ownership integration can involve either cooperatives or proprietary agribusiness firms.

Proprietary ownership integration (frequently referred to as **corporate integration**), which is not as common as might be gleaned from the popular press, accounted for only 8 percent of the production in 1993–1994. It is concentrated in most of the same commodities where contract integration is prevalent (Table 12.1). There is a definite tendency for corporate integration to increase. For example, integration in fruits and nuts has persistently increased from 15 percent in 1960 to 25 percent in 1993–1994. Market integration of eggs has increased at an even faster pace than that of vegetables for the fresh market and turkeys. Freese indicates that, in 1995, the 28 largest pork contractors produced about one-fourth of the market hogs sold in the United States,[4] substantially higher than the numbers reported by Manchester in Table 12.1.

Between contract integration and ownership integration, the following commodities are nearly *totally integrated:* vegetables (fresh and processed), potatoes, sugar beets, sugarcane, seed crops, eggs, broilers, and turkeys.

Cooperative integration refers to farmer ownership of marketing and/or processing facilities. Although overall cooperative integration (Table 12.2) is more common than corporate contracting and ownership integration (Table 12.1), it is not as tightly controlled. That is, whereas corporate contract integration controls production, the most common forms of cooperative contract integration only commit the farmer to market through the cooperative. In this case, the cooperative contract is generally referred to as a marketing agreement. This is the case for cooperative integration in rice, cotton, soybeans, and milk. For much cooperative

---

[4]Betsy Freese, "What a Difference a Year Makes: Pork Powerhouses 1995," *Successful Farming* (October 1995, p. 21).

**TABLE 12.2   Extent of Cooperative Integration in 1993
by Commodity and Extent of Producer
Commitment (percent of total sales)**

Commodity	Cooperative Market Share	Extent of Commitment		
		No commitment	*Marketing agreement*	*Production agreement*
Crops				
Feedgrains	30	28	2	0
Wheat	50	49	1	0
Rice	45	2	43	0
Vegetables for fresh market	5	0	5	0
Vegetables for processing	10	1	0	9
Dry beans and peas	20	10	10	0
Potatoes	7	1	3	3
Citrus fruit	45	0	28	7
Other fruit and nuts	25	5	20	0
Sugar beets	35	0	0	35
Sugarcane	45	0	0	45
Cotton	35	4	31	0
Tobacco	10	3	7	0
Soybeans	30	20	10	0
Seed crops	16	8	0	8
Livestock				
Fed cattle	2	0.5	0.5	0.5
Other cattle	5	3	1	2
Sheep and lambs	16	14	1	1
Hogs	9	3	1	5
Milk	82	0	82	0
Eggs	0	0	0	0
Broilers	8	0	0	8
Turkeys	5	0	0	5
Total farm output	26%	9%	15%	2%

Source: Estimates by Ron Knutson with the assistance of professional staff in
the Agricultural Cooperative Service/USDA.

integration, such as feedgrain and wheat, the commitment to market does not exist; the cooperative is just another competitor. Therefore, the total cooperative share in Table 12.2 is divided into three components:

- No commitment means that the cooperative simply operates in a spot market.
- Marketing agreement integration means that members agree to market production through the cooperative.
- Production agreement integration means that members agree to produce to specification, frequently involving input supply arrangements with the cooperative.

Overall Integration

Tables 12.1 and 12.2 suggest that the combination of corporate integration and cooperative integration is fairly substantial but certainly not overwhelmingly so. It is certainly not correct to charge that agricultural production is dominated by large agribusiness firms as is stated or implied by some advocates of doing away with farm programs.[5] From a farmer perspective, cooperative integration is certainly of less concern than corporate integration. Yet, quite clearly, total integration is substantial and increasing in certain sectors of agriculture but not in major crops such as wheat and feedgrains. Furthermore, there are questions as to whether the development of biotechnological designer crops combined with the decoupling or elimination of farm subsidies might lead to integration of the crops sector.

SOURCES OF CONCERN ABOUT STRUCTURE

Concern about the structure of agriculture arises from economic, social, and political sources. The economic concerns relate to the efficiency of production and the concentration of market power. Social concerns relate to the survival of family farms, impacts on the environment, and the impact on the quality of life in rural America. Political concerns relate to the balance of political power within agriculture and controversy over the role of agriculture in maintaining stability in a democratic system.

Economic Concerns

The economic concerns about the structure of agriculture relate to issues of concentration of market power, efficiency, and prices.

Concentration of Market Power

Market power refers to the ability of a firm to influence the terms of trade in a market. The terms of trade may be either the prices for products or inputs. Concentration of market power indicates the extent to which market control is centered in the hands of a few firms. **Concentration** is *measured by the proportion of production or marketings controlled by the largest firms in a market.* From an economic perspective, concentration is important because the more highly concentrated the market, the greater the potential impact of a firm or group of firms on price. Concentration of production, input markets, and product markets is an important consideration when analyzing the structure of agriculture.

---

[5]V. James Rhodes and Glenn Grimes, *U.S. Contract Production of Hogs: 1992 Survey,* Ag. Eco. Report 1992–2 (Columbia, Mo.: University of Missouri, 1992).

**TABLE 12.3   Proportion of Gross and Net Cash Income and Direct
Government Payments by Farm Size, 1995**

Gross Sales	Number of Farms (Thousands)	Percentage of Farms	Percentage of Gross Cash Income	Percentage of Net Cash Income	Percentage of Payments
Less than $40,000	1469	70.97	12.07	6.09	23.53
$40,000–$99,999	260	12.56	10.98	9.95	15.02
$100,000–$249,999	219	10.58	21.03	20.35	29.17
$250,000–$499,999	75	3.62	14.72	14.35	18.92
$500,000–$999,999	30	1.45	12.33	14.18	9.46
$1 million and over	17	0.82	28.87	35.08	3.90

Source: Economic Research Service, *Agricultural Income and Finance:
Situation and Outlook*, AIS-62, (Washington, D.C.: USDA/ERS, September
1996), p. 16.

Concentration of production in *agriculture is generally low but highly variable
from commodity to commodity.* Overall decline in farm numbers has been accompanied by growth in the number and size of the largest farms. Concentration of production has occurred to the point at which 17,000 farms having sales of more than
$1 million (about 0.8 percent of all farms) control 29 percent of the production and
35 percent of net cash income (Table 12.3). Although the image is that large family farms tend to be concentrated in the West and South, over one-third of those
farms with more than $1 million sales are located in traditional family farming
areas of the Midwest and Great Plains.[6]

Wide variation is seen in the extent of concentration of production from commodity to commodity. For example, in 1995, there were 39 beef feedlots in the 13
major cattle feeding states that reported cattle inventories of more than 50,000
head (0.1 percent of the total 43,332 feedlots) producing 21 percent of the fed cattle. The 89 largest feedlots produced 38 percent of the fed cattle.[7]

Concentration in cattle feeding has received attention because a number of
the largest feedlots are owned by the biggest packers (IBP, ConAgra, and Excel).
In 1994, the four largest packers accounted for 68 percent of U.S. fed cattle slaughter and 86 percent of U.S. boxed beef production (Figure 12.1). ConAgra owns
Monfort of Colorado; Cargill owns Excel and Caprock Industries. Monfort and
Caprock are two of the largest cattle feeders.

Captive supplies refer to cattle which are committed to a specific buyer two
weeks or more in advance. In reality, captive supplies take three forms:

■ Packer feeding in packer-owned and commercial feedlots.

[6]Donn A. Reimund, Thomas A. Stuckey, and Nora L. Brooks, *Large Scale Farms in Perspective,*
Agric. Inf. Bul. 505 (Washington, D.C.: ERS/USDA, February 1987), p. 3. The states included are North
Dakota, South Dakota, Kansas, Nebraska, Minnesota, Iowa, Missouri, Wisconsin, Illinois, Indiana,
Michigan, and Ohio.
[7]National Agricultural Statistics Service, *U.S. Cattle on Feed* (February 23, 1996).

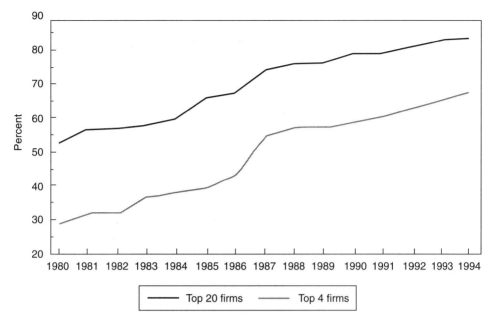

FIGURE 12.1 Cattle slaughter concentration, 1980–1994.

Source: Grain Inspection, Packers and Stockyards Administration, *Packers and Stockyards Statistical Report: 1991–94 Reporting Years* (Washington, D.C.: USDA/GIPSA, October 1996), p. 62.

- Fixed price and basis forward contracts.
- Exclusive marketing and purchasing agreements with individual cattle feeding firms.

In 1985, the only form of captive supplies was packer-owned feedlots, which accounted for 4 percent of the supplies. By 1989–1994 captive supplies in all three forms had increased to 17 to 23 percent of steer and heifer slaughter with substantial year-to-year variability. Studies have indicated minimal impact on prices, although the issue of captive supplies continue to receive widespread attention.[8]

In 1990, the 10 largest broiler integrators controlled about 70 percent of the production, whereas the 4 largest controlled 47 percent.[9] Comparable concentration levels for turkeys are 62 percent for the 10 largest integrators and 32 percent for the 4 largest.[10] In eggs, the 10 largest integrators have 24 percent of the production and the 4 largest have 15 percent.[11] In each of these three segments of the poultry industry, trends suggest continued increases in concentration.

[8]Clement E. Ward, Stephen K. Kootz and Ted C. Schroeder, *Short-Run Captive Supply Relationships with Fed Cattle Transaction Prices* (Washington, D.C.: GIPSA/USDA, May 1996).
[9]*Feedstuffs,* 1992 reference issue, p. 18.
[10]Bernard E. Heffernan, *Turkey World* (December 1991), p. 32
[11]*Egg Industry* (November/December 1991), p. 26.

Such high levels of concentration are limited largely to livestock, poultry, and selected vegetable products. Concentration is still very low for most of crop agriculture. In fact, relative to other U.S. industries, where the market share of the four largest manufacturers frequently exceeds 50 percent, concentration in agricultural production is quite low. But attention is drawn to agriculture because of the rapidity with which certain industries, such as poultry, fed cattle, and hogs, have gone— or are going—from a diffused to a concentrated and integrated structure.

Four types of concern arise from concentration and integration in production:

- *Buyers may have the ability to depress prices for producers outside* the integrated system. These are issues on which little light has been shed except from a theoretical perspective. Extensive study of pricing in beef indicates that much of the slaughter beef is sold with the producer receiving only one bid.[12] Yet when prices received by producers are analyzed, scant evidence of price suppression is found. Ward noted that during June 1989, the big three packers, as a group, paid significantly lower prices than did their competitors.[13] However, individual packers from the big three paid different prices, all of which were not significantly lower. Arguably, paying lower prices would not only increase packer margins on cattle purchased but also put members of the big three in a more favorable competitive position relative to other packers. However, much more research of a longer term nature is needed to validate such theoretically plausible relationships.

- *Integrators may be able to suppress the returns to growers* under a production contract. Here, the evidence is purely theoretical and anecdotal. It is clear that the integrator has substantial market power with regard to its growers. Instances in which integrators have attempted to stifle the formation of grower bargaining groups have surfaced. Such bargaining groups are designed to obtain more favorable terms of trade for growers. This issue is discussed in a policy context later in the chapter.

- *Increasing concentration of agricultural production could lead to higher food prices,* if extended over a period of time.[14] This would result from increased merchandising and marketing costs, potential unionization of agricultural workers, the institutionalization of grower bargaining structures, and the lack of effective competition.[15] The basis for this conclusion, once again, is the economic theory of imperfect competition. That is, as an industry progresses

---

[12]Clement E. Ward, "Structural Change: Implications for Competition and Pricing in the Feeder-Packer Subsector," in *Structural Change in Livestock: Causes, Implications, Alternatives,* Wayne D. Purcell ed. (Blacksburg, Va.: Research Institute on Livestock Pricing, February 1990), P. 83.

[13]Clement E. Ward, "Inter-Firm Differences in Fed Cattle Prices in the Southern Plains." *Am. J. Agric. Econ.* (May 1992), pp. 481–485.

[14]Harold F. Breimyer and Wallace Barr, "Issues in Concentration versus Dispersion," in *Who Will Control U.S. Agriculture?* North Central Reg. Ext. Publ. 32 (Urbana, Ill.: University of Illinois, August 1972), pp. 13–22.

[15]James V. Rhodes and Leonard R. Kyle, "A Corporate Agriculture," in *Who Will Control U.S. Agriculture?* North Central Reg. Ext. Publ. 32–3 (Urbana, Ill.: University of Illinois, March 1973).

TABLE 12.4   Concentration in Farm Input Markets, 1992

	Market Share Percentage	
Input	*Four largest firms*	*Fifty largest firms*
Nitrogen fertilizer	48	99
Phosphate fertilizer	62	99+
Farm machinery	47	71
Petroleum refining	30	97
Tires	70	99+
Agricultural chemicals	53	95

Source: Bureau of the Census, Census of Manufacturer's, *Concentration Ratios in Manufacturing* (Washington, D.C.: U.S. Department of Commerce, 1996).

toward higher levels of concentration, competition among the remaining firms would be expected to assume higher cost forms of nonprice competition and increased profit margins. This conclusion assumes that there are relatively few economies to be gained from increased size and from integration of production and marketing functions. We will see later that this assumption appears to be suspect. Moreover, it assumes that competition from international sources is insignificant—an assumption that is questionable in an era in which the competitive forces of freer trade are evident.

■ *The perception that agriculture is becoming integrated and controlled by a few large agribusiness firms* makes it more difficult for crop and dairy producers to convince Congress and the administration of the need for subsidies.

Concentration in purchased inputs *used in agricultural production is generally high in absolute and relative terms.* Concentration in inputs such as machinery, energy, seeds, credit, and chemicals is high at all levels of the market channel, with the four largest firms approaching or exceeding 50 percent market shares (Table 12.4). For example, in 1992 the four-firm concentration ratio for nitrogen and phosphate fertilizer was 48 and 62 percent, respectively. The four largest firms sold 53 percent of the agricultural chemicals and 70 percent of the tires. The four largest farm machinery manufacturers sold 47 percent in 1992, while an earlier, more detailed study indicated that in tractors and combines, the market share approaches 80 percent.[16]

It should be clear that in addition to potential monopolistic abuse from high concentration in industries such as agricultural chemicals, there is substantial nonprice competition, which may have limited social value. This suggests substantial costs associated with advertising and product differentiation as well as the

---

[16]Robert F. Leibenluft, *Competition in Farm Inputs: An Examination of Four Industries* (Washington, D.C.: Federal Trade Commission, February 1981).

potential for excessive profits. However, there is a lack of evidence of excessive profits. Two reasons may exist for this lack of evidence:

- Cooperatives have experienced substantial success in penetrating farm supply markets in terms of direct farmer sales. Lower prices may result from cooperatives being able to centralize purchases and bargain for lower prices. In addition, sales to farmers may be taken away by competition between cooperatives and direct farmer purchases from input suppliers.
- Many of the advances in agricultural technology have been the products of publicly supported research. The results of this research have been generally available for adoption, and continuous infusion of advances has made it difficult to maintain high profits on inputs over the long term. In addition, the Extension Service has generally maintained extensive trial or test plot demonstrations of the effectiveness of a wide range of purchased inputs. This has tended to provide incentives for competition among input suppliers on the basis of objective measures of product quality and performance.

Recent research policy changes allowing the patenting of new life forms and computer software could change the competitive relationships within the farm input industries. Discoveries involving the products of biotechnology are being patented as rapidly as allowed, even by the land-grant universities. This has precipitated an extensive number of acquisitions within the seed business. However, the pace of technological change in biotechnology could be so fast that the danger of a firm capturing a market and reaping monopoly profits may be minimal.

Concentration in farm product markets *also contains substantial evidence of imperfect competition.* The concern here involves the extent to which the market power of first handlers (buyers) of farm products may be sufficient to suppress producer prices.[17]

Farmers generally sell in markets where there is a relatively small number of buyers (Table 12.5). For example, as has been seen previously, the national market share of the four largest beef packers is 68 percent of the cattle slaughtered. Table 12.5 indicates relatively high concentration in major agricultural product markets. The four largest sheep and lamb packers had 76 percent of the slaughter. In hogs, the concentration is lower, with the four largest packers having a 44 percent market share. Regional livestock market concentration is considerably higher, often running more than 90 percent.[18] It is not unusual for a seller to be in a position of having only one buyer bid—essentially a monopsony situation.

High concentration is not limited to livestock. The four largest textile mills for cotton had a market share of 39 percent in 1992; in wool, it was 55 percent. The four largest soybean mills crushed 71 percent of the beans; and in flour, 39 percent.

[17]Issues of concentration in consumer products and retail food markets are addressed in Chapter 14.

[18]Clement E. Ward, *Meat Packing Competition and Pricing* (Blacksburg, Va.: Research Institute on Livestock Pricing, 1988), p. 17.

TABLE 12.5   Percentages of Concentration in Agricultural Product Markets Measured by the U.S. Market Share of the Four Largest Firms

Product	Year	Four-Firm Market Share
Cattle slaughter	1994	68
Boxed beef	1994	86
Pork	1994	44
Sheep and lamb	1994	76
Broilers	1991	47
Turkeys	1991	32
Eggs	1991	15
Cheese	1992	42
Flour	1992	39
Cereal and breakfast foods	1992	85
Canned fruit and vegetables	1992	27
Corn wet milling	1992	73
Soybean oil milling	1992	71
Cotton weaving mills	1992	39
Wool	1987	55

Major sources: Grain Inspection, Packers and Stockyards Administration,
Packers and Stockyards Statistical Report: 1991–94 Reporting Years
(Washington, D.C.: USDA/GIPSA, October 1996); Bureau of the Census,
Concentration Ratios in Manufacturing: 1992 Economic Census (Washington,
D.C.: U.S. Department of Commerce, 1996).

Distance limits the number of alternative outlets available for most grain and cotton producers. Fruit and vegetable producers have long been recognized as facing a highly concentrated buyer market for their products.[19]

The theoretical implications of monopsony pricing are clear (Figure 12.2). In a competitive market, quantity Q_c would be purchased at price P_c with supplies S and demand D. However, because the monopsonist faces an upward sloping supply curve (S), its marginal cost (MC) yields a profit-maximizing quantity purchased at Q_m, and the price paid producers is reduced to P_m.

Specific economic evidence of the extent of monopsony influence on farm prices is sketchy. Miller and Harris found that buyer concentration was a statistically significant factor having a negative influence on the price of hogs.[20] It has also been found that as the volume of commodities moving through terminal markets declines, prices fall relative to other, higher volume markets.[21] Extensive research has been conducted on potential monopsonistic abuse in fed beef resulting from increased concentration associated with mergers and consolidations of meat packers.

[19]Leon Garoyan, "Thin Markets in the Fruit and Vegetable Industry," in Marvin L. Hayenga, ed., *Pricing Problems in the Food Industry*, N.C. 117, Monogr. 7 (Madison, Wis.: University of Wisconsin, February 1979), pp. 105–114.

[20]Steven E. Miller and Harold M. Harris, *Monopsony Power in Livestock Procurement: The Case of Slaughter Hogs*, (Clemson, S.C.: Clemson University, October 1981).

[21]William G. Tomek, "Price Behavior on a Declining Terminal Market," *Am. J. Agric. Econ.* (August 1980), pp. 434–444.

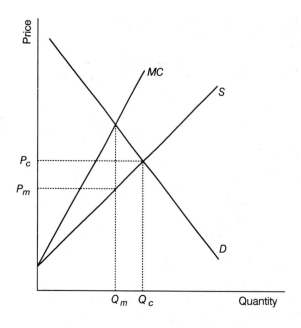

FIGURE 12.2   A monopsonist reduces the quantity purchased and the price paid producers.

This research, as indicated previously, has been less than decisive, indicating only that the largest three packers, as a group, pay lower prices.[22] Although this evidence is far from decisive, it lends support to the general proposition that first handlers of farm products generally have sufficient market power to affect the terms of trade.

Efficiency

Efficiency, as used here, *is defined as minimum cost.* The issue of the most efficient size farm was discussed in Chapter 9 and is not repeated. Suffice it to say that the least-cost farm size extends far beyond a family-size farm having up to $250,000 in sales.

Table 12.6 provides a rough comparison of the size farm where most of the economies appear to have been realized. The most efficient size is not just a function of pure technical economies. The pecuniary and technology economies may be at least as substantial as technical economies. These economics may push the optimal size beyond the levels indicated in Table 12.6. That is, the ability to buy inputs at lower prices and sell products at higher prices is a key to efficiency gains embodied in management skill. Moreover, the most progressive firms in adopting technology tend to be on the front of the treadmill lowering costs. Thus, in California and New Mexico, progressive dairies frequently put together multiple units of 2,000 dairy cows per unit. The same phenomenon has happened in hogs, poultry and beef feedlots.

[22]Ward, "Inter-firm Differences."

TABLE 12.6   Efficient Farm Size Where Most Economies Have Been Realized Compared with $250,000 Gross Sales Family Farm

Product Produced	Family Farm	Efficient Size
Iowa corn	950 acres	2,200 acres
Kansas wheat	1,500 acres	3,300 acres
Texas cotton	1,700 acres	4,000 acres
Arkansas rice	500 acres	1,300 acres
New York dairy	100 cows	1,200 cows

Source: Agricultural and Food Policy Center (College Station, Tex.: Texas A&M University, December 1996).

The efficiency implications of vertical integration are much debated. It has long been recognized that a principal problem in agriculture has been the difficulty of coordinating production with market needs. Vertical integration, to the extent that production is influenced by the integrator, makes substantial contributions to satisfying this need. For example, in broilers and turkeys, vertical integration has contributed to the uniform size and quality of poultry sold in the supermarket and served in restaurants and fast-food outlets.

Equally important contributions are made by vertical integration to increased efficiency and reduced costs. This is seen again in the case of broilers, where increased efficiency in feed conversion and reduced costs in real terms have been evident.[23] In addition to technological breakthroughs, vertical integration eliminates multiple handling of products.

These same questions logically can be asked for beef and hogs for which the combined forces of market power and efficiency gains have been studied extensively. For beef, the following conclusions and trade-offs on structure issues appear to be warranted based on the research of Ward[24] and Purcell[25]:

- Economies of size in meat packing plants extend throughout the full range of existing plants, perhaps to more than one million head annually.
- Vertical integration provides increased assurance that adequate supplies of cattle are available to keep packing plants operating efficiently.
- Improving the uniformity of product quality holds the longer run potential for improving the consumer appeal of red meat.

[23]Lee F. Schrader and George B. Rogers, "Vertical Organizations and Coordination in the Broiler and Egg Subsectors," in *Vertical Organization and Coordination in Selected Commodity Subsectors*, N.C. 117, W.P. 20 (Madison, Wis.: University of Wisconsin, August 7, 1978); and Ronald D. Knutson, "The Structure of Agriculture: An Evaluation of Conventional Wisdom," in *1980 Agricultural Outlook* (Washington, D.C.: Committee on Agriculture, Nutrition, and Forestry, U.S. Senate, December 23, 1979), pp. 135–143.

[24]Ward, *Meatpacking Competition and Pricing.*

[25]Wayne D. Purcell, *Structural Change in the Livestock Sector: Causes, Implications, Continuing Issues* (Blacksburg, Va.: Research Institute of Livestock Pricing, February 1990).

■ High levels of concentration obviously create the potential for monopsonistic and monopolistic exploitation, although neither has been decisively proven.

Prices

It is frequently suggested that in the long run, market power in integrated agriculture will become sufficiently highly concentrated that the consumer will pay higher prices for food. Although such an idea can be supported by economic theory, most conclusions of this type fail to take into account efficiency gains from integration. Alternatively, it is assumed that such gains could be realized without the development of tightly knit vertically integrated systems. The extent to which this would, in fact, happen can be hotly debated. Advances in the convenience of having broilers cut up and sold in ready-to-cook consumer packages are evident, but one can logically question the costs in terms of paying prices for chicken that are nearly equivalent to those for beef. Would the highly efficient broiler, fresh mushrooms, or processed fruit and vegetable industries have been possible without the development of integrated production systems? Are the growers receiving a share of the benefits of their efficiencies?

Social Concerns

Many of the arguments against structural change in agriculture, whether in the form of increased concentration or integration, have been of a social or a socioeconomic nature. These arguments place emphasis on the impact of concentration and integration on the "family farm" institution, rural communities, and rural institutions.

Major concern has been expressed that continuously increasing concentration and integration will lead to the demise of the family farm as an institution. As defined earlier in this chapter, the term *family farm* has been associated with the existence of an independent business and social entity sharing the responsibilities of ownership, management, labor, and financing. The family farm system is perceived as being equivalent to dispersion of economic power. It has also been associated with the perpetuation of basic American values and the family as an institution. Although the importance of these roles is debated (see Chapter 1), it can be persuasively argued that concentration and integration tend to destroy the family farm institution.

Several potentially adverse consequences of integrated agriculture have been identified. Contract integration with corporations and sometimes cooperatives radically changes the role of the traditional independent farmer. More often than not, the farmer loses control—if not legal title—to the commodities grown under a production-integrated arrangement. Payment to the grower is largely on a per unit of product grown or piece-wage basis and not necessarily related to product value. In the process, most of the management functions traditionally associated with the family farm are removed. The farmer thus takes on more of the characteristics of a laborer and loses many of the characteristics of a small business owner.

An additional social concern that has been raised is that concentration and ownership integration reduce the number of farms and make the integrator less dependent on the local community. In the process, smaller rural towns and their social institutions decline or vanish. It is further argued that this potentially reduces the cohesiveness of family life. It is also suggested that towns where corporate agriculture dominates have few services, lower quality education, and less community spirit.[26]

Although these arguments are plausible, they rely on little supporting evidence. The most frequently cited study involves a comparison of two California communities, one dominated by corporate ownership-integrated agriculture and the other by family farm agriculture. Study of these communities in two different time periods supports the social relationships and hypotheses suggested here.[27] However, studies in contract-integrated agriculture fail to clearly establish these relationships and, in some instances, point out distinct contradictions.[28]

Political Concerns

One of the longest standing arguments supporting the maintenance of a decentralized agriculture is political in nature. Jefferson visualized the merits of a decentralized political system in which power was highly diffused and every individual had the opportunity for input into public decisions. His agrarian philosophies placed a high value on independent farmers and landowners as a means of maintaining a democratic system of government.

Without question, there has been a marked departure in the United States from the decentralized power structure ideal visualized by Jefferson. What political influence farmers have today is undoubtedly related to a combination of numbers, the tendency to vote as a bloc, and a basic belief in farming as a way of life. Some of the more powerful agricultural political lobbies, such as milk and cotton, have come to rely more on political strategies that are characteristic of the corporate and labor union sector than on the ideals visualized by Jefferson. As U.S. agriculture continues the trend toward increased industrialization, power relationships in agriculture will increasingly take on the characteristics of politics in the nonfarm sector. Some will interpret this trend as progress; others will interpret it as a step backward.

[26]*A Time to Choose: Summary Report on the Structure of Agriculture, Preliminary Report* (Washington, D.C.: USDA, January 1981), pp. 38–39.

[27]Community Service Task Force, *The Family Farm in California, Report of the Small Farm Viability Project* (Berkeley, Calif.: November 1977); and D. MacConnell and Jerry White, *Agricultural Land Ownership and Community Structure in California Central Valley* (Davis, Calif.: University of California, December 1980).

[28]William D. Heffernan, "Agriculture Structure and the Community," in *Can the Family Farm Survive?* Spec. Rep. 219 (Columbia, Mo.: University of Missouri, 1972), pp. 481–499; and Louis A. Plock, *Social and Family Characteristics of Main Contract Broiler Growers,* Bull. 569 (Orono, Maine: Maine Agric. Exp. Stn., August 1960).

STRUCTURE POLICY OPTIONS

Numerous farm bills have been enacted to "preserve family farms," but no general structure policy can be clearly identified (see box). Instead, bits and pieces of policy influence structure, but the nature of that influence is not well understood. The remainder of this chapter discusses the policy options for dealing with structure.

Targeting Farm Program Benefits

Although farm programs were designed to help all farmers overcome the instability, price, and income problems that confront agriculture, it is clear that large farmers receive more total benefits than small farmers. This occurs because most farm program benefits, such as deficiency payments, are allocated on the basis of volume of production. This is still the case under the 1996 farm bill where decoupled payments are allocated on the basis of base acres or historical payments. Yet relative to sales, moderate-size farms receive a larger proportion of the benefits (Table 12.3). For example, midsize farms with $100,000 to $249,999 sales accounted for 21 percent of the gross income but received 29 percent of the payments. Farmers with more than $1 million sales accounted for 29 percent of the gross income but received 4 percent of the payments. Some of the lower percentage of payments is accounted for by large farms, such as cattle feeders, not growing pro-

---

Government Programs without Policy

I remember those years now (as a Congressman from Minnesota) as one crisis after another, a seemingly endless debate on agriculture bills with little or no discussion of agricultural policy . . .

We didn't know who exactly was being helped or who was being hurt by the measure before us. The problems were seldom clearly defined. If they were, they were cast as narrow but immediate crises that needed patches quickly. Other than a dime a bushel here or a few pennies more a pound there, the remedies presented were either politically unacceptable or simply made no sense.

We thought—we hoped—that if we helped the major commercial farmers, who provided most of our food and fiber (and most of the political pressure), the benefits would filter down to the intermediate-sized and then the smallest producers.

I was never convinced we were anywhere near the right track. We had symbols, slogans and superficialities. We seldom had substance.

Source: Bob Bergland, *A Time to Choose: Summary Report on the Structure of Agriculture* (Washington, D.C.: USDA, January 1981), pp. 5–6.

gram crops. Still, it is not correct to assert that the benefits of farm programs go primarily to the largest farms in America.

Yet, on an individual commodity basis, as indicated previously, large farms are able to avoid payment limits. For example, GAO found that 18,716 rice farms had 54,311 payment recipients.[29] It found that two farms received $1.4 million in deficiency payment benefits. Ten percent of the rice farms received 26 percent of the benefits or an average of $169,000 per farm. GAO recognized that despite the magnitude of this concentration of payments, they had not accounted for all of the benefits received by large farms because many rented additional units on a cash rent basis. By utilizing a cash rent, a portion of the benefits could be encompassed within the rental rate, but not accounted for by FSA.

As noted in Chapter 10, all farm bills since 1970 have contained payment limits. The 1990 farm bill contained two levels of limits:

- A $50,000 per person limit on deficiency payments.
- A $250,000 limit per family under the three-entity rule for deficiency payments and marketing loan gains.

Under the 1996 farm bill the $40,000 per person limit applies to the combination of transition/contract payments. A $230,000 limit is applied to the family unit under the three-entity rule for transition payments and marketing loan gains.

Despite these limits, or perhaps because of them, payment limits have never been truly effective in reallocating benefits toward moderate-size farms. Lawyers and farmers, perhaps, are smarter than politicians. Alternatively, politicians, realizing the convenience—and perhaps necessity—of being able to say that there are limits, never really intend limits to be so tight as to realize the adverse consequences of having no loopholes.

The adverse consequences of effective or ineffective payment limits lie in the efficiency arena. If, as contemporary research clearly indicates, there are economies of size in agriculture that extend substantially beyond the point where effective payment limits would take effect, it is inefficient to target farm program benefits toward moderate-size farms. One estimate suggests that a structure of only family-size crop farms could increase cost of production by as much as $2.3 billion.[30] This magnitude of costs could be borne under even the current payment limit program, despite its ineffectiveness. The reason lies in the legal, accounting, and organizational costs that result from a large number of sham legal entities and rental arrangements established by large farms to avoid the limit, which they must do if they are to survive.

---

[29]General Accounting Office, *Rice Program: Government Support Needs to Be Reassessed* (Washington, D.C.: U.S. Congress, May 1994).
[30]Ronald D. Knutson *et al., Economic Impacts of Farm Program Payment Limits* (College Station, Tex.: Texas Agricultural Experiment Station, Texas A&M University, October 1987).

Some argue that the inefficiency costs of effective payment limits are worth paying as a means of encouraging the adoption of low-input sustainable farming methods in place of input-intensive conventional farming. Effective payment limits, they argue, would bring other societal benefits of decreased pollution, reduced soil erosion, and enhanced habitat for wildlife.[31] In other words, they see effective payment limits as a means of internalizing the costs of externalities and sustaining family farm agriculture.

One method suggested for targeting benefits involves limiting payments to the first 40,000 bushels of grain harvested.[32] Such a proposal has been studied for its feasibility and impacts.[33] It was found that such a limit would have substantial differential regional impacts, depending on the crop and the extensiveness of farming conditions. It would, therefore, be very difficult to establish an equitable set of rules. And, even if such a set were established, problems would be encountered with farmers moving production from farm to farm or dividing farms to avoid hitting the 40,000 bushel limit.

Research has also indicated that if farm programs were dropped, moderate-size farms would suffer the most. Without government programs, large farms were found to have twice as high a probability of surviving as moderate-size farms.[34] The reason lies in the greater ability of large farms to manage risk and the economies of size realized by large farms.

Perhaps of greater concern is the inability of government programs to target assistance to those farmers needing it the most. Harrington found that in 1985, the least financially stressed farms (55 percent of the total number of farms) received three-fourths of the farm program benefits. On the other hand, the most stressed farms (45 percent of the total number) received only one-fourth of the benefits.[35] These conditions have caused Willard W. Cochrane, a former advocate of mandatory production controls (see box), and the Office of Technology Assessment (OTA), the congressional research agency, to argue that farm programs ought to be specifically targeted toward the problems of moderate-size farms. The OTA study made the following structure-related recommendations:[36]

- Eliminate payments to farms having more than $250,000 in sales.
- Provide income protection to farms having less than $250,000 in sales through the marketing loan.

---

[31]"A New Interest in Targeted Payments," *Food Fiber Lett.* (McLean, Va.: Sparks Commodities, July 13, 1992), p. 4.

[32]Edward G. Smith, Daniel R. Sechrist, and Ronald D. Knutson, *Targeting Proposal Evaluation,* AFPC Working Paper 90–6 (College Station, Tex.: Texas A&M University, June 1990).

[33]Ibid.

[34]Edward G. Smith, James W. Richardson, and Ronald D. Knutson, "Impacts of Alternative Farm Programs on Different Size Cotton Farms in the Texas Southern High Plains: A Simulator Approach," *West. J. Agric. Econ.* (December 1985), pp. 365–374.

[35]David H. Harrington, *Agricultural Programs: Their Contribution to Rural Development and Economic Well-Being, Rural Economic Development in the 1980s* (Washington, D.C.: ERS, USDA, July 1987).

[36]Office of Technology Assessment, *Technology, Public Policy, and the Changing Structure of Agriculture,* OTA-F-285 (Washington, D.C.: U.S. Congress, March 1986), pp. 287–290.

Making the Game Fair

The 400,000 moderate-size family farmers struggling to survive represent a different kind of problem. These farmers have been slow to adopt new managerial, commercial, and technological practices. Hence, they have been slow to get their cost structures down. As a result, they are continually in financial difficulty, except in periods of rapidly rising farm prices. Maintaining the present level of price and income support helps them some, but it helps their large, aggressive neighbors a lot more. And raising the level of price and income support to assist these struggling farmers would create a profit bonanza for the large, aggressive farmers. Either way, the support programs provide a price and income shelter under which the large, aggressive farmers can expand the size of their operations by gobbling up their struggling moderate-sized neighbors. And either way, a few more of our 400,000 fall by the wayside.

The moderate-size family farmer struggling to survive needs two things: (1) long-run managerial guidance and (2) subsidized credit. These farm families need guidance in adopting the right practices at the right time and in using them effectively once they are adopted. Such families need subsidized credit to compete with their aggressive neighbors, who know how to obtain favorable financial deals, in acquiring and operating modern, productive, but often expensive, physical inputs. Some might argue that government should not be doing these kinds of things. But I do not agree. Providing guidance and special credit to the weak and struggling is one way of making the competitive game tolerable. It is comparable to letting the team with the worst win record in the National Football League draft first in the search for player talent. It is a way of saving a high proportion of the 400,000 and maintaining a high degree of competition in the production of the nation's food supply.

In sum, I am not arguing to get government out of agriculture as a general ideological proposition. I am arguing to get government out of agriculture where it is helping one group of farmers do in another group. I am arguing to keep government in agriculture where it operates to make the competitive game of farming more fair to all concerned, hence, a more acceptable game for all concerned.

Source: Willard W. Cochrane, "The Need to Rethink Agricultural Policy and to Perform Radical Surgery on Commodity Programs in Particular," *Am. J. Agric. Econ.* (December 1985), pp. 1008–1009.

- Make nonrecourse loans available only to farms having less than $250,000 in sales. All larger farms would be provided only a recourse loan.
- Target educational programs in risk management, future markets, contracting, and cooperative marketing to moderate-size farms.
- Increase testing of new agricultural inputs for their comparative efficiency and intensify extension of results to moderate-size farms.
- Make sure ample credit is available to moderate-size farms.

Note that the consequence of these proposals would be to remove the floor on prices provided by the nonrecourse loan because the volume of production by large farmers would be under a recourse loan. The success of the program would then depend on the ability to teach moderate-size farm operators to become more efficient and technologically up to date in terms of effectively utilizing the tools of risk management.

After doing a detailed review of various targeting options, Sumner concluded that other policy tools such as credit or welfare programs "using commodity programs as distributional tools is like using a pipe wrench to drive in a nail."[37] When the Clinton administration took office, it investigated a number of options for targeting farm program benefits, including tighter payment limits. After the dust settled, it recommended to Congress that farmers having off-farm adjusted gross income of more than $100,000 not be eligible for farm program benefits. Once again, it was uncertain as to how to prevent farm operations from being divided in order to fall within this limit. However, when the focal point of the debate shifted to Freedom to Farm, decoupled payments, and potentially ending payments in year 2002, the notion of more effectively limiting payments was lost.

Animal Rights and Plant Rights

While seldom recognized as a structure policy, proposed animal rights laws are exactly that. They are also specifically targeted toward large-scale integrated agriculture, although they may have spillover effects on conventional family farm agriculture in the case of veal calves.

Wunderlich notes that at least 200 U.S. animal rights organizations press for stricter government control over the way animals are housed, used in laboratories, transported, and slaughtered.[38] These are basically ethical issues with important economic consequences and, in the case of research, health consequences.

---

[37]Daniel A. Sumner, *Targeting Farm Programs*, FAP90–1 (Washington, D.C.: Resources for the Future, October 1989), p. 44.

[38]Gene Wunderlich, "The Ethics of Animal Agriculture," *Food Rev.* (Washington, D.C.: ERS/USDA, October-December 1991), pp. 24–27.

One of the major concerns of animal rights advocates is confinement production of livestock and poultry, which is practiced throughout integrated agriculture—50,000 head feedlots, 18,000 sow hog production units, and stacked batteries of 200,000 caged laying hens, for example. Integrated, confined production contributed to the highly efficient, low-cost, uniform-quality, nutritious broilers that have overtaken beef in per capita consumption. Would America be better off with range chickens produced at three or four times the cost? The vegetarian population may think so but they do not have to pay the price! If animal rights laws are imposed with the effect of putting farms/corporations out of business, should they be compensated for their losses?

As Tweeten notes (see box), the rights argument logically extends far beyond vertically integrated agriculture. Family farm agriculture, where hogs may wallow in the mud as opposed to being confined on a slab of concrete, is affected. Likewise, plants grown in greater than natural populations per acre of land arguably are being abused. Are insects rights infringed when they are sprayed with chemicals?

What is the logical stopping point for protection of animal and plant rights? These are very real policy questions with important economic and structural implications.

Information Policy

Access to and utilization of information on production, markets, and prices are two of the keys to survival in today's agriculture. Without government assistance, information is available only to large farmers and agribusiness firms that have the expertise and can afford to gather it or purchase it. Information made available free leads to a more equitable production and marketing environment for all who are part of the food system, but it is of particular benefit to farmers who cannot afford their own sources of information.

One of the primary goals of USDA's programs has traditionally been to improve the quantity and quality of information on agriculture and its markets that is available to farmers. This information covers a wide spectrum including current market prices by the Agricultural Marketing Service (AMS), product utilization information and estimates of crop and livestock production in the United States by the National Agricultural Statistics Service (NASS), and foreign country production, consumption stock and trade data by the Foreign Agriculture Service (FAS). Information on U.S. market prices is made more meaningful by the establishment of commodity grades that facilitate reporting prices on a uniform quality basis.

Although USDA data are frequently criticized, they are recognized throughout the world as being the most comprehensive and reliable in the world. The USDA is also recognized as having the most reliable information on the agricultural industry that is available from any source in the United States.

The Logic of Animal and Plant Rights

Moral rights philosophers have erected a superstructure of conclusions from logic that goes something like this:

- *First premise:* All sentient animals (capable of feeling pleasure and pain) have rights.
- *Second premise:* All domestic farm animals are sentient.
- *Conclusion:* Therefore, all domestic farm animals have rights.

The soundness of the logic may entice acceptance of the conclusion. However, it is well to scrutinize especially the first premise. It has no empirical base and, therefore, the conclusion is a value judgment that must be accepted on faith.

Activists know full well that animal rights are not bestowed by God. Instead, animal rights will be precisely what society through the political process dictates them to be although animals do not possess intellects capable of abstract reasoning, higher order communication skills, or emotions such as feelings of injustice. One line of reasoning is that human rights extend to infants and the mentally handicapped who possess no more sentience, intelligence, interests, or reasoning capabilities than many animals. Human rights extend to comatose and other people incapable of feeling pleasure or pain.

The logic is extended as follows:

- *First premise:* Moral justice requires that all living things with the same sentience possess the same rights.
- *Second premise:* Animals have as much sentience as comatose humans.
- *Conclusion:* Therefore, justice requires that all animals have the same rights as humans.

By the logic that rights of humans with no sentience ("vegetables" in the vernacular) should be extended to all life with no sentience, the plant kingdom deserves plant rights. It is not too farfetched to extend the same rights to inanimate forms, leading to an Earth First! paradigm. The success of animal rights in the policy process will depend partly on how many Americans accept such thinking. But one network television expose with a particularly graphic illustration of farm animal abuse probably would go farther than the above sophistry to further the animal rights political agenda.

Source: Luther Tweeten, *Public Policy Decisions for Farm Animal Welfare* (Columbus, Ohio: Ohio State University, June 10, 1991), p. 5.

The Value of Information

USDA information policy is rooted in the following principles:

- Improved information helps everyone involved in the agriculture and food system make more economically rational decisions. For example, information on prices, price trends, and forecasts is crucial to producers and marketing firms when making decisions on what to produce and when to market.

- Improved information makes markets more competitive. One of the basic assumptions of the economic concept of a perfectly functioning market is perfect knowledge by everyone in the market regarding prices, production, costs, and so forth. Absolute perfection cannot be achieved. However, the more reliable the information available to decision makers, the more rational the decisions and the more competitive the market.

- Improved information results in a more equal balance of power among firms in a market. Information is market power. This is probably most widely recognized in the extensive and highly sophisticated information systems that have been developed by the multinational grain companies. These systems bring together all available public as well as privately generated information as a basis for dealing with governments, other market intermediaries, producers, and consumers throughout the world. Highly complete and complex information systems are by no means limited to the grain companies. Every firm that is doing an effective job of purchasing and marketing strives to have more complete information than its competitors or the firms with which it is dealing. Most farmers and smaller businesses, however, do not, as individuals, have the resources to develop or purchase information that even approaches that of the larger firms. The provision for public information bridges that gap and thus results in more balanced bargaining relationships within markets.

- Improved information results in more efficient market coordination. Information on the quantities of products available and expected to be produced provides signals to decision makers up and down the market channel. Information thus becomes at least a partial substitute for vertical integration.

Information policy issues.   With all its far-reaching beneficial effects, one might think that the marketing information policies of USDA would be relatively uncontroversial. Such is not the case. Five interrelated issues are at the center of these concerns:

- Questions are persistently raised about the *accuracy of USDA data* and their influence on the level of prices. Crop and livestock production estimates have been a major target of this criticism. NASS crop production estimates require a combination of acreage estimates and yield estimates. Once crops are planted, acreage estimates are very accurate. Yield estimates are based primarily on actual statistical sampling and field counts. The basic goal of NASS is to be within 2 percent of actual production. In reality, they are generally within 1 percent of actual production. The human nature of farmers is such that when NASS estimates are high and thus tend to suppress prices, extensive criticism frequently results. Livestock numbers are more difficult than crop production to estimate, but NASS's record of accuracy stands well against the criticism.

- Charges are sometimes made of *political manipulation of numbers* to make the food and agricultural situation look better or worse than the facts would suggest. On the critically important estimates of crop and livestock production numbers, every precaution is taken to see that there is no political influence. A procedure is used whereby once the data have been pulled together by the states, those working on the estimates are literally locked in a room until the final estimate is reached. During the 1973–1974 food crisis, CIA and White House officials were surprised to find that even the secretary of agriculture did not have prior access to the crop production estimates.

  This is not to suggest that all USDA data are completely free of political influence. At times, particularly in the heat of a legislative battle, GATT trade negotiations, or a political campaign, numbers do appear to be generated by USDA that are designed to prove a point with little basis in fact. This is particularly true of numbers relating to anticipated expenditures on programs and food price inflation rates. These numbers may be influenced by the office of the secretary. However, those numbers that are most crucial in affecting market prices are the most universally respected for their accuracy and objectivity.

- Questions consistently arise regarding the *appropriate division of responsibility* for information generation, analysis, and forecasting between USDA and the private sector. Private firms have been generating price and production information for years. In commodities such as eggs and red meat, the private sector quotations from sources such as Urner-Barry and the Yellow Sheet are recognized to be more widely used by the industry than comparable USDA price quotations.

- Computer technology and increased price instability have brought on a new wave of private information, analysis, and forecasting services. These profit-oriented firms appear increasingly to be viewing USDA as an effective source of competition for analyzing, forecasting, and formatting statistics so that they could be sold privately at a profit. The result has been increasing advocacy by these interests that USDA handle only data collection, with the private sector handling the analysis, forecasting, and distribution. Budget pressures and private sector initiatives have resulted in a significant reduction in the Economic Research Service (ERS) economic situation and outlook reports on a commodity basis. Although ERS was once looked to as the training ground for the best commodity analysts in the world, this is no longer

Agriculture and the Information Superhighway

Someday I am sure, your children or grandchildren will ask you,"What was life like before the Internet?" We are on the threshold of a new age of information exchange. Just as the telegraph, radio, and television impacted life in the past, the Internet will impact our lives in ways which we cannot even imagine.

Today the Internet is known mostly for its entertainment value, but it was originally conceived so that research institutions and government entities could exchange information. Presently, it is a medium for all types of entities to exchange information: governments, corporations, universities, agencies, businesses, and individuals.

Few disciplines have as superb an Internet presence and abundance of quality information as agriculture. Most USDA reports and data sets are available via the Internet. Many universities, foreign governments, trading blocs, and agribusiness firms have web sites relating to agriculture. The majority of the tables and figures in this textbook were compiled from information gathered over the Internet. Furthermore, the Internet was utilized extensively for checking facts in this edition. Some Internet sites that might be of interest to the student of agricultural and food policy:

- The Agricultural and Food Policy Center at Texas A&M University, *http://afpc1.tamu.edu*
- The U.S. Department of Agriculture, *http://www.usda.gov/* National Agricultural Statistics Service, *http://www.usda.gov/nass/* Economic Research Service, *http://www.econ.ag.gov/* Foreign Agricultural Service, *http://ffas.usda.gov/ffas/*
- The Food and Agriculture Organization of the United Nations, *http://www.fao.org/*
- AgriNet, *http://agrinet.tamu.edu/*
- Resources for Economists on the Internet, *http://econwpa.wustl.edu/EconFAQ/EconFAQ.html*

the case. This change in policy represented a departure from persistent efforts by USDA to see that all participants in production and marketing have equal access to its information and analyses. It gives larger farmers and agribusiness firms an advantage that, over time, is reflected in increased concentration of market power.

- Most of the data collected by USDA are obtained voluntarily from individuals and firms. The increased complexity of agriculture, integration, size of firms, and the proprietary nature of the information have made collection of certain types of data, such as prices and production information, more difficult to obtain. As a result, increased requirements for mandatory reporting of information have been suggested.[39] There seems little doubt that long-term trends toward a more concentrated agriculture continue to undermine the ability of USDA to collect data voluntarily. Thus, if the quality and quantity of food and agricultural data are to be maintained, more reporting requirements may be necessary.

Countervailing Power Policy

Countervailing power *offsets market power with market power.* When market power develops on one side of the market, initiatives frequently develop to build a neutralizing power on the other side of the market. **Countervailing power policy** is *designed to assist producers in balancing the monopsony power of buyers in the markets where producers sell their products.* Countervailing power is important to structure because it provides farmers with a means to form and utilize cooperatives either to provide a marketing and bargaining interface with the corporate agribusiness sector or to integrate forward in competition with corporate agribusiness firms. Because all sizes and types of farmers are eligible to join cooperatives, small farmers can gain some of the advantages enjoyed by both large farmers and integrated corporate firms.

Specific programs designed to accomplish these objectives include the Capper-Volstead Act, marketing orders, commodity promotion programs, agricultural bargaining legislation, and the Packers and Stockyards Act.

Capper-Volstead Act. The Capper-Volstead Act gives producers the right to jointly market their products. Marketing has been interpreted broadly to cover any activity beyond the production level, including joint preparation for market, processing, and pricing. The act also gives cooperatives the right to form a common marketing agency for the purpose of carrying out the same functions encompassing a market share that could presumably extend to a monopoly.[40] It is still unclear whether cooperatives can legally control the production of their members,[41] but it is clear that without the Capper-Volstead Act, many of the marketing activities currently engaged in by cooperatives would be a violation of either the Sherman Antitrust Act or the Clayton Act provisions restricting mergers.[42]

---

[39]Kirby Moulton and Daniel I. Padberg, "Mandatory Public Reporting of Market Transactions," in *Marketing Alternatives for Agriculture* (Washington, D.C.: Committee on Agriculture and Forestry, U.S. Senate, April 1976), pp. 28–37.

[40]*Cape Cod Food Products v. National Cranberry Assn.*, 119 F. Supp. 900 (D. MA, 1954).

[41]Practically, this may be a moot point because it is highly unlikely that cooperatives can effectively control the production of its producer-members without the help of a government program.

[42]*U.S. v. Borden Co.*, 308 U.S. 188 (1939).

Although the Capper-Volstead Act gives cooperatives substantial latitude for organizing to exercise countervailing power against buyers, the exemption from antitrust prosecution is not unlimited. Cooperatives do not maintain their exemption when they engage in predatory market conduct to exclude competition.[43] In addition, cooperatives cannot maintain their exemption if they either have non-producer members[44] or if they conspire with nonproducer corporations.[45]

The Capper-Volstead Act has given producers substantial latitude for offsetting buyer market power. There are, however, only certain commodities for which producers and their cooperatives have taken anything like full advantage of their antitrust exemption. Examples of cooperatives that have acquired sufficient market power to test the limits of the Capper-Volstead exemption include those in milk, rice, navel oranges, lemons, grape juice, cranberries, walnuts, almonds, raisins, and prunes. Among these, only milk and rice could be classified as major agricultural commodities.

Cooperatives have been sufficiently successful to be the dominant force influencing prices in the dairy industry. This influence ranges from procuring, processing, and marketing milk and its products to organizing effective legislation to support milk prices as well as raising prices above the minimum prescribed in federal milk marketing orders. This influence has created sufficient concern to result in several antitrust suits against milk cooperatives for engaging in predatory trade practices and monopolizing markets for milk. Rice cooperatives have been nearly as effective as those for milk, but with considerably less public visibility and with a smaller overall market share. In other words, rice cooperatives have relied more on marketing skill than on the raw market and political power tactics that have been typical of some milk cooperatives.

The influence of cooperatives in grain, livestock, and poultry is limited to their role as competitors in the markets where they operate. In these commodities, the balance of market power still lies with the major corporate grain and food marketing firms. Having said this, there is considerable controversy over the potential advantage that the Capper-Volstead Act gives cooperatives over their proprietary competitors.

Cooperative Policy Issues

Issues of cooperative policy evolve largely from public and competitor concerns about cooperatives obtaining and exploiting a truly dominant market position. Specific concerns include the following:

- USDA has been charged with not enforcing the provisions of the Capper-Volstead Act that prohibit undue price enhancement by cooperatives. The

[43]*Maryland and Virginia Milk Producers Assn. v. U.S.*, 363 U.S. 459 (1960).
[44]*Case-Swayne Co. Inc. v. Sunkist Growers, Inc.*, 389 U.S. 384 (1967).
[45]*U.S. v. Borden Co.*

undue enhancement provision is viewed by many as the final public interest protection against potential monopolistic cooperative abuse. Concern exists that the USDA has neither formally defined what is meant by the term *undue enhancement* nor conducted a hearing under the undue enhancement provisions. USDA has, however, conducted several investigations of potential undue enhancement abuses. The secretary of agriculture invariably finds that while, on the one hand, USDA is charged with promoting cooperatives and the related producer interest, on the other hand, it is charged with regulating potential cooperative monopoly price exploitation. One alternative for removing this conflict of interest is to move Capper-Volstead enforcement from USDA to either the Department of Justice or the Federal Trade Commission.[46]

■ Substantial concern also exists regarding the apparent *right of cooperatives to merge* with one another to the extent of obtaining a *dominant market share*. Although some sympathy might exist to curb mergers among cooperatives having a large market share, the court always finds itself caught in the dilemma that if the merger is denied, the farmer-members of the two cooperatives could simply disband and form a single new cooperative.[47] If restrictions are to be placed on marketing cooperative mergers, it will probably have to be done by Congress. Congress has not, however, been inclined to further restrain cooperative market power.

■ As agriculture continues its trend toward increased scale of production and vertical integration, cooperatives are sometimes being formed by large proprietary agribusiness firms that are also producers of agricultural products. This creates the potential *for large corporations to utilize the Capper-Volstead Act to* **achieve monopolistic objectives.** This issue may once again need to be resolved by Congress or might be resolved by the courts ruling that large corporate membership is not consistent with the original intent of the Capper-Volstead Act. A recent Supreme Court ruling with regard to broilers tended to have this effect when it was held that the contracting growers as opposed to the proprietary integrators were the producers within the meaning of the Capper-Volstead Act.[48]

■ As cooperatives continue to grow and acquire market power, questions will arise as to *whether domestic farm programs* **are needed any longer.** These questions have already arisen in milk. In milk, people question whether cooperatives have already acquired sufficient market power to no longer need the assistance of price supports and marketing orders.[49]

[46]An excellent discussion of the undue price enhancement issue is contained in E. V. Jesse, A. C. Johnson, Jr., B. J. Marion, and A. C. Manchester, "Interpreting and Enforcing Section 2 of the Capper-Volstead Act," *Am. J. Agric. Econ.* (August 1982), pp. 431–443.

[47]*Sunkist Growers, Inc.*, et al. *v. Winkler and Smith Citrus Products Co.*, 370 U.S. 19(1962).

[48]*U.S. v. National Broiler Marketing Association*, 98 S. Ct. 2122, Trade Cases, para. 62074 (1978).

[49]Robert Masson and Phillip Eisenstat, "The Pricing Policies and Goals of Federal Milk Order Regulations Time for Reevaluation," *S.D. Law Rev.* (Spring 1980), pp. 662–697.

Marketing orders.   Another specific area of concern is that of marketing orders. **Marketing orders** *provide another mechanism by which producers may initiate programs to regulate the marketing of their commodity to achieve orderly marketing through unified action.*[50] Marketing orders for certain fruits, vegetables, nuts, and milk can be requested by producers and implemented after a hearing if two-thirds of the producers favor the proposed order. Once approved by producers, an order is binding on all producers.

The order defines the commodity and market area to be regulated. Orders give producers a degree of control over product quality, quantity, or price, depending on the specific provisions of the order. Because federal orders are quite different for milk than for fruits and vegetables, each is described separately.

Milk marketing orders set minimum prices that processors must pay for fluid-grade milk. Under marketing orders, milk is priced according to the use processors make of the milk. That is, milk used for fluid purposes is priced at a higher level than milk used for making manufactured products, such as butter, cheese, or ice cream. The higher price results from the more inelastic demand for fluid milk. The result of this price discrimination on the basis of end use is higher producer returns. Because orders prescribe only minimum prices, cooperatives are free to negotiate premiums over federal milk order prices.

Fruit and vegetable marketing orders rely primarily on control of quality, market flow, and volume management to influence the level and stability of producer returns. Quality control measures specify minimum grades and sizes of products that can be marketed. Market flow regulations limit the quantities of products that can move into fresh markets on a weekly basis during the heaviest marketing season. Volume management regulations restrict the supplies of storable dried fruits and nuts going into fresh markets through holding supplies in reserve until the next marketing season. The effect is to restrict the quantity a producer can market or to divert excess supplies to alternate markets such as exports (two-price plan). Additional provisions of fruit and vegetable marketing orders may specify varieties of products that can be planted; establish standard packs and containers; and fund research, development, commodity advertising, and promotion.

Marketing Order Issues

Several marketing order issues have surfaced in recent years. The center of concern in these issues involves the extent to which order provisions operate in the public interest. A related concern is whether orders are really needed to protect producer returns. The following are among the major issues:

- Pricing milk used for fluid purposes on the basis of Eau Claire, Wisconsin, plus transportation has been charged with discriminating against Upper

---

[50]This discussion is limited to federal marketing orders. State marketing orders for milk, fruit, vegetables, and nuts have similar provisions but tend to be used mainly in the western and eastern states.

Thin Markets

In an era of increased integration of production and marketing, there is a need for a constant flow of products through the market to large processing plants. The result is increased contracts with prices specified by formula off from some spot market or from a market news report issued by AMS/USDA.

The National Cheese Exchange (NCE) was such a market. Trading on NCE occurred every Friday. Often there was little or no trading on NCE. Yet the NCE price for 40-pound blocks of cheddar cheese was said to be the formula pricing base for the entire U.S. cheese industry. More important, the cheese price is the driver for all farm level milk prices.

Needless to say, NCE was a focal point of attention for dairy farmers, processors, and the AMS/USDA, which administers federal milk marketing orders—the price setting regulatory agency for milk. A University of Wisconsin study concluded that the NCE price was controlled by a couple of cheese manufacturers and cooperatives. One of the cooperatives as much as admitted that it refrained from marketing on NCE to prevent the price of cheese from falling and purchased from NCE to support the price of cheese. A University of Maryland study indicated that NCE was not rigged but was simply operating as a residual market absorbing excess supplies and demands.

Farmers in Wisconsin became sufficiently riled up by the NCE rigging controversy that they drove their manure spreaders to NCE in protest. AMS looked into the alternative of pricing milk by collecting actual 40-pound block transaction prices between manufacturers and buyers. Farmers argued that if NCE prices are rigged the transaction prices would effectively also be tainted. In 1997, after pressure from dairy farmers and the Wisconsin legislature, the NCE was closed and moved to the Chicago Mercantile Exchange where butter is traded in an equally thin market. The movement of the NCE is unlikely to accomplish anything in terms of the underlying thin market problem.

Such policy controversies are increasingly being played out in agricultural markets. In the process, market information becomes more important at a time when there are less public monies to support collecting and providing this information.

Commodity Advertising: A Zero-Sum Game?

Farmers have gotten into the advertising and promotion business in a big way. They advertise milk, beef, grapes, oranges, peaches, cotton, apples, potatoes, and eggs. The total cost to producers and government is about $1.0 billion and growing. The money is largely collected under check-off programs. Some of the money is used for market research, but most of it goes for generic advertising. It is generally recognized that brand advertising is more effective, but farmers, USDA, and the Congress do not want to support food and fiber agribusiness firms in their marketing efforts.

USDA initially resisted getting into the advertising business because total demand would not be enhanced. The stomach is only so large, they argued. Therefore, one farm commodity group is just taking market share from another. The winner is Madison Avenue, NFL football players, and media personalities. From the perspective of any commodity, advertising can be justified, but does agriculture as a whole benefit? Probably not. But in America, advertising is like baseball, hot dogs, apple pie, and California fruit. Some producers do not think so! They view a check-off from the buyer's payment for their marketings as a gross injustice. Why should the government mandate that producers participate in financing an advertising system? Proponents assert the benefits of generic advertising in expanding demand at a time when government is discontinuing its subsidies and price supports.

Midwest producers and in favor of farmers located in the South and Southeast. Reform of the milk order system to consolidate orders and modernize its pricing system was mandated by the 1996 farm bill. In the reform process, the basis for pricing milk used for fluid products, soft products, and manufactured products will change. Controversy extends to the use of the NCE as the adjuster/mover of milk prices over time (see box).

- Pricing milk on the basis of use has been attacked by consumer interests for *raising the price of milk above competitive levels* to the detriment of consumers. Without question, consumers pay a higher price for fluid milk as a result of the federal milk order program.[51] Milk producer interests suggest that the intent of marketing orders was to raise producer return but that the orders benefit consumers by providing increased stability.

- Fruit and vegetable marketing order provisions that limit the supply of products moving to market have similarly been attacked as being *detrimental to consumer interests*. It is argued that grade, flow to market, and volume management regulations all hold edible products off the market, to the detriment of

---

[51]Roger A. Dahlgran, "Welfare Costs and Interregional Income Transfers due to Regulation of Dairy Markets," *Am. J. Agric. Econ.* (May 1980), pp. 288–299.

consumers.[52] Those critical of fruit and vegetable marketing orders do not consider the consumer benefits in terms of uniform product quality over a longer market period sufficient to offset consumer losses in terms of higher prices.

■ Fruit and vegetable marketing orders are administered by an administrative committee made up of largely industry interests. USDA has made efforts to get a public representative appointed to each administrative committee, but consumer interests have suggested that two or three *bona fide consumer representatives ought to be required.* Producers are opposed to any consumer involvement in orders.

■ It is also suggested that *orders unduly favor cooperatives,* by giving them the right to vote for orders on a bloc basis for all their members. As a result, a cooperative having two-thirds of the producers is in a position to control the order, subject, of course, to the discretion of USDA.[53]

Commodity advertising.   Extensive use of nonprice competition in the form of advertising by large agribusiness marketing firms has led farmers to advocate the establishment of commodity advertising programs for farm products. These programs, totaling about $1 billion, are financed by producer check-off funds. That is, when farmers market their products a given amount per unit is deducted from their receipts. This money goes into the promotion fund, which is managed by a producer promotion board appointed by the secretary of agriculture.

Commodity promotion programs exist for a wide range of commodities; the largest of which are for milk, orange juice, beef, and cotton. Although their programs have been authorized in farm bills as separate titles, many promotion programs operate under either state or federal marketing orders.

Promotion programs are essentially competitive advertising designed to persuade consumers away from substitute products or to attract foreign customers to buy from the United States. For example, milk advertising is designed to encourage youth, older Americans, and women to consume milk rather than soft drinks. Orange juice advertising has a similar objective. Cotton advertising is designed to attract customers who might otherwise buy synthetic fabrics. The effect is to expand demand and raise the farm price.

Returns to generic advertising have been extensively studied. They range from $1.85 in revenue per advertising dollar invested for milk to $14 in soybean export promotion (Table 12.7). Most of the results show returns in the $4 to $6 range per dollar invested. Supporters of generic advertising view these returns as evidence of its success.

Agricultural bargaining.   Providing producers bargaining rights comparable to those enjoyed by organized labor has been the goal of several major agri-

[52]Richard Heifner *et al., A Review of Federal Marketing Orders for Fruits, Vegetables, and Specialty Crops: Economic Efficiency and Welfare Implications* (Washington, D.C.: AMS, USDA, October 15, 1981).
[53]Ibid.; see also Robert T. Masson and Phillip M. Eisenstat, "Welfare Impacts of Milk Orders and Antitrust Immunities for Cooperatives," *Am. J. Agric. Econ.* (May 1980), pp. 270–278.

TABLE 12.7 Returns to Commodity Promotion Programs in the United States

Commodity	Revenue per Dollar Invested
Fluid milk	$1.85–$7.04
Fluid and manufacturing grade milk	$2.04–$4.77
Milk and cheese	$11.29
Beef	$5.74
Soybeans and products (for export)	$14.00
Orange juice	$2.28
Grapefruit juice	$10.44
Apples	$6.74
Cotton	$5.38–$5.95

Source: Oral Capps, Jr., et al. *Economic Evaluation of the Cotton Checkoff Program*, (College Station, Tex.: Market Research and Development Center, Texas A&M University, October 1996).

cultural movements. The National Farmers Organizations's major goal is to obtain good faith collective bargaining rights for producers. These efforts have been strongly resisted by the organizations most directly affected, such as fruit and vegetable processors and poultry contractors.

The greatest bargaining benefit for producers is in contract-integrated agriculture, where producers have lost control of many production and marketing decisions. This resulted in the passage of the Michigan Agricultural Marketing and Bargaining Act. This act gives certified Michigan cooperatives the right to bargain with processors of specified fruits and vegetables on behalf of all producers.

No comparable law exists at the federal level. The closest is the Agricultural Fair Practices Act, which prohibits processors from discriminating against producers because they are members of a cooperative. This law has done little to assist farmers who desire good faith bargaining rights equivalent to labor.

To be successful in bargaining, the following organizational and product characteristics must exist:

- *The demand for the product involved must be inelastic.* Without an inelastic demand, the percentage decline in the quantity demanded will be greater than the percentage increase in price. The result will be reduced producer revenue. Since most farm products have an inelastic demand, this condition is normally satisfied.

- *The bargaining group must control a sufficient volume of the product.* The larger the volume of product controlled, the more bargaining power. Generally speaking, at least two-thirds of the production must be controlled to be a significant bargaining force.

- *Unity and discipline must exist within the bargaining group.* Unless the members of the group stick together, the bargaining effort will be lost.

- *There must be recognition that the bargaining group can exercise control.* This recognition must be established at an early stage in the bargaining process. Once a bargaining group fails to establish recognition, it is much more difficult to negotiate the second time around. Bargaining laws can assist in obtaining recognition for the bargaining group.

- *If successful, the bargaining group must be able to control production.* Producers normally respond to higher prices by increasing output. This normal response must be controlled.

- *The bargaining group must be efficient in performing its functions.* Unless it is efficient, alternative means of production and marketing will be found by the buyers of the product—even to the extent of integrating to produce the product themselves.[54]

Packers and Stockyards Act. In the early 1900s, major concern developed over the potentially dominant position of the major meat packers. In 1910, a Federal Trade Commission (FTC) study of the meat industry concluded that the five largest meat packers were manipulating livestock markets, controlling prices of dressed meat, and extending their control to fish, poultry, and eggs, as well as to fruit, vegetables, and staple groceries.

As a result of this investigation, the Department of Justice brought action against Swift, Armour, Morris, Wilson, and Cudahy for violating the Sherman Act. In 1920, these packers signed a consent decree that prohibited them from

- Owning public stockyards, public cold storage facilities, or stockyard rail terminals
- Handling or letting others use facilities to handle 114 nonmeat products, including fish, vegetables, fruits, canned goods, and cereals
- Handling fresh cream or milk except for sale in processed form
- Operating retail meat markets in the United States

The consent decree has had a major impact on the structure of the meat industry. Its greatest effect has been to encourage the rise of the supermarket chain as a major force in meat marketing. It prevented major packers from being a competitor in retailing meat. In late 1981, the consent decree was dropped. However, its effects on the industry continue to be felt.

The consent decree formed the basis for the passage of the Packers and Stockyards (P&S) Act in 1921. It was, in a very real sense, designed to prevent the basic abuses discovered in the FTC investigation that led to the consent decree. Specifically, the P&S Act was designed to accomplish the following:

[54]Leon Garoyan and Eric Thor, "Observations on the Impact of Agricultural Bargaining Cooperatives," *Agricultural Cooperatives and the Public Interest,* N.C. 117, Monogr. 4 (Madison, Wis.: University of Wisconsin, September 1978).

- Prohibit unfair, deceptive, or unjustly discriminatory practices.
- Prohibit practices that would give particular persons or localities undue competitive advantages.
- Prohibit practices that would have the effect of apportioning supplies, manipulating or controlling prices, or restricting competition.
- Authorize the regulation of services and rates at public stockyards.
- Provide for honest weights as well as prompt and full pay in marketing livestock and poultry.

Much of the initial regulatory emphasis of P&S was on the maintenance of fair competition and reasonable charges in central public markets for livestock. P&S pursued a policy that was, in effect, designed to maintain the industry structure in its original mold, but pronounced structural changes occurred. Increased direct farmer-packer sales reduced the role of public markets, which were a major proportion of P&S regulatory activity. In addition, many new aggressive firms entered the meat packing business. These firms felt less restrained by the provisions of the consent decree. They were more willing to challenge the regulatory authority of P&S at a time when there was structural change occurring in the meat industry, including the development of large-scale feedlots and integrated systems of production and marketing.

In the late 1960s, P&S attempted to gain control of structural change in the industry by prohibiting packer ownership of cattle on feed. It was, however, unable to demonstrate adverse effects on competition and eventually dropped legal action. Subsequent investigations into packer joint ventures with feedlots, formula pricing of meat, vertical integration by grain companies, and potential monopoly abuses also ended in a lack of decisive action by P&S.

In the late 1970s, Congressman Neal Smith (Dem.–Iowa) began actively investigating potential monopoly abuses in the meat industry. After hearings, Smith proposed major regulatory changes, ranging from prohibition of formula pricing of cattle off the Yellow Sheet[55] to placing an absolute 25 percent limit on the market share of livestock purchased by any single packer in any state.

Implicit in the action of Congressman Smith was a concern that P&S has not been sufficiently active in dealing with structural and pricing issues in the meat industry. This point was reemphasized in the late 1980s when four packers gained control of more than 70 percent of the steer and heifer slaughter capacity with no concern being expressed by P&S.[56] In 1995 the Secretary authorized university studies and a special task force to investigate pricing issues in meat. This led to P&S litigation as well as new initiatives in price and volume reporting. Obviously, this is not the end of the livestock pricing controversy.

---

[55]The Yellow Sheet is a private market news service that provides daily quotes on wholesale meat prices. These quotas are used extensively as a basis for pricing meat in the wholesale meat trade.
[56]Ward, Structural Change in Livestock.

Over time, it has become increasingly apparent that USDA's ability and/or will to regulate markets is weak.[57] As a result, the ability of farmers to countervail the power of input and marketing firms is also weak.

Tax Policy

Tax laws have a significant impact on the structure of agriculture. The 1986 Tax Reform Act was designed to reduce the incentives for farms to get larger solely for tax reasons. Available research suggests that it may not have accomplished that objective in crop agriculture, although the incentives for tax-avoidance investments in breeding stock should be materially reduced.

The following are among the major changes in tax provisions introduced by the 1986 Tax Reform Act:

- Lower tax rates and fewer brackets
- Capital gains treated as ordinary income
- Lengthening of depreciation schedules
- Limiting prepayment of expenses
- Requirement of capitalization of preproductive expenses exceeding two years
- Elimination of income averaging
- Prohibition of passive investors deducting losses from wages and salaries

USDA contends that these tax changes help small farms more than large ones,[58] but other research indicates the opposite. Research at Texas A&M and Minnesota indicates the following:[59]

- Crop farms with more than $40,000 in income will pay less taxes, while those with less than $40,000 will pay more taxes.
- Taxable incomes for dairy and cow/calf farms increase dramatically. However, the percentage is more for small operators than for large operators.
- The ability to defray risk is reduced by the elimination of tax averaging. Farmers who are best at managing risk will have a greater advantage.
- Incentives increase to cash rent because share rent is considered a passive investment.
- Incentives for greater labor intensity result from reduced capital write-offs.

[57]Ronald D. Knutson, L. Leon Geyer, and John W. Helmuth, Trade Practice Regulation, Fed. Agric. Market. Prog. Leaf. 8 (College Station, Tex.: Texas Agricultural Extension Service, 1988).

[58]Gregory D. Hanson and Diane R. Bertelsen, "Tax Reform Impacts on Agricultural Production and Investment Decisions," paper presented at the American Agricultural Economics Association Annual Meetings, Michigan State University, East Lansing, Michigan, August 3, 1987.

[59]Clair J. Nixon and James W. Richardson, "Tax Act Signal to Commercial Farmers: Get Larger or Get Out," and Thomas F. Stinson and Michael D. Boehlje, "Dramatic Tax Rule Changes: Significant But Not Immediate Effects," *Choices* (2nd Quarter, 1987), pp. 12–16.

Tax policy does influence structure. The trend toward a bimodal distribution of farming is accelerated by the tax policy changes. Tax policies could be constructed to favor moderate-size farms, requiring more progressive property and income taxes and more restrictions on the ability to transfer estates from one generation to the next. Changes in tax policies that have occurred tend to go in the opposite direction.

Research and Extension Policy

It was recognized previously that technological change is one of the factors that has made U.S. agriculture highly productive. One of the side effects of technological advance has been structural change. That is, new technology generally shifts the cost curve downward and to the right. For example, the self-propelled combine not only reduced the cost of harvesting corn but also made it possible for a farmer to harvest a much larger number of acres. The result of a large number of virtually continuous technological advances has been fewer but larger farms.

In the early 1970s, Ralph Nader's Agriculture Accountability Project became interested in the contribution of public research to the process of technological change. Specifically, this project investigated the extent to which the land-grant university and USDA agricultural research complex emphasized research that primarily benefited large farmers and agribusiness firms. The study concluded:

> America's land grant college complex has wedded itself to an agribusiness vision of automated, vertically integrated and corporatized agriculture. . . . Had the land grant community chosen to put its time, money, its expertise and its technology into the family farm rather than into corporate pockets, then rural America today would be a place where millions could live and work in dignity.[60]

The publication *Hard Tomatoes, Hard Times* was extensively criticized by researchers and officials in the land-grant university system. Although the conclusion of the Nader project was obviously overdrawn to make the point, the fact remains that technology is seldom structurally neutral. In fact, the question exists whether technology could be made structurally neutral and still realize its benefits in terms of lower costs.

In 1979, Secretary Bergland picked up on the *Hard Tomatoes* theme as an outgrowth of the structure of agriculture project. He questioned the use of federal funds for research projects having the objective of producing large-scale, labor-saving technology.[61] He set up a special task force to investigate the impact of research and extension on the structure. During this period, Congress earmarked research and extension funds for increased work with small farms and for projects involving direct marketing from farmers to consumers.

---

[60]Jim Hightower, *Hard Tomatoes, Hard Times* (Cambridge, Mass.: Schenkman Publishing Company, 1973), pp. 138–139.

[61]Bob Bergland, *The Federal Role in Agricultural Research*, USDA 262–80 (Washington, D.C.: USDA, January 31, 1980).

The Bergland initiative on research appeared to die with the election of the Reagan administration. It was, however, rekindled by the announcement of joint initiatives between private sector companies and universities in high-priority genetic engineering research. Questions arose as to whether the primary beneficiaries would be the private sector firms or the initial farmer adopters of the resulting new technology.

In 1997 the National Academy of Sciences became concerned about the impacts of new technologies and agricultural research on structure. A trigger for this concern was precision farming techniques utilizing sensors and computers on farm equipment to get the highest yields with the least application of fertilizer and pesticides.

The question remains as to whether public research could, or should, be oriented more toward smaller family farms, as suggested by Cochrane and OTA. The long-run cost and competitive implications of such a strategy have not been adequately evaluated. Similarly, the impact of patenting agricultural research discoveries has not received adequate attention. Structure policy has been largely ignored in the farm policy debate because farmers have not put it on the agenda. Farm organizations tend to be controlled by large farmers who benefit most from technology.

THE FUTURE OF STRUCTURE POLICY

There is a distinct possibility that agriculture has moved so far and decisively toward an integrated industrialized structure that there is little chance of reversing the trend. Government policies, although not designed to foster these trends, clearly contribute to them. That is, income supports, credit, research, and tax policies all tend to put the large farmer in a relatively better position to grow and expand than the smaller farmer. The larger the farmer, the more the apparent benefits.

At the same time, few policies are designed overtly to stem the trend toward agricultural integration and industrialization. Only the antitrust laws appear to have that potential, but even there, not at current relatively low levels of concentration.

The trend toward industrialization will adversely affect farmers' ability to convince the Congress of the need for traditional farm programs. The ability of the family farm to survive is in question. Family farm agriculture may be limited in the future to part-time farming. Commercial agriculture will, at that time, be primarily integrated, industrialized agriculture. That eventuality is not a pleasant prospect to the believer in Jeffersonian agrarianism and the contribution of family farms to the productive success of American agriculture.

ADDITIONAL READINGS

1. USDA's most comprehensive report on the structure of its agriculture project was *A Time to Choose: Summary Report on the Structure of Agriculture,* by Bob

Bergland (Washington, D.C.: USDA, January 1981). When President Reagan took office in January 1981, there was an overt effort to suppress this report and ignore the whole structure of agriculture issue.

2. The OTA report titled *Technology, Public Policy, and the Changing Structure of Agriculture* (Washington, D.C.: U.S. Congress, March 1986) is must reading for anyone interested in the structure issue. OTA initiated the study after it became apparent that USDA, under President Reagan, was not going to pursue the issue.

3. A second OTA report, *A New Technological Era for American Agriculture,* OTA-F-474 (Washington, D.C.: U.S. Congress, August 1992) does an excellent job of addressing the potential technological impacts on structure.

4. Clement Ward's study of the beef industry developments toward concentration and integration, *Meat Packing Competition and Pricing* (Blacksburg, Va.: Research Institute on Livestock Pricing, July 1988) is must reading as is Wayne Purcell, ed., *Structural Change in Livestock: Causes, Implications, Alternatives* (Blacksburg, Va.: Research Institute on Livestock Pricing, February 1990).

5. Olan D. Forker and Ronald W. Ward have authored an excellent book on commodity promotion programs titled *Commodity Advertising* (New York: Lexington Books, 1993).

RESOURCE POLICY

In a world of scarcity, it is impossible to implement freedom for everyone. One person's freedoms and opportunities are a cost to another.

—*A. Alan Schmid*

Economics may be viewed as the mechanism by which limited resources are allocated to obtain maximum satisfaction for society as a whole. For most of the first two centuries following the signing of the Declaration of Independence, government policies and programs related to agriculture and food were designed to assist in the exploitation of resources.

America was founded on principles of property rights, profit-motivated enterprise, market economies, and technological change. In this institutional setting, individuals who, through productive enterprise, *utilized resources* in a manner that met consumer needs were rewarded. During the settlement of America's West, resources such as land were given away for purposes of settlement and/or exploitation. Water resources were developed, given away, and/or sold for less than their cost.

The combination of private property rights, profit motivation, technological change, and the development of transportation, public utilities, and irrigation infrastructure has been tremendously successful in fostering agricultural production. Although the political rhetoric was that of free enterprise and market economy, for farmers much of the risk on prices and incomes was removed by government programs. Elementary economic principles indicate that this was a prescription for overproduction, which has received most of the attention of agricultural policy since the 1950s.

Excessive use of resources by agriculture was of minor and only sporadic concern during the majority of the twentieth century. Often the word *conservation* was used in policymaking to politically justify programs that were designed more to remove excess capacity or, at times, even foster production (drainage projects) than to conserve resources.

ORIGIN OF RESOURCE POLICY ISSUES

Resource policy issues arise from four interdependent sources:

- Scarcity
- Differences in values placed on goods
- Property rights
- Public goods

Scarcity

Goods have value because they are scarce. If goods were not scarce, they would be free—like air. Even with air, people are probably willing to pay more to live in cleaner environments distant from the smells of industrial pollution, pulp mills, or confined animal-feeding operations (feedlots).

The more scarce resources are, relative to their demand, the greater their value. Thus, as water becomes more scarce in the West, the higher the price that is placed on it or the greater the need for governments to allocate its use between agriculture and industrial, residential, and recreational sources of demand. Likewise, as the demand for wood increases, more political pressure is exerted to harvest trees from public lands. The result is more conflict over endangered species, lower water quality in our national forests, and grazing on public lands.

It can be argued that scarcity is the origin of all resource policy issues. According to the following discussion, the scarcity of resources creates the conflict and pressure for changes in public policy regarding resource use.

Differences in Values Placed on Goods

What is water worth? Its value depends on its availability and the use to which it is put. In the aftermath of a hurricane or if a city's water supply is contaminated by a parasite, potable water is valuable. Generally speaking, most consumers and industry are *willing and able* to pay more for water than farmers. This *difference* in demand creates an inherent conflict among interest groups over a limited water supply. It creates conflict among farmers, some of whom might decide to sell their water rights to industry or to cities.

These differences in value permeate resource policy issues. Farmers and ranchers almost certainly place a different value on coyotes, wolves, or other endangered species than the wildlife advocates. Ironically, erosion may be considered a more important problem to those interested in water quality than it is to many farmers who depend on topsoil for their production. Such conflicts in values often lead to laws restricting how resources are used.

Property Rights

The rights of property were imbedded in the Constitution only to be impeded by the power of eminent domain. These property rights were, from the very beginning, identified with land and water. In this economic and political environment, a philosophy developed that the role of government was to help when the need existed but otherwise to leave people and their businesses alone—particularly their land and related resources. Encompassed within the land resource was the right to water, which was available for landowners to use in conjunction with property under law.

A *right* is a legally enforceable expectation. It involves an assurance that an activity will be protected by the legal system. The rights of private property generally include the right to possess, use, and dispose of that property and to prohibit others from interfering with that right.[1] Some rights are public in nature and are held in common by all people, such as the right to use a national park.

But no right, public or private, is absolute. For example, zoning laws establish the use of specific land areas. The Fifth Amendment of the U.S. Constitution provides that no government may take private property without just compensation. The government may therefore exercise its power of eminent domain to acquire land for roads.

Important legal questions arise when government regulations designed to protect health, safety, and welfare result in an infringement of private property rights that may be considered illegal. Under what conditions must just compensation be paid? This issue has become an important focal point for laws that could, for example, severely restrict the use of property to protect an endangered species for the benefit of the public welfare.

Not until the 1990s did farmers and ranchers fully recognize that the sanctity of their property rights could be jeopardized by government regulation. This reality is the result of two significant developments:

- A large subset of society, perhaps a majority, began to realize that the resources on which agriculture is based—land and water—are becoming scarce in significant parts of the United States.
- Society began to hold agriculture accountable for its externalities. In Bromley's words, farms were expected "not only to produce food and fiber but also to produce a particular bundle of countryside and community attributes."[2]

Acceptance by farmers of an obligation to society to do more than produce food and fiber has been a difficult proposition. Farmers and ranchers once fought

---

[1]William Goldfarb, *Water Law* (Chelsea, Mich.: Lewis Publishers, 1988), p. 10.

[2]Daniel Bromley, "A Modern Proposal for Increasing Equity through Agricultural Policy," in Steven B. Lovejoy and Kenda M. Resler, eds., Resource Police Consortium, *Implementing Agricultural Policies to Achieve Environmental Objectives: Lessons for the 1990s* (West Lafayette, Ind.: Purdue University, July 1990), p. 4.

independently against nature to prevent wolves from killing their sheep and calves, but now they are expected to protect the wolves! These societal demands have been more difficult for farmers and ranchers to accept than for other businesspeople for the following reasons:

- Nonfarm businesspersons became sensitized many years ago to the need to adjust because of their interaction with labor (particularly with labor unions) even though they did so grudgingly.
- Nonfarm businesses were located in population centers where societal and environmental pressures are more evident. "In a world of scarcity, it is impossible to implement freedom for everyone. One person's freedoms and opportunities are a cost to another. Rights defining opportunities can be understood by looking at the reciprocal relationships of people with incompatible preferences."[3] These powerful words by Schmid merit rereading and studying because they convey the sources of conflict inherent in property rights. They also explain why agriculture is experiencing a watershed of conflicts over environmental issues that farmers view as infringements of their property rights. That is, scarcity inherently implies increased infringements of freedom. As resources (land, plentiful water, clean water, clean air, wildlife, wetlands, and open spaces) become more scarce, the freedoms that farmers and property owners have enjoyed are more likely to be infringed on by society and, in turn, by government.

Public Goods

Public goods are provided or protected by government for society in the interest of social welfare. Public goods are unique in that it is physically impossible or prohibitively expensive to exclude consumption by any individual.[4] Public lands and natural amenities such as mountains or wildlife, clean air, and clean water are public goods.

As a result of having no competitively determined market price, public goods tend to be exploited, misallocated, and/or inefficiently allocated. Public goods may have either a positive or a negative value. Thus, some public goods such as pollution, have undesirable attributes and are at the root of many environmental problems.

Public goods have become a major source of resource policy controversy. Society logically chooses to exercise greater influence over public goods than over private goods. Thus, more direct government control is exercised over grazing on

---

[3]A. Alan Schmid, "The Idea of Property: A Way to Think About Soil and Water Issues," in *Water Quality and Soil Conservation: Conflicts of Rights Issues,* Special Report 394 (Columbia, Mo.: University of Missouri, Agricultural Experiment Station, November 1988), p. 14.

[4]Maynard M. Hutschmidt *et al., Environment, Natural Systems and Development* (Baltimore: The Johns Hopkins University Press, 1983), p. 60.

public lands, the allocation of rights to water originating from public lands, or the harvesting of timber from public lands than for the same issues arising on private lands. Moreover, many public goods directly interact with private property. Efforts to protect endangered species, maintain a clean water supply, or prevent air pollution are designed to protect public goods but have the potential for infringing on private rights.

The concept of public goods is closely related to **common property resources** *for which numerous owners have equal rights.*[5] An example of a common property resource is a groundwater aquifer whose rights are not lost through nonuse; neither is there a restriction on the quantity or timing of its use. With no restrictions on use, the resource can be overexploited. Users independently optimize their profit assuming that if they do not make maximum use of the common property resource, others will. As a result, there is no market mechanism to curtail exploitation. Common property resources are public goods because it is impossible to exclude consumption by any individual until government regulation of use takes over. This is precisely why government has regulated water use from aquifers in some instances.

SHIFTING POLICY FOCAL POINT

Conflicts over public goods, property rights, the allocation of scarce resources, and differences in the values to be placed on goods by markets, individuals, and groups are taking up an increasing share of the time of policymakers. As a result, the centerpiece of agricultural and food policy may be shifting from price and income support to resource policy. In the resource policy arena, although the issues are different, they involve basic conflicts in values and trade-offs that must be weighed by policymakers. These conflicts are similar to those used to justify farm programs, but with different interpretations:

- To what extent should farmers be free to exploit resources by virtue of their property rights, even though others, including future generations, may be adversely affected? Recall that freedom also was a key issue in farmers' attitudes toward price and income policy options such as production controls.

- To what degree can the market be relied on to allocate scarce resources valued differently by various interest groups? Recall that commodity programs have made decisive moves in the direction of market determined pricing.

- Are market systems sustainable or will exploitation be at the expense of future generations or time periods? Recall that food reserve (stocks) programs evolved because of questions regarding whether the private sector could be relied on to store grain for the future. Contemporary questions exist over the adequacy of government stocks.

[5]S. V. Ciriacy-Wantrup and R. Bishop, "Common Property as a Concept in Natural Resource Policy," *Natl. Res. J.* (December 1975), pp. 713–737.

- If farmers consider themselves as being close to nature, how could they possibly be viewed as violating the interests of nature? Their very survival, to a degree, was determined by their ability to control or effectively deal with nature. Recall that this identity with nature can be related to the agricultural fundamentalist views discussed in Chapter 1.

- If farmers' livelihoods and that of future generations of their families depend on the existence of their land and water resources, can it be assumed that farmers will protect these resources for future generations?

TYPES OF RESOURCES

Resources are generally classified into two categories:

- **Nonrenewable resources** have a limited supply in the relevant decision time frame. There is only a limited stock of nonrenewable resources. For example, underground aquifers that regenerate very slowly are nonrenewable resources. Once the well is dry, there is no more resource.

- **Renewable resources** are capable of being replenished. They are, in essence, a flow resource. The water in the Mississippi River may be viewed as a renewable resource in that it is continuously regenerated.

As implied by these definitions, the time frame is important in determining whether a resource is nonrenewable. Resources that are often considered to be renewable may be nonrenewable in a limited time period. For example, much of the West depends on snow melt for its water supply. Within each year, there is essentially a given stock of water—a nonrenewable resource until the next snow melt. Before population and irrigation pressures developed in the western states, water might have been thought of as a renewable resource. Recently, scarcity has made it seem more like a nonrenewable resource.

In another sense, *it may be argued that all resources are nonrenewable.* Once used, many resources will never be the same again. In other words, the quality of the resource changes. While a virgin forest will regenerate, it will never again be virgin. Alternatively, the cost of restoring a resource to its original state may be viewed as being prohibitively high. Therefore, *the issue of limited nonrenewable resources is not only one of quantity but also one of quality.*

ROLE OF TECHNOLOGY

Historically, heavy reliance has been placed on technology as a substitute or means to avoid the consequences of resource limitations. This seemed to work exceedingly well in the case of land whereby the Malthusian consequences of exponential population growth and a fixed land area were avoided by technological change. That is, limited land areas were substituted for by increased yields resulting from public and private investments in research and development that also involved the substitution of chemicals, energy, water, and management resources for land.

Investment in technology was a matter of overt government policy that, at least in relative terms, was unique to agriculture. As early as 1862, land-grant universities were created in every state. These universities were created with a specific charge in agricultural research and education through the creation of agricultural experiment stations and in adult and youth education through the creation of an extension service, in addition to the traditional university function of conferring undergraduate and graduate degrees in agriculture. Thus, many states have more than one state-supported university, but each state has only one land-grant university with unique roots in the generation of agricultural technology; Penn State University, Kansas State University, Purdue University, Michigan State University, the University of California, Auburn University, Oregon State University, and Texas A&M University are but a few examples. In 1890, a separate set of predominantly black universities was established in the South with comparable missions; Prairie View A&M University and Alcorn State University are examples.

Initially, the federal government research and extension support provided to these universities was allocated by a formula on the basis of the farm population of the states. Pressures from states having larger farms and higher sales, such as California, eventually resulted in some weight being given in the formula to agricultural sales (California being the largest agricultural state in terms of dollar sales). An equally important and controversial move involved allocating a portion of the federal agricultural research dollars to competitive peer review proposals open to scientists in both land-grant and non-land-grant institutions (Harvard, Yale, Stanford, University of Michigan, University of Texas, University of Kansas, etc.).

The land-grant university system has been the model for agricultural research and technology transfer throughout the world. However, in recent years, questions have arisen as to whether the land-grant universities are continuing to serve the public interest adequately. Like USDA, the agriculture component of the land-grant universities has been challenged on a number of bases:

- Not looking beyond agriculture for solutions to problems
- Having research and education programs that are biased toward agriculture (particularly large farms and agribusiness) as opposed to consumer and environmental concerns
- Maintaining an outdated county extension structure.

These charges, while fascinating topics for discussion, are too complex to explore here. In part, *they reflect increasing frustration over the ability of science and government to solve the contemporary problems of resource scarcity.* Perhaps problems are simply more difficult to solve in an environment of scarcity. The obstacles to increasing food production were considerably fewer when we had ample land, water, and energy resources without the government imposing restrictions on environmental degradation. *Science has a very important role to play, but that role will involve scientists operating within a different set of societal constraints just as business and farmers operate within a different set of constraints.* Technology can play an important role in solving resource problems, but it does not forestall the need for an adjustment in how resources are used. Technology can complement our ability to adjust to resource policy issues.

VALUE PLACED ON RESOURCES

Every resource has value. Who determines that value and the constraints placed on the process of value determination are central concerns of resource policy. Resource policy decisions involve consideration of both market values and nonmarket values.

Market Value

Market value is the price that a resource will receive in the marketplace. The market value is determined by the conditions under which a resource is sold. For example, land located near a metropolitan area may have a considerably different market value if it can be sold only for agricultural purposes than if it can also be used for development. Likewise, wetlands may have greater market value if they can be drained and used for agriculture. Then, too, the price of agricultural land is influenced by the availability and cost of water from either natural rainfall or irrigated sources.

Nonmarket Value[6]

Nonmarket value is the value of a resource or activity that is not revealed in the marketplace. It is the value placed on a resource that is not traded in markets.[7] What, for example, is the value of maintaining a wetland when draining it would affect not only the populations of waterfowl, fish, and invertebrates, but also its availability and scenic value and that of related remote areas? It is possible for the nonmarket value of wetlands to be greater than either its market value for agriculture or development? Informed policy decisions require an estimate, understanding, or judgment of this nonmarket value relative to the market value of wetlands for agricultural or developmental purposes.

[6]This section draws on the following sources: Teofilo Ozuna, Jr., and John Stoll, "The Significance of Data Collection and Econometric Methods in Estimating Nonmarket Resource Values," Am. Fisheries Soc. Symp. (1991), pp. 328–335; John R. Stoll, "Recreational Activities and Nonmarket Valuation: The Conceptual Issue," *So. J. Agric. Econ.* (December 1983), pp. 119–125; and Alan Randall, "Total and Nonuse," in J. B. Braden and C. D. Kolstad, eds., *Measuring the Demand for Environmental Quality*, (Amsterdam: North-Holland Publishers, 1991), pp. 303–321.

[7]John C. Bergstrom, John R. Stoll, and Alan Randall, "The Impact of Information on Environmental Commodity Valuation Decisions," *Am. J. Agric. Econ.* (August 1990), pp. 614-621.

Nonmarket values originate from three primary sources:

- **Current use value** involves the benefits derived from the consumptive or nonconsumptive benefits or costs derived from a good not valued by the market. Current use value may be either direct on-site use, such as hunting in wetlands, or off-site use, including activities such as reading about the natural environment or seeing films about it or watching sport fishing or nature television programs about wetlands.
- **Optional use value** involves the willingness to pay a premium to ensure the future availability of an amenity, even if it is not currently used.
- **Existence value** involves benefits derived from seeing that a resource is available to future generations or that a resource is left undisturbed regardless of its current or optional use value.

Nonmarket values are much more difficult to determine than market values because they cannot be measured directly in the market. Economists use three techniques to determine nonmarket values:

- **Property value** approaches provide an indirect measure of nonmarket values by comparing the market value of property having different environmentally relevant characteristics. This technique is also referred to as the *hedonic price* in that the value is implied from the characteristics of property. Examples include placing a value on soil erosion by comparing the price of land in a given area having various levels of topsoil or by placing a value on clean air by comparing the value of comparable housing located varying distances from large confined animal-feeding operations.
- **Travel cost** approaches to nonmarket valuation involve tabulating the amount of money spent visiting a site. Because different individuals live various distances from the site, travel costs differ, thus providing an indication of the price people are willing to pay for the use of the site. Therefore, from a survey of travelers to a site and the amount of money they spend, it is possible to construct a demand curve for the site. Of course, travel costs can be used only to establish the current use value of a recreational site.
- **Contingent value** approaches utilize survey techniques, which, in effect, ask consumers to place a value on certain nonmarket attributes such as wetlands. Contingent valuation methods may be used as alternatives to travel costs for on-site valuation. However, contingent valuation is the only means by which optional use values and existence values can be measured. In addition, contingent valuation methods can be used as an alternative to hedonic property value comparisons to value the impact of resource policy issues such as how much consumers would be willing to pay to avoid the location of a solid waste landfill near their community.[8] Knowing this value is impor-

---

[8]Roland K. Roberts, Peggy V. Douglas, and William M. Park, "Estimating External Costs of Municipal Landfill Siting Through Contingent Valuation Analysis: A Case Study," *So. J. Agric. Econ.* (December 1991), pp. 155–165.

tant in making comparisons with solid waste policy alternatives, such as recycling, that avoid a portion of the actual costs and nonmarket values associated with operating landfills.

Nonmarket values are becoming increasingly important to decision making in the resource policy arena. Often resource policy decisions are made without knowing anything but the market effects on values. The extent to which nonmarket values are recognized is often limited to a marginal comment. For example, the externalities associated with a large integrated hog operation having five units of 18,000 sows each include the disposal of animal wastes and related odors. Until recently, this fact was frequently noted only in passing to rural communities by industry recruiting committees. But such externalities are the origin of many resource policy conflicts. Therefore, much of the controversy surrounding resource policy involves the existence of nonmarket values, their magnitude, the openness with which they are discussed, and the weight to be placed on them in the political process. Environmental interest groups exist because of these nonmarket values. These interest groups devote a major share of their resources to studying nonmarket values and explaining their magnitude to the electorate and to policymakers.

RESOURCE POLICY CHOICES: WHO PAYS

The broad resource policy choices involve the means by which resource allocation decisions are to be made. That choice determines who pays and who benefits. Each specific resource issue has its own detailed list of options, but the choices basically involve four broad policy options:

- **Market determination** uses market prices to allocate resources. Those prices are set under actual market conditions. As such, they reflect only the values of those who are willing and able to pay. They reflect basic self-interest economic motives of profit and utility maximization. Except to the extent that future profit and asset values are known to be impacted, market-determined prices depend on concepts of stewardship, what is right, concern about future generations, and peer pressure for decisions relating to resource conservation and sustainability. Therefore, nonmarket values are not considered in a market-determined resource allocation system in which even services provided by the government (public goods) are valued on an at-cost basis predominantly through user fees. For example, western farmers and ranchers would pay the full cost of water for irrigation purposes. In an even more pure market-determined system, farmers would bid for available water supplies in competition with consumers and industrial users.

- **Subsidies** pay people for engaging in certain specific practices. Subsidies may also be referred to as *incentive payments* or green payments. Subsidies come in many forms. Public goods such as water may be provided at less than cost. Farmers may be paid for engaging in certain practices such as putting in terraces to curb runoff and soil erosion. Attached to subsidies such as target or support prices may be a condition that certain practices must be pursued. Therefore, a payment by government for an easement not to develop agricultural land is a form of a subsidy as is a payment to retire highly erodible land from production. The effect of the subsidy is to reduce the marginal cost of engaging in a certain practice so that the marginal revenue from the practice will be higher than the marginal cost. Sometimes the subsidy may not cover all of the costs of engaging in the specified practice by all producers, but it must be sufficient to entice the desired level of participation. The subsidy is designed to reflect the marginal cost of engaging in the subsidized practice, good, or service. Subsidies are often preferred politically because they are voluntary and involve no penalty, although, from an economic decision perspective, the practice may not be voluntary at all. Subsidies, therefore, utilize an economic carrot to accomplish certain desired end objectives.

- **Taxes** are designed to internalize the costs of externalities associated with a resource problem. That is, the tax is designed to approximate the cost borne by society and/or to offset the benefits (profits) to be derived by the producer in engaging in a specific practice. For example, a tax may be placed on agricultural chemicals to discourage their use because of the perceived or actual adverse affects on the environment. The effect of the tax is to increase the cost of the input, resulting in its use in reduced quantities. Taxes utilize an economic stick to accomplish the desired end objectives. How effective a tax is at accomplishing its intended objective depends on the elasticity of demand for the good. If the tax is placed on a good whose demand is highly inelastic, the tax is more effective at raising revenue for the government than at reducing the use of the input.

- **Regulations** are mandated government actions. They utilize the stick to ensure compliance in that penalties for noncompliance are imposed. Regulations shift the burden of cost to the abusing producers and ultimately to the consumer or, more accurately, to a combination of the producer and the consumer. Therefore, the notion that the polluter pays for the cost of pollution is a regulatory concept. With increased problems of budget deficits (less willingness to pay producers incentives for desirable practices), regulation is becoming a more common means to deal with resource policy issues.

Major constitutional questions have arisen over the conditions under which the government must compensate a property owner if the effect of the regulation is to reduce the value of the property. For example, if a new wetlands regulation makes it impossible to profitably farm certain lands, is this an unconstitutional "taking"? Does the landowner need to be compensated? If it eliminates the potential for development of the land, is that "taking"? Does it make a difference whether the motive for land ownership is speculative (development) or the landowner's way of earning a living (farming)? Who pays?

A recent Supreme Court decision, *Lucas v. South Carolina*, surprised many when it held that South Carolina's regulation of construction on its environmentally fragile Atlantic coastline amounted to taking property rights from Lucas, who had bought two beachfront lots on which he planned to build two houses.[9] Questions have arisen regarding the extent to which the *Lucas* case could be used to limit the government's power to regulate in areas such as wetlands and endangered species.

Regardless of its unpopularity with farmers, regulation is being advocated to solve environmental problems as an alternative to subsidies because regulation is viewed as being more effective. For example, Ken Cook, an environmental advocate, has warned farmers and ranchers that their failure to be more environmentally conscious voluntarily would result in the use of mandatory approaches.[10]

RESOURCE POLICY ISSUES AND OPTIONS

Soil Conservation Policy

Efforts to encourage soil conservation are among the oldest of federal farm programs. Current programs have their roots in the Dust Bowl days of the 1930s when Congress declared soil erosion a national emergency. As a result, in 1935, the Soil Conservation Service (SCS), now the Natural Resources Conservation Service (NRCS), was created as an agency of USDA.

Since its creation, the degree of emphasis placed on soil conservation has changed from a major national concern, as indicated by its status as a national emergency in the 1930s, to nearly being eliminated as a USDA agency during the 1970s and early 1980s. The two factors that saved SCS were its grass-roots structure of conservation districts and the emergence of the environmental movement linking soil erosion and pesticides to issues of water quality. Now NRCS, supported by the regulatory authority of the EPA, is one of the more powerful USDA agencies.

The Land Resource Base

According to the Census of Agriculture, the total land area of the United States was 2.263 billion acres in 1992.[11] Of the 460 million acres classified as cropland, only about 338 million were actually cropped. The remaining 122 million acres were either held idle by the government or were in pasture. An additional 591 million

---

[9]Paul M. Barrett, "Supreme Court Supports Rights of Landowners." *The Wall Street Journal* (June 30, 1992), p. A3.

[10]*The Food and Fiber Letter* (Washington, D.C.: Sparks Companies, March 1, 1993, p. 3, and March 29, 1993, p. 3); and Congressional Record (Washington, D.C.: U.S. Congress, July 17, 1991).

[11]National Agricultural Statistics Service, *Agricultural Statistics* (Washington, D.C.: USDA/NASS, 1995), p. IX-5.

acres of pastureland (26 percent) were used for grazing cattle and sheep. Approximately 648 million acres were in forest land (not including parks), of which 40 percent are owned by the government. With the new Conservation Reserve Program provisions in the 1996 farm bill, in 1996, an estimated 346 million acres were used for crops and an estimated 34.4 million acres were idled by government programs.[12]

Historically, the major concern with the land resource base has been that of soil erosion. The societal value placed on soil conservation relates to the maintenance of agricultural productivity and the adverse impacts of soil erosion on water quality. For most U.S. cropland, the majority of the available plant nutrients is contained in a few inches of topsoil. This productive topsoil, which takes many years to regenerate, can be eroded rapidly by either wind or water. Water erosion, along with the related chemicals (including pesticides and commercial fertilizer) used in food production and animal wastes, has been identified as the largest polluter of inland streams, lakes, and coastal waters.

From a policy perspective, the standard for measuring erosion is a soil loss tolerance value referred to as T. The T value is the maximum level of soil erosion that will permit a high level of crop productivity to be sustained economically and indefinitely.[13] Depending on the soil's characteristics, the T value allows a loss of two to five tons of soil per acre per year. The 1985 farm bill effectively established a goal of reducing soil erosion on all lands to T. However, special emphasis was placed on highly erodible acreage having erosion rates of more than eight tons of soil (a large dump truck load) per acre per year. In 1987, 29 percent (123 million acres) fell into this highly erodible category.[14]

NRCS has completed three 5-year inventories of the status and trends in land, soil, and water in the United States. The results of the latest survey, the 1992 National Resources Inventory (NRI), indicate tremendous success in reducing soil erosion. Sheet and rill erosion (water erosion) on cropland fell from 4.1 tons per acre per year in 1982 to 3.1 tons per acre per year in 1992. Similarly, wind erosion was reduced from 3.3 tons per acre per year to 2.5 tons per acre per year. Combined, more than one billion tons of soil are saved per year.[15]

Much of the success can be attributed to the conservation provisions of the 1985 farm bill and CRP. In 1992, the year of the latest NRI survey, approximately 34 million acres were under contract in CRP. Of the 16.5 million CRP acres classified as highly erodible, in a 10-year period, sheet and rill erosion decreased from

---

[12]Personal correspondence with the USDA/ERS, Natural Resources and Environment Division, Natural Resource Conservation and Management Branch, November 1996.

[13]Russell C. White, "Alternative Conservation Systems in Conservation Compliance," in *Water Quality and Soil Conservation: Conflicts of Rights Issues,* Special Report 394 (Columbia, Mo.: University of Missouri, Agricultural Experiment Station, November 1988), p. 65.

[14]*Agricultural Resources,* AR-23 (Washington, D.C., Economic Research Service, USDA, September 1991), p. 36.

[15]Natural Resources Conservation Service, *Graphic Highlights of Natural Resource Trends in the United States between 1982 and 1992* (Washington, D.C., Natural Resources Conservation Service, USDA, April 1995).

8.6 to 0.6 tons per year while wind erosion decreased from 10.7 to 1.3 tons per year.[16]

Even with the before-mentioned successes, there is still work to be done to achieve *T* on all lands. The 1992 NRI indicated that an average of 19 percent of cropland experienced a water erosion rate greater than *T*, and 14 percent experienced a wind erosion rate greater than T.[17]

Soil Conservation Policy Options

Many different policy options have been utilized to reduce soil erosion:

- Market prices and stewardship
- Agriculture conservation payments
- Conservation compliance
- Retirement of highly erosive land
- Regulation

Market prices and stewardship were the primary means to deal with soil erosion issues prior to the 1930s and through much of the surplus production period of the 1970s and early 1980s. In a market setting, the effectiveness of soil conservation is determined by the following criteria:

- **Its short-term impact on the profitability of farming.** The market provides incentives to employ soil conservation practices if their effect is to increase returns by more than the cost of employing soil conservation practices or, stated alternatively, if the immediate returns from practicing conservation are greater than the costs. An Iowa study indicates that reductions on water-caused soil erosion from 20 tons per acre (roughly 5*T*) to 4 tons per acre (roughly *T*) require investments of $250 to $300 per acre.[18] The investment to achieve *T* erosion resulted in a negative return of $67.90 per acre. A representative farm study in Texas found that reductions in wind erosion to *T* would reduce net cash income of a representative farm by $12.79 per acre and decrease the chances of the farm surviving over a five-year planning horizon to near zero.[19] *Therefore, in a current cash flow context, studies indicate that conservation practices designed to achieve T erosion values (sustained productivity) do not pay.*

---

[16]Natural Resources Conservation Service, 1992 Natural Resources Inventory Highlights, Internet version (Washington, D.C., Natural Resources Conservation Service, USDA).

[17]Natural Resources Conservation Service, 1992 Natural Resources Inventory Summary Tables, Internet version (Washington, D.C., Natural Resources Conservation Service, USDA).

[18]J. F. Timmons and O. M. Amos, *Economics of Soil Erosion Control with Application to T- Values,* Iowa Agric. Exp. Stn. J. Pap. J-1 625 (Ames, Iowa: Iowa State University, 1979).

[19]James W. Richardson *et al., The Economic Impacts of Conservation Provisions in the 1985 Food Security Act on a Representative Dawson County, Texas Farm,* AFPC Policy Research Report 89-1 (College Station, Tex.: Texas A&M University, June 1989).

- **Its long-term impact on the price of land.** If soil erosion discernibly affects the soil's productivity, the price offered for eroded or eroding land should decline proportionately.[20] Some studies have found that soil depth and erosion impact land values.[21] Others have not found a statistically significant relationship between erosion and the market value of land with the implication of a market failure.[22] Still other economists point out that although a theoretical basis exists for establishing a relationship between erosion and land values, the empirical results are mixed.[23] Based on their analysis, Gardner and Barrows point out that "if the value of soil conservation investments is not generally capitalized into land prices, the rational, well-informed, profit-maximizing farmer will underinvest in soil conservation because the value of land at the end of the planning horizon (sale date) will not reflect the value of the required investment."[24] Libby explains the erosion control strategies of farmers in the following terms:

  > Farmers have many motives influencing their behavior. They have a sense of responsibility for the land. . . . They acknowledge the community responsibility. . . . But farmers respond to economic incentives of price and cost. . . . Primary attention must be given to economic incentives that affect the relative attractiveness of conservation.[25]

The consequences of utilizing market prices and stewardship to control erosion include:

- Increased soil erosion
- Reduced water quality
- Reduced future productivity
- Little or no consideration of the off-site nonmarket costs of erosion on water quality, sedimentation of lakes and waterways, and reduced future productivity

The implication may be drawn that *if the desired goal is to have soil erosion of amount T or less, government investment in soil conservation is necessary*—implying the need for conservation subsidies. Such a conclusion implies that society places

---

[20]K. E. McConnell, "An Economic Model of Soil Conservation," *Am. J. Agric. Econ.* (February 1983), pp. 83–89; David E. Erwin and John W. Mill, "Agricultural Land Markets and Soil Erosion: Policy Relevance and Conceptual Issues," *Am. J. Agric. Econ.* (December 1985), pp. 938–942.

[21]R. B. Palmquist and L. E. Danielson, "A Hedonic Study of the Effects of Erosion Control and Drainage on Farmland Values," *Am. J. Agric. Econ.* (February 1989), pp. 55–62.

[22]Kent Gardner and Richard Barrows, "The Impact of Soil Conservation Investments on Land Prices," *Am. J. Agric. Econ.* (December 1985), pp. 943–947.

[23]Gregory L. Poe, Richard M. Klemme, Shawn J. McComb, and John E. Ambrosius, "Commodity Programs and the Internalization of Erosion Costs: Do They Affect Crop Rotation Decisions," *Rev. Agric. Econ.* (July 1991), pp. 224–235.

[24]Gardner and Barrows, "The Impact of Soil Conservation Investments," p. 947.

[25]Lawrence W. Libby, *Developing Agricultural Policy to Achieve Lower Rates of Erosion on Fragile Lands* (East Lansing, Mich.: Michigan State University, 1986), p. 5.

a higher value on soil conservation than on the market. Tweeten notes that if non-market costs are large, reliance on the market alone will not optimize conservation even if farmers do a good job of equating private costs with private benefits at the margin.[26] He points out studies by Clarke *et al.,*[27] which estimate nonmarket costs of $6.1 billion associated with erosion damage, and estimates of $7.1 billion by Ribaudo.[28] Tweeten concludes that the environmental problems of agriculture are very real and will not be resolved by the market alone.[29]

Green payments are made to farmers for engaging in environmentally sound practices—often referred to as best management practices (see p. 375-376). Such payments could be made for practices related to sources of environmental degradation due to either crops or livestock enterprises.

The **Agriculture Conservation Payments (ACP)** program was the first green payment program. ACP was the congressional reaction to the Dust Bowl conditions of the 1930s. The ACP program, paid for such important soil conservation practices as terraces, diversions, and windbreaks. ACP payments were complemented by technical assistance involving the design of soil conservation practices such as contour farming and crop rotation systems by SCS professionals. Interestingly, decisions on what practices were to be given payments were the responsibility of the Congress and ASCS administrators, who wrote the checks to the farmers.

Over time, the definition of ACP-eligible conservation practices was broadened to encompass cropland- and output-enhancing practices apparently unrelated to the mainstream objective of controlling erosion. Included were payments for the application of lime to reduce soil acidity, the construction of livestock water tanks, and drainage programs allegedly to control floods and to plan and protect watersheds. The drainage and watershed programs may actually have facilitated erosion and runoff. In the process, waterfowl habitat was destroyed,[30] much to the dismay of contemporary wetlands advocates. A 1981 study found that fewer than 19 percent of the soil conservation practices had been installed on highly erosive land.[31]

The result of such findings was to reduce support for ACP payments and reorient the program toward practices designed to reduce erosion, targeting highly erodible lands, while maintaining wetlands. In 1994, the Clinton administration proposed moving full responsibility for ACP payment decisions to NRCS, located

[26]Luther Tweeten, "The Economics of an Environmentally Sound Agriculture," in *Research in Domestic and International Agribusiness Management* (Greenwich, Conn.: JAI Press, Inc., 1992), p. 55.

[27]E. H. Clarke, J. Haverkamp, and W. Chapman, *Eroding Soils: The Off-Farm Impacts* (Washington, D.C.: Conservation Foundation, 1985).

[28]M. O. Ribaudo, "Reducing Soil Erosion: Off-site Benefits," Agric. Econ. Rept. No. 561 (Washington, D.C.: ERS, USDA, 1986).

[29]Tweeten, "Environmentally Sound Agriculture," p. 19.

[30]Earl O. Heady, *Agricultural Policy under Economic Development* (Ames: Iowa State University Press, 1962), pp. 558-560.

[31]USDA, *National Summary Evaluation of the Agricultural Conservation Program: Phase I* (Washington, D.C.: ASCS, USDA, 1981).

under the newly created assistant secretary for natural resources. Farm organizations strongly opposed this move. The 1996 farm bill changed the ACP program into the Environmental Quality Incentive program (EQIP), which provides green payments for reducing or eliminating crop- and livestock-created environmental problems.

The following are some consequences of an environmentally sensitive EQIP or green payment program:

- Reduced soil erosion
- Improved water quality
- Enhanced farmland productivity
- Enhanced farm income
- Improved waterfowl and wildlife habitat

Conservation compliance provisions of the 1985 farm bill required farmers to develop and file with the USDA a conservation plan on all highly erodible cropland by January 1, 1990, with a goal of reducing the level of erosion to T by January 1, 1995. Farmers not filing and implementing satisfactory conservation plans become ineligible for farm program benefits, including deficiency payments and access to price support loan provisions.[32] The conservation plan specifies production practices that are to be implemented within a specific time period. These practices may range from relatively low-cost conservation tillage to higher cost systems such as the construction of terraces or the planting of alternative strips of row crops and cover crops. Conservation compliance is credited with resulting in farmers adopting the practice of no-till or mulch-till technology on one-third of the cropland.

In setting the T erosion goal, Congress indicated that the secretary of agriculture should apply standards of reasonable judgment and consider the economic consequences in determining what practices should be included in the mandated conservation plans. In other words, the Congress desired not to cause undue hardship on producers. As a result, the local NRCS soil conservationists were given latitude in developing conservation plans that deviate from the achievement of the T value goal when economic conditions warrant. However, it soon became apparent that some farmers and/or local NRCS officials may have developed conservation plans that could be implemented only with substantial expense and economic hardship. Therefore, although the 1990 farm bill extended and even strengthened the conservation compliance provisions of the 1985 bill, substantial questions have arisen as to the conditions under which the conservation plans might be enforced and/or modified.[33] Although environmental advo-

---

[32]Food Security Act of 1985, Conference Report, Report No. 99-447 (Washington, D.C.: U.S. House of Representatives, December, 1985), pp. 459–460.

[33]Ken Cook, quoted in *The Food and Fiber Letter* (Washington, D.C.: Sparks Commodities, Inc., March 29, 1993), p. 3.

cates favor stiff enforcement and revocation of farm program benefits in the event of violations, farmers and rural congresspeople are pushing for a more pragmatic approach.

Closely related to conservation compliance is the sodbuster provision, which requires that highly erodible land brought into production have an approved conservation plan before becoming eligible for program benefits. The combination of farmers operating under frozen base acres with the sodbuster provision substantially reduces or eliminates the incentives for bringing new highly erodible land into production.

The consequences of the conservation compliance and sodbuster provisions include the following:

- Reduced soil erosion
- Reduced water runoff
- Improved water quality
- Increased cost of production associated with the costs of implementing the conservation plan
- Lower producer returns
- Reduced participation in farm programs

With regard to the last item, conservation compliance utilizes farm program benefits as the carrot for encouraging farmer participation. If, from the farmers' perspective, the costs are greater than the benefits, they will drop out of the program. If farm price and income support programs were eliminated, cross-compliance could not be implemented. Alternatives such as green payments or regulation would become necessary to achieve the program goal.

Retirement of highly erodible land was established as national policy in the 1985 farm bill. Referred to as the **conservation reserve program (CRP),** its original goal was to remove 40 to 45 million acres of highly erodible land from production for a 10-year period. Farmers bid an annual rental payment that they would be willing to accept to retire their land from production for 10 years. Once accepted, the farmer is required to plant a grass, legume, or tree ground cover when the establishment cost is shared by the USDA. Grazing or harvesting hay is not allowed on CRP land except in declared emergency situations. Of course, the trees would eventually be harvested for timber. The 1990 farm bill modified the CRP program to put greater emphasis on water quality by authorizing a wetland reserve and extended the achievement of the 40 to 45 million acre goal in either CRP or the wetland reserve to 1995. The budget authorization process subsequently modified this goal to less than 40 million acres.

From an economic perspective, the farmer's bid price for participation in the CRP or wetland reserve must be equal to or above the return from farming. The average accepted CRP bid for 36.4 million acres enrolled in CRP was about $50 per acre.[34]

---

[34]*Agricultural Resources,* AR-23, p. 27.

CRP may be looked on as a successor to the earlier Soil Bank program enacted in 1956, which also removed cropland from production and encouraged the planting of grass and/or trees. However, there were some substantial differences. The Soil Bank program, while touted for its conservation benefits, was truly a land retirement program designed as a voluntary production control program and was not targeted toward any particular land type. While CRP received the support of farmers because of benefits in curbing production, its targeting of highly erosive land gained the support of environmentalists. This environmental support arguably was necessary to enact the 1985 farm bill.

Another difference between the Soil Bank program and CRP is that whole farms were retired under Soil Bank. This feature drew considerable criticism from rural communities whose agricultural economic base and related living expenses were removed from the local economy as participating farmers moved to favorite retirement sites such as Florida, California, and Arizona. CRP, likewise, has been subject to the same criticism, although whole farms are not retired. An effective 30 percent limit[35] was set on the amount of cropland that can enter CRP in any county. In addition, the application of payment limitations to CRP land curbed the quantity of land enrolled in the program by any one person.

A major policy issue faced in the 1996 farm bill involved what to do with CRP land at the end of the 10-year retirement period. Environmentalists wanted to see this highly erosive land kept in retirement or regulated from going back into crop production.

While farm organizations wanted to see CRP continued, export-oriented agribusiness wanted to see most CRP land put back into production. The 1996 farm bill extended CRP but passed the buck for this controversial decision to Secretary Glickman. NRCS interests favored enrolling CRP lands that were sensitive to water quality problems. This upset Great Plains farmers and their representatives whose lands were subject to high levels of wind erosion. They argued that greater consideration needed to be given to the wildlife preserving effects of CRP.

The following are some of the major consequences of the land retirement option for controlling soil erosion:

- Reduced soil erosion; USDA estimates that the average reduction in erosion on CRP land is 19 tons per acre annually, about $4T$[36]
- Improved water quality because of erosion reduction, runoff reduction, and less chemical use
- Increased habitat for wildlife
- Reduced crop production; American agriculture perhaps is the closest to a relative supply–demand balance since the 1930s

---

[35]Actually, the statutory limit is 25 percent but, practically, that limit has been waived to allow up to 30 percent of county cropland into the CRP.
[36]*Agricultural Resources*, AR-23, p. 29.

- Increased farm price levels and variability; as land is retired from production, commodity stocks decline, raising the level and variability of prices
- Increased food prices because of reduced crop production and increased farm prices
- Fewer farm input suppliers in areas of high CRP participation
- Increased farm program costs

Easements involve the acquisition of the right to limit the use of property for specific purposes. While CRP may be looked on as a type of easement, the objective of an easement may be more limited. For example, as an alternative to CRP, an easement may be purchased on land to prevent it from being used for row crops or to limit its use to haying and grazing. Allowing the use of the land for some productive purpose would reduce the cost of controlling erosion.

Regulation mandates that certain farming practices (best management practices) be employed as a condition for farming. While regulation is implied by policy options such as conservation compliance, it uses the carrot of farm program payments to accomplish the conservation objective. If farm program payments vanish in times of tight budgets, trade negotiations, or dissatisfaction with farm programs, pure regulation is an alternative means to conserve land resources. We will see subsequently that such requirements are already being employed to achieve water quality goals—a component of which is the reduction of soil erosion. A barrier to utilizing regulation to keep CRP land out of production could involve issues of property rights and unconstitutional taking. There is increasing agitation in farm organizations and Congress over regulations that limit the use of land and, therefore, reduce its value.

Water Supply Policy

Water supply policy deals with the legal principles that govern the rights to and allocation of an increasingly limited supply of water. Water rights are imbedded in complex interrelationships of constitutional law, statutory law, administrative regulations, executive orders, and common law court decisions. In statutory terms, they are also interrelated with the issues of water quality, wetlands, soil erosion, and endangered species, which are treated separately in this chapter.

This section is not intended to represent a comprehensive treatment of water rights issues.[37] It certainly should not be relied on as the basis for private decisions regarding water rights. Rather, the purpose of this section is to convey the content and importance of the rapidly developing water supply issues. These issues evolve from competition for an increasingly scarce resource and foretell the potential for increased federal intervention in the allocation of limited water supplies.

---

[37]Such a comprehensive treatment of water law is contained in Goldfarb, *Water Law.*

The Water Supply Base

Agriculture is the single largest user of water, accounting for more than 70 percent of the consumptive water use in the world. In the United States, agriculture's share of consumptive water use is more than 80 percent, although in the western United States nearly 90 percent of water use is agricultural. Of the total U.S. crop-land area, 13.5 percent (over 46 million acres) is irrigated. Of the area actually planted to crops, more than 14 percent is irrigated.[38] *Groundwater* and *surface water* each supply about half of the irrigation needs in the United States. Of the 32 million irrigated acres in the 17-state arid West region, 57 percent is partially or fully irrigated with *surface water*. Much of this water is captured and flows through the publicly financed western reservoir, canal, and irrigation system. The central United States relies on groundwater for more than 70 percent of its irrigated water supply.[39] Although the distinction between surface water and groundwater generally is made and frequently has legal significance, the two sources are often closely interrelated because surface water influences the level of the water table and recharges the groundwater aquifers.

Irrigated acreage accounts for all of the rice production, about 70 percent of the orchards, more than half of the vegetables, and about one-third of the cotton produced in the United States.[40] Because of the concentration of high-valued crops such as fruits and vegetables on irrigated acres, the value of output per acre is about 2.5 times as high as on dryland acres, whereas the volume of production tends to be 50 percent or more higher than on dryland acres.[41] In California, irrigation accounts for more than 80 percent of the output.

The water supply problem is considerably more severe and pervasive west of the Mississippi than east of it. This is the case because western agriculture depends much more on water for irrigation than eastern agriculture and because of population pressures brought on by migration to the Sunbelt. Intensive farming in the West developed largely because federal and/or state governments saw fit to subsidize the construction of dams that served as a source of both hydroelectric power and irrigation water. Water from these projects has seldom been priced at a level that reflects its cost.[42] Since agriculture's ability to pay for water generally is more limited than that of either other businesses or consumers, agriculture is in an unfavorable competitive position. In addition, because farm prices are determined in national and/or international markets, farmers have no power to pass through water cost increases.

---

[38]Economic Research Service, *Irrigation Water Use*, 1994 AREI Update No. 8 (Washington, D.C.: ERS/USDA, August 1996), p. 2.
[39]Kenneth P. Frederick, "Irrigation under Stress," *Resources* (Washington, D.C.: Resources for the Future, Spring 1988), p. 1.
[40]Ibid.
[41]Ibid.
[42]Ibid.

Accordingly, any increase in the price of water to farmers places them at a severe competitive disadvantage—potentially forcing them to materially change their crop mix or to discontinue farming.

Postel notes that for the United States, "One-fifth of the total irrigated land is watered by excess pumping of groundwater: roughly half of the western rivers are over-appropriated: to augment supplies, cities are buying farmers' water rights."[43] Accordingly, questions are being raised regarding the adequacy of existing water rights and policies to deal with present and future supply problems. Temporary shortages are being dealt with by water allocation systems that have the potential for starving agricultural production and/or changing cropping patterns. Longer run policy options are being experimented with, some of which hold the potential for abrogating assumed water rights, leading to extensive litigation. The case is being made for a new set of institutions to deal with water issues that could potentially lead to more federal involvement in an area of policy that has traditionally been left to the states and the courts.

Underscoring the complex water issues in the United States is the fact that the water scarcity problem is even more pervasive internationally and it is a problem which is projected to become much worse. Rosegrant predicts that the control of water resources may well be the defining issue of the twenty-first century. Presently, 28 countries with an aggregate population of 228 million people are subject to water shortages. It is projected that 46 to 52 countries with an aggregate population of 3 billion will be water stressed within 30 years. Even with the before mentioned scarcity issues in the United States, on a continental basis, North America has water wealth. North America has 17,500 cubic meters of water per person, South America has 28,300, Africa has 5,100, Europe has 4,100 and Asia has only 3,300.[44]

Rather than treat water as a scarce good, many countries, including the United States, subsidize water for urban and rural use. Irrigation water is also frequently underpriced. Rosegrant identified poor water policies, such as water subsidies, as the single greatest force behind water scarcity. Only 25 to 40 percent of the water used for irrigation in the developing world is actually used beneficially. Furthermore, often 50 percent of the water usage in major metropolitan areas of the developing world is lost, or simply goes down the drain.[45] The Rosegrant study concludes that, "new strategies for water development and management are urgently needed to avert severe national, regional, and local water scarcities that will depress agricultural production, parch the household and industrial sectors, damage the environment, and escalate water-related health problems."[46]

[43]Sandra Postel, "Saving Water for Agriculture," State of the World 1990, ed. Lester R. Brown et al. (New York, N. Y.: W.W. Norton and Co., 1990), p. 50.
[44]Mark W. Rosegrant. *Water Resources in the Twenty-First Century: Challenges and Implications for Action.* (Washington, D.C.: International Food Policy Research Institute, March 1997), p. 2.
[45]Ibid. p. 4.
[46]Ibid. p. 1.

Water Supply Policy Options

The water policy options for the United States fall into the following categories:

- Water rights doctrines promulgated out of state statutory law and common law that are different for surface water than for groundwater
- Water development involving the construction of irrigation systems
- Water pricing involving the sale of water from development projects or the sale of existing water rights; included in water pricing are payments for water rights resulting from the employment of conservation practices
- Water management and conservation systems involving a wide range of licensing-mandated management practices and water allocation systems that could, in the extreme, involve government appropriation of water rights

Water rights.   Riparian and prior appropriation doctrines have historically been used to allocate U.S. **surface water rights.** The riparian doctrine applies to all states east of the Mississippi (except for the state of Mississippi) as well as Arkansas, Iowa, and Missouri. The prior appropriation doctrine prevails in the West.

The **riparian doctrine** holds that only persons holding land on natural waterways have the right to divert water for their use on riparian land. *Riparian land* is actually adjacent to a waterbody or watercourse (stream). However, states vary in their interpretation of the extent to which riparian rights can be applied to land that is not contiguous to a watercourse. Riparian rights are generally subject to a rule of reasonable use whereby a riparian cannot unreasonably interfere with the legitimate use of other riparians. Riparianism, therefore, makes interbasin transfer of water difficult.

The **prior appropriation doctrine** applies the concept of "first in time is first in right." Therefore, those who were first to appropriate (use) water have the right to use it indefinitely in the future as long as that use is beneficial. Goldfarb draws four crucial distinctions between prior appropriation and riparianism:[47]

- Appropriative rights are limited only by the economics of applying water to a beneficial use.
- A right holder can utilize all unappropriated water in a watercourse unless the state imposes a minimum flow.
- An appropriative diversion right is transferable apart from the land as long as there is no injury to other right holders.
- An appropriation right is of indefinite duration as long as it is used. Generally, nonuse must be for a period of three to five years.

The potential for expiration of the appropriation right is actually an incentive for use. Because the concept of beneficial use has been applied liberally,

---

[47]Goldfarb, *Water Law,* pp. 33-34.

California utilizes a legal standard of "reasonable beneficial use," a concept that is gradually expanding to other states. In times of shortages, the priority of use has normally been as follows:

1. Domestic households
2. Agriculture
3. Industrial and hydroelectric power
4. Fish, wildlife, and recreation

However, there has been increasing pressure to change this priority because, in effect, no rights are granted for water to be left in stream to preserve and protect fish and wildlife habitat, recreational whitewater, and other ecological or aesthetic values. This issue has come to a head in protecting the salmon population in the Columbia River system of the Great Northwest—a region where industry is built on low-cost hydroelectric power and agricultural irrigation. Here environmental interests argue that farmers should pay more for water to reduce their consumption and reflect the cost of pollution by pesticides. Thus, under both riparian rights and prior appropriation doctrines, states have provided methods to permit (approve), adjudicate, and distribute water. In an increasing number of instances, even these systems do not appear to be adequate.

Groundwater rights evolved independent of the surface water doctrines of riparianism and prior appropriation. Because of the dangers of overpumping the limited supply of aquifers and the potential for long-lasting pollution, groundwater sources have in some states succumbed to administrative management systems sooner than for surface water. However, these management systems may be no more effective and, therefore, just as temporary as those for surface water.

The **absolute ownership doctrine** creates an absolute right for each landowner to pump groundwater located under his or her land without limitation or consideration of the needs of or effects on others, except pumping for spite or revenge. Such absolute rights are vanishing.

The **reasonable use doctrine** for groundwater is the same as absolute ownership except that waste is prohibited and water must be used on overlying land unless it can be transported without injury to other overlying owners. Thus, under the reasonable use doctrine, a groundwater landowner may pump indefinitely as long as the water is used on overlying land without waste. For example, the construction of a catfish farm on overlying lands may be considered reasonable use even though the availability of groundwater to other users is limited.

The **restatement rule doctrine** has its origin in the law of torts and lists three criteria for determining an unreasonable interference in the use of groundwater:

■ Causing unreasonable harm by lowering the water table or artesian pressure
■ Exceeding the owner's reasonable share of supply
■ Having unreasonable effects on surface supplies.

According to Goldfarb, this is the same reasonable use test as for surface water.

A **correlative rights doctrine,** applied in California to groundwater, effectively states that in times of shortages:

- Overlying users are entitled only to their just and fair proportion for on-site uses.
- Overlying users have priority over transfers out of the aquifer.
- Transfers out are based on first in time, first in right.

The consequences of the correlative water rights doctrine include the following:

- The protection of property values for the holders of water rights
- The potential overuse of available water supplies and, therefore, future shortages
- Allocation of water to those having the greatest ability to pay
- High levels of short-run production of agricultural products but potential future reductions
- Potential long-run water shortages

Water development.   Historically, the major source of increased water supplies has been water development activities undertaken by agencies. Many of these are huge multipurpose projects involving federal and state production of hydroelectric power, irrigation, recreation, and tourism, as well as other products (see box). The major federal agencies involved in these water development projects include these:

- The Bureau of Reclamation of the Department of Interior was created to facilitate settlement of the western lands. The Bureau has responsibility for irrigation projects that include as much as 20 percent of the irrigated land area. In 1987 the bureau underwent substantial reorganization, shifting its emphasis from large multipurpose projects to environmental protection on existing projects. The original Reclamation Act of 1902 limited the provision of irrigation water to 160 acres (320 for husband and wife). In the late 1970s, there was an effort to enforce this largely ignored law. The result was the Reclamation Reform Act of 1982, which increased the irrigation acreage limitation to 960 acres but contained provisions for leasing, making the limitation largely ineffective.
- The Corps of Engineers of the Department of Army is the oldest water resources program. The corps is responsible for planning, constructing, and managing many major navigation, flood control, irrigation, and hydroelectric projects. These projects are individually authorized through the House and Senate Committees on Public Works after extensive study with subsequent appropriation action. The Water Resources Development Act of 1986 placed requirements for cost sharing with states or other entities.

New Era for Water Rights

About 350 farmers from states that included Nebraska, Kansas, Colorado, and Wyoming met in Rapid City, South Dakota. They talked about the growing pressure on them to share with communities and wildlife some of the water farmers have relied on for decades.

For years, pressure has been applied by environmentalists. Now the U.S. Bureau of Reclamation, a farmer friend for decades, is joining the effort to change farmers, too.

The bureau is insisting that more than 136,000 U.S. farmers deploy bureau water more efficiently in irrigating nearly 12 million acres in 17 states.

In a nine-state region—Nebraska, South Dakota, North Dakota, Colorado, Wyoming, Montana, Kansas, Oklahoma, and Texas—the bureau's tough new stance affects nearly 16,000 farmers irrigating three million acres.

"We have to recognize and address the needs of all water users, including urban, recreation, fish and wildlife, and other instream uses," Dan Beard, commissioner of recreation, told the farmers.

The agency built more than 300 major dams, 16,000 miles of canal and 1,460 miles of pipeline during the past 90 years to help farmers and power users at a U.S. taxpayer cost of $9.9 billion.

"Many irrigators believe, 'This water is mine, by God, and nobody is going to get a piece of that,'" said J. Michael Jess, Nebraska's water resources director. "Beard was saying, 'You've got to set that view aside. This is a new era.'"

Source: The Associated Press, "Farmers Meet to Outline Threats to Water Usage," *The Eagle*, Bryan-College Station, Tex., December 26, 1993, p. C7.

- The Soil Conservation Service administers only small watershed projects largely for flood control, although irrigation may be involved.
- The Tennessee Valley Authority (TVA) is a federal government corporation with a primary initial mission of flood control on the Tennessee and Mississippi river basins. In addition to the navigation and hydroelectric power, TVA utilizes its system of more than 30 dams as a source of irrigation water and electricity for the production of nitrogen fertilizer.

Since the 1980s, the initiation of new water projects has been severely limited by environmental concerns, largely involving the protection of habitat for endangered species. However, quite clearly, other nonmarket values such as the destruction of natural scenic areas are also involved.

Water development options include the following consequences:

- Increased water availability

- Increased agricultural production and reduced food prices
- Reduced endangered species
- Reduced scenic areas

Water pricing. Pricing water is becoming an increasingly important policy alternative for dealing with issues of water scarcity. The price of a resource obviously influences its use. Except for the cost of pumping, in many areas water is essentially a free good. Pricing water below its marginal value fosters overuse and mismanagement of resources. This basic economic principle applies to water as well as it does to other commodities. Or does it? If domestic and industrial users are willing and able to outpay agriculture, should they not be allowed to—even if the result is reduced food production, higher food prices, and fewer farmers? These are difficult questions to answer, but they are being asked and, in some cases, are being answered—particularly in the water-starved West.

The origin of the problem in the eyes of some was that dams, canals, and water systems were built, operated, and maintained with users being charged much less than cost for the resulting water supply. California's Central Valley Water Project (CVP) is the Bureau of Reclamation's largest water resource project. Construction began 60 years ago and continues to this day. In the case of the CVP, roughly 50 percent of the reimbursable costs were allocated to agriculture with the remainder appropriated to municipalities, industry, and hydroelectric power generation (nonreimbursable costs include flood control, recreation, fish and wildlife and highway improvement).[48] As of 1994, it has been estimated that farmers benefitting from the CVP (85 percent of the water is allocated to agriculture) have paid only 12 percent of their allocated costs over the span of the project—$186 million out of $1.6 billion in costs allocated to irrigation. The problem is not agriculture's alone—municipal and industrial users have effectively paid nothing toward project costs while adding additional expenses in the form of unpaid operation and maintenance costs.[49] Of the total estimated project cost of $3.8 billion, only 42 percent of costs are currently scheduled to be repaid (90 percent of these costs in the future) with agriculture responsible for 94 percent of those repayments.[50]

Farmers object strongly to increases in water costs. In the short run, increases in the price of water cut directly into farmers' profit margins and in the long run adversely impact their competitive position relative to other production regions. In the process, asset values (the price of land) are reduced.

However, increasing the cost of water encourages conservation. When the price of pumping water rose substantially with the oil crisis in the late 1970s, it

[48]General Accounting Office, *Information on Allocation and Repayment of Costs of Constructing Water Projects* (Washington, D.C.: General Accounting Office, July 1996).

[49]General Accounting Office, *Central Valley Project Cost Allocation Overdue and New Method Needed* (Washington, D.C.: General Accounting Office, March 1992).

[50]General Accounting Office, *Information on Allocation and Repayment of Costs.*

fostered increased interest in water conservation systems.[51] For example, techno-logical advances such as low-energy precision application (LEPA) and drip irri-gation systems deliver water close to the ground and cut evaporation losses. Laser leveling, recycling water, and soil moisture monitors likewise help to con-serve water resources. Reductions in water usage of about 30 percent have been experienced in broad-based commercial farming (as opposed to experimental) applications. In the future, expanded use of urban wastewater for irrigation pur-poses is likely.

Water marketing involves the sale of water rights to municipal and indus-trial users. Water marketing facilitates voluntary transfer of water from one use and/or location to another. In 1982 California declared water marketing a bene-ficial use and, therefore, not a cause for the loss of water rights. Several states such as Oregon and Nebraska have followed this lead. One source indicates that prices varied from $200 per acre foot (325,850 gallons) in Salt Lake City to over $3,000 on the Colorado Front Range.[52] Trades of this kind have the potential to threaten crop production, although they clearly represent the economic forces of supply and demand.

In some instances, payments for water conservation have been demonstrat-ed to be the lowest cost alternative source of water. Such payments are a means of pricing water. For example, the Metropolitan Water District of Southern California (half of California's residents) has been financing agriculture water conservation projects in the Imperial Valley Irrigation District at a cost of $128 per acre foot—less than the cost of any alternative supply source.[53]

The following are some of the consequences of water pricing options:

- Increased water costs
- Potentially reduced agricultural production
- Increased water conservation
- Water availability as needed

Water allocation.   Water management, conservation, and allocation sys-tems involve government mandates regarding the use of water. Water manage-ment systems are supervised by some form of water commission, agency, or individual with the right to regulate groundwater withdrawal in several states. Overlapping the riparian system in a majority of eastern states is an adminis-trative permit system that may be utilized to allocate water independent of the riparian rights. The permit contains information on the volume, flow rate of the diversion, location, and nature of water use. Permits are generally issued for 10 to 20 years.

---

[51]At the same time, there was concern about depletion in the supply of water from the Ogallala aquifer with an increasing number of wells going dry.
[52]Postel, "Saving Water for Agriculture," p. 49.
[53]Ibid.

Management systems are more sophisticated in western states that have a longer history of water problems, although states such as New Jersey and Florida have also instituted water management systems. Even states such as Texas, which has been one of the last states to employ an absolute ownership doctrine, recently reversed a long-standing Texas policy that landowners could pump virtually unlimited quantities of groundwater. A court ruling that the Edwards aquifer, which serves the city of San Antonio as well as major vegetable-growing regions, is really an underground flowing river over which the Texas Water Commission has greater regulatory power triggered the Texas legislature to modify its absolute ownership doctrine.[54] Limits were then placed on how much water various users, including agriculture, could pump. These limits are being challenged in the courts. The Ogallala aquifer supplies groundwater for irrigation to the semiarid agricultural regions of eight states (14.3 million acres, 25 percent of the nation's irrigated farmland, 12 percent of the corn, cotton, sorghum and wheat, and 50 percent of the beef cattle). The type and degree of state management of groundwater use in the region served by the Ogallala varies widely from state to state. Such variations in regulatory systems and standards, along with ever-increasing scarcity, are likely to lead to increased federal preemption (regulation) of water use.

According to Goldfarb, "Federal preemption is rare in water diversion law because the federal government has traditionally deferred to state lawmaking when it comes to allocating rights to divert water, even on federally-owned land."[55] Most cases of federal intervention in the allocation of water relate to Indian reservations, national parks, and national forests. However, the potential for substantially increased federal intervention lies in efforts to implement the provisions of the Endangered Species Act discussed later. For example, when ranchers holding Nevada appropriation permits for irrigation pumping reduced the water table in a national monument cavern to the point where a unique species of desert fish was endangered, an injunction was issued to prohibit pumping.[56] Likewise, in a Colorado case, an irrigation district was denied the right to build a reservoir that would reduce instream flow sufficiently that the habitat for whooping cranes 250 to 300 miles downstream would be adversely affected.[57] Such instances of federal involvement can be anticipated to become increasingly common.

One of the incentives for increased use of management systems involves the reality that much of the irrigation water has its origin in other states, such as the mountainous regions of the West. Disputes over water having its origin in interstate waterways where federal influence predominates have three primary methods of resolution:[58]

[54]*New Waves* (College Station, Tex.: Texas Water Resources Institute, Texas Agricultural Experiment Station, June 1992), p. 12.

[55]Goldfarb, *Water Law*, p. 49.

[56]Ibid., p. 50.

[57]*Riverside Irrigation District v. Andrews*, 568 f. Supp. 583 (1983).

[58]Goldfarb, *Water Law*, pp. 53–55.

- The **U.S. Supreme Court** resolves conflicts among states in the utilization of water from interstate waterways. The doctrine utilized by the Supreme Court is that of "equitable apportionment" from international water law. Equitable apportionment begins with the historical priority of appropriation on the basis of a mass allocation to each state.

- **Interstate compacts** to allocate available water supplies are entered into under the provisions of the Constitution with the approval of the Congress. More than 30 interstate compacts now exist to deal with various water problems, including those involving allocation. Some compacts, such as the Upper Colorado River Basin Compact or the Delaware and Susquehana River Compacts, create a commission that has the power to allocate supplies and curtail use in times of shortages.

- **Congressional apportionment** is rarely used but has been applied to the allocation of water supplies between the upper and lower basins of the Colorado River.

As water becomes increasingly scarce and disputes more prevalent, pressures for federal legislative and administrative remedies can be anticipated. Such a federal administrative remedy could build on the existing system of water rights and management allocation systems to achieve greater uniformity in policy among states as well as between surface water and groundwater sources.

The following are examples of the consequences of water management, conservation, and allocation systems:

- Increased water conservation
- Allocation of use more consistent with need
- Reduced water scarcity
- Reduced freedom
- Reduced values of water rights

Water Quality Policy

Contemporary water quality policy has its origin in the Clean Water Act of 1972 (CWA). The CWA rejected the notion of an acceptable level of pollution. Prior to the CWA, each state was allowed to set its own clean water standards based on the use of the water, from swimming to agricultural and industrial use. In one state, water for navigation was not considered legally objectionable until it caught fire.

Although the CWA did not immediately do away with such measures of water quality, it gradually but progressively tightened the effluent limitations until the **zero discharge goal** was reached. Second, it set a goal that all waters be fishable and swimmable regardless of their use. Concurrent with the CWA, the Congress enacted the Coastal Zone Management Act of 1972 (CZMA) authorizing states to voluntarily develop comprehensive programs to protect and manage coastal resources (including the Great Lakes). The CWA represented a change in policy because it focused on utilizing technology to eliminate the causes of pollution rather than its tolerable effects. Both the CWA and the CZMA require reauthorization every five years. Contemporary CWA debate centers on three primary issues:

- Whether the cost of installing continuously improving technology is worth the benefits of improving water quality
- The degree to which pollution control should be extended beyond point sources to nonpoint sources, of which crop production agriculture is a part
- The weight of private property rights in water quality decisions; one proposal would require compensation for reductions in land value resulting from water quality initiatives.

Point sources of pollution involve identifiable single-discharge sources such as an industrial plant or a municipal sewage water treatment plant from a pipe or a ditch into a stream or body of water. Large confined animal-feeding operations are point pollution sources in that animal wastes must be contained for up to 24 hours for a flood event that happens every 25 years.[59] **Nonpoint sources** are diffuse sources of water pollution that do not discharge at a single location but whose pollutants are carried over or through the soil, ground cover, and storm water runoff. "It is caused by rainfall or snow melt moving over and through the ground and carrying natural and manmade pollutants into lakes, rivers, streams, wetlands, other coastal waters, and groundwater."[60] Crop production, forestry, storm water, mineral extraction, construction, and landfills are the principal sources of nonpoint pollution.

Every five years since 1972 in the CWA reauthorization process, Congress has tightened the reins on water quality regulations by initially concentrating on point sources of pollution and then gradually shifting emphasis to nonpoint sources. Amendments and reauthorizations to the CWA and the CZMA have narrowed the distinctions between point and nonpoint sources of pollution. The Water Quality Act of 1987 amended the CWA and declared: "It is the national policy that programs for the control of nonpoint sources of pollution be so developed and implemented in an expeditious manner so as to enable the goals of this Act to be met through the control of both point and nonpoint sources of pollution."

The 1990 reauthorizations of the CZMA expressed major concern about the impacts of nonpoint pollution on coastal waters. In reauthorizing amendments, Congress made it clear that nonpoint pollution, having its origin substantially inland in major river basins, may be regulated by the provisions of the CZMA.

---

[59]What constitutes a large confined animal feeding operation has become a moving target. At the federal level, farms having more than 700 mature cattle have been required to obtain a discharge permit. In some states, there are considerably lower size requirements. See Joe L. Outlaw *et al.*, *Impacts of Dairy Waste Management Regulations*, Agricultural and Food Policy Center Working Paper 93-4 (College Station, Tex.: Texas A&M University, May 1993.)
[60]Environmental Protection Agency, *Proposed Guidelines Specifying Management Measures for Sources of Nonpoint Pollution in Coastal Waters*, WH-553 (Washington, D.C.: Office of Water, May 1991), p. 1–1.

These amendments mandate that the state provide management measures designed to achieve the act's objectives. Agriculture has become more uncomfortable with each reauthorization and amendment.[61]

The CWA was due to be reauthorized in 1992. It has been debated each year since with no action on reauthorization. The status of agriculture is a contributor to the deadlock.

The Water Quality Base

Agriculture is the largest nonpoint source of water pollution, which accounts for about half of all water pollution.[62] Agricultural nonpoint pollution impacts both surface water and groundwater, each of which has different treatment problems.

Surface water pollution results largely from runoff from land (nonpoint pollution) and from livestock production facilities (point pollution if large size). Land runoff is closely, but not exclusively, related to the previously discussed problems of soil erosion. Precipitation- and irrigation-induced runoff carries sediment, minerals, soil nutrients, and pesticides into streams, rivers, estuaries, and lakes. These effects may be more adverse than the impact of erosion on soil productivity.[63] Sediment in water resulting from erosion decreases the light available for aquatic vegetation, which reduces the fish population and related aquatic species. Water-soluble nitrogen readily ends up in surface water runoff. Increased nutrient levels stimulate algae growth and reduce available oxygen in the water supply, once again reducing the fish population. Pesticide runoff likewise creates concerns about chemical residues in fish and the quality of the surface water supply. Many chemicals are difficult or impossible to remove in the drinking water treatment process.

Groundwater pollution results largely from contamination of one-fourth of the precipitation that falls on the United States, infiltrates the soil, and recharges the aquifers.[64] Groundwater pollution is of major concern because of its importance as a source of drinking water. Groundwater provides drinking water for more than 50 percent of the total U.S. population and for almost all of the rural population.[65] Consequently, rural people are very concerned about groundwater contamination.

---

[61]Rewriting the Clean Water Act—Can National Economic Policy Get Any Worse Than S.1081, Econ. Rev. (Chicago, Ill.: American Farm Bureau Federation, April 1992).

[62]G. Chesters and L. J. Schierow, "A Primer on Nonpoint Pollution," J. Soil Water Conservation (1985), pp. 14–18.

[63]Soil Conservation: Assessing the National Resources Inventory (Washington, D.C.: National Academy Press, 1986); and USDA, Agricultural Resources—Cropland, Water and Conservation, AR-8 (Washington, D.C.: Economic Research Service, 1987).

[64]David W. Moody, "Groundwater Contamination in the United States," J. Soil Water Conservation (March–April 1990), pp. 170–179.

[65]Ibid., p. 172.

Groundwater pollution is a difficult problem to deal with because once contamination occurs, it is often very difficult to eliminate. Moody described the problem in the following terms:

> Because of the slow rate of groundwater movement in deeper aquifers and the greater travel distances of contaminants, a contaminant may not be detected in water from wells until years after it has entered an aquifer. Similarly, it may take years for improvements in groundwater quality to be observed after remedial action has been taken.[66]

Therefore, prevention of groundwater pollution may be considerably easier than remedying it.

The major sources of groundwater pollution include the disposal of solid and liquid waste (septic tanks, landfills, etc.), storage and handling tanks, mining, oil and gas production, agriculture, and saline intrusion. Because agriculture occurs on at least 465 million acres of cropland and 591 million acres of pastureland, it is an important but much debated source of groundwater pollution. The major sources of agricultural pollution of groundwater include commercial fertilizer (particularly nitrogen), pesticides, animal wastes, and irrigation.

Agriculture became a focal point for concern about groundwater pollution when pesticides were discovered in the groundwater of California, Florida, Iowa, and Minnesota during the mid-1980s.[67] These findings precipitated a national drinking water survey, which concluded that "10 percent of the nation's community drinking water wells and four percent of the rural domestic drinking water wells have detectable residues of at least one pesticide. But less than one percent of the wells have pesticide residues above levels considered protective of human health."[68] The most common pesticides found in the groundwater included dacthal (primarily a broadleaf *lawn* weed killer), atrazine (a herbicide), alachlar (a soil fumigant no longer approved for use by EPA), FBDC (a fumigant canceled by EPA), and gamma lindane (an insecticide). EPA concluded that "the findings of the survey indicated that the vast majority of the drinking water wells in this country do not have levels of pesticides or nitrates that would pose a risk to public health."[69] Yet concern for groundwater quality has remained on the agenda as a viable public policy issue.

Water Quality Policy Options

The four general approaches to dealing with water quality issues are the following:

- Free market
- Tax inputs
- Performance standards
- Prescribe practices

---

[66]Ibid., p. 171.

[67]D. W. Moody *et al.*, *National Water Summary*, Water Supply Paper 2325 (Reston, Va.: U.S. Geological Survey, 1986).

[68]Environmental Protection Agency, *National Survey of Pesticides in Drinking Water Wells*, R-187 (Washington, D.C.: Office of Water, November 1990), p. 1.

[69]Ibid.

Free Market

The free market in a water quality context involves no government regulation of agricultural pollution of the water supply. The free market integrates into farmers' decisions only those costs and revenues to which farmers are subject. These include the costs of productions, receipts from products, and changes in asset values such as land. As indicated previously, the free market does not consider the nonmarket values and costs derived, for example, from externalities or situations in which land markets fail to reflect changes in productivity due to soil erosion. Consequently, from an economic theory perspective, the adverse externalities of chemical use or animal wastes may not be adequately considered from a public interest perspective in farmers' decisions unless they are treated as point pollution sources. However, farmers and farm families are perhaps more adversely affected by pesticides than the general public, suggesting that considerable care would be taken by farmers in handling chemicals. Such a generalization may hold for pesticides, but it may not hold for other water quality-related externalities such as soil erosion or animal waste runoff. However, it must be realized that confined animal-feeding operations are much easier to identify and control as a source of pollution than either chemicals or soil erosion.

Input Taxes

Input taxes have been proposed as a means to reflect at least a portion of the costs of the externalities associated with the use of environmentally sensitive inputs such as fertilizer and pesticides. The effect of an input tax is to raise the cost of the input. Increased input costs mean lower input use as farmers equate the higher marginal cost of the input with marginal revenue. However, the magnitude of the reduction is influenced by the elasticity of demand for chemicals. Large reductions in yields associated with reduced chemical use suggest a highly inelastic demand for chemical inputs. This means that the tax would need to be large, perhaps several times the value of the pesticide, to substantially reduce chemical use.

Some states have imposed input taxes. California, Iowa, South Dakota, and Wisconsin each place a tax on fertilizer, although these taxes may have been designed more to generate revenue than to reduce use.[70] Revenues from the Iowa input tax are designated for research designed to find substitutes for chemicals.

Performance Standards

Performance standards prescribe the quality of water that must be the end result of farming activity. It does *not* prescribe the farming practices that must be pursued. Stated differently, it gives the farmer flexibility in deciding methods by which the performance standards are achieved. As a result, such standards facilitate and encourage technological advance and innovation in achieving the desired goal at the least cost to the farmer.

---

[70]Marc Ribaudo and Danette Woo, *Summary of State Water Quality Laws Affecting Agriculture,* AR-23 (Washington, D.C.: ERS/USDA, September 1991), pp. 50–54.

Implementation and enforcement of performance standards require regular monitoring, which may be costly. Monitoring may be more difficult with nonpoint pollution of surface water. With groundwater, the problem may be more difficult to discover before actual pollution occurs, although analyzing soil samples at various depths may provide a satisfactory monitoring approach.

Prescribing/Proscribing Practices

Prescribing and/or proscribing certain practices includes a broad range of regulatory activity designed to improve water quality. Because farmers have voluntarily and willfully adopted the current set of practices, it is virtually inevitable that prohibiting current practices will increase unit costs of production for those farmers who must, as a result, change farming methods. Proscribing specific practices may foster technological change and innovation, although that will take time. Prescribing specific practices may delay technological change and innovation because of the requirement that certain technologies be employed. The following are examples of this regulatory approach.

Proposals to severely restrict chemical use have been high on the agenda of policymakers since the publication of Rachel Carson's *Silent Spring* in 1962.[71] The book postulates a future world without birds and other wildlife if the use of chemicals such as pesticides continues unrestrained. The book is credited with having led to the establishment of the Environmental Protection Agency (EPA) in 1970. Federal policy regulating the sale and application of pesticides is contained in the Federal Insecticide, Fungicide, and Rodenticide Act (FIFRA) first enacted in 1947 and periodically reauthorized and updated since.[72] The authority for registering pesticides was transferred from USDA to the newly created EPA in 1970. Agricultural chemicals must be registered by EPA before they can be sold. The criteria used to deny the registration of a chemical involve determining whether it does what it purports to do, is an oncogene (causes benign or malignant tumors), harms the environment, and affects agricultural productivity. In other words, FIFRA utilizes a cost-benefit standard for chemical registration.[73] To meet these criteria, EPA can require up to 70 different types of tests investigating its chemistry, toxicity, environmental fate, ecological effects, and impacts on productivity.[74] Each chemical is ruled on individually, based largely on data supplied by chemical companies developed under EPA-specified procedures. In 1988, FIFRA was amended to require that reregistration of all agricultural chemicals be completed by 1997. Because of the cost of reregistration, concern arose that some chemicals used on minor crops such as fruits, vegetables, and nuts would not be

---

[71]Rachel Carson, *Silent Spring* (Boston, Mass.: Houghton Mifflin Publishing Co., 1987). (Originally published in 1962.)

[72]Provisions of the Pure Food and Drug Act as discussed in Chapter 14 are also relevant to the regulation of agricultural chemicals.

[73]National Research Council, *Regulating Pesticides in Food: The Delany Paradox* (Washington, D.C.: National Academy Press, 1992), p. 30.

[74]Leonard P. Gianessi and Cynthia A. Puffer, "Registration of Minor Use Pesticides: Some Observations and Implications," *Agricultural Resources*, AR-25 (Washington, D.C.: ERS/USDA, February 1992), pp. 52–60.

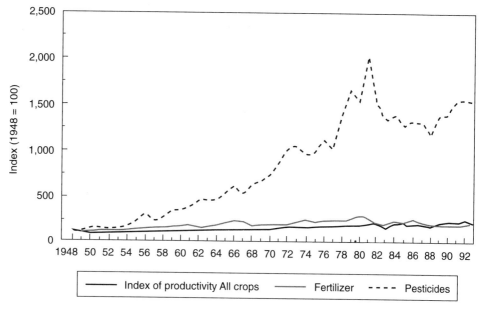

FIGURE 13.1 **Relationship between crop productivity and chemical use, 1948 base, 1948–1993.**

Source: Economic Research Service, *Increases in Agricultural Productivity, 1948–1993* AREI Update No. 6 (Washington, D.C.: USDA/ERS, July 1996) p. 3.

dropped; this is referred to as the minor use pesticide problem. Some of these pesticides have been dropped. Decisions not to produce certain chemicals have created consternation among producers—particularly in cases where no substitutes exist.

Many view agricultural chemicals as a symbol of modern commercial agriculture. Crop production per acre tracks fertilizer use in agriculture (Figure 13.1)—more closely than sales of pesticides.

Despite an extensive system of regulation, agricultural chemicals have become a major target of environmental interests extending back to the publication of *Silent Spring*. As indicated previously, finding residues of chemicals in the groundwater certainly heightened support for eliminating chemical use in agriculture.

The impacts of severely restricting the use of agricultural chemicals is a highly contentious issue. The only major study of this issue, based on the estimates of plant scientists, concluded that substantial reductions in crop yields and increases in unit costs of production would be associated with complete curtailment of chemical use on the major farm program crops (Figures 13.2 and 13.3).[75] A companion study[76] of the economic impacts of reduced chemical use

---

[75]Edward G. Smith, Ronald D. Knutson, C. Robert Taylor, and John B. Penson, *Impacts of Chemical Use Reduction on Crop Yields and Costs* (College Station, Tex.: Agricultural and Food Policy Center, Texas A&M University and Tennessee Valley Authority, 1991).

[76]Ronald D. Knutson, C. Robert Taylor, John B. Penson, and Edward G. Smith, *Economic Impacts of Reduced Chemical Use* (College Station, Tex.: Knutson & Associates, 1990); and C. Robert Taylor et al., "Economic Impacts of Chemical Use Reduction in the South," *So. J. Agric. Econ.* (July 1991), pp. 15–24.

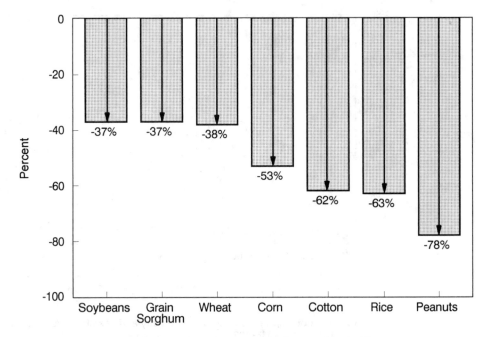

**FIGURE 13.2   Estimated percentage reduction in yields associated with
no chemical use by crop.**

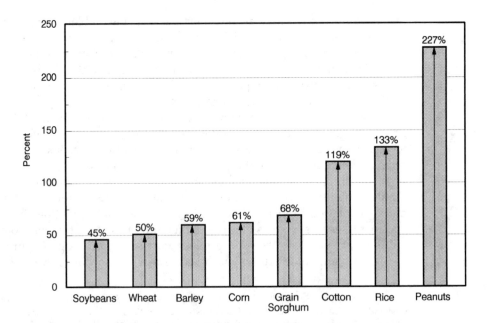

**FIGURE 13.3   Percentage increase in unit costs associated with no chem-
ical use by crop.**

concluded the following effects would result from the absence of chemicals on major crops:

- Most adversely affect producers having the greatest comparative disadvantage in production—corn in South, wheat in West, etc.
- Sharply reduce agricultural exports.
- Reduce stock levels to the point where shortages could become a problem.
- Require import quotas if U.S. farmers and consumers were to be protected from foreign products produced with chemicals.
- Raise net income to crop producers but reduce income to livestock producers.
- Increase food prices by as much as 12 percentage points annually, the level at which export embargos were imposed in the early 1970s.

These studies were extensively discussed and evaluated.[77] They were criticized for having been financed by chemical companies and farm organizations, for studying only the zero- use option, and for not considering potential adjustments to zero chemical use such as sustainable agriculture.[78] Tweeten found that studies of yields on organic farms support the results of the Texas A&M study.[79] Moreover, Tweeten points to studies indicating that more than 95 percent of the pesticides ingested by humans are naturally occurring in the food supply.[80]

The consequences of reduced chemical use include a number of nonmarket values, such as reduced residues of chemicals in water and/or in the products on which they are used and increased wildlife. This issue perhaps provides one of the clearest illustrations of the trade-off between market values and nonmarket values inherent in many resource policy issues. The extent of the controversy and the cross-current of facts and opinions clearly indicate the sensitivity of chemical use issues in modern agriculture.

A **best management practice (BMP)** is a farming method, measure, or practice designed to prevent or reduce pollution from nonpoint sources.[81] While

[77]Tweeten, "Environmentally Sound Agriculture," pp. 66–69; John E. Ikerd, "Applying LISA Concepts on Southern Farms," *So. J. Agric. Econ.* (July 1991), pp. 43–52; and Michael R. Dicks, "Applying LISA Concepts on Southern Farms or Changing Farm Philosophies," *So. J. Agric. Econ.* (April 1991), pp. 53–56.

[78]Harry Ayer and Neilson Conklin, "Economics of Ag Chemicals: Flawed Methodology and a Conflict of Interest Quagmire," *Choices* (4th Quarter, 1990), pp. 24–30; Ronald D. Knutson, "Economic Impacts of Reduced Chemical Use," *Choices* (4th Quarter, 1990), pp. 24–30; Harry Ayer and Neilson Conklin, "No Chemicals, No Pesticides," *Choices* (1st Quarter, 1991), pp. 40–41; Otto Doering, "No Chemicals, No Pesticides," *Choices* (1st Quarter, 1991), p. 41; Ronald D. Knutson *et al.*, "No Chemicals, No Pesticides," *Choices* (2nd Quarter, 1991), p. 41; David Zilberman and Andrew Schmitz, "No Chemicals, No Pesticides," *Choices* (2nd Quarter, 1991), p. 40; and C. Robert Taylor and John Penson, "No Chemicals, No Pesticides," *Choices* (2nd Quarter, 1991); Tweeten, "Environmentally Sound Agriculture," p. 41.

[79]Tweeten, "Environmentally Sound Agriculture," p. 69.

[80]Ibid., p. 52.

[81]Terry J. Logan, "Agricultural Best Management Practices," in *Ground Water and Public Policy* (University Park, Pa.: ES/USDA, Soil and Water Conservation Society, Freshwater Foundation, July 1991).

originally conceived as voluntary, BMPs have become the regulatory means of implementing nonpoint pollution control under federal and state policy.

A most interesting issue involves the economic content of BMPs. That is, do BMP requirements have to be economically feasible to have the force of law? It appears that they do. The Coastal Zone Act Reauthorization Amendments of 1990 require that specified management measures be "economically achievable measures . . . which reflect the greatest degree of pollutant reduction achievable through the application of the best available nonpoint pollution control practices. . . . Thus, the management practices must be economically achievable."[82] Yet it is unclear how this economic achievability criterion will either be interpreted or administered. Economic feasibility is important because if the proposed BMP is economically infeasible, the following are possible:

- Farms could be put out of business.
- Regional competitive relationships among farms could be upset.
- Administration of the program would be more difficult.
- Political backlash to the extent that the program becomes politically infeasible may result.

The following are examples[83] of specific alternative practices that are acceptable to EPA by pollution problem area:

- *Erosion and sediment control.*—vegetative cover, crop rotation, conservation tillage, filter strips, strip cropping, and terraces
- *Confined animal-feeding operation management.*—runoff control such as dikes and diversions, manure and runoff storage such as treatment lagoons, utilization of manure on a specified number of acres of cropland per animal unit, or marketing manure
- *Nutrient management measures.*—nutrient testing of waste material, testing for heavy metals and organic toxic substances, testing soil, proper timing and application methods, and use of cover crops
- *Pesticide management practices.*—pesticide inventory, evaluating soil characteristics, integrated pest management systems, and maintaining a history of pesticide use on each field
- *Range and pasture management practices.*—deferred grazing (not overgrazing), rotational grazing, fencing off streams or lakes, and grassland reestablishment
- *Irrigation management practices.*—irrigation scheduling, irrigation measuring, water conservation, field leveling, and filter strips

---

[82]Environmental Protection Agency, *Proposed Guidelines Specifying Management Measures.*
[83]These are only examples and are certainly not inclusive. Adopted practices must be turned specifically to the soil and water condition on the site for which BMPs are specified.

If BMPs are to be utilized, they can be implemented either by economic incentives or by regulation. Economic incentives may include either conservation compliance or green payment. **Conservation compliance** makes receipt of farm program benefits contingent on applying BMPs in the framework of a conservation plan. **Green payments** would offset at least a portion of the cost of applying the BMP. Regulation would make the BMP a requirement for farming. An administrative regulatory structure must exist for seeing that the management practices are properly implemented. Under the EPA guidelines, implementation would be performed by a state government employee.[84] Within the USDA the implementing agency would be NRCS, including its state and local affiliates.

The 1996 farm bill took significant steps in implementing a green payment policy by the establishment of EQIPS. These payments can be made either for crop production ($100 million authorized) or livestock waste management ($100 million authorized). The $50,000 limit on payments raises questions regarding the efficiency of this program initiative in reducing problems of animal waste pollution.

The BMP option does not single out agriculture. Separate guidelines have been proposed for other nonpoint pollution sources including forestry, urban runoff, marinas, and recreational boating.

The consequences of a BMP policy include the following:

- Improved water quality
- Reduced farmer freedom in production decisions
- Increased cost of production
- Reduced conflict with environmental interests

Sustainable agriculture is defined in the 1990 farm bill as follows:

> Sustainable agriculture is an integrated system of plant and animal production practices having a site-specific application that will, over the long term, satisfy human food and fiber needs; enhance environmental quality and the natural resource base upon which the agricultural economy depends; make the most efficient use of nonrenewable resources and on-farm/ranch resources and integrate, where appropriate, natural biological cycles and controls; sustain the economic viability of farm/ranch operations; and enhance the quality of life for farmers/ranchers and society as a whole.[85]

The sustainable agriculture movement has its origins in the organic agriculture and low input, sustainable agriculture (LISA) concept. According to John Ikerd, a sustainable agriculture disciple, the LISA concept reflects two perspectives:[86]

- The low input perspective is that farmers must reduce their use of commercial chemical inputs as a means of reducing environmental and ecological risks.

---

[84]Environmental Protection Agency, *Proposed Guidelines Specifying Management Measures*, p. 2–83–84.

[85]Richard M. Kennedy, "Title XVI—Research," Provisions of the Food, Agriculture, Conservation, and Trade Act of 1990, Agr. Inf. Bul. 624 (Washington, D.C.: ERS/USDA, June 1991).

[86]Ikerd, "Applying LISA Concepts," p. 43.

■ The sustainable agriculture perspective is that long-run productivity and utility of agriculture depend ultimately on their ability to keep farms both ecologically and economically viable. Reduced reliance on commercial inputs is seen as one means of addressing the ecological risk that could threaten long-run sustainability.

The sustainable agriculture movement got a boost of support from the National Academy of Sciences' report *Alternative Agriculture.* This report, based on a case study of 11 "sustainable" farms, was strongly criticized for its lack of objectivity and economic content. Tweeten's analysis of the farms led to the conclusion that there were insufficient data to determine whether most of the case study farms were economically sustainable and that many of the practices employed were typical of conventional commercial farms.[87] However, the sustainable agriculture movement, perhaps unfortunately, has drawn a line between itself and conventional commercial agriculture. Ikerd, for example, notes that sustainable agriculture is distinguished from conventional agriculture by differences in goals and objectives. He notes that conventional agriculture has primary goals of increased agricultural productivity and short-run profitability with environmental protection and resource conservation being viewed as constraints in achieving these goals.[88] In contrast, according to Ikerd, sustainable agriculture farms strive to achieve both ecological and economic goals in the interest of maintaining their productivity and usefulness to society indefinitely.[89]

Sustainability from the perspective of prescribed farming practices is difficult to define. The following are some characteristics that appear to cut across the sustainable agriculture movement:[90]

■ Crop rotation to mitigate pests, reduce the need for commercial fertilizer, and reduce soil erosion
■ Integrated pest management
■ Soil conserving tillage
■ Animal systems that avoid high-density confinement and utilize animal waste as a commercial fertilizer substitute
■ Farms on which both plants and animals are grown in a synergistic fashion

As a result of employing sustainable practices, it is generally concluded that sustainable farms will tend to be more diversified and smaller than their counterparts.[91] However, this conclusion may be wishful thinking since management may be as important in a sustainable system as in conventional agriculture.

---

[87]Tweeten, "Environmentally Sound Agriculture," p. 60.
[88]Ikerd, "Applying LISA Concepts," p. 40.
[89]Ibid.
[90]National Research Council, *Alternative Agriculture,* 1989.
[91]Ikerd, "Applying LISA Concepts," p. 47.

Because good management is one of the most limiting resources in agriculture, large sustainable farms may have substantial advantages over smaller ones. In fact, some of the more successful sustainable farms in the *Alternative Agriculture* project were clearly larger than average size and used the techniques of commercial agriculture.[92]

The 1996 farm bill has been looked on by sustainability advocates as another step forward in promoting their cause. This is the case for the following reasons:

- The removal of the target price reduces incentives for increased production and increased use of purchased inputs.
- Full flexibility allows for greater diversity in production including the opportunity for more use of crop rotation and less specialization.
- The EQIP program provides the potential for sustainable farms participating in green payments to their objectives.

On the negative side, the 1996 farm bill increases price variability, which could create difficulty for usually smaller sustainable farms.

Ahead on the agenda of the sustainability movement is the reauthorization of the research and extension title of the farm bill. The 1996 bill extended the research title through 1997. Sustainability advocates want to see relatively more emphasis on applied research—and less emphasis on biotechnology and basic research.

The consequences of sustainable agriculture are debatable, including the most obvious and confirmed ones:

- Reduced chemical use
- Higher costs of production
- Higher food prices
- Improved water quality
- Reduced soil erosion

Integrated pest management (IPM) involves making decisions to apply and utilize pesticides based exclusively on approved scouting techniques and economic thresholds. Economic thresholds involve making application decisions only when the economic benefits exceed the cost of pesticide application. IPM also emphasizes the utilization of biological pesticides, crop rotation, and related control techniques.

The Clinton administration, in designing its policy to modify U.S. pesticide registration policy and reduce pesticide use, indicated a goal of 70 percent IPM adoption by 2000. In the mid-1990s, IPM was most extensively utilized on fruit and vegetable crops, cotton, and rice.

---

[92]Tweeten, "Environmentally Sound Agriculture," p. 60.

Treating all of agriculture as point source pollution would involve higher levels of regulation than is currently proposed for nonpoint pollution. Point pollution sources are subject to requirements to obtain permits that require proof of compliance whenever farmers make improvements or expand the size of their operation. Farmers would be required to engage in specific practices and employ specific remedies that remove the sources of pollution. Farmers would be subject to testing requirements to ensure that specified standards are being met. Prescriptions that are very specific, with little flexibility, are designed to internalize the costs of externalities.

In some cases, agriculture is already regulated as point sources. For example, large confinement animal-feeding operations (more than 700 mature cattle or their equivalent animal units) are regulated by EPA and by several states as point pollution sources. Any animal-feeding operation that discharges waste into streams and/or against whom a complaint is registered regardless of size may be regulated as a point pollution source.

Regulation of agriculture as a point source would require that water from a farm be channelized and monitored for pollutants. Implementation of techniques to remedy the source of pollution would be required. Permits would be required to change production practices and production enterprises to eliminate adverse environmental impacts.

An important issue in environmental regulation of agriculture involves whether farms and ranches are judged according to the **practices** (BMPs) they employ to control externalities or according to their **performance** as measured by the quality of the runoff. If regulated by performance, farmers would have more flexibility in adjusting their practices than if regulated by the practices themselves. However, performance measures require more data on the quality and quantity of runoff.

The following are some consequences of treating agriculture as a point source of pollution:

- Improved water quality
- Reduced soil erosion
- Increased costs of production
- Increased food prices
- Reduced freedom of farmers to farm

Wetlands Policy

Wetlands are "those areas that are inundated or saturated by surface or groundwater at a frequency and duration to support, and that normal circumstances do support, a prevalence of vegetation typically adapted for life in saturated soil conditions."[93] Wetlands generally include swamps, marshes, bogs, and similar areas

[93]Environmental Protection Agency, *Proposed Guidelines Specifying Management Measures*, pp. 7–12.

where water periodically and regularly stands. Associated with wetlands are riparian vegetative areas around a body of water. Riparian areas or zones have many of the characteristics of wetlands and act as a filter for water flowing into wetlands.

This definition is not without controversy, nor is it completely definitive in terms of what constitutes wetlands. Substantial debate exists between farmers and NRCS/USDA and other related government agencies over the definition of wetlands and whether particular land areas have this designation. This debate led to a compromised administrative interpretation that would require land to be flooded at least 15 consecutive days during the growing season or saturated for at least 21 days.[94] If so designated, substantial restrictions exist on the ability to drain and/or farm these land areas.

The Wetlands Resource Base

Until 1977, government policy was to encourage the drainage of wetlands, often with the benefit of direct government assistance. In the two centuries from 1780 to 1980, more than half of the 221 million acres of continental U.S. wetlands were drained and converted to other land uses.[95] In recent decades, the rate of loss of wetlands declined dramatically from 458,000 acres annually during the mid-1950s to the mid-1970s, to 290,000 acres annually from the mid-1970s to the mid-1980s, to 150,000 acres annually during the 1980s.[96] Agriculture's share of conversion likewise has declined from 87 percent in the mid-1950s to the mid-1970s to about half in more recent periods.

Much of this decrease in the rate of loss of wetlands was due to a change in government policy. As noted previously in the discussion of soil conservation policy, for the first seven decades of the twentieth century, USDA provided financial and technical assistance to farmers for wetland drainage. In 1977, concern over the effects of wetland loss on both wildlife and water quality led to the issuance of Executive Order 11990 directing all government agencies to "minimize the destruction, loss, or degradation of wetlands" and to "avoid direct and indirect support of new construction in wetlands whenever there is a practical alternative."[97] Since the 1977 executive order, the policy toward wetlands has not wavered, consistently moving in the direction of no net loss and even restoration of wetlands.

The interest in preserving wetlands results from both wildlife (particularly waterfowl) preservation and water quality policy goals. From a water quality perspective, wetlands serve as a buffer or filter between uplands and the water body into which runoff flows, such as the Mississippi River and, ultimately, the Gulf of

---

[94]*Agricultural Resources*, AR-23, p. 27.

[95]Ralph E. Heimlich, "The Policy Context," in Ralph E. Heimlich, ed. *A National Policy of "No Net Loss" of Wetlands* (Washington, D.C.: ERS/USDA, August 1991), p. 4.

[96]*Agricultural Resources*, AR-23, p. 26.

[97]Ibid.

Mexico. Without wetlands, nonpoint pollutants flow more directly from the farmland to the river and into the Gulf. The same relationship exists for all major rivers and the ultimate water bodies into which they flow. As a result of the loss of wetlands, excessive plant nutrients, pesticides, sediment, oils, greases, and heavy metals (from landfills) may be rapidly discharged into rivers and, ultimately, into the coastal waters. Likewise, along coastal areas, the removal of wetlands and riparian areas exposes the associated water bodies directly to the effects of erosion and other forms of environmental pollution resulting from agricultural and developmental activities. These coastal areas then become directly subject to the erosive effects of storms and tidal activity, including the destruction of saltwater habitat for fish and shellfish.[98]

The protection and restoration of wetlands has been a major goal of the Department of Interior's Fish and Wildlife Service, supported by private interest groups such as Ducks Unlimited.[99] Their interest arises from a concern about the impacts of drainage and development projects on habitat. Wetlands serve as a nursery ground for frogs, fish, and invertebrates. More obvious is the importance of wetlands and related riparian areas for habitat of various types of wildlife and waterfowl, particularly ducks and geese. The major waterfowl flyways on the east and west coasts, as well as in the central United States, depend on the maintenance of wetlands on both ends for migration, wintering, and reproduction purposes.[100] Benefits to fish, waterfowl, and wildlife represent a haven for economists interested in nonmarket valuation techniques.

Wetlands Policy Options

Wetlands policy has rapidly evolved into six main categories:

- Market policies
- No net loss
- Swampbuster
- Best management practices
- Wetlands reserve
- Compensation

The manner and extent to which each of these will be implemented is open to considerable debate.

Market policies in the wetlands area have taken some interesting twists in that private sector interest groups have become involved in the purchase of wetlands for purposes of preservation and habitat. Therefore, the issue of what would

---

[98]Environmental Protection Agency, *Proposed Guidelines Specifying Management Measures*, p. 7–1.
[99]Heimlich, "The Policy Context," p. 6.
[100]John Bergstrom and Richard Brazee, "Benefit Estimation," in Ralph E. Heimlich, ed., *A National Policy of "No Net Loss" of Wetlands*, (Washington, D.C.: ERS/USDA, August 1991), pp. 18–22.

happen to wetlands in a market economy involves more than just evaluating the economic motives of farmers and developers.

Farmer development (drainage) of wetlands for agricultural purposes in the past has been very much influenced by government subsidies and by the provisions of commodity programs. Drainage subsidies provided considerable incentive for conversion of wetlands. Without these subsidies, conversion would have occurred more slowly and some may never have occurred. Commodity program provisions provided incentives for drainage in that they generated higher returns to farmers and fostered higher land values, the benefits of which extended to lands that were drained. Set-aside programs also encouraged the destruction of riparian areas because, if cropped, these acres became eligible for a farm program base. These acreages could then be set aside to meet the acreage reduction requirements of the program. Thus, it is not unusual to see set-aside lands being classified as wetlands. Recall from the farm program chapters that the least productive lands will be set aside in acreage reduction programs.

As a result of past programs encouraging conversion, fewer agricultural acres would be converted from wetlands to farmland in a free market context than historically was the case. Yet, clearly, in a free market economy, some agricultural conversion would be made by farmers who conclude that the marginal benefits of conversion are greater than the marginal cost. A larger threat to the environment from conversion lies in development for residential, recreational, or industrial purposes. Economic development will continue to occur as population grows. Pressures will exist to develop in areas classified as wetlands because of their proximity to coastal waters, including their industrial and recreational value.[101]

A private sector free-market twist on the wetlands issue involves the purchase of wetlands by organizations such as Ducks Unlimited and the Nature Conservancy. Lovejoy expands on this idea by suggesting that the interests of society could best be served by publicly auctioning off the rights involved in the preservation of wetlands.[102] Thus, a farmer currently restricted by regulatory policy could go ahead and drain wetlands if no private interests were willing to pay an acceptably high price for the farmer not to drain the wetland.

The consequences of a market wetlands policy include

- Continued conversion of agricultural land but at a slower rate than in the past
- Continued conversion for residential, recreational, and industrial purposes at a faster rate than in the past
- Reduced water quality
- Reduced wildlife numbers and related habitat.

---

[101]Leonard Shabman, "Integrating Agricultural Reconversion of Wetlands into Achieving Environmental Goals in Urbanizing Regions," in Ralph E. Heimlich, ed., *A National Policy of "No Net Loss" of Wetlands,*(Washington, D.C.: ERS/USDA, August 1991), pp. 23–28.

[102]Stephen B. Lovejoy, "Wetlands: Sell to the Highest Bidder," *J. Soil Water Conservation,* (November–December 1991), pp. 418–419.

The **no net loss policy** specifies that if an individual decides to convert wetlands to farming or development, the same acreage must be replaced as wetlands elsewhere.[103] In other words, under a no net loss policy, a farmer who has a pothole that must be drained, would be required to create or expand another wetland by the same acreage.

The no net loss policy option is controversial to both wetlands advocates and to farmers. Wetlands advocates argue that it is not possible to recreate a natural wetland.[104] Farmers view the no net loss constraint as being an unreasonable and costly restriction on their operations, if not an unconstitutional "taking" of their property rights and values.

The following are examples of the consequences of the no net loss policy:

- Reduced farmer and developer freedom
- Higher cost of production
- Increased stable quantity of wetlands
- Reduced quality of wetlands
- More wildlife
- Improved water quality

The **swampbuster policy** denies program benefits for wetlands brought into production. However, in the case of minimal conversions, swampbuster allows the application of the no net loss concept. That is, if one wetland is drained, another prior converted wetland must be restored. Farm program benefits, such as transition/contract payments, are not reinstituted until the converted wetlands are restored.

The consequences of swampbuster include the following:

- Reduced farmer freedom
- Stable quantity of wetlands
- Restoration of converted wetlands
- Increased number of wildlife
- Higher cost of production
- Improved water quality

Of course, the swampbuster policy does nothing about wetland conversion resulting from economic development.

Wetlands BMPs specify farming methods, measures, or practices that must be pursued on wetlands and riparian properties regardless of whether they are farmed. EPA has set forth generally defined BMPs for wetlands and

---

[103]Heimlich, "The Policy Context," p. 7.
[104]Shabman, "Reconversion of Wetlands."

riparian areas under the CZMA. These are examples of specified management practices[105]:

- Consider wetlands and riparian areas as a continuum of filters before waters enter an estuary.
- Do not alter the wetlands to achieve other functions.
- Use upland vegetation or forested buffers around existing wetlands when necessary to prevent nonpoint source pollution.
- Restore native species of diverse plant types.
- Maximize habitat.
- Plant vegetative filter strips adjacent to water bodies subject to sediment, suspended solid, and/or nutrient runoff.

Such BMPs have significant implications for agriculture and for farmers who own wetlands. This policy option could place an obligation on farmers owning wetlands to maintain the wetlands culturally. Not only could farmers not run cattle on wetlands (which was not listed as a BMP but could certainly be implied), but the farmer would also have to expend money to maintain the wetlands and the riparian area and to convert adjacent cropland to grassland or forestry land as a vegetative filter. By implication, farmers may be required to fence off cattle from the wetlands as well as the vegetative filter. An important policy issue involves whether farmers should be paid for carrying out such regulatory obligations, or is that part of the "privilege" of owning wetlands and land adjacent to wetlands? Part of that privilege may include the property right to lease wetlands for hunting or fishing, an increasingly important source of income to some farmers. Or would that privilege be lost also?

The **wetland reserve program (WRP)** pays farmers a rental fee per acre for converting farmland into wetlands. This program, enacted in the 1990 farm bill, has a modest goal of enrolling only 975,000 acres during the five-year life of the bill. In addition to farmed wetlands, noncropped wetlands or what are referred to as *water bank lands, riparian corridors,* and *critical wildlife habitat* may be entered in the wetlands reserve.

In contrast with the CRP, which provides for a 10-year government lease, the WRP participant must agree to a permanent or 30-year easement. The cropland base on the land, if any, is forfeited to the government. The farmer is paid an annual rental rate per acre, as in the CRP, and the government shares in the cost of restoration. The farmer still retains values associated with hunting rights, fishing rights, managed timber harvest, and, when warranted, periodic haying and grazing.

The consequences of the wetlands reserve program include

---

[105]Environmental Protection Agency, *Proposed Guidelines Specifying Management Measures,* pp. 7–1–20.

- Preservation of wetlands
- Improved water quality
- Option for farmers to enter the program
- Increased habitat for fish, wildlife, and waterfowl
- Reduced production to the extent that existing cropland is put in the program

Compensation policies relate to the property rights issue of whether land owners whose property value is adversely affected by wetlands decisions should be compensated. This issue arises as part of the *Lucas* case, which allowed compensation in a case in which the value of land was reduced to zero. Compensation policies could extend the *Lucas* principle to all wetland regulatory land value reductions. Such a policy, if implemented, would be a severe deterrent to wetland regulation.

Endangered Species Policy

Endangered Species Resource Policy Base

Biological diversity is represented by the total variety of life on earth. Biodiversity is analyzed at three levels[106]:

1. The ecosystems within which organisms live
2. The species
3. The genetic variety within those species

Much of the environmental sciences is concerned with the degradation of ecosystems, species, and genetic diversity. The enactment of the Endangered Species Act of 1973 recognized the explicit importance of curbing the trend toward fewer species of plants and animals and, therefore, less genetic diversity.

Genetic diversity is of substantial concern to agriculture. One of the contributing factors to the world food crisis of the early 1970s was the susceptibility of nearly all varieties of corn to a blight. The preservation of species may be viewed as having special significance for an era of biotechnology because of the potential for utilizing genetic material for fostering productivity and disease resistance in both plants and animals.

E. O. Wilson, a prominent Harvard scientist, estimates that more than 500 species and subspecies of plants and animals have become extinct in North America since European settlement.[107] The concern is that the rate of extinction is accelerating to a level of one species per hour in the world as a whole.

---

[106]John C. Ryan, "Conserving Biological Diversity," in Lester R. Brown, ed., *State of the World 1992,*(New York: W. W. Norton and Company, 1992), pp. 9–26.

[107]Edward O. Wilson, *The Diversity of Life* (Cambridge: Harvard University Press, 1992).

Endangered Species Policy Options

The policy options for endangered species include the following:

- Market allocation, no regulation
- Provision of the Endangered Species Act
- Consideration of economic costs relative to benefits
- Compensation

Market allocation treats species only in terms of their productive value. Predators are a threat to livestock even if the predators are endangered. Sufficient hunting or destruction of predators puts them on the threatened or endangered species list. Likewise, weeds, even if an endangered species, reduce crop yields. If an endangered plant species happens to be in or near cropland, it may be killed either by herbicides or by cultivation.

The following are some consequences of market allocation of endangered species:

- An increased number of threatened or endangered species
- Reduced biodiversity
- Low production costs
- Low food prices.

The **Endangered Species Act** has as one of its purposes preserving endangered and threatened species of plants and animals and the ecosystems on which they depend. It provides for the listing of endangered species and prohibits "taking" actions. The term *"taking"* is interpreted broadly to include activities ranging from harassment and harm to killing. Preserving the species requires the preservation of habitat critical to the species. As of 1992, 681 plant and animal species were listed under the Endangered Species Act. Five states (California, Texas, Tennessee, Alabama, and Florida) each had more than 50 endangered species.

The Fish and Wildlife Service of the Department of Interior has the prime responsibility for administering the Endangered Species Act, but all agencies are required to prevent farmers and ranchers from jeopardizing either the lives or habitat of threatened or endangered species. Therefore, EPA is required to consider the effects on endangered species in licensing chemicals. Under the Endangered Species Act, chemicals may be registered for use only in areas that do not jeopardize an endangered species. Thus, it may be illegal to spray specific herbicides in or near areas where endangered plants or endangered crops exist. Ranchers may be required to fence off waters where endangered fish exist. Agriculture and irrigation, as it is known, could be jeopardized by efforts to preserve several genetic varieties of salmon in the Columbia River Basin.

One of the unique features of the Endangered Species Act is that there is no requirement for the Fish and Wildlife Service to weigh the costs imposed by protecting endangered species against the benefits. In other words, a listed endangered species must be preserved at all costs. The only remedy is a so-called "god squad" made up of the secretaries of agriculture, army, and interior; the administrators of the EPA and the National Oceanic and Atmospheric Administration; the chairperson of the Council of Economic Advisers; and the president. A majority vote by the members of this committee is required for action to be taken that allows a species that is endangered to die out. The combination of the potential impacts and the lack of a feasible remedy has created great consternation among farmers and ranchers.

Economic cost–benefit analyses, as applied to endangered species, would require that the Fish and Wildlife Service weigh the increased cost and reduced productivity imposed by protecting endangered species against the benefits. The benefits could be determined by contingent valuation techniques and by the specific perceived benefits of biodiversity. This process would not be easy, which may explain why the Congress provided for no cost–benefit valuation procedures. Yet difficult decisions will be involved in many endangered species issues under current policy, with the god squad likely playing an important role in decisions.

The consequences of cost–benefit valuation procedures in determining endangered species include the following:

- Consideration of the economic costs imposed on agriculture and consumers
- Less adverse impacts on productivity and costs of production
- Less protection of endangered species and less biodiversity.

Compensation policies would require that the landowner be compensated for reductions in land values resulting from the declaration of the need to protect an endangered species. As in the case of wetlands, such a policy would put a severe curb on endangered species initiatives.

FARMLAND RETENTION AND PRESERVATION

Anyone who has traveled in California from Santa Rosa to San Diego (about 600 miles) or from Portland, Maine, to Williamsburg, Virginia (about 1,000 miles), may readily observe that urban encroachment on farmland has proceeded to the point where one hardly leaves metropolitan America. Even from Chicago, Illinois, to Madison, Wisconsin, it is becoming possible to envision an urban corridor eating up considerable quantities of prime farmland.

These trends have become public policy issues because of the effects on agricultural production and congestion, and a desire to retain the aesthetic value of rural America. From a production perspective, conversion of farmland to nonfarm

Are Farms Public Goods?

The value recognized here suggests a national interest in preserving the unique character of rural lifestyles, rural landscapes, and historically significant rural settlements and structures. If the national interest is served by protecting these "public goods" as aspects of the rural environment, then the cost should be shared by urban people. Indeed, many urban residents are the strongest advocates for some kinds of rural preservation and will often be the major consumers of the preserved rural countryside. The question of how these public goods are paid for and valued is important, since most of the "things" that people want to protect cannot be readily priced in the marketplace. Thus, the only way to ensure their survival may be to intervene with public policy. Also, in many cases there may be trade-offs between rural economic development and rural preservation strategies. Protecting rural quality of life, especially if that means maintaining the rural environment in a pristine and unaltered state, may come at the expense of rural people's economic well-being.

Collectively, these concerns about rural public goods have led to an interest in rural residents' role as "scenekeepers," tending to the national interest in rural places, and being paid for that task. In Europe, particularly Switzerland and the United Kingdom, the importance attached to protecting the aesthetics of rural places has become a central concern in rural policy discussions. Where the rural landscape is largely man-made and has been "gardened" for hundreds of years, it is not surprising that there is such a strong interest in maintaining it. Furthermore, in many European countries, the most important rural growth industry is tourism, which depends heavily on maintaining the picturesque and tranquil character of rural places.

Source: Kenneth Deavers, *Rural Vision—Rural Reality* (Madison, Wis.: University of Wisconsin, Department of Agricultural Economics, April 20, 1990), p. 3.

uses is unlikely to jeopardize our food supply. Viewed from the perspective of the total 465 million acres of available cropland, conversion of an estimated 1 to 5 million acres of cropland to nonfarm uses annually does not appear to be a major concern. Despite common misperceptions, most of this conversion of farmland has not been to concrete; as much as half of the land was converted from agriculture to trees—much of that under government programs such as CRP. At this point in time, from a supply–demand perspective, the need for trees may be considered to be at least as great as the need for food.

On the other hand, one can realistically be concerned about the loss of the agricultural production base in Napa, Sonoma, and Santa Clara counties in California or the loss of the Shenandoah Valley of Virginia in the middle Atlantic states to urban encroachment. What is lost is more than just agricultural production. The congestion of San Francisco and Washington, D.C. has expanded to some of the most valuable farmland in the world. At the same time, the aesthetic value of both the California wine country and the Shenandoah Valley is materially reduced by industrial and residential development.

From another perspective, aesthetic value is also reduced by the replacement of family farms with large feedlot dairies and integrated poultry operations. Interestingly, the potential loss of the pastoral dairy farms with cows grazing in the scenic mountainous valleys prompted the Vermont legislature to enact a lump-sum subsidy of $2,500 per dairy. It was hoped that this subsidy might restore the economic viability of these farms—an ill-fated venture that was soon abandoned. In light of such concerns, Deavers suggests that it may be time to treat parts of agriculture as a public utility (see box).

The problem may be viewed as being two dimensional:

- **Farmland preservation** involving the prevention of farmland conversion to nonagricultural uses
- **Farmland retention** involving keeping farmland actively in cultivation until an orderly transfer occurs to higher valued uses

The land that is sought to be preserved or maintained in agricultural use varies from prime farmland to quaint images of rural America. Prime farmland is often the first chosen to be subdivided because of the low cost of development. Preservation of the images of rural America as amenities, for example, places a high value on maintaining small family dairy farms or roadside markets.

Preservation motives involve issues of the desired structure of agriculture, as discussed in Chapter 12. If the goal of preservation or retention is to maintain agriculture as a healthy component of the economy, preference may be given to larger commercially oriented operations or even industrial farms. If, on the other hand, the goal relates to maintaining rural ambiance, preference may be given to small family farm structures. Cultural issues such as the maintenance of ethnic communities might also be involved in preservation policy decisions. Increasingly, the issue of farmland retention involves protection of the natural environment designed to maintain water quality and wildlife habitat or control soil erosion.

Sifting through this set of potentially conflicting policy objectives and goals, Gardner identified the following potential benefits that society may desire to gain from farmland preservation or retention policies:

- Maintenance of food and fiber productive capacity
- Maintenance of a healthy local agricultural economy
- Maintenance of environmental amenities associated with open space
- Maintenance of a sound, orderly, and efficient urban development strategy[108]

---

[108]B. D. Gardner, "The Economics of Agricultural Land Preservation," *Am. J. Ag. Econ.* (1977), pp. 1026–1036.

Success in developing land use policy is very dependent on care and consensus in the specification of goals. As in any well-designed policy, clear distinctions need to be made between objectives and the exact relationship of policies to objectives.

Farmland Retention and Preservation Policy Options[109]

Options for retention and preservation inherently involve issues of the level of government at which action is taken and the substance of the regulation itself. Considerable political energy has been spent to develop a national farmland retention policy, but the product of this action has been described as symbolic but ineffectual.[110] The 1981 Farmland Protection Policy Act was signed into law to reduce the effect of federal activities on the conversion of farmland. The act provides a framework for USDA to develop criteria for the identification of the effects of federal programs and agencies on the availability of farmland. It then requires that all agencies (except defense) evaluate and propose revisions in rules, procedures, and programs to reduce adverse impacts on farmland availability. For example, a federally funded highway might be rerouted to avoid consumption of prime farmland. While enacted with good intentions, such impact analyses are frequently better at keeping bureaucrats employed than at effectuating policy. The thrust of meaningful policy development, therefore, remains at the state and local level, where six main options are employed to influence agricultural land use.

Agricultural zoning laws *provide exclusive zoning of land for farming in 22 states.* The main purpose of agricultural zoning appears to be the preservation of open space. The most frequent use of this technique is in the Northeast where there is relatively intense competition among nonagricultural uses of farmland. While agricultural zoning is a low-cost option to the public, farmland can be reduced in value by agricultural zoning because of limitations on its development potential. Wisconsin blends tax breaks with zoning to offset at least a portion of the resulting land value reduction.

Purchase of development rights *involves public action to compensate farmland owners for benefits of development separate and distinct from the farm value of land.* This leaves the landowner the right to use the land for agricultural purposes while transferring the development rights to the local or state government.

About a dozen states authorized the purchase of development rights which compensate farmland owners for income forgone from development. However, the purchase of development rights is costly. For example, King County, Washington, was projected to pay $240 million for development rights on 30,000 acres ($8,000 per acre). One of the important effects of the purchase of develop-

[109]This section draws heavily on the work of E. Jane Luzar and Sandra S. Batie, *Improving Land Use Policy Analysis in the Southeast: Lessons from Virginia's Agricultural and Forestry District Act* (Mississippi State, Miss.: Southern Rural Development Center, April 1986), p. 14.
[110]Ibid., p. 27.

ment rights is to reduce the supply of land available for development. This increases the price of land with development rights.

Right-to-farm laws *protect farmers against legal actions taken against accepted agricultural practices.* These politically popular (with farmers) laws protect farmers from nuisance lawsuits that might otherwise drive them from agricultural pursuits. For example, a hog farm might be protected against an air pollution suit. Critics charge that such laws have little potential for achieving specific land use objectives and constitute a license to pollute.[111] Governmental costs are low, but such laws will have less chance for passage and/or survival in the future, as suggested by the previous discussion of agricultural pollution.

Preferential assessment of agricultural land *requires that land for property tax be assessed on the basis of its agricultural use value rather than its market value.* In its pure form, no penalty is associated with conversion to nonfarm uses, although deferred taxation in cases where development takes place, or restrictions on development, may be included in preferential taxation contracts. The theory is that without a property tax break, farmers located near a city may not be able to afford a farm and may be prematurely forced to sell their land to developers. However, experience suggests that differential assessment may be more of an income transfer to landowners than a mechanism for retaining land in agriculture. As one might expect, most of the benefits are concentrated near metropolitan areas. In Texas, 16.4 percent of the farmland located in metropolitan counties received 38.3 percent of the tax benefits. Often, these "farms" had only a few cows or cultivated acres with low productivity. Conversely, 83.6 percent of the rural land received 61.7 percent of the benefits, which, for them, amounted to property tax relief.[112]

Agricultural districts *designate specific areas of land for long-term agricultural use.* Membership in the district is voluntary and is linked to differential assessment and/or protection from nuisance ordinances. Pioneered by New York, agricultural districts are used by about 12 states. Districts are effective only if there is a substantial economic incentive to attract participation, districts are formed in most regions, and the contract is over a sufficiently long period. The cost of agricultural districts is the revenue forgone from differential assessment.

GLOBAL WARMING

The topic of global warming is at the cutting edge of contemporary scientific debates but is largely ignored in the policy arena. Arguably, this topic brings together all of the resource policy issues discussed in this chapter. Ruttan effectively does this by segmenting resource and environmental concerns into three periods[113]:

---

[111]Ibid., p. 35.
[112]David P. Ernstes *et al. The Effects of Differential Assessment on Tax Incidence in Texas Counties* Agricultural and Food Policy Center Working Paper 96-7, (College Station, Tex.: Texas A&M University,, Nov. 1996) p. 15.
[113]Vernon W. Ruttan, "Sustainable Agriculture and the Environment" in Vernon W. Ruttan, ed., *Perspectives on Growth and Constraints* (Boulder, Colo.: Westview Press, 1992), pp. 3–4.

1. The first wave of concerns was the *limits of growth* era of the late 1940s and early 1950s when Malthusian-type questions were resurrected. The adequacy of resources including land, water, energy, forests, and other natural resources was concluded to be inadequate to sustain growth.[114] Utilizing the induced innovation theory, Ruttan suggests that the response was increased research and technological change, which, while relieving the symptoms of surpluses, may have added to the problems of the second and third waves. The **induced innovation theory** states that problems in an education and research environment induce discoveries and institutional innovations designed to solve the problems.[115]

2. The second wave of concerns may be termed the **limits of the environment era** of the late 1960s and the early 1970s, which questioned the capacity of the environment to assimilate the multiple forms of pollution generated by growth.[116] A collision course developed between the externalities of rapidly increased production (e.g., asbestos insulation, agricultural chemicals, and smog) and rapidly growing effective demand for natural environments, freedom from pollution, and freedom from congestion. The response was the organization of institutions such as EPA designed to internalize the costs of externalities. Of course, research designed to deal with environmental issues expanded dramatically.

3. The third wave may be termed the **global era** of the mid-to late 1980s and the 1990s when the problems of the second wave transcended to the international arena and became considerably more complex. Global warming, acid rain, ozone depletion, and deforestation have international dimensions, making them much more difficult.[117] The types of institutional and research innovations required to deal with the problems of the global era are considerably more difficult to foresee.

The Global Warming Base of Knowledge

The knowledge base for the extent of global warming and its probable impact on agriculture is subject to considerable uncertainty and debate. Experts seem to agree on some useful generalizations. However, there is less certainty regarding timing, magnitude, and cause–effect relationships. The following discussion reflects this uncertainty.

---

[114]Report of the U.S. President's Water Resources Commission (Washington, D.C.: White House, 1950); and Report of the U.S. President's Materials Policy Commission (Washington, D.C.: White House, 1952).

[115]Vernon W. Ruttan, *Agricultural Research Policy* (Minneapolis, Minn.: University of Minnesota Press, 1982).

[116]Vernon W. Ruttan, "Technology and the Environment," *Am. J. Agri. Econ.* (December 1971), pp. 707–717.

[117]Committee on Global Change of the Commission on Geoscience, Environment and Resources, *Research Strategies for the U.S. Global Change Research Program* (Washington, D.C.: National Academy Press, 1990).

There appears to be an expert consensus opinion that the accumulation of carbon dioxide, methane, nitrous oxide, and chlorofluorocarbons will result in some rise in global average surface temperature during the next 30 to 70 years.[118] The amount of the rise in temperature and when it will occur are not agreed on. The range of potential average global temperature increase is generally from 1.5 to 5.50° C.[119]

Sources differ on the importance of agriculture as a contributor to global warming. USDA estimates that agriculture's contribution is 25.6 percent—the sum of carbon dioxide (10 percent), methane[120] (13 percent), and nitrous oxide (2.6 percent). The main agriculture sources of carbon dioxide include biomass burning, cultivated soils, natural soils, and fertilizer. The main agriculture sources of methane include ruminant animals, rice fields, wetlands, and biomass burning.[121] However, Ruttan notes considerable debate over the estimates for methane and nitrous oxide. Cattle and dairy interests have been particularly upset with cows being identified as the major source of methane. Questions arise as to whether agriculture should get the rap for methane gas produced from wetlands.

At least one source[122] sharply disagrees with these USDA estimates. It argues that the proper way to measure agriculture's contribution is on a net basis considering, for example, the CO_2 released from agriculture, less that stored through photosynthesis. Using this measure, one can contend that agriculture contributes 1 percent to the global warming potential and 5 percent of the U.S. warming potential.

Estimates of the impacts of global warming on production depend on the amount of adjustment that takes place in agriculture in response to the change in climatic conditions. For example, without adaptation of agricultural practices such as irrigation and crop rotation, world cereal production is estimated to decline 1 to 7 percent.[123] With moderate adaptations, global production reductions range from 0 to minus 5 percent. With the most extensive adaption, global production could decrease as little as 2.4 percent.[124] Developed countries are less adversely affected by global warming trends than developing countries.

Global Warming Policy Options for Agriculture

Policy options for dealing with global warming can be viewed from the perspective of needed reforms in world industrial production and living customs or from

---

[118]Ruttan, "Sustainable Agriculture and the Environment," p. 178; Cynthia Rosenzweig, Martin Perry, Gunther Fischer, and Klaus Frohberg, *Climate Change and World Food Supply* (Oxford, U.K.: University of Oxford, Environmental Change Unit, May 1992); and Paul E. Waggoner *et al.*, *Preparing U.S. Agriculture for Global Climate Change* (Ames, Iowa: Council for Agricultural Science and Technology, June 1992).

[119]Richard M. Adams *et al.*, "Global Climate Change and U.S. Agriculture," *Nature*, 345 (May 17, 1990), pp. 219-224; and Dean E. Abrahamson, "Greenhouse Gas and Climate Change," in Vernon W. Ruttan, ed., *Sustainable Agriculture and the Environment* (Boulder, Colo.: Westview Press, 1992).

[120]J. Reilly and R. Bucklin, *World Agriculture Situation and Outlook Report*, WAS-55 (Washington, D.C.: ERS/USDA, June 1989).

[121]Ruttan, "Sustainable Agriculture and the Environment," p. 178.

[122]Waggoner *et al.*, *Preparing U.S. Agriculture*, p. 39.

[123]Rosenzweig *et al.*, Climate Change, p. iv.

[124]Ibid., Table 9.

the perspective of what can be done solely within agriculture. The latter approach is adopted here. Within agriculture, the approaches for dealing with global warming can be divided into those designed to prevent or substantially reduce global warming and those designed to facilitate adjustment. The following discussion groups the global warming policy options into one preventionist option and four adaptation options:

- The **preventionist option** would reduce fossil fuel use, the intensity of production, ruminant livestock production, and biomass burning and increase the production of crops relative to livestock. This approach would mean a substantial change in the nature of farming and of consumption patterns. It would, for example, mean significantly less beef consumption and more cereal, fruit, and vegetable consumption. Nutritionists might argue that the result could be a more nutritious diet.

- **Adjusting farming practices** to increased temperatures involves shifts in planting dates, utilizing new varieties and perhaps crops, new irrigation systems designed to conserve water, and increased fertilizer applications. Such adjustments imply increased educational efforts and more highly applied research in an attempt to determine the optimal mix of practices as climate changes progress regionally.[125]

- **Increased agricultural research and extension education commitment** is implied by climatic change. Specific research needs involve the development of water-conserving irrigation systems, analyses of the interactions between crop production patterns and water needs, discovery of new biological fuels, and the means to stash more carbon in trees, plants, and soil, as well as to reduce energy use in agriculture and determine new sources of genes to adopt crops and animals to crops. Generally, these changes involve a broadening of the agricultural research agenda to encompass the adaptation to climate change. It also implies the modernization of extension education information exchange systems. The basic requirement is for a reversal of the trend toward lower real expenditures on agricultural research and extension.

- **Changes in the allocation of water rights** in terms of the priority given to agriculture and the rules governing water use may be necessary.[126] As indicated previously in this chapter, there has been a strong tendency to reallocate water rights from agriculture to industry. In observing this trend, the National Research Council emphasized in a recent study the importance of having the flexibility in water allocation to meet changing demand.[127] Climatic change could be a primary factor creating the need to reverse this trend if agriculture productivity is to be sustained. It may also provide the impetus for increased federal involvement in the allocation of water rights.[128]

[125]Ibid., Climate Change, p. iii.

[126]Waggoner *et al., Preparing U.S. Agriculture,* p. 29–32.

[127]National Research Council, *Water Transfers in the West: Efficiency, Equity and the Environment* (Washington, D.C.: National Academy Press, 1992).

[128]Ibid.

- **Trade liberalization and increased flexibility** in farm programs to shift cropping patterns may be required to expedite changes in product flows and encourage production efficiency. While countries and areas below about 40°N latitude (in the United States, Kansas and south) would experience less favorable growing conditions (reduced yields), the states of Nebraska, Iowa, and above would experience more favorable growing conditions (increased yields).[129] The income benefits of trade, therefore, would primarily accrue to the northern developed countries as producers. As consumers, the benefits in terms of increased food production and food availability would accrue primarily to the developing countries—assuming that they have the ability to buy or that the developed countries are willing to give it to them. With increased demand for food and liberalized trade relationships, making farm programs more flexible could become more politically feasible.

THE TRAUMA OF ADJUSTMENT

The adjustments required in farming and the economy to resource limitations in agriculture have been more the subject of romantic philosophical thought than of empirical analyses. Some advocates of sustainable agriculture contend that movements away from a chemical-reliant specialized agriculture would lead to smaller, more diversified family farms. On the other hand, the implications of virtually all of the policy options covered in this chapter foretell a more complex agriculture requiring higher levels of management expertise. It is easy to visualize pork, dairy, and hog production moving its large-scale confinement operations to the drier Great Plains climate with its sparcer population. It is easy to be romantic about smaller family farms, but there is also great need for realism.

Likewise, there is need for realism concerning the seriousness of the problems agriculture faces in adjusting to scarcities from a farmer, agribusiness, environmentalist, and consumer trade-off perspective. These are problems that policymakers will come to grips with and make decisions on, often much to the dismay of farmers.

It makes far more sense for farm, environment, and consumer interests to sit down together and evaluate the trade-offs than to fight these issues in the political arena. Sitting down at the same table requires a level of understanding, communication, and good faith compromise that has not heretofore existed. It also requires a deescalation of the level of controversy among these conflicting interests. The alternative is to leave that evaluation process to the Congress and to congressional staff, who, more often than not, will not know how to deal with the consequences and trade-offs in a balanced manner across the farm, agribusiness, environment, and consumer sectors.

---

[129]Reilly and Bucklin, *World Agriculture Situation*, p. 42.

ADDITIONAL READINGS

1. The publication by Anum S. Malik, Bruce A. Larson, and Marc Ribaudo, *Agricultural Nonpoint Source Pollution and Economic Incentive Policies: Issues in the Reauthorization of the Clean Water Act*, Staff Rpt. No. AGES 9929 (Washington, D.C.: ERS/USDA, November 1992) is an excellent treatment of the nonpoint pollution issue.

2. William Goldfarb's *Water Law* (Chelsea, Mich.: Lewis Publishers, 1988) does a really good job of tracing the differences in water policy among the states.

3. The book *Measuring the Demand for Environmental Quality* is a comprehensive compilation of articles edited by J. B. Braden and C. D. Kolstad (Amsterdam: North-Holland Publishers, 1991).

4. The National Academy of Sciences completed a comprehensive study of the land- grant system titled *Colleges of Agriculture at the Land Grant Universities: Public Service and Public Policy* (Washington, D.C.: National Academy Press, 1996).

5. The publication *The Economist's Role in the Agricultural Sustainability Paradigm*, edited by Steve Ford (Ames, Iowa: Amer. Agr. Economics Assn., July 27, 1996), provides a good update on the status of the sustainability movement.

6. The report, *Water Resources in the Twenty-First Century: Challenges and Implications for Action*, written by Mark W. Rosegrant (Washington, D.C.: International Food Policy Research Institute, March 1997) is an excellent analysis of global water supply and demand. It identifies strategies to avert projected severe water scarcities worldwide.

CONSUMERS, RURAL DEVELOPMENT, AND AGRIBUSINESS

American consumers are an ultimate beneficiary of farm policy. They enjoy a plentiful supply of food at relatively low prices. Yet, a significant segment of the U.S. population does not have sufficient income to fulfill food, shelter, and health care needs. In addition, questions are being raised continuously regarding the safety and nutrition of the food supply. These issues are not just short-run concerns but encompass the utilization of resources in a long-run context.

The brunt of consumer and related special-interest criticism of the U.S. food supply is not aimed just at farmers. Food processors have provided consumers with a host of highly processed, table-ready foods that often contain a host of additives and preservatives that are the subject of increasing health and nutrition questions. Issues of sanitation and the potential for tampering with the food supply continue to haunt the mass distribution-oriented food industry.

The purpose of the next four chapters is to provide an overview of consumer, rural development, and agribusiness policies and issues. Chapter 14 reviews policies and issues relating to the price of food. Chapter 15 discusses food safety and nutrition policies and issues. Chapter 16 surfaces the policies and issues involved in rural development. Although this discussion may seem out of place, 25 percent of the U.S. population lives in rural areas. It is surprising to many that rural poverty is higher than urban poverty, rural unemployment is higher than urban unemployment, and social conditions in many rural areas are deteriorating. This deterioration is not necessarily related to conditions in agriculture.

Having completed an analysis of the impact of agricultural and food policy on farmers and consumers, Chapter 17 analyzes the remaining group, the agribusiness firm.

FOOD PRICE POLICY

Inflation has become so bad it has hit the price of feathers. Even down is up.
—*Maynard Speece*

Food prices are a major policy concern throughout the world. Because of political sensitivity of consumers to the price of food, governments sometimes directly control both food prices and farm prices.

U.S. consumers are not seriously in danger of experiencing food shortages. The average consumer spends only 12 percent of income on food. But those with low incomes spend a much higher percentage. General concerns about the price of food periodically arise during inflationary periods. These concerns result both from the impact of food prices on expenditures of individual consumers and from the contribution of food prices to the overall rate of inflation.

At the same time that consumers express concerns about food prices, farmers frequently contend that the government pursues a "cheap food policy." The purpose of this chapter is to attempt to explain these apparently conflicting perspectives and discuss government options for dealing with food price issues.

ECONOMIC PERSPECTIVES AND ISSUES

Despite the inherent instability of farm commodity prices, food prices were rarely major considerations in national economic policy except during wars and during the world food crisis of the 1970s. The reasons for this follow.

- Food price increases have generally been less than the rate of increase in income, resulting in a decline in the share of consumer income spent on food. The proportion of consumer-disposable personal income spent on food declined from 23.9 percent in 1929 to 11.4 percent in 1994. Since World War II the only significant rise was during the world food crisis in the early 1970s (Figure 14.1). Since 1975, personal income spent on food resumed its decline from 14 to 11.4 percent in 1990. It is important to note

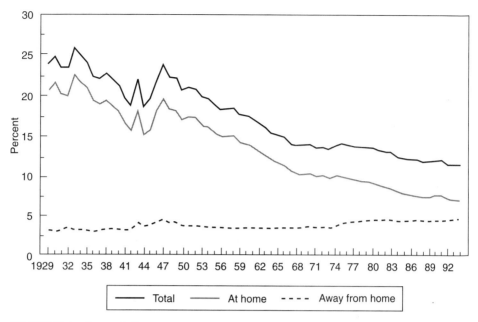

FIGURE 14.1   **Share of disposable personal income spent on food by families and individuals, 1929–1994.**

Source: Alden C. Manchester, *U.S. Food Expenditures* electronic database (Washington, D.C.: USDA/ERS, January 1996), Table 7.

that although the proportion of income spent on food at home is declining, away-from-home expenditures as a share of income are increasing. This reflects a higher proportion of food consumed away from home. It also reflects the higher level of services contained in food consumed away from home. As away-from-home expenditure continues to increase, it may be more difficult to reduce the proportion of income spent on food in total.

■ Farm programs and related surplus conditions have generally kept farm prices low and relatively stable. In the 1950s and 1960s, technological advances increased the level of output and reduced the unit cost of producing food. Price supports were forced lower and remained relatively stable because of these increases in output and efficiency, combined with the tendency for government stocks to remain high. Stable to declining farm prices offset food price inflation pressures resulting from inflation and consumer demands for increased convenience. In the 1980s, target prices guaranteed farmers a certain level of return, fostering higher production and lower market prices. The result was further downward pressure on the proportion of income spent on food. That source of downward price pressure disappeared when payments were decoupled from market prices in the 1996 farm bill.

The 1970s were an aberration in the long-run trend toward a lower share of income spent on food. While average annual food price increases of 3.0 percent were the rule during the 1950s and 1960s, food prices rose rapidly after sharp increases in domestic farm product prices occurred in 1972. Food prices rose more than 14 percent in 1973 and 1974 and accounted for nearly 50 percent of the overall inflation experienced in the economy during those two years. Food price inflation averaged 8 percent in the 1970s and accounted for an average of 26 percent of the overall inflation for the decade. These comparatively large increases in food prices resulted in only a 0.6 percent increase in the share of income spent on food, from 13.5 percent in 1973 to 14.1 percent in 1975. Food prices increased sharply during this world food crisis period, but inflation in the rest of the economy was substantially less than food price inflation (6.1 percent in 1973 and 9.8 percent in 1974).

During the 1970s, great concern arose among consumer activists and policymakers over the contribution of food cost to inflation. President Ford placed an embargo on wheat exports because of the alleged potential bread price of $1 per loaf. Price controls were extended to food despite inevitable adverse consequences for production and prices. Those adverse consequences occurred when cattle producers refrained from marketing during a government-imposed price freeze, which drove beef prices even higher. Broiler chicks were killed because integrators concluded that controlled broiler prices and rising feed prices meant that they could not even cover their variable cost of production—an economic shutdown situation.

A lesson from the 1970s was that when food price increases rise to double-digit levels (10 percent or more) and substantially exceed the inflation rate for the rest of the economy, sharp consumer reaction can be anticipated. This reaction occurs even though the average share of income spent on food rises by less than one percentage point. Three potential reasons for this consumer and policymaker sensitivity to food price increases are as follows:

- Consumers shop for food more frequently and regularly than for any other commodity group. As a result, consumers are more aware of food price levels—they develop a sense of a "fair price." For other commodities purchased less frequently, price changes are not as noticeable, and the concept of a fair or reasonable price is not as clear in the consumers' mind.

- The average household spends about 14 percent of income on food, but poor households spend as much as 87 percent (Table 14.1). As a result, poorer people tend to be more sensitive to price changes than wealthy persons.

- Food and shelter are two vital necessities of life. Annual consumer spending on each is nearly identical. Because of its overall importance, it is only natural for both consumers and policymakers to react with vigor to food price inflation. Because of the level of government involvement in agriculture, it has been easier for government to drive down the price of food than the price of housing.

TABLE 14.1 Household Expenditures for Food in Relation to Income, After Taxes, by Income, 1993

Income group	Percentage of Total Households	Average Number of Persons in Household	Food Expenditures as a Percentage of Income After Taxes
Under $5,000	5.6	1.7	86.9
$5,000–$9,999	12.6	1.9	34.8
$10,000–$14,999	11.6	2.1	24.2
$15,000–$19,999	10.1	2.2	21.0
$20,000–$29,999	15.6	2.5	17.3
$30,000–$39,999	12.5	2.7	14.7
$40,000–$49,999	9.6	3.0	13.6
$50,000–$69,999	11.8	3.1	11.9
Over $70,000	10.5	3.1	8.5

Source: Food Consumption, Prices, and Expenditures electronic database (Washington, D.C.: USDA/ERS, February 1996), Table 99.

Why Food Prices Increase

Food price increases create news and consumer awareness; food is basic to life, and its purchases are frequent. Fundamentally, it is the variability in food prices, as much as the level of prices, that makes food prices a public concern. Changes in farm commodity prices, marketing costs, and consumer demand combine to cause year-to-year changes in food prices. Certain of the factors that create fluctuations in commodity prices, such as input costs, farm policies, and trade policies, are controllable to a degree. Those stemming from weather conditions and/or the biological nature of the food production process are much less controllable.

In the popular press, statements are sometimes made implying that farmers set the price of their products. The writer of such statements *erroneously implies that farmers have the power to set prices or to pass on increases in production costs.* In reality, neither is the case. With no market power, farmers can only wait for reductions in supply or increases in demand to boost prices in response to cost increases. This is a distinguishing feature of agriculture.

Commodity and Retail Price Variability

Since 1970, year-to-year changes in farm commodity prices have been dramatic. Prices nearly doubled from 1970 to 1974 and then remained relatively stable from 1975 to 1977. Commodity prices rose another 34 percent from 1977 through 1980 and then pursued a general downward trend in the 1980s through the early 1990s.

Variability in commodity prices is the major factor causing variability in retail food prices. Still, retail prices are less variable than commodity prices. That is, in almost all cases, the percentage change in retail prices is less than the percentage change in the price of the agricultural products used to produce them. This is true for two primary reasons:

- *The value of farm commodities represents only 21 cents of each food dollar spent on food.* The other 79 cents pays for the costs of processing, transporting, and selling (Figure 14.2). This means that each 5 percent increase (decrease) in commodity prices results in an approximate 1 percent increase (decrease) in consumer food prices if other things remain unchanged.
- *A lag exists between the time commodity price changes are reflected in retail price changes* because of the time involved in transporting, processing, and distributing food products. This time lag is different among commodities. In part, this lag reflects food retailers' pricing practices, which reflect the overall trend in prices rather than the price of particular foods at the wholesale level. Thus, when meat prices rise at the farm and packer levels, retail margins normally tighten. On the other hand, a fall in farm and packer prices results in wider retailer margins.

In the short run, the lagged retail price response tends to increase the variability in farm commodity prices. Suppose that beef slaughter is larger than expected. The relative increase in supply will put downward pressure on cattle prices. If retail meat prices fall quickly to reflect the lower raw product cost, con-

FIGURE 14.2   What a dollar spent on food paid for in 1994.

Source: Office of Communications, *Agricultural Factbook 1996* (Washington, D.C.: USDA, 1996), p. 9.

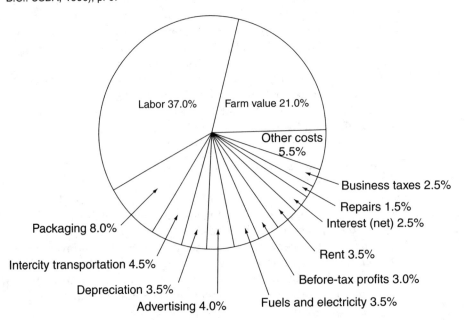

sumer meat purchases will increase. *Any stickiness in the retail price decline will cause cattle prices to fall even more because of sluggish demand.* This greater price variability occurs because a farm price reduction due to higher supplies is not reflected at the retail level. As a result, there is no price incentive for consumers to increase purchases. With short-run production being largely predetermined, a glut of supplies develops, driving the price down even further.

This stickiness has, from time to time, created sufficient political frustration that the secretary of agriculture has found it expedient to "get on the stump" and "jawbone" retailers to reduce meat prices. The result is a commonality of the producer and the consumer interest in making the market system work better.

Marketing Costs

Annual changes in the cost of transporting, processing, and selling domestically produced food products are monitored by the USDA. These data, referred to as the *marketing bill,* exclude the costs of marketing imported foods and fish, which account for nearly 20 percent of total food expenditures.

The marketing bill has increased each year since World War II. In 1994, the marketing bill was nearly $401 billion, 5.3 times higher than in 1970.[1] Inflation alone accounts for 71 percent of this increase. The remainder is due to increases in quantity marketed, increases in the cost of marketing, and increases in the proportion of products consumed away from home.

Changes in the relative importance of the various components of the marketing bill indicate that prices of some marketing inputs are changing more than others. In addition, there have been important changes in the types of foods being consumed and increased services provided.

Labor, packaging, and transportation costs have all increased as a percentage of the total food cost. The annual percentage of change in costs of these inputs has tended to exceed the percentage of change in retail food prices. Labor is the single most important component of consumers' food costs, accounting for 37 percent of the total (Figure 14.2). Packaging, the second largest nonfarm cost component, accounts for 8 percent of the total food cost. Transportation costs account for about 4.5 percent of the total.

The relative size of a cost component has not been a good indicator of its potential policy interest. Profits and advertising are the most controversial components of marketing costs. Together, in 1994, these components accounted for 7 percent of consumer expenditures, up from 5.6 percent in 1980. Changes in advertising and food industry profits have frequently been taken as indicators of the need for concern about the structure of the food industry, particularly in food processing and retail distribution.

[1]Howard Elitzak, *Food Cost Review 1995* (Washington, D.C.: USDA/ERS, April 1996), p. 29.

Eating Out

One of the major changes in food consumption patterns impacting marketing costs is the change in the proportion of the food dollar spent on consumption away from home—at fast-food outlets, restaurants, and so on. In 1994, consumers spent 40 percent of their food dollars on away-from-home consumption, up from 27 percent in 1974 (Figure 14.3). This trend has significance because of the higher cost of eating out. That is, while the farmers' share of the food dollar is $0.25 for at-home consumption, it is only $0.16 for away-from-home consumption (Figure 14.4). Because restaurant prices are rising faster than grocery store prices, the farmer's share can be expected to continue to decline for away-from-home consumption relative to at-home consumption. The primary factors leading to increased away-from-home consumption appear to be higher consumer incomes, an increased proportion of women working outside the home, and a larger proportion of single-person households.

Who Needs the Intermediary?

The alleged economic exploitation of farmers and consumers by food industry "intermediaries" is perhaps the oldest food policy issue. During the settlement

FIGURE 14.3   Proportion of the food dollar spent at home and away from home, 1929–1994.

Source: Alden C. Manchester, *U.S. Food Expenditures* electronic database (Washington, D.C.: USDA/ERS, January, 1996), Table 7.

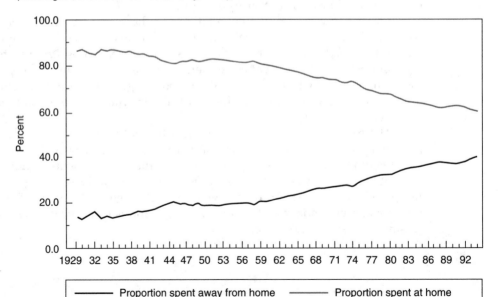

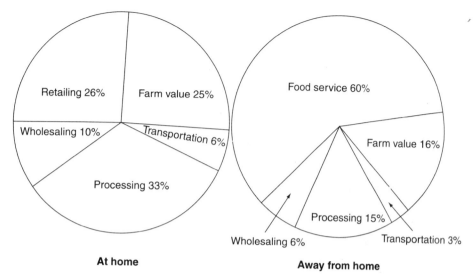

FIGURE 14.4   Where the food dollar goes at home versus away, 1994.

Source: Office of Communications, *Agricultural Factbook 1996* (Washington, D.C.: USDA, 1996), p. 10.

period, farm groups lobbied hard for policies that reduced rail freight rates and increased farmers' relative bargaining power. The interest of the consuming public was first aroused in a major way when it became apparent at the turn of the century that some food firms mislabeled products or used unsanitary processing practices to deceive the public.[2]

Modern-day concerns about intermediaries are more subtle. But the public policy issue—economic exploitation—remains the same. From a political–economic perspective, the issue has two related aspects: the farmers' share and industry structure.

Farmers' Share

Farmer and consumer interests have, from time to time, argued for policies that would increase the farmers' share of the food dollar. Farmer interests typically contend that their 21 cents of each dollar spent on food is "too small." The implication is drawn that food firms use monopoly power to the disadvantage of farmers, with the result that prices for farm commodities remain low. To dramatize the costs of food marketing, in the late 1970s, members of the National Farmers Organization sold meat, milk, and eggs directly to consumers at prices approximating their "farm value." The practice ended presumably because the farmers found they could not do it any more cheaply than the modern food processing-

[2]Ben Senauer, Elaine Asp, and Jean Kinsey, *Food Trends and the Changing Consumer* (St. Paul, Minn.: Eagan Press, 1991), p. 242.

distribution system. Farmer cooperatives have consistently encountered problems challenging the competitive position of major food companies. Once again, this suggests that the farmers' share may be about right.

Yet the fact that farmers receive, on average, one-fourth of the food expenditure dollar is used to help support legislation that would raise farm prices. Senator Jesse Helms (Rep.–N.C.), then chair of the Senate Agriculture Committee, made the following statement as he introduced his version of the Agriculture and Food Act of 1981: "Consumers have nothing to fear from higher farm prices . . . the cost of labor in food products exceeded the share of the cost that went to farmers for growing the food in the first place."[3]

Those representing consumer interests frequently use the "farmer share" statistics as well. For example, they sometimes argue that food processing costs are excessive. Consumer activists have cooperated with farmers to encourage programs to promote direct farm-to-consumer marketing. However, while reducing processing costs would increase the farmer's share of the food dollar, it would not necessarily improve prices or profits for farmers. In a competitive economic system, lower processing costs could simply be passed on to consumers in the form of a lower price.

Data from USDA show a very close relationship between the farmers' share of each at-home food expenditure dollar and the amount of processing required to bring it to the consumer (Figure 14.5). Foods that require considerable processing and are bulky (such as bakery products) return a considerably smaller portion to the farmer than foods that require almost no processing (such as eggs). But egg pro-

FIGURE 14.5   Farmers' share of retail food prices, 1995.

Source: Data supplied by USDA/ERS, Food and Consumer Economics Division, November 1996.

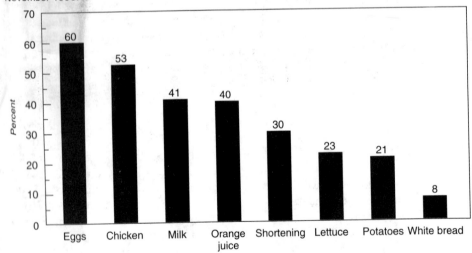

[3]Statement by Senator Jesse Helms, Congressional Record—Senate, S-3484/S-3508 (Washington, D.C.: U.S. Senate, April 7, 1981).

ducers with a 1995 farmers' share of 60 percent are not necessarily any better off than wheat producers with a farmers' share of only 8 percent because margins say nothing about bottom line profits. *In other words, the importance of the farmers' share tends to be blown out of proportion. Although it may be possible to reduce the number of intermediaries and thereby increase productivity in the marketing sector, the consumer would probably be the primary beneficiary of productivity changes.* In fact, this is exactly what has happened as agriculture has become more highly integrated and the share of income spent on food has declined.

Industry Structure

The declining number and increasing size of firms in the food marketing industry surface from time to time as policy issues. They could surface as central policy issues in the future if trends toward increased concentration continue and/or food prices sharply increase relative to the overall inflation rate. The concern is with the potential exercise of monopoly power by food manufacturers and grocery retailers.

There have been two concerted, rather widely publicized efforts to determine whether monopoly abuses exist in the food industry. In the mid-1960s, a National Commission on Food Marketing conducted a comprehensive study of the food industry, evaluated its competitive nature, and made recommendations for policy changes. The technical study on food retailing concluded that although there have been substantial changes in the organization of the industry, "Net profit has not contributed to the steadily rising gross margins."[4] A second study was conducted for the Joint Economic Committee by a land-grant university research group in the mid-1970s; it was part of a broad congressional examination into the causes of inflation. The study and subsequent related efforts attracted considerable policy attention because of its conclusion that monopoly overcharges "are likely in retail food markets that are dominated by one or two firms and/or where sales are highly concentrated among the largest four firms."[5] Increasingly, this structural characterization is becoming appropriate for the entire food industry. Parker and Connor subsequently concluded that monopoly overcharges in U.S. food manufacturing industries run at least $10 billion but could possibly run as high as $15 billion.[6] This is as much as 25 percent of the value added to food products by food manufacturers. This conclusion was, however, sharply rebutted by O'Rourke and Greig and also Bullock.[7]

---

[4]*Organization and Competition in Food Retailing,* Tech. Study 7 (Washington, D.C.: National Commission on Food Marketing, June 1966).

[5]*The Profit and Price Performance of Leading Food Chains 1970–1974* (Washington, D.C.: Joint Economic Committee, April 1977).

[6]Russell C. Parker and John M. Connor, "Estimates of Consumer Loss due to Monopoly in the U.S. Food Manufacturing Industries," *Am. J. Agric. Econ.* (November 1979), p. 639.

[7]A. Desmond O'Rourke and W. Smith Greig, "Estimates of Consumer Loss due to Monopoly in U.S. Food Manufacturing Industries: Comment," *Am. J. Agric. Econ.* (May 1981), pp. 284–289; and J. Bruce Bullock, "Estimates of Consumer Loss due to Monopoly in U.S. Food Manufacturing Industries: Comment," *Amr. J. Agric. Econ.* (May 1981), pp. 290–292.

Monopoly concerns increased again in the late 1980s and early 1990s when the trends toward increased concentration in food processing appeared to accelerate sharply. The focal point of attention initially focused on the beef trade for which the sales of boxed beef cuts by the four largest packers rose from 53 percent in 1980 to 86 percent in 1994.[8] During this same time period, steer and heifer slaughter by the four largest packers increased from 36 to 81 percent. Concern about structure spread in the early 1990s with the discovery of extensive bid rigging on school milk involving more than 50 criminal cases.[9]

The practical net effect of increasing concentration in the food industry remains in dispute. Substantial research indicates large economies of size in food processing.[10] The efficiency gains that come from consolidating transportation, processing, and merchandising functions are doubtless substantial. These gains provide most of the incentives for consolidation. Questions increasingly appear to be arising as to whether these gains may be offset by the exercise of monopoly power. Events such as those in milk serve to heighten these concerns. Yet the food industry is highly dynamic with continuous changes occurring in the composition and structure of the food industry. For example, the entrance of Sam's Clubs in the food retailing business substantially changed the structure, balance of power, and competitive position of the major chains. The food industry is highly dynamic, but farmers, the consuming public, and many economists continue to link farm problems and rising food prices to changes in food industry structure. That presumed linkage alone is sufficient to make structure a continuing food policy issue—particularly with the extensive integration and consolidation that is occurring in the food processing industries.

FOOD PRICE POLICY OPTIONS[11]

Prices serve as the primary rationing device in a market economy. When a larger quantity is available than people want to purchase at a particular price, the price will fall and more will be consumed. When less is available than people want at the current price, the price will rise and less will be consumed.

If allowed to do so, prices will always rise or fall enough to balance production with consumption. But the functioning of the pricing system has a human effect. When cattle or corn prices fall, farmers see their personal income positions decline. When meat or bread prices rise, consumers see the purchasing power of

[8]*Packers and Stockyards Statistical Report*, SR-96-1 (Washington, D.C.: GIPSA/USDA, October 1996), pp. 60 and 66.

[9]*Dairy Foods Newsletter* (Chicago, Ill.: Delta Communications, August 3, 1992), p. 1.

[10]Purcell, Structural Change, p. 14; Clement E. Ward, *Meatpacking Competition and Pricing* (Blackburg, Va.: Research Institute on Livestock Pricing, 1988), pp. 21–23; Larry A. Duewer and Kenneth E. Nelson, "Beefpacking Costs Are Lower for Larger Plants," *Food Rev.* (Washington, D.C.: ERS/USDA, October–December 1991), pp. 10–13; and Ron Knutson and Charles E. French, *Structural Change in the Fluid Milk Industry* (Washington, D.C.: Milk Industry Foundation, February 1990).

[11]This section draws heavily from John T. Dunlop and Kenneth J. Fedor, *The Lessons of Wage and Price Controls—The Food Sector* (Cambridge, Mass.: Harvard University Press, 1977).

their incomes decline. Some consumers must make do with less. The extent of this human effect makes price movements in the food system a public policy concern.

Six alternative government programs have been utilized in attempting to directly or indirectly control food price increases:

- Price controls
- Export embargoes
- Reduced import restrictions
- Marketing order controls
- Farm program provisions
- Antitrust restraints

The first five of these six programs were used in the early 1970s to deal with what were considered to be extraordinarily high rates of inflation combined with high unemployment. Initially, the inflation problems were not indigenous to agriculture; they were, instead, general economic problems. However, beginning in 1973, food became a major source of inflation, with reduced worldwide food production and sharply increased prices. In the 1980s, although food price inflation was not a major concern, government farm programs were a significant factor holding down food prices. In 1996, food price increases surfaced once again as tight stocks drove up grain prices. The 1996 farm bill made farm programs a less significant factor influencing prices. In the 1990s, antitrust restraints applied to the food industry have the potential to be revived.

Price Controls

Historically, price controls have developed under two conditions:

- In times of military conflict, such as during and immediately after World War II, substantial excess demand is evident. Shortages of goods require a method to ration supplies between domestic needs and war requirements. Under shortage conditions, strong pressures exist to control prices as a means to prevent undue inflation and windfall profits.
- In peacetime, price controls have been imposed during a broad-based inflationary surge when there is no other perceived politically acceptable means to bring these problems under control. Price controls are, under these conditions, generally looked on as a "quick fix" alternative to the restraint required by more effective, longer term policies such as reduced government spending, restricted monetary growth, and market price incentives for increased production.

For example, in the early 1970s there was the political perception of the need for a quick fix on inflation. Total economic output had declined during the

1969–1970 recession. The unemployment rate was about 6 percent in 1971, nearly 60 percent higher than in 1966–1969. The consumer price index was rising at a then dramatic 6 percent annual rate. The political importance of these economic woes was heightened by the 1972 presidential election campaign. It was perceived that something had to be done to put things back on course.

President Nixon announced his new price control policy in August 1971. The program was designed to deal with inflation, unemployment, and the growing balance-of-payments deficit. The 90-day freeze on nearly all prices and wages was perhaps the most dramatic for its effect on the average citizen. Retail food prices were frozen, but farm prices were not.

Food prices were, in fact, stable during the freeze. But lower farm commodity prices were the primary reason for the observed food price stability. The reason was not as important as the fact. Food price stability during the period had an important influence on the popular perception that the price freeze had been a success.

In November 1971, President Nixon lifted the price freeze and focused on controlling profit margins. Retailers were allowed to pass through raw material and labor cost increases as long as there was no increase in profit margins.

After the freeze was lifted, food prices rose at a rate that stood in stark contrast to the stability experienced during the freeze. In fact, rising farm commodity prices, not larger retailer profit margins, were responsible for the 4.3 percent rise in food prices in 1972. The cause of the farm price increase was the result of increased exports and had little or nothing to do with the price controls. Prices received by farmers rose 30 percent during the margin control period. This dramatic price increase was particularly noticeable in the meat sector, highlighting the important role that commodity prices play in influencing food prices.

Theory of Price Controls

Food price controls are relatively ineffective in fighting food price inflation. The principal purpose of such controls—keeping prices from rising—is itself a cause of their failure. In a market-based economy, price changes are relied on to correct supply and demand imbalances. High prices both discourage consumption and encourage production. This is particularly true of a competitive industry in which economic forces operate continuously to eliminate excess profits.

Suppose in Figure 14.6 that the market for meat without price controls is in equilibrium at price P_1 and quantity Q_1. If income increases because of a wage increase, the consumer's demand for meat will shift to D_2, which means that, in the aggregate, consumers are willing to purchase more meat than previously at each price. The first reaction in the market is to notice that the meat counter is empty sooner. That is, consumers are willing to buy Q_3 pounds of meat at the old price. Noting the shift in demand, the retailer attempts to buy more carcass beef. To do that, given a fixed supply, it must pay higher prices for cattle. Those higher prices are then passed on to consumers. The rising price has two effects. It encourages

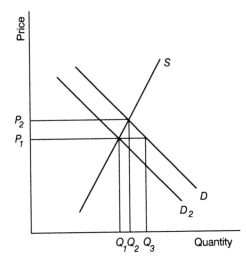

FIGURE 14.6
Impact of price controls.

producers to expand output. But it also discourages consumption, resulting in an eventual equilibrium price and quantity, P_2 and Q_2. The result eliminates the "shortage" $(Q_3 - Q_1)$.

The effect of an arbitrary price freeze at P_1 is immediately obvious. Unless the price is allowed to rise, there will be a shortage of meat—consumers will want to purchase more than producers are willing to sell at the frozen price. In that case, some other system must be used to ration the available meat (Q_1). First-come, first-served is a frequently used rationing device. When the meat case is empty, there is no more to sell. Thus, in 1972 on the East and West Coasts, to the amazement of many, lines developed at meat counters, and butcher shops operated with shortened hours—an effect identical to the oil embargo in 1973. Although rising prices are unpopular, price controls are even more unpopular. More important, price controls do not help solve the underlying supply problem. Indeed, using arbitrary price controls to solve economic problems is like trying to cool the temperature of a warm room by covering the thermometer. *The eventual result is the formation of a black market for the product in which products are sold at or even above the equilibrium price (P_2) to those buyers who have the money to pay for the product.*

Uniqueness of Food Price Controls

Price controls on food pose even more problems than for most other commodities for three major reasons:

- At the farm level, competitive market forces operate. Only a small profit margin exists against which the effects of price controls can be cushioned. This contrasts with other industries, where monopolistic profit margins and reduced costs associated with nonprice methods of competition can be used to offset the pressures of price controls.

■ The effects of farmers' responses to price controls are frequently lagged or modified by the biological nature of food production. Production cannot be expanded or contracted quickly in response to changes. In fact, price changes in the livestock sector tend to result in short-run production changes that seem perverse. When cattle prices are held down relative to the cost of production, producers begin to sell their breeding stock. Such sales reduce their costs but add to current production, which tends to make prices fall. In turn, the lower prices intensify the extent of the liquidation and prices fall even more. When prices begin to rise, producers are encouraged to add to the breeding herd, which reduces production and forces prices higher. Thus, prices initially fall as farmers liquidate and then they go up sharply once controls are lifted, as farmers attempt to rebuild their herds. Actions that discourage rebuilding, such as importing beef, aggravate efforts to bring prices back to equilibrium.

■ These production/consumption distortions are the predictable and inevitable result of a policy doomed to failure by its inherent approach. In many respects, the only cure for high retail food prices is high farm prices, which inevitably result in increased production.

Export Embargoes

Export embargoes to reduce inflationary pressures, like price controls, are political solutions to perceived economic problems. Political pressure for export controls in the United States increased abruptly in 1973 with the threat of worldwide food shortages. Preventing shortages and controlling prices required either an increase in supply or a reduction in demand. In agriculture, short-term increases in supplies are impossible. The only political option available in the short run becomes one of reducing demand by restricting exports. Export shipments of agricultural products were restricted for purposes of domestic price control three times during the 1970s. The first, in June 1973, was the most extensive and had the greatest effect on prices. The other two, in October 1974 and August 1975, were precautionary in nature and were directed primarily at blocking large sales of grain to the USSR.

Prices fell dramatically after the first embargo was imposed. Soybean prices, for example, fell from their high of about $12.25 per bushel to about $6.25 over a period of about two weeks.[12] This full drop cannot be attributed to the embargo. For one thing, the $12.25 price of soybeans was higher than could be sustained— it resulted from a lack of good information about supply and demand conditions.[13] Then, too, the 1973 crop was developing nicely and became a record 1.6

---

[12]*Fat and Oil Situation* (Washington, D.C.: ERS, USDA, December 1975), p. 10.

[13]Subsequently, an export sales reporting system was installed to improve the quality of information.

million bushels—22 percent greater than the 1972 crop.[14] These two factors, the lack of quality information about actual exports and the improving crop prospects, undoubtedly contributed in a major way to the lower prices. But farmers blamed the lower prices on the embargo, just as consumers became convinced that the embargo had done its job.

It is naive to argue that export controls will never again be used to control prices. When commodity exports begin to threaten domestic food security or when grain prices increase so rapidly as to threaten a major liquidation of livestock, it is almost a certainty that ways will be found to reduce demand. An export embargo remains a possible short-run policy option.

Reduced Import Restrictions

Recall from Chapter 6 that two major agricultural commodities—milk and beef—utilize quotas as a means to maintain domestic prices above world prices. The president has used the ability to increase these quotas as a means to control prices. Increasing imports, of course, increases supplies available to the domestic market and, thus, lowers price.

After the price freeze in 1973, domestic production dropped and prices rose. With the potential for shortages, import quotas were lifted for both milk products and beef at the behest of food processors. Imports of nonfat dry milk increased from 1.6 million pounds in 1972 to 267 million pounds in 1973, cheese imports from 179 to 230 million pounds, and butter imports from less than 1 million pounds to 43 million pounds.[15] Fresh and frozen beef imports increased 16 percent.[16]

The relaxation of import restrictions—applauded by consumers, food processors, and retailers—was strongly condemned by producers on the grounds that it reduced incentives for domestic production. Interestingly, when U.S. supplies are short, world supplies are also frequently short. Thus, even though import restrictions are raised, the supplies may not be readily available, or they are available because only U.S. consumers are able to pay a higher price and bid supplies away from foreign consumers.

In the 1980s, reduced import restraints were suggested by market-oriented economists as a means to curb the market power of marketing boards in Canada. Provincial Canadian producer boards had sufficiently restricted output to raise the price of milk and poultry to what some considered to be excessively high levels. A relaxation of import restrictions by the Canadian government would be an outside source of competition restraining the boards' exercise of monopoly powers. In negotiating the Canadian–U.S. Free Trade Agreement (CUSTA), which is part of the overall NAFTA umbrella, the United States agreed to allow Canada to

---

[14]*Agricultural Statistics* (Washington, D.C.: SRS, USDA, 1974), p. 133.
[15]*Agricultural Statistics* (Washington, D.C.: SRS, USDA, 1977), p. 392.
[16]Ibid., p. 355.

continue its marketing board activities in return for U.S. continuation of supports for peanuts, milk, and sugar. This U.S. concession has since haunted U.S. efforts to secure true free trade with Canada.

Marketing Order Controls

Recall from Chapter 10 that marketing orders either directly establish prices (milk) or provide varying degrees of control over the quantity and quality of products that move to markets (fruits and vegetables). Beginning in 1972, White House agencies such as the Office of Management and Budget, the Council of Economic Advisers, and the Cost of Living Council began to take a great deal of interest in how marketing order decisions of the secretary of agriculture raise commodity prices and thus exacerbate the inflation problem. To prevent this occurrence, these agencies began to closely monitor USDA marketing order decisions. Particular emphasis was placed on milk marketing order price increases and flow-to-market regulation for navel oranges and lemons. In fact, White House scrutiny of orders became so intense that changes in order provisions by the secretary required White House approval.

White House interest in orders continued through the 1980s, as indicated by the designation of fruit and vegetable marketing orders as one of the candidates for deregulation by the Reagan administration.[17] Subsequently, policy guidelines were issued by Secretary John Block to prevent orders from being used as a means of controlling production and artificially raising prices.

In the late 1980s, concern about the adverse impact of marketing orders on prices and resource misallocation spread from fruits and vegetables to milk. In 1988, the General Accounting Office issued a report calling for the elimination of milk marketing orders. Orders were charged with unduly raising milk prices in the Northeast and South, lowering prices in the Upper Midwest, and preventing the use of substitute milk forms such as reconstituted milk. Surprisingly, USDA did not contest these charges but stated that Congress would have to act if policy was to be changed. Subsequently, however, in 1992, USDA issued a decision recommending that federal milk marketing order provisions be changed so as not to discriminate against the use of reconstituted milk. This decision was hailed by both consumers and Upper Midwest milk producers as a major change in dairy policy, in effect, breaking the "Southern monopoly" on milk prices. However, this decision did not appear to have any perceptible impact on fluid milk price levels.

Although the reform mandates for marketing orders contained in the 1996 farm bill were not specifically designed to curb the price-enhancing efforts of milk orders, there existed an undertone of concern that higher prices for milk used for fluid consumption (bottling) in the Southwest had provided incentives

---

[17]Richard Heifner *et al.*, *A Review of Federal Marketing Orders for Fruits, Vegetables and Specialty Crops*, Agric. Econ. Rep. 477 (Washington, D.C.: AMS, USDA, November 1981).

for increased production that led to lower prices for milk used for manufacturing. Midwest interests urged a more flat price surface for milk used for fluid consumption.

Farm Program Provisions

Farm programs do affect the price of food. The target price program reduced the price of food by providing incentives to produce more than would otherwise be the case. Other domestic farm programs, such as sugar import quotas, raise the price of food. Sugar producers deny this effect by arguing that a reduction in the price of sugar would not result in a reduction in the price of candy bars. Such an argument results from lags in price responsiveness through the marketing channel. But, surely, over time, farm prices and consumer prices are related.

Cheap Food Policy

If farm prices and consumer prices were not related, there would be no foundation or purpose to the frequently heard farmer charge that the United States pursues a cheap food policy. A farm policy economist responds to such a charge with mixed emotions.

A **cheap food policy** *involves the government overtly pursuing policies that hold down the price of food below the competitive equilibrium price.* We saw earlier that the consequence of the government setting the price of food below the competitive market equilibrium and paying farmers that price is a shortage. Many developing and centrally planned economies pursue this type of cheap food policy.

When the United States maintained the target price program it was fair to argue that a cheap food policy was being followed. The effect of the target price was to stimulate production, which lowered the market price. With decoupling, contained in the 1996 farm bill, this argument vanished. The only arguments remaining to support cheap food are flimsy at best. Government support for technology creation (research) and education might be considered to be support for cheap food. Its main justification however, is progress, efficiency, and competitiveness internationally. Allowing the development of integrated agriculture might be considered cheap food. Allowing the use of pesticides might be considered cheap food at the expense of the environment. But these arguments are most certainly stretching the reality of the reasons for such policies.

Grain reserves also affect the price of food and may be considered by some to be part of a U.S. cheap food policy. Yet reserves both raise and lower prices. When reserves are being acquired, farm prices rise. In fact, most publicly held stocks of farm commodities are acquired when the government is trying to support (raise) the price. Once reserves are acquired, they become a part of the potential supply with the effect of lowering and stabilizing price.

Reserves are probably the only means by which government can moderate price increases without potentially severe adverse impacts affecting the production decisions of farmers. Even here, however, the timing and method of release become important to determining the nature of the impact. For example, a large quantity of CCC stocks of grain placed on the market just before planting could adversely affect farmers' decisions on what crops to plant as well as the level of fertilizer applications. As in many other public policy decisions, the appearance and timing of the action may be as important as the action itself in determining its effect. Moreover, the 1996 farm bill makes it quite clear that the CCC stocks are not to be used to manage prices and that grain reserves are to be released only under emergency conditions.

Expensive Food Policy

Expensive food policies raise the price of food above the competitive market equilibrium price. For example, price supports on sugar are combined with import controls to raise their prices above the competitive equilibrium. In addition, milk marketing orders raise the price of milk for fluid consumption above competitive levels. Meat import controls raise prices. Land retirement programs support prices by reducing production. Questions about the desirability of these programs are continuously debated and are dwindling in number and importance as U.S. farm policy moves toward fewer subsidies and freer trade.

Some countries have more overtly pursued an expensive food policy. The most notable examples have been Japan and the European Union (EU). Higher food prices have been imposed only on their domestic consumers. Such a policy may be sold to consumers on grounds such as food security or preserving a decentralized farm structure (larger number of smaller farms).

Expensive food policies often result in surplus production that must either be curbed by production controls, held in reserve, or dumped on the world market. If surpluses are dumped on the world market, consumers in other countries receive short-run benefits while unprotected farmers in other countries receive lower prices. This is why the Japanese and EU policies have been so strongly objected to in the GATT negotiations. It is also the reason why Japan has begun to provide market access for rice and why the EU has reformed its policies to look more like those of the United States.

Antitrust Restraints

Antitrust policy was discussed in Chapter 12 as structure of agriculture policy. It is also a food price policy. The degree of emphasis on antitrust in the food industry has changed dramatically during the past 30 years. In the 1960s, mergers among milk processors and food retailers were severely restricted. In the 1970s, the emphasis in antitrust shifted to hard-core antitrust, such as price fixing and predatory market practices designed to drive competitors out of markets.

In the 1980s, concern about market structure and antitrust almost vanished. Mergers and acquisitions within and among both agricultural input and product markets mushroomed. Market shares for the four largest firms increased sharply. Although not as profound, concentration also increased sharply in the seed corn, meat packing, farm machinery, and poultry businesses. Such changes in structure could once again make antitrust a focal point for fostering competition and influencing food prices. If a change in antitrust policy were to be pursued in the food industry, it could, in certain highly concentrated industries, require the application of the Sherman Antitrust Act policies designed to break up existing firms where it is perceived that monopoly exists, as opposed to preventing further mergers. Alternatively, it could be argued that interproduct and international competition has increased sufficiently that current high levels of product concentration are no longer meaningful measures of the potential for abuse.

A RED HERRING?

To some, it may seem ironic that while U.S. consumers spend less than 12 percent of their income on food, they are still concerned about food prices. It also appears ironic that while most of this book has been concerned with low farm prices and incomes, high food prices are also an issue. As has been pointed out previously, the root of the U.S. food price concern lies in the inherent price instability of agriculture. Prices become a concern only when the rate of increase exceeds income growth. Consumers are particularly conscious of food price increases because they buy almost daily. Government policy cannot curb that instability unless it supports farm prices—an expensive food policy that reduces the competitiveness of U.S. agriculture internationally.

Such ironies point up the complexity of agriculture and the forces affecting it. They are what make the farm and food policymakers' jobs so extremely difficult. They also point out the role economists can play in evaluating the trade-offs from various policy options.

15 NUTRITION, FOOD SAFETY, AND FOOD QUALITY

If the government was as afraid of disturbing the consumer as it is of disturbing business, this would be some democracy.

—*Frank McKinney Hubbard*

While consumer interests in food price and structure issues appeared to die in the 1980s, concerns about nutrition, diet, food safety, quality, and health grew. These concerns were fostered by the continuing flow of research results on the relationships between diet and health; the findings of persistent poverty, hunger, and malnutrition; the willingness of consumers to buy increasingly highly processed foods; the improved ability to detect chemical residues in the food supply; and the detection of chemical residues in the water supply. This chapter is designed to reflect the dimensions of these concerns and government options for dealing with them.

THE FOOD SAFETY-NUTRITION CONTINUUM

Issues of food safety and nutrition are becoming increasingly intertwined as more information becomes available on the relationships between diet and health. Additives such as salt or sodium nitrite (used to cure bacon, ham, and other processed meats) were once considered completely safe. Now salt has been linked to high blood pressure and sodium nitrite has been tagged as a potential cancer-causing substance. Pesticide residues have become a major health and environmental concern.

Foods themselves, or natural substances within them, are being implicated in specific food safety issues. For example, cholesterol, a naturally occurring substance in animal fat, has become nearly as suspect as a cause of heart disease as salt has as a cause of high blood pressure. Similarly, obesity has been implicated as an unsafe physical condition, suggesting the need to avoid certain foods and

---

Rare Hamburgers with Farmers' Market Cider

Should consumers who eat out have the right to order their hamburgers rare or medium rare? Should restaurants be allowed to cook them to fit consumers' desires? Note that any pathogens on steak are killed by surface heat. Grinding hamburger incorporates the pathogens into the meat, thus increasing their chances of survival if not cooked well-done. Should farmer markets be allowed to sell apple cider made by their press from apples produced in their orchard? This cider, if allowed to sit in the sum for a short time, garners tang and effervescence not obtained when pasteurized.

Some consumers believe they should be able to take the risk of *E. coli* food poisoning to enjoy a juicy hamburger with cider meal. Should restaurateur who serves rare or medium-rare hamburgers and cider ordered by customers be liable in a court of law for resulting food poisoning? Should they be required to post a sign or notice on the menu that alerts consumers to the dangers of consuming such products?

Note that policies restricting the sale of such products reduce consumer choice. Is that restriction justified? Where is the line to be drawn on regulatory policy and liability involving consumer choice?

---

reduce food intake—a message that certain producers did not want to hear. In fact, scientists generally rate diet-related factors such as cholesterol and fat as being a greater problem than pesticides as a cause of health problems.[1]

Looked at in this context, issues of food safety and nutrition may be viewed as a continuum of health-related food concerns. (Figure 15.1). That is, food safety concerns range from a series of natural substances in trace amounts found in particular foods, to larger quantities of substances such as cholesterol and fat believed to cause health problems, to chemical residues and additives having primarily long-term adverse health implications, and, finally, to various food-borne pathogens that have the potential to cause serious illness or even death in a short time. Policy concerns may likewise be treated as a continuum ranging from a buyer beware option with no government involvement to government-authorized consumer education, to the requirement for warning labels, to self-regulation by industry under government authorization and supervision, to a tightly regulated government inspection system, and ultimately to zero tolerance of any disease-causing substances. Any one of the options could theoretically be utilized to deal with any of the food safety concerns. The policy issue involves the desired matching of the safety concern with the policy alternative.

---

[1]Ben Senauer, "The Impact of Reduced Agricultural Chemical Use on Food: A Review of the Literature for the United States," staff paper (St. Paul, Minn.: Center for International Food and Agriculture Policy, University of Minnesota, October 1992).

Safety concerns

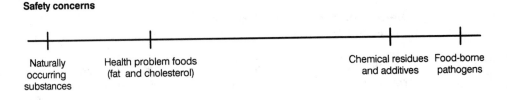

Policy alternatives

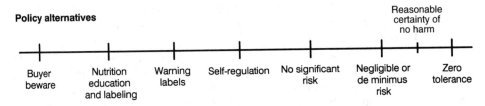

FIGURE 15.1   Conceptual continuum of food safety concerns and policy approaches.

FOOD SAFETY

The safety of the food supply has long been considered a responsibility of the federal government. That safe food is an important consumer concern is hardly debatable. But it is also a producer, processor, and retailer concern—just ask the producers of canned mushrooms, apple juice, or cranberries who have had products withdrawn from the market because of a health threat. Even the otherwise highly independent meat industry has lobbied for federal meat and poultry inspection to protect against pathogens in its products. But with the development of chemical-based technologies and, more recently, biotechnologies in agriculture and the manufacture of increasingly highly processed foods, the problem of food safety has expanded to chemical residues of pesticides and a wide array of additives—many of which have been used for years and were generally recognized as safe.

Food Safety Concerns, Policies, and Programs

Table 15.1 indicates the major federal agencies involved in food safety regulations, the laws under which they operate, and their general responsibilities. Note that with each statutory policy, regulatory agencies at the federal, state, and often local levels enforce these regulations.

Food-Borne Illness

Scientists concerned with food safety issues consider microbiological contaminants to pose the most serious food-related health risk.[2] Accordingly, the perisha-

[2]Ibid., p. 2.

TABLE 15.1 Federal Agencies Primarily Responsible for Food Safety

Agency	Principal Statutory Authority	Responsibilities
Food and Drug Administration	Federal Food, Drug, and Cosmetic Act	Safety/quality/effectiveness of animal feeds and drugs, and all foods except meat and poultry
USDA–Food Safety and Inspection Service	Federal Meat Inspection Act and the Federal Poultry Products Inspection Act	Safety/wholesomeness/ accurate labeling of meat and poultry products
USDA–Agricultural Marketing Service	Egg Products Inspection Act	Safety/quality of egg products and shell eggs
Environmental Protection Agency, USDA–Food Safety and Inspection Service, Food and Drug Administration	Federal Insecticide, Fungicide, and Rodenticide Act; Federal Food, Drug, and Cosmetic Act	Safety of pesticide products and pesticide residue tolerance in foods/feeds
National Marine Fisheries Service and Food and Drug Administration	Agricultural Marketing Act	Voluntary seafood inspection

Source: Office of Technology Assessment, 1992.

bility of food and its proclivity for carrying bacteria and transmitting disease historically have been the central public policy concerns with food safety. Perhaps the most feared food safety health risk is botulism. This often fatal disease is caused by a widespread organism that is found in both terrestrial and marine environments. Although feared, its incidence in the United States during the past century has been rare. Of the 688 outbreaks from 1899 to 1973, 72 percent were traced to home-processed foods.[3] Yet virtually every year, incidents of botulism food poisoning occur somewhere in the United States. Botulism, while the most deadly, is not the most common of the food-borne pathogens. *Salmonella* is a major cause of food-borne illness, particularly for chicken, eggs, and even cheese. Its origin, like *Escherichia coli* (*E. coli*), lies primarily in fecal material, reflecting careless sanitary procedures in the plant.[4] *Staphylococcus* (staph) is also a common food-borne illness. The staph bacteria is spread by unsanitary conditions, with incubation being common in salads at room temperature.

The increased occurrence of food-borne illness appears to be largely the result of improper handling of food, a deterioration in cooking skills, a lack of knowledge of the need to cook, a lack of realization of the dangers of pathogens, and the time pressures resulting from a fast-moving society. Processors, retailers, restaurants, and consumers are all responsible for these failures. When bacteria have their origin in processing plants, the cause of the contamination is a lack of care to prevent contamination. In fast-food outlets and restaurants, the origin is a lack of care in

[3]Center for Disease Control, *Botulism in the United States, 1899–1973*, HEW Publ. 77-8279 (Washington, D.C.: HEW, June 1974).
[4]Ben Senauer, Elaine Asp, and Jean Kinsey, *Food Trends and the Changing Consumer* (St. Paul, Minn.: Eagan Press, 1991), p. 258.

refrigeration and handling; inadequate training, and the failure to cook to a sufficiently high temperature to kill the bacteria. In 1992, this failure to cook food properly resulted in both death and illness from hamburger sold through fast-food outlets in the Pacific Northwest. In 1996 six people died in Scotland from *E. coli* food poisoning traceable to the shop of an award-winning butcher. Likewise in 1996, many people encountered food poisoning from drinking unpasteurized apple cider.

Regulations of conditions leading to food-borne illnesses have their origin in unsanitary conditions in food processing plants and in blatant cases of adulteration of the food supply. On January 20, 1879, Congressman Wright of Pennsylvania introduced a law in the 45th Congress, H.R. 5916, "for preventing the adulteration of articles of food and drink." In 1904, Upton Sinclair first arrived in Chicago to chronicle the life of immigrant meat packing company workers. His book, *The Jungle,* went beyond the immigrant worker theme to highlight the unsanitary conditions in America's meat packing industry.[5] Shortly after its publication, President Theodore Roosevelt ordered a special report on conditions in the stockyards. The report confirmed Sinclair's account. Before the end of the year (1906), Congress enacted the Meat Inspection Act and the Pure Food and Drug Act. In the following year, the meat inspection law was amended as the Meat Inspection Act of 1907; it was not rewritten until 1996.

Initially, both the Food and Drug Administration (FDA) and the Meat Inspection Service were part of the USDA. In 1938, the Federal Food, Drug, and Cosmetic Act was enacted to strengthen the power of the secretary to regulate. Subsequently, FDA was transferred to the Federal Security Administration and then to the Department of Health, Education, and Welfare—and, most recently, to Health and Human Services. However, in 1957, the authority for poultry inspection was placed with the USDA, but it was not until 1970 that egg inspection was authorized for USDA.

The transfer of FDA out of USDA represents a specific case of perceived conflict of interest within USDA that was referred to earlier as making it difficult for USDA to become a Department of Food. An early administrator of USDA's Department of Chemistry (Dr. Harvey Wiley) was initially in charge of enforcing the Food and Drug Act. He enforced the law so vigorously that he antagonized large food manufacturers and was forced to resign.[6] The resignation was a clear signal to subsequent FDA administrators in USDA that tough enforcement was not conducive to survival in USDA. In 1940, FDA was transferred out of USDA.

Another interesting policy issue involves the separation of inspection responsibilities between USDA and FDA. FDA handles all inspection responsibilities except for meat and poultry. This opened the door for conflicting regulations as well as duplicatory inspection requirements. This potential has become more likely as the amount of food processing has increased. It will exist as long

---

[5]Upton Sinclair, *The Jungle* (New York: Doubleday, Page & Co., 1906; republished by Viking Press 1946).

[6]Ross B. Talbot and Don F. Hadwiger, *The Policy Process in American Agriculture* (San Francisco, Calif.: Chandler Publishing Co., 1968), p. 260.

as responsibility for meat and poultry inspection is housed in USDA, like processed food inspection, or until USDA recognizes that inspection has primarily a consumer constituency—like FDA. The Clinton administration proposed remedying this potential conflict by transferring the meat and poultry regulatory authority to FDA, a policy change that was effectively repelled by the meat industry.

From 1907 through 1996 the basic meat inspection policy involved a veterinarian using senses of sight, smell, and touch to determine if meat was uncontaminated by pathogens. In 1996 it began the switch to a science-based system of hazard analysis critical control point (HACCP) procedures designed to minimize and detect pathogens. This procedure identifies critical points in the slaughter process where contamination may occur such as when the gut is removed from the slaughtered animal. Yet even a well-designed HACCP program cannot guarantee elimination of pathogens. If one virulent *E. coli* survives it can subsequently multiply into many more.

Whether HACCP needs to be applied at all levels of the food system is now evolving into a major policy issue. Another issue involves whether the origin of meat should be traceable so that the source of contamination can be identified. How would ranchers react to a requirement to develop and implement a HACCP plan? Yet tracing the cause is critical to controlling pathogen contamination.

Agricultural Chemicals

One of the major objectives of the sustainable and organic agriculture movements is to reduce the use of agricultural chemicals. From a health perspective, the major concerns relate to the impacts of nitrogen fertilizer on the nitrate content of water and pesticide residues as a cause of diseases such as cancer.[7]

Nitrogen is a critical nutrient to nonleguminous crop production, particularly for corn and high-protein wheat. However, when nitrates from either organic or inorganic sources leak into the water supply, health problems can result. These problems occur when the body converts nitrates to nitrites and to nitrosomines. Nitrosomines are themselves carcinogens, and nitrites are a danger to babies resulting from their tendency to tie up oxygen through the oxidation of hemoglobin in bodies. The result is the so-called "blue baby" syndrome, or methemoglobinemia, which can be fatal.[8] Of course, the higher the nitrate concentration in the water, the greater the potential problems.

Concerns about the health impacts of pesticides relate primarily to the effects of direct exposure by farmers and farm laborers, the effects of residues of chemicals appearing in both fresh and processed agricultural products, and the effects

---

[7]Recall from Chapter 13 that an additional concern, apparently unrelated to health and safety, is the runoff and permeation of phosphorus into lakes and streams, which promotes excessive vegetative growth; thus damaging the environment.

[8]H. Vogtmann and R. Biedermann, "The Nitrate Study—No End in Sight," *Nutrition and Health* (1985), pp. 203–216.

of contamination of the water supply due to leaching or improper handling. Most of the potential impacts relate to pesticides as a cause of cancer. The resulting causal relationships, therefore, are the focal point of pesticide regulation.

Regulation of agricultural chemicals at the federal level began with the Federal Insecticide Act of 1910, a farmer-oriented statute that prohibited only the manufacture or sale of adulterated or misbranded insecticides and fungicides. The current system of pesticide regulation originated with the enactment in 1947 of the Federal Insecticide, Fungicide, and Rodenticide Act (FIFRA). FIFRA set up a system for registration of chemicals for specific uses. The requirements for registration involve weighing the benefits of the pesticide against the costs while requiring proof of efficacy and safety.

- **Efficacy** requires demonstration that the chemical will indeed control the pests for which the pesticide is registered.
- **Safety** requires no unreasonable adverse effects on human health or the environment.

Although the federal government initially had the responsibility for proving that the efficacy and safety requirements were being violated, the burden of proof was shifted to the manufacturer in 1954. In addition, it was required that tolerances be set for residues of pesticides in foods. FDA was transferred out of USDA in 1953, and not until 1970 was the regulatory authority for pesticides transferred to the Environmental Protection Agency (EPA). However, FDA is still a factor in the regulatory equation for pesticides because it retains the regulatory authority for additives in processed foods. FDA rulings on additives by implication include pesticide residues in processed foods.[9]

Many pesticides were already on the market when EPA was given the responsibility for regulating them. Because of the belief that many of the old pesticides could not meet the standards required for a new pesticide, in 1988 Congress directed the EPA to reregister all old pesticides by 1997. This reevaluation process made many chemicals unavailable. This reality, along with public interest advocates' concerns about the use of agricultural chemicals, has heightened interest in the future of pesticide regulatory policy.

USDA is not completely left out of the loop when it comes to pesticide regulatory policy. Its extension state specialists and county agents conduct educational programs that lead to certification for a pesticide applicator as having a knowledge of the risks and regulations relating to handling and applying pesticides. Land-grant universities also conduct research to discover and develop safer and more effective pesticides.

In the Congress there has always been a struggle over whether the agriculture or health committees should have jurisdiction over pesticide policy issues. In

[9]National Research Council, *Regulating Pesticides in Food: The Delaney Paradox* (Washington, D.C.: National Academy Press, 1992), p. 25.

the initial phases of the recent debates the health committees initially took charge. However, with the election of the Republican majority the balance of power began to shift to agriculture. Perhaps this was because the agriculture committees had more at stake in getting something done.

Food Additives

"Food additives are substances added to foods in minor amounts, either intentionally to improve nutrition, quality, or shelf life, or unintentionally as a result of production, processing, storage, or packaging."[10] Actually, the Food, Drug, and Cosmetic Act defines an additive even more nebulously to include "any substance the intended use of which results, or can reasonably be expected to result . . . in its becoming a component of food."[11] From this statutory definition, it can easily be seen how pesticides, animal drug residues, and packaging material residues can be interpreted as being an additive and within the purview of FDA's regulatory standards for food safety. However, originally, the main concern of the 1958 Food Additive Amendment was with artificial sweeteners, preservatives, food colorings, cultures, and various other chemical processing aids.

As part of the 1958 additive act, many additives that had traditionally been used in food processing were designated as generally recognized as safe (GRAS). This was consistent with the original FDA philosophy of products being safe unless proven otherwise by the federal government. With the reversal of this philosophy, FDA began requiring the testing of GRAS substances as well as new additives.

Critical Policy Issues and Options

One of the most prominent and highly qualified FDA commissioners, Donald Kennedy, identified two requirements for sound food safety decisions[12]:

1. An objective method of estimating the level of risk, an issue that science can resolve

2. A judgment about the acceptable level of risk using that measure, an issue that only the policy process can resolve

The critical policy issues in food safety exist largely because science moved faster than policies have in two main areas:

- The ability to measure or quantify the existence of minute residues
- The ability to adopt regulations for the development or manufacture of new life-forms through genetic engineering

[10]Senauer *et al., Food Trends,* p. 261.
[11]21 USC Sec 32.(s) (1984).
[12]Senauer *et al., Food Trends,* p. 244.

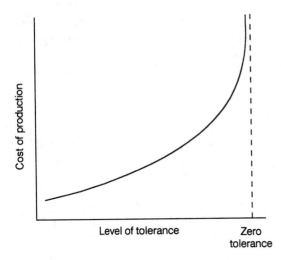

Level of tolerance   Zero   **FIGURE 15.2**
                                  tolerance   **Hypothetical cost of achieving zero tolerance.**

Acceptable Level of Risk

The notion of an acceptable level of risk due to food-borne illness, pesticide residues, or food additives implies that there is some standard or tolerance against which the risk is to be judged. Tolerance refers to the amount of a particular substance allowed in the food supply. Any tolerance level higher than zero implies the willingness to accept some level of risk.

How much risk people are willing to accept is a political decision. The level of risk that is accepted has economic content, because the cost of achieving the desired level of purity increases at an increasing rate as the level of tolerance approaches zero (Figure 15.2). The elimination of the use of an additive or a pesticide can mean either higher costs of production or the removal of the resulting processed food product from the market. Alternatively, it can mean the substitution of another risk. For example, banning the use of sodium nitrate in curing meats would either increase the risk of food poisoning or eliminate cured meats from the supermarket.

The application of the zero tolerance standard is complicated for four reasons:

- There are practical limits on the ability to remove specific substances, particularly when they exist in very small, yet measurable, amounts. Therefore, for most food supply contaminants, some level of tolerance is provided under law. For example, food inspection procedures allow small amounts of insect parts, rodent hair, and so forth in finished products because, practically, they are impossible to remove or eliminate from the food supply. Interestingly, if a decision is made to reduce significantly the use of pesticides on fruits and vegetables, tolerances for insect parts in processed foods may need to be increased. These are very interesting and important trade-offs.

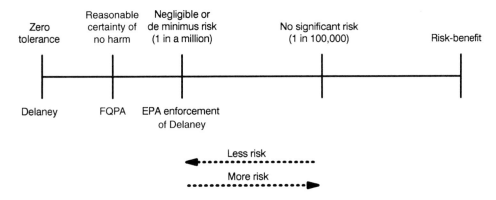

FIGURE 15.3 Continuum of tolerance options for food safety

Source: Adapted from Sondra Archibald, B. Hurd, and R. Marsh, "Perspectives of Chemical Manufacturers," in *Regulating Chemicals: A Public Policy Quandry* (Davis, Calif.: Agric. Issues Center, University of California, 1988).

- Many substances that may be carcinogens occur naturally in the food supply in very minute quantities. For example, potatoes contain arsenic; lima beans contain hydrogen cyanide; and spinach and broccoli contain nitrites.[13] Raw cabbage, lettuce, and spinach contain small amounts of 3,4-dibenzopyrene, which has been linked to gastric cancer.[14]

- The development of proven safe substitutes for potential disease-causing substances is very costly. For example, it costs more than $4 million for a pesticide manufacturer to complete the research needed to demonstrate the safety and efficacy of labeling (registering) a pesticide for use on an individual crop. If a pesticide is to be used on several crops, $4 million must be spent to obtain the label (register) for each crop. If a chemical is found to be unsafe, finding an acceptable substitute and proving that it is safe can easily cost more than $100 million.

- Detection instruments can now measure parts per trillion. Paarlberg uses the analogy that a part per trillion is like a grain of sugar in an Olympic-size swimming pool.[15] Our ability to measure potentially dangerous substances exceeds our ability to eliminate them from the food supply.

Archibald represented tolerance levels or acceptable levels of risk on a continuum from the zero-tolerance option to the risk-benefit option used under FIFRA before the adoption of the Food Quality Protection Act (FQPA) of 1996 (Figure 15.3).

---

[13]Reay Tannahill, *Food in History* (Briarcliff Manor, N.Y.: Stein & Day Publishers, 1973), pp. 379–380.

[14]Senauer *et al., Food Trends,* p. 261.

[15]Don Paarlberg, *Farm and Food Policy: Issues of the 1980s* (Lincoln, Neb.: University of Nebraska Press, 1980), p. 91.

Regulatory Irony

One of my vivid recollections is of sitting around a table in Washington with a roomful of government regulators. Their purpose was to get the last molecule of diethylstilbestrol out of beef liver, although scientists could not show that anyone had ever been harmed by consuming residues of this feed additive. The air was thick with cigarette smoke (325,000 deaths a year attributable to smoking). After the discussion was adjourned to cocktails (25,000 deaths a year from drunken driving, many of the victims nondrinkers). The inconsistency of regulating others and indulging oneself went unremarked by these people and, so far as I could see, unobserved.

Source; Don Paarlberg, *Farm and Food Policy: Issues of the 1980s* (Lincoln, Neb.: University of Nebraska Press, 1980), p. 82

Zero tolerance. A standard of zero tolerance means that products should be completely devoid of scientific evidence indicating specific harmful substances have been directly or indirectly added to the food supply. Zero tolerance was included in the 1958 Food Additives Amendment to the Food, Drug, and Cosmetic Act, hereinafter referred to as the *Delaney clause.* The **Delaney clause** stated that "no additive shall be deemed to be safe if found to induce cancer when ingested by man or animal, or if it is found after tests that are appropriate for the evaluation of the safety of food additives, to induce cancer in man or animal." Congressman Delaney was the chairman of the House Select Committee to Investigate the Use of Chemicals in Foods and Cosmetics.[16] Delaney, whose wife died of cancer, took a special interest in food safety issues.[17]

Zero tolerance is appealing to the public, but it created insurmountable problems for FDA enforcement, for the food industry, for the courts, and, in turn, for Congress. The problem is that the zero-tolerance standard became less workable as the ability to detect residues increased from 100 parts per billion in 1958 to 1 part per trillion in the early 1990s. Because "zero" cannot truly be measured, the standard of zero tolerance became a moving target determined by the continuously improving scientific ability to measure residues.

Delaney is, by admission, an unworkable standard, but it is the standard. This reality was determined by both a federal district court and by the U.S. Court of Appeals for the Ninth Circuit (San Francisco), which stated "If there is to be a change, it is for the Congress to direct," not the FDA.[18] Yet, consumer advocate support for zero tolerance remains strong today.

---

[16]National Research Council, *Regulating Pesticides in Food,* p. 163.
[17]Senauer *et al., Food Trends,* p. 246.
[18]Sparks Companies *Food and Fiber Letter,* (Washington, D.C.: July 20, 1992).

Reasonable certainty of no harm.[19]   A tolerance of reasonable certainty of no harm requires the establishment of a threshold at which there are discernable health effects. This is a point between some risks and zero risk/tolerance. This standard was established by the FQPA in 1996 and remains to be precisely interpreted.

De minimus or negligible risk.   The *de minimus* tolerance option suggests that extremely small risks (residues) can be ignored.[20] Because of the unworkable nature of the Delaney clause, the FDA and EPA adopted *de minimus* tolerance for enforcement of Delaney issues involving food additives. The specific *de minimus* tolerance level was established as 1 in 1 million—meaning that an additive or residue cannot cause more than one additional death per million people over their lifetime.[21]

No significant risk.   A tolerance of no significant risk involves *a tolerance level of 1 in 100,000*. That involves one additional death per 100,000 people over their lifetime. This standard was adopted by California's Safe Drinking Water and Toxic Enforcement Act (Proposition 65). It implies a higher level of risk being allowed than the current FDA *de minimis* or negligible risk standard but less risk than was allowed under FIFRA.

Risk–benefit.   The risk–benefit approach requires a pesticide to accomplish its intended effect without "any unreasonable risk to man or the environment, taking into account the economic, social, and environmental costs and benefits of the use of any pesticide."[22] It is the cost–benefit balancing standard applied under FIFRA prior to amendments of pesticide regulations in 1994 and 1996. This standard recognized that pesticides have both risks and benefits. The benefits to be considered under FIFRA included the impacts on production, commodity prices, food prices, and the agricultural economy. As noted in Chapter 12, one of the major issues in contemporary regulation involving the environment is whether economic benefits should receive any consideration in policy decisions.

Food Quality Protection Act

The FQPA does more than establishes the reasonable certainty of no harm criteria. Pesticides for which it has not been possible to show the point or threshold at which adverse health effects begin may be permitted slightly higher levels of tolerance if they confer benefits such as control of a potentially harmful substance such as aflatoxins (toxic and carcinogenic compounds produced by certain strains of fungi), or increase production of a valued crop. Equally important, FQPA expands the coverage beyond carcinogens to all possible health harms. Because

[19]James D. Wilson, "Resolving the Delaney Paradox," *Resources* (Washington, D.C.: Resources for the Future, Fall 1996).
[20]Senauer *et al.*, *Food Trends*, p. 247.
[21]National Research Council, *Regulating Pesticides in Food*, p. 39.
[22]7 USC Sec. 136(a) (1978).

children consume more food per pound of body weight, a special safety factor was added to recognize their higher exposure to additives.

Ability to Adapt to New Life Forms: Biotechnology

A major new challenge to the regulations of the safety of the food supply involves the ability to adapt contemporary regulatory procedures to the products of genetic engineering. Theoretically, this should present no greater problem than that of testing new pesticides or food additives. This appeared to be the case with the genetically engineered substitute for the enzyme rennin used as the culture for making cheese. It was approved by FDA for sale in March 1990.

A different set of conditions developed with respect to the approval to sell recombinant bovine somatotropin (rBST), a genetically engineered substitute for the hormone that allows cows to produce milk. The fact that cows having higher milk production secrete more natural BST provided the incentive for splicing the BST gene into the DNA of the *E. coli* bacteria. Fermentation causes the resulting bacteria to multiply rapidly. The synthetic BST, which is indistinguishable from the natural, can then be separated and purified.

FDA approved the milk produced with synthetic BST as being safe in the late 1980s. After years of testing by four licensed companies plus several independent studies, and expenditures approaching $500 million, FDA did not approve the sale of rBST until 1994. In the 1993 budget resolution, Congress inserted provisions for a 12-month moratorium on its sale. The slowness of FDA approval and negative congressional response appeared to result from protests by less progressive farmers and the fear of a consumer backlash precipitated by public interest groups. The negative positions of these public interest groups on biotechnology appeared to have a plant and animal rights overtone. Specifically, these groups appear to be opposed to any technology that disturbs the purity of the species and their natural evolution. They argued in the case of rBST that health could be jeopardized by increased antibiotics used to control mastitis in milk. Their theory was that higher producing cows had more mastitis. But it is illegal for a farmer to sell milk from cows treated for mastitis with antibiotics until after the residues disappear. In an era of potentially large productivity gains and the potential for the development of chemical substitutes, *the challenge for FDA, EPA, and USDA is the ability to make decisions on a scientific basis for individual products, given the politically established standards for risk tolerance levels.*

The Office of Technology Assessment (OTA) concluded that the rBST issue encompasses the whole sphere of biotechnology developments and regulations.[23] According to OTA, "Biotechnology-related risk assessment focuses on the planned introduction of genetically modified organisms into the environment

---

[23]Office of Technology Assessment, *A New Technological Era for American Agriculture*, OTA-F-474 (Washington, D.C.: U.S. Congress, August 1992).

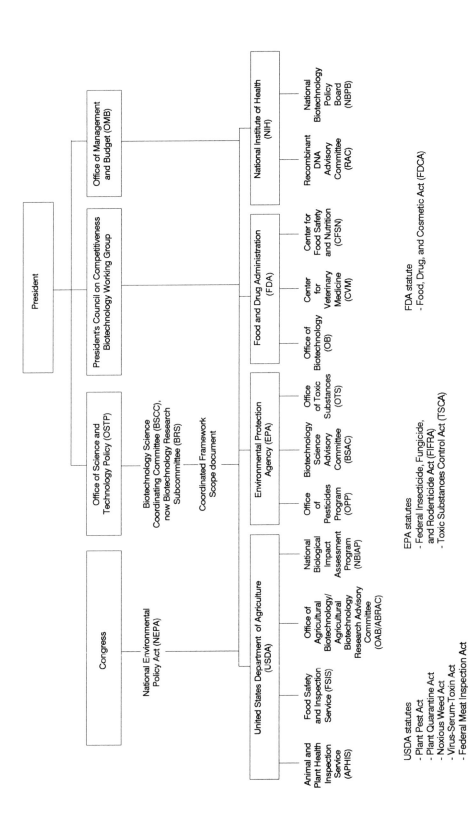

FIGURE 15.4 Jurisdiction and coordination of environmental policy for biotechnology-derived agricultural products.[a]

[a]OSTP, the Council on Competitiveness, and OMB do not have direct oversight over the federal agencies; the connections shown here are those of influence through law, key documents, or review.

Source: Office of Technology Assessment, 1992.

(environmental safety) and on the consumption of products derived from biotechnology."[24] Figure 15.4 indicates the regulatory maze that has developed in this relatively new arena of scientific endeavor. The policy issues identified by OTA that needed to be addressed by the Congress and other regulatory agencies include the following:

- Should the products of biotechnology be treated in the same science-based, risk-based regulatory paradigm as pesticides and food additives? For example, should the products of biotechnology be treated as new products, requiring protracted approval procedures, even though they are synthetically produced replicas of naturally occurring substances?
- Should disease or pest-resistant plants that are the product of biotechnology be treated any differently from a regulatory perspective than such plants produced by conventional plant breeding procedures?
- Should transgenic plants (genetic material from another plant or animal inserted) be treated as a food additive and require the complete new additive approval process?
- Should the same residue tolerances set for pesticide residues in general be applied to pest-resistant plants? For example, should Round-Up–resistant soybeans be required to go through the same regulatory maze for approval as the Round-Up herbicide?
- What role should the public play in resolving issues of biotechnology regulation faced by the various regulatory agencies?
- Do consumers have a right to know whether they are buying products of biotechnology? For example, should food produced from transgenic plants and animals be so labeled?
- Should current laws regulating chemical applications to agriculture be applied to biotechnology where the control mechanism is microorganisms?
- What steps should be taken to protect the natural gene pool from the genetically modified gene pool when, for example, a new biotechnology-derived plant or animal is produced?

Safety Information and Labeling

As an alternative to regulating product availability, the private sector could be required to provide safety information regarding their products, leaving consumers the choice of whether to consume them. For products demonstrated to be potentially dangerous, warning labels could be required. Presumably, different levels of food safety would be available and consumers could choose among them, considering the level of risks involved. This approach has been used on cigarettes,

[24]Ibid., p. 12.

alcohol, and saccharin. Consumption of cigarettes and some alcoholic beverages has declined significantly, and many saccharin consumers switched to other synthetic noncaloric sweeteners when it was labeled a potential health hazard.

The information and labeling approach may not be applicable to all food safety situations. The risks from disease-causing substances are by no means equal. It is unlikely that a labeling system can accurately distinguish between the degree of risks involved in different substances. The amount of information required to make an objective decision would be vast.

In certain instances, it may be impossible to translate safety information effectively. For example, it might be impractical for fast-food outlets or restaurants to convey information on safety to the customer. Where children are involved, parents would be making food safety decisions. In situations such as school lunches, the government would still be making food safety decisions on behalf of individuals.

NUTRITION AND HEALTH

As noted at the beginning of this chapter, the nutrition issue may be viewed as a continuum of the food safety issue. In other words, the health of people is influenced by what they eat. The effects of poor nutrition are not as violent and visible as food poisoning, but poor nutrition has been implicated in many of the same diseases as pesticide residues or food additives. Poor nutrition implicates other health concerns such as cardiovascular diseases, which may also be contributed to by additives or pesticides. A related dimension of the nutrition and health issue involves the problems of hunger and malnutrition. These issues are treated after discussion of diet, health, and nutritional labeling.

Sources of the Nutrition-Health Problems

The specific nature of the relationship between diet and health is much disputed. Yet it is undisputed that there is a diet–health relationship. Three main sources of the diet–health problem can be identified:

- Eating the wrong foods
- Having an unhealthy lifestyle
- Being obese.

Eating the Wrong Foods

Individual choice, custom, food availability, cost, habits, health, geographic location, peer pressure, ease of preparation, age, religion, and ethnic considerations all influence what people eat. Food selection and consumption are related in a complicated way to other aspects of life. For example, the per capita consumption of

TABLE 15.2 Influence of Selected Preventable Factors on Mortality

	Cause of Death				
	Major cardiovascular-renal diseases	Malignant neoplasms	Accidents: motor vehicle and other	Respiratory diseases	Diabetes mellitus
	Level of Influence[a]				
Smoking	VH	VH	L	VH	VL
Nutrition	VH	VH	VL	VL	VH
Occupational hazards	VL	L	VH	H	VL
Alcohol abuse	L	L	VH	L	L
Drug abuse	VL	VL	H	VL	VL
Radiation hazards	VL	L	VL	VL	VL
Air and water pollution	VL	VL	VL	L	VL
	Number of Premature Deaths				
In 1973	395,000	90,000	44,000	16,000	24,000
In 2000[b]	595,000	127,000	71,000	33,000	30,000

[a]VH, very high (30%); H, high (20 to 30%); M, medium (10 to 20%); L, low (below 5%).

[b]If current trends remain unchanged.

Source: R. B. Gori and B. J. Richter, "The Macroeconomics of Disease Prevention in the United States," *Science* (June 1978), p. 1125. Copyright 1978 by the American Association for the Advancement of Science.

dairy products among black Americans may be low because some blacks have had difficulty digesting lactose. Teenagers may choose not to participate in the school lunch program because they prefer to spend their lunch hour away from school, because of the quality of the lunch, or because they simply prefer to eat "junk food." Highly processed food consumption has increased, in part, because of the working mother and higher preferences placed on leisure time.

Available studies indicate an increasing consensus that the American diet is contributing to chronic diseases, such as heart disease and cancer, which generally afflict people in later life.[25] In addition, they assert that cutting down on fats, calories, sugar, and salt would be positive steps toward reducing heart disease, certain cancers, and strokes.[26] In drawing this conclusion, they cite comprehensive studies such

[25]Public Health Service, *Healthy People: The Surgeon General's Report on Health Promotion and Disease Prevention*, Pub. 79-55071 (Washington, D.C.: U.S. Department of Health, Education, and Welfare, 1979); Public Health Service, *The Surgeon General's Report on Nutrition and Health*, Pub. 88-50210 (Washington, D.C.: U.S. Department of Health, Education, and Welfare, 1988); National Research Council, *Toward Healthful Diets* (Washington, D.C.: National Academy Press, 1980); National Research Council, *Diet, Nutrition, and Cancer* (Washington, D.C.: National Academy Press, 1982); and National Research Council, *Diet and Health* (Washington, D.C.: National Academy Press, 1989).

[26]Ibid.

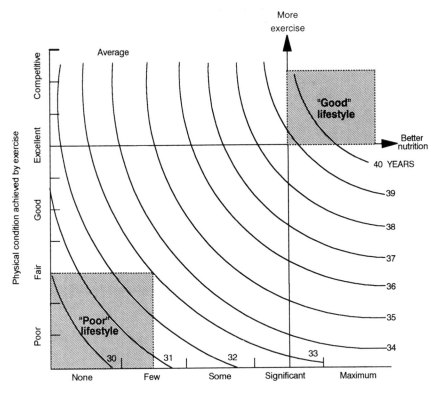

FIGURE 15.5 **Hypothetical relationship between physical condition from exercise, and diet restriction and expected number of years free from chronic disease for a 40-year-old nonsmoking adult.**

Source: C. Peter Timmer and Malden C. Nesheim, "Nutrition, Product Quality and Safety" in Bruce L. Gardner and James W. Richardson, eds., *Consensus and Conflict in U.S. Agriculture* (College Station, Tex.: Texas A&M University Press, 1979), p. 167.

as those of Glueck and Conner[27] and Stamler,[28] which recommend that alteration of dietary practices to reduce saturated fat and cholesterol consumption would be desirable for the public. There appears to be general agreement by a number of scientists that healthy Americans should avoid too much fat, saturated fat, and cholesterol.

These views were highly controversial and remain so in many agricultural circles, but the consensus of evidence has become overwhelming. Agricultural organizations involving dairy and beef producers appear to have begun to adjust their strategy, as they recognize the problems of selling a product containing substantial amounts of fat and cholesterol.

[27]C. J. Glueck and W. E. Conner, "Diet-Coronary Heart Disease Relationship Reconnoitered," *Am J. Clin. Nutr.* (1978), pp. 727–737.

[28]J. Stamler, "Life Styles, Major Risk Factors, Proof and Public Policy," *Circulation* (1978), pp. 3–19.

Poor Lifestyles

Without question, diet–health relationships are more complex than requiring a simple nutrition-health prescription. Gori and Richter effectively summarized the results of research on the controllable factors in mortality prevention (Table 15.2). Nutrition is only one of seven factors affecting mortality.

In summarizing the causal relationship for various environmental causes of death, Gori and Richter conclude that the government is devoting too many resources to lower priority causes such as air and water pollution to the detriment of making substantial progress on higher priority causes such as smoking, diet, and alcohol. Yet recent evidence on the extent of water pollution, including pesticide residues, suggests to environmentalists that all environmental problems deserve greater attention.

The complexity of the relationship between nutrition and health is probably best illustrated by the relationships suggested in Figure 15.5, which points out the interactive effects of diet, physical exercise, and smoking on life expectancy. A "good" lifestyle involves a high level of physical exercise, significant dietary restrictions to maintain low levels of serum cholesterol and total caloric intake, few highly processed foods, and no smoking. With somewhat less exercise and fewer dietary restrictions, the risk of chronic disease increases. The individual can consciously control these variables. The trade-off is one of giving up some of the "pleasures of life" in the short run to live longer.

Obesity

Obesity, a body weight that is more than 20 percent above the ideal weight, is increasingly becoming recognized as a major health problem. An increasing body of evidence points to the undesirability of the average trend to accumulate weight gradually from age 25 and onward. Public health data suggest that at least 25 percent of the population is obese.[29] The principal cause of obesity is overeating, the intake of food in excess of requirements. Available evidence suggests a persistent increase in the extent of obesity.

Sources of excess calories are related primarily to excess intake of fats, oils, and caloric sweeteners, all of which have tended to increase in consumption.[30] High fat intake has been associated with increased risk of both heart disease and cancer.[31] Yet total food intake also increased 22 percent from 1970 to 1986 and then declined by 5 percent from 1986 to 1989.[32] Americans are eating too much of the wrong foods—a factor contributing to health problems. Yet there are some posi-

---

[29]Nancy J. Schmidt, Chatham College, Pittsburgh, Pa., Personal correspondence.

[30]"Food Consumption," *National Food Review* (Washington, D.C.: ERS, USDA, 1987 Yearbook), p. 4.

[31]Cronin *et al.*, "Developing a Food Guidance System to Supplement Dietary Guidelines," *J. Nutr. Educ.* (1987), p. 285.

[32]K. Bunch, "Highlights of 1985 Food Consumption Data," in *National Food Review* (Washington, D.C.: ERS, USDA, Winter–Spring 1987), p. 1.

tive signs. Consumers are consuming fewer animal fats, obtaining a larger proportion of their food supply from plants, increasing intake of complex carbohydrates, and eating more fruits and vegetables. But they are still becoming more obese. The lack of physical exercise is clearly a factor.

Nutrition Education Policy

A key to improved nutrition is education. People need to have the information necessary to make intelligent decisions on both what to eat and how much. The key policy issues involve who should provide that information and in what form. Put more directly, what is the role for government in nutrition education?

The Nutrition Message

The USDA has always played a key role in nutrition education. In the federal government, it is designated as the lead agency in the establishment of nutrition education policy. To implement this role, it has established the Human Nutrition Information Service, which operates through and with key public and private agencies such as the Extension Service, National Institute of Health, Food and Drug Administration, American Red Cross, American Medical Association, American Heart Association, and the American Dietetic Association.[33]

Nutrition education: Prior to 1970.   Prior to 1970, nutrition education programs were tuned to a generally higher level of physical activity than exists today. *The main message was to eat a variety of the basic food groups.* From the 1940s through much of the 1950s, there were seven basic food groups: green and yellow vegetables; oranges, tomatoes, and grapefruit; potatoes and other vegetables and fruits; milk and milk products; meat, poultry, fish, and eggs; bread, flour, and cereals; and butter and margarine. In the mid-1950s, the basic seven were condensed to four basic food groups: fruits and vegetables; meat, poultry, fish, and eggs; milk and milk products; and bread and potatoes. The food groups were generally depicted as a "food wheel," the form of which gave no preference to any particular food group—only to variety.

The role of controlling food intake was not emphasized. As traditionally conducted, these programs did not threaten anyone's interests because they did not give any order of preference among the groups and were not intended to restrict food consumption behavior.[34] It is interesting to note that neither fat nor sugar was categorized as a food group in either the basic seven or four food groups.

---

[33]Cronin *et al.,* "Developing a Food Guidance System."
[34]C. Peter Timmer and Malden C. Nesheim, "Nutrition, Product Quality and Safety," in *Consensus and Conflict in U.S. Agriculture,* ed. Bruce L. Gardner and James W. Richardson (College Station, Tex.: Texas A&M University Press, 1979), p. 175.

Nutrition education: 1970 and beyond. The whole approach to nutrition education began to change in the 1970s. A focal point for this change was the formation and activities of the Senate Select Committee on Nutrition and Human Needs. The initial charge of this committee, chaired by George McGovern (Dem.–S.D.), was to define the extent of malnutrition, examine government feeding programs, and make recommendations.[35] Out of the committee's hearings and related activities grew the White House Conference on Food, Nutrition and Health in 1969. The conference did not focus solely on malnutrition but on the broader concept of food and nutrition policy.[36]

The next step in the developing change in nutrition policy was the formation of the National Nutrition Consortium in 1974. The result of its work was an outline of recommendations for a national nutrition policy. The work of the National Nutrition Consortium was debated extensively before the Senate Select Committee on Nutrition and Human Needs. The result of the committee's deliberations was the publication of a set of **dietary goals,** which were as follows:

- To avoid being overweight, consume only as much energy (calories) as is expended; if overweight, decrease energy intake and increase energy expenditure.

- Increase the consumption of complex carbohydrates and naturally occurring sugars from about 28 percent of energy intake to about 48 percent of energy intake.

- Reduce the consumption of refined and processed sugars by about 45 percent to account for about 10 percent of total energy intake.

- Reduce overall fat consumption from approximately 40 percent to about 30 percent of energy intake.

- Reduce saturated fat consumption to account for about 10 percent of total energy intake; and balance that with polyunsaturated and monounsaturated fats, which should account for about 10 percent of energy intake each.

- Reduce cholesterol consumption to about 300 milligrams a day.

- Limit the intake of sodium by reducing the intake of salt to about 5 grams a day.[37]

Needless to say, the dietary goals caused great consternation in American agriculture. Livestock producers, dairy farmers, sugar producers, and related agribusiness interests saw the potential for substantially reduced demand and, thus, lower prices and incomes from their producers. This adverse reaction did not discourage

---

[35]K. Schlossberg, "Nutrition Policy in the United States," in B. Winkoff, ed., *Nutrition and National Policy,*(Cambridge, Mass.: The MIT Press, 1978), p. 350.

[36]White House Conference on Food, Nutrition, and Health, Final Report (Washington, D.C.: 1969).

[37]*Dietary Goals for the United States*, 2nd ed. (Washington, D.C.: Select Committee on Nutrition and Human Needs, U.S. Senate, December 1977). An earlier version of these goals was published in February 1977. The revision lowered recommendations on carbohydrate intake and raised the recommended salt intake.

either Congress or the Carter administration from further pursuing the issue. Congress included in the 1977 farm bill authorization for USDA to be the lead agency in nutrition policy. In doing so, it recognized increasing evidence of a relationship between diet and many of the leading causes of death and health problems. In 1980, USDA and the Department of Health, Education and Welfare (now Health and Human Services) issued the report *Nutrition and Your Health, Dietary Guidelines for Americans.* The **dietary guidelines** were not as specific as the dietary goals:

- Eat a variety of foods.
- Maintain ideal weight.
- Avoid too much fat, saturated fat, and cholesterol.
- Eat foods with adequate starch and fiber.
- Avoid too much sugar.
- Avoid too much sodium.
- If you drink alcohol, do so in moderation.[38]

Despite their lack of specificity, the guidelines drew a negative reaction from farmers at least as strong as the dietary goals. In this case, the reverberations extended through the countryside as employees of the Extension Service began to adopt new educational materials based on the guidelines. These educational materials expanded the four basic food groups to five: fruits and vegetables; bread and cereal; milk and cheese; meat, poultry, fish, and beans; and fats, sweets, and alcohol. *The basic educational message switched from variety to variety, moderation, and avoidance.* No servings of the fats, sweets, and alcohol group are recommended.[39] Serving sizes were reduced throughout, with less meat used in the menus and recipes.

When the new educational approach was taken to the country by extension nutrition specialists, it was not unusual for farm groups, particularly cattle raisers and dairy farmers, to question them and their administrators regarding the use of the guidelines and related educational materials (see box). Some educational programs were stopped—at least temporarily. Timmer and Nesheim put the controversy in the following perspective: "When nutritional educators begin to mount programs aimed at reducing the intake of foods thought to be associated with major public health problems such as coronary heart disease, cancer, and diabetes, then opposition and widespread controversy erupted quickly."[40]

More fuel was added to the nutrition–health, nutrition education controversy in 1982 when a National Academy of Sciences scientific advisory committee found that foods such as fats and smoked foods appear to be linked to certain types of cancer.[41] The committee recommended reduced salt- and smoked-cured

---

[38]*Nutrition and Your Health, Dietary Guidelines for Americans* (Washington, D.C.: USDA, HEW, 1980).
[39]C. A. Davis *et al.,* "Food," Home Garden Bull. 228 (Washington, D.C.: SEA, USDA, 1980).
[40]Timmer and Nesheim, "Nutrition, Product Quality and Safety," p. 175.
[41]National Research Council, *Diet, Nutrition, and Cancer* (Washington, D.C.: National Academy Press, 1982).

Economic Impacts of Dietary Guideline Adoption on Farmers

A policy question before us is whether USDA should adopt and promote quantitative dietary goals.

The effect on farm and food prices and on farm income was analyzed under a scenario in which all U.S. consumers would adopt a diet that followed the dietary goals between 1981 and 1985.

Consequences for Producers

Following dietary goals strictly would have led to a decrease in the consumption of fat, sugar, and sodium. If all Americans had strictly adopted such a diet between 1981 and 1985, the domestic demand for red meats would have decreased 8 percent; eggs, 23 percent; cheese, 15 percent; sugar, 27 percent; fats and oils, 38 percent; and nuts 27 percent. The demand for poultry would have increased 5 percent; fish, 9 percent; yogurt, 9 percent; vegetables, 14 percent; rice, 60 percent; wheat, 50 percent; and lentils, 108 percent. Net farm income would have increased for wheat farmers by 17 percent, for rice farmers by 31 percent, and for broiler and turkey producers by about 11 percent.

Net income of egg and pork producers would have declined by 51 percent and 38 percent, respectively. Corn, soybean, and beef producers would have seen small decreases in income: 3, 11, and 6 percent, respectively. The dairy sector was largely unaffected, due to government price supports.

Source: Jean Kinsey, "Food Quality and Prices," in *Agricultural and Food Policy Issues for the 1990s* (Urbana, Ill.: Department of Agricultural Economics, University of Illinois, April 1990), p. 2.

---

food intake; reduced fat intake; increased fruit, vegetable, and whole-grain cereal consumption; and alcohol consumption in moderation. The National Cattlemen's Association called the panel's findings "inconclusive and premature."[42]

Politicians, however, had difficulty arguing against the moderation message in a time of increased obesity and concern about cancer. Even the Reagan administration, which had made a campaign issue in opposition to the dietary goals and guidelines, did not repudiate them. In 1982 Secretary of Agriculture Block appointed a committee of scientists to review the dietary guidelines published in 1980. In its 1985 report, the committee recommended only minor changes in the

---

[42]Bart Schorr, "Certain Foods May Increase Cancer Risk Committee Says, Urging Changes in Diet," *The Wall Street Journal* (June 17, 1982), p. 4, col. 2.

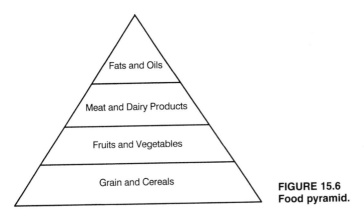

FIGURE 15.6
Food pyramid.

dietary guidelines and some clarifying statements in the accompanying text.[43] The second edition of the dietary guidelines formed the basis for federal dietary guidance policy.[44]

The dietary goals and guidelines were followed by a host of government and private organization reports, all of which essentially confirmed the message contained in the original dietary goals report. After making a detailed comparison of the dietary recommendations of 15 different studies, Senauer, Asp, and Kinsey drew the following conclusion: "Overall, the recommendations of different groups are remarkably similar. Differences concern intended audience, that is, whether the recommendations are for the general public or professionals—whether they address the risk of a particular chronic disease, or whether they are made for everyone to improve health and decrease a 'whole spectrum of chronic diseases.'"[45]

Yet from a political perspective, how to convey the message to the public in an educational context was not yet over. A new version of the food wheel had been created that integrated fats, sweets, and alcohol into the wheel, but the implication was that each should be consumed, in addition to other food groups. USDA began to look for a new educational tool that would convey a proper visual representation of the new widely accepted nutritional message better than the food wheel. They came up with a food pyramid (Figure 15.6). The pyramid was designed to convey the notion that the diet should be based on grains and cereals. Next in the order of preference was fruits and vegetables, followed in smaller quantity by meat and dairy products. The smallest part of the diet is fats and oils. Note that alcohol and sugar are implicitly not included in the diet.

Unfortunately for Secretary Edward Madigan, the food pyramid was released just days after he became agriculture secretary and on a weekend when the National Cattlemen's Beef Association (NCBA) was in town. Madigan first

---

[43]*Report of the Dietary Guidelines Advisory Committee on the Dietary Guidelines for Americans,* Admin. Rep. 100 (Washington, D.C.: USDA, 1985).

[44]*Nutrition and Your Health: Dietary Guidelines for Americans,* 2nd ed., Home Garden Bull. 232 (Washington, D.C.: USDA, HEW, 1985).

[45]Senauer *et al., Food Trends,* p. 35.

read about the pyramid in *The Washington Post,* just like the cattlemen. When confronted on Monday by the NCBA leadership, Madigan stopped the release of the pyramid pending further study. However, after $543,000 worth of study, the pyramid was vindicated.

This example of public policy teaches an extremely important lesson. *Despite much controversy surrounding the dietary goals and guidelines, the scientific base of their content, combined with the persistence of their professional advocates, allowed them to survive and gain general acceptance.* Clearly, the guidelines and the food pyramid are still controversial, but they are based on science.

Effectiveness of the Guidelines in Influencing Consumption

Since their inception in the mid-1970s, the guidelines do appear to be having some impact on American food consumption habits. Studies of American consumers indicate that a health trend is becoming part of the American lifestyle. Most Americans (87 percent) know that their choice of foods has long-term effects on health, including heart problems, weight gain, cholesterol buildup, high blood pressure, diabetes, and cancer. Up to two-thirds of the respondents knew that salt was associated with high blood pressure, eggs were associated with cholesterol, and fats contributed to heart disease.[46]

As indicated previously, some improvements in diets have been observed. Consumption of nutritious foods such as lowfat milk, yogurt, fresh fruit, fresh vegetables, chicken, and fish are up. Fat is down but sugar is up.[47] Average cholesterol intake is 440 milligrams per day while the guidelines recommend 300 milligrams per day—but down from 500 milligrams in 1968.[48] Therefore, the guidelines appear to be having some impact, but there are still substantial changes in diet that would have to take place if the dietary goals are to be reached.

Food Labeling

The most recent policy innovation in the nutrition–health arena involves nutritional labeling. The National Labeling and Education Act of 1990 was a logical sequel to the dietary goals and guidelines controversy. That is, although the goals and guidelines have made people more conscious of nutrition health issues, they alone do not provide consumers the product information needed to implement them. Until May 1994, when the new labeling requirements went into effect, there was a mixture of largely voluntary, mandatory, and certainly unstandardized regulations. For example, manufacturers were allowed to establish their own

---

[46]A. E. Sloan, "Educating a Nutrition-Wise Public," *J. Nutr. Educ.* (1987), p. 303.
[47]"Food Consumption," p. 3.
[48]Nancy Raper, "Nutrient Content of the U.S. Food Supply," *National Food Review* (Washington, D.C.: ERS/USDA, July–September 1991), p. 14.

serving size. Therefore, even if two companies had the same basic nutritional information, it could not readily be compared. A company that made a low-fat sour cream was not allowed to call it sour cream under the standards of identity for sour cream. Instead, the low-fat product had to be labeled an imitation or a substitute. The new regulations have even cracked the health claim barrier that required companies making health claims to have the scientific evidence to back them up. Now, certain health claim links are recognized as being scientifically valid and, in such instances, can be utilized on the label without product-specific scientific support.

The following are major areas covered by the labeling law for processed foods:[49]

- A standard serving size is specified as the amount normally consumed by an average person over age four in a standard unit of measure such as a cup.
- The nutrients to be listed include those having both a favorable and unfavorable nutrition connotation, as follows:
  - Calories
  - Calories from total fat
  - Total fat
  - Saturated fat
  - Cholesterol
  - Total carbohydrates
  - Complex carbohydrates
  - Sugar
  - Dietary fiber
  - Protein
  - Sodium
  - Vitamin A
  - Vitamin C
  - Calcium
  - Iron
- A dictionary of product descriptors is uniformly applied. For example, *low calorie* means less than 40 calories per serving, *reduced calorie* means twenty five percent less, and *light* means one-third fewer calories. *Reduced cholesterol* means 50 percent less. *Freshly prepared* means not having been frozen, thermal processed, or preserved.
- Health claims regarding established relationships between the following can be made:
  - Fat and cancer

[49]Judith E. Foulke, "Food Labeling," *FDA Consumer* (January–February 1992), pp. 9–13.

- Fat and heart disease
- Calcium and osteoporosis
- Sodium and hypertension

Scientific evidence establishing health-improving relationships for fiber and heart disease and fiber and cancer is expected.

This is not the end of the road for the nutritional labeling controversy. Labeling for fresh meat, poultry, fish, fruits, and vegetables is voluntary. The important away-from-home market presents even more difficult labeling problems.

A futuristic alternative for labeling involves the potential for reflecting in a single number the total nutritional quality of a particular food, combination of foods, or meal. Padberg suggested an approach in which dieticians are asked to rank the relative importance of nutrients.[50] The consensus ranking then becomes part of a computerized formula whereby any food, combination of foods, or meal could be converted to a single number indicating the nutritional ranking on a scale of, say, 0 to 100. Conceptually, all foods in a supermarket or a restaurant would be required to have the nutritional number displayed.

Despite such grandiose potential innovations in labeling, there remains a host of standards of identity that could be roadblocks to nutritious product innovation, depending on the flexibility with which the new FDA labeling regulations are applied. Standards of identity require certain product composition and physical characteristics for a product, for example, to be called cheddar cheese. **Standards of identity** were put in place to facilitate informed consumer choices. In an era of biotechnology and food processing innovations and with a policy of encouraging more healthy foods, standards of identity stand as a barrier to progress. Should vegetarian chili be allowed to be called chili? Would the alternative of simply allowing consumers to make comparisons, using the nutritional label, be sufficient?

HUNGER IN AMERICA

Hunger from a worldwide perspective may be defined as the sustained failure to consume nutrients in sufficient quantity and in the proper proportion to sustain normal body functions (see Chapter 5). When it comes to U.S. citizens, the tendency is to use a considerably more charitable definition of hunger, which is in itself a source of considerable controversy and debate. For example, hunger statistics are often based on a lack of purchasing power rather than actual consumption or nutrition levels.[51]

---

[50]D. I. Padberg, Karen S. Kubena, Teofilo Ozuna, Heaseon Kim, and Lacye Osborn, "The Nutrition Quality Index: An Instrument for Communicating Nutrition Information to Consumers," AFPC Research Report 93-10(College Station, Tex.: Agricultural and Food Policy Center, Texas A&M University, October 1993).

[51]Robert Rector, "Food Fight. How Hungry Are America's Children?" *Policy Rev.* (Fall 1992), pp. 38–43; and Jean Y. Jones, *The Hunger Issue and Federal Food Assistance* (Washington, D.C.: Congressional Research Service, April 14, 1993).

Hunger in America became a major issue in the mid-1960s. The National Advisory Commission on Rural Poverty was created to study, evaluate, and make recommendations on means of removing poverty in America. Its report, *The People Left Behind*, focused increased attention on the problems of the poor—only one of which is hunger.[52] This and subsequent studies identified hunger in America as a major national problem and concern.

A late-1960 USDA study found that one-fifth of the households in the United States had "poor" diets. Thirty-six percent of the low-income households were found by the same study to subsist on poor diets. A team of medical doctors made on-site visits in several poor counties and shocked the nation by its conclusion that people in 280 counties were living in such distressed conditions "as to warrant a Presidential declaration naming them as hunger areas."[53]

In the late 1960s and early 1970s, the elimination of hunger and poverty in America became a major policy goal. A plethora of programs was established as part of President Johnson's Great Society plan. Many of these—including the food stamp; the women, infants, and children program; free or subsidized school lunch; and expanded nutrition programs—emphasized food assistance and education.

In the 1970s, the hunger problem appeared to dissipate as the economy expanded and the Great Society programs were implemented. A 1979 study by the Field Foundation, an organization dedicated to the elimination of hunger, came to about the same conclusion: "Our first and overwhelming impression is that there are far fewer grossly malnourished people in this country today than there were ten years ago . . . tremendous progress has been made."[54]

In the 1980s, the hunger problem reappeared, not with the vigor of the 1950s and 1960s, but nonetheless a serious concern.[55] As the unemployment rate increased from 5.8 percent in 1979 to 9.7 percent in 1982, the poverty rate increased from 11.7 to 15.0 percent.

In spite of these economic adversities, the Reagan administration held the line on food stamp expenditures and actually cut them in 1982 (Figure 15.7). Poverty rates remained above 14 percent through 1985.

In reaction to the deteriorating situation, a Physicians' Task Force on Hunger in America was appointed. The Physicians' Task Force concluded that of 35 million people in poverty, 20 million go hungry.[56] These results became the subject of a major controversy.[57] In June 1988, *Insight* magazine published an investigative report on the hunger issue that raised several questions about the reliability of the

---

[52]National Advisory Commission on Rural Poverty, *The People Left Behind* (Washington, D.C.: NACRP, 1967).

[53]Citizens Board of Inquiry into Hunger and Malnutrition in the United States, *Hunger USA* (Boston: Beacon Press, 1968).

[54]Nick Kotz, *Hunger in America: The Federal Response* (New York: The Field Foundation, April 1979).

[55]Jones, *The Hunger Issue*, p. 1.

[56]J. L. Brown, "Physicians' Task Force on Hunger in America," in *Poverty and Hunger in America*, Serial 99-4 (Washington, D.C.: Committee on Ways and Means, U.S. House of Representatives, April 30, 1985), p. 11.

[57]Carolyn Lockhead, "Hunger Hype," *Insight on the News* (June 27, 1988), pp. 8–15.

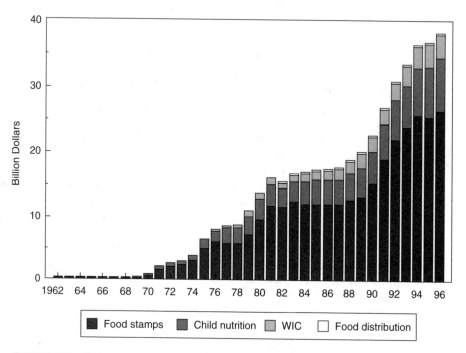

FIGURE 15.7   Federal expenditures on food programs, 1962–1996.

Source: USDA Historical Budget Outlays, electronic database. (Washington, D.C.: USDA/ERS, March 1995).

Physicians' Task Force. It pointed out that no standard definition of hunger was used; that no statistics on hunger exist; that proxies for hunger such as anemia, underweight, and infant mortality may not be hunger in many or even most cases; and that counties pointed out by the study to have high hunger rates had not, in fact, experienced hunger problems.

The physician's study served to highlight the difficulty of defining hunger in an American context. It was followed by a 1991 FRAC study, which concluded that 5.5 million children under age 12 are hungry and 11.5 million under age 12 are hungry or at risk of hunger.[58] In 1993, Bread for the World concluded that 30 million Americans (10 percent) were hungry.[59]

It is quite clear that these studies had a different concept of hunger than the presentation of starving people in Africa seen on television. That is not to say hunger is not a problem in America. The difference in hunger estimates only points up the difficulty of defining the problem. Is the existence of more people in

[58]Food Research Action Center, *Community Childhood Hunger Identification Project. A Survey of Child Hunger in the United States* (Washington, D.C.: Food Research Action Center, March 1991).
    [59]Bread for the World, *Fourth Annual Hunger Report* (Washington, D.C.: Bread for the World Institute, November 1993).

soup lines, more homeless, or more food stamp recipients evidence of hunger or poverty, or both?

In the heat of the 1996 presidential campaign with Republicans in the majority and strong economic growth, awareness of the problems waned as momentum built for welfare reform. What appeared to be never-ending growth in food assistance and other welfare expenditures took its toll in the form of welfare reform legislation, a key feature of which was to require work in return for benefits.

Causes of Hunger in America

How is it possible for hunger to be a problem in a food-rich nation? Most public policy discussions on that topic focus on three factors:

- Lack of resources
- Lack of access
- Lack of knowledge

Lack of Resources

The American food system is market oriented. The available foods, like other goods and services, are rationed in the market to those with the resources to purchase them. To put it simply, it is frequently argued that a large portion of the population does not earn enough to purchase nutritionally adequate diets.

For example, households earning between $5,000 and $10,000 annually spend 35 percent of their income on food, whereas those earning $30,000 to $40,000 spend 15 percent. This difference, obviously, is the center of the hunger problem.

How to deal with unequal income distribution as a cause of hunger has been a matter of public debate. Some believe that *a lack of income indicates a lack of initiative*. Others argue that *access to food is a basic human right, not something to be "earned."* The Rome World Food Conference in 1974 and, subsequently, the U.S. Congress, passed such "right-to-food" resolutions. Policy solutions usually fall somewhere between these two views.

Lack of Access

Even if sufficient income is available, consumers must have access to food to avoid malnutrition. Natural disasters, like floods or earthquakes, can cause hunger if food supply lines are cut off. Disasters and the coverage they receive readily command emergency relief, thus avoiding temporary hunger. Access to food supplies is more difficult for those living in remote areas, such as on Indian reservations. Lack of access has also been identified as a problem in the city ghettos, among the elderly, and for some children.

The question of access to food has been at the center of many food policy debates. Congressional inquiries from time to time have focused attention on the exodus of grocery stores from the inner city. The issue was hotly debated when the food stamp program replaced the direct government distribution of surplus foods.

Lack of Knowledge

Educational level has been identified as one of the most important factors influencing food choice. Data from the Household Food Consumption Surveys indicate that, on average, the highly educated spend more and buy a better mix of food per person. After a thorough analysis of the data, one researcher concluded: "Regardless of the amount of money spent per person for food, among households with less education, there was a larger proportion with poor diets . . . the percent of poor diets increased as education decreased."[60] Therefore, education has been at the center of public policy debates about food. Nutrition education programs have been widely supported because the data are so convincing.

Food Assistance Policy and Program Alternatives

Like the commodity programs, the national commitment to eliminate hunger is an outgrowth of the Great Depression. The early food assistance programs were designed to dispose of surpluses—they were producer oriented. Beginning in the 1960s, the emphasis shifted from farm income support to the elimination of domestic hunger and improvements in nutritional health within the general population and among target groups such as school children and pregnant women. The program emphasis is very likely to shift some over time, but the policy emphasis on improving nutrition is likely to continue. The reality is that food assistance programs are a major part of the USDA budget—64 percent in 1995. The focal point of the debate centers around the options of no food assistance, food distribution, food stamps, and welfare reform.

No Federal Public Assistance

The free market distributes food and all other goods and services on the basis of ability to pay. Those with adequate incomes are able to make the needed purchases. Those with less income are not capable of purchasing as much. The poor become dependent on their families and on private organizations such as churches and the Salvation Army for assistance and care. Families of the poor are also fre-

---

[60]T. K. Meyers, "Can a Case Be Made for Nutrition Education?," address at the Third International Congress, Food Science and Technology, Washington, D.C., August 9, 1970. For more recent confirmation of the relation between education and nutrition, see Jean-Paul Chavas and K. O. Keplinger, "Impact of Domestic Food Programs on Nutrient Intake of Low Income Persons in the United States," *South. J. Agric. Econ.* (July 1983), pp. 155-162.

quently poor. Dependence on private organizations does not spread the burden of unemployment and low pay over the general public. It is precisely this result that has led, over time, to the development of public food assistance programs. *Markets allocate food based on the ability to pay; the nutritional needs of people depend primarily on biological factors. Private relief operates on a highly imperfect basis.*

Publicly financed food assistance programs have generally lacked the support of free-market advocates. They have argued that such programs tend to reward the lazy, unproductive, and fertile. There is an element of truth in that charge. However, certain members of all societies have limited resources through little or no fault of their own, at least for a time. Examples include the involuntarily unemployed (the steel worker whose plant is closed down because its technology is obsolete), those physically or mentally unable to work, the elderly, and children.

Adequate nutrition is critical to breaking the cycle of poverty. Neither adults nor children are able to perform up to their physical or mental capabilities when they do not receive nutritionally adequate diets. An adequate and balanced distribution of the food supply has also been an important element in political stability in almost all organized societies. For all these reasons, it is difficult to provide even a single example of a country that relies exclusively on free-market allocation of food.

Food Distribution Programs

The Federal Surplus Relief Corporation (FSRC) was organized in October 1933. It had two primary purposes:

- To relieve the national economic emergency in agriculture by expanding markets for farm products
- To relieve the hardship and suffering caused by unemployment by purchasing, handling, storing, and processing surplus agricultural commodities

This government agency, in cooperation with the Agriculture Adjustment Administration, purchased surplus agricultural commodities for direct distribution to the unemployed and their families.

In 1935, the FSRC became the Federal Surplus Commodities Corporation. The name change reflected a change in program emphasis from general help for the unemployed to specific help for the farm sector through surplus disposal and the encouragement of domestic consumption. This change in program thrust was accompanied by an amendment to the Agricultural Adjustment Act of 1933 that for 30 years would serve as the cornerstone for funding food assistance programs—Section 32.

Section 32 appropriates 30 percent of the gross receipts from duties collected under the customs laws (tariffs) for exclusive use by the secretary of agriculture to encourage exports and the domestic consumption of "surplus" commodities. Surplus food distribution programs were developed to provide commodities such as butter, cheese, flour, cornmeal, and dry beans directly to the poor and school lunch programs.

The most serious problem with direct food distribution was the lack of choice on the specific commodities received. As a result, waste was noticeable. In addition, there were complaints that donated foods were displacing regular market purchases. By the late 1930s, the need for a new system was clear.

Yet food distribution programs continue. One explanation lies in the existence of surpluses. Quite clearly, this was an origin of the Temporary Emergency Food Assistance Program (TEFAP), which was enacted with the concurrent events of the Physicians' Task Force and substantial dairy product surpluses. Another explanation lies in the compatibility of direct distribution with other programs including Indian reservation nutrition, school lunch, summer food service, and elderly feeding programs.

Food Stamps

A novel food stamp idea grew from this dissatisfaction with direct food distribution. The plan generated substantial food industry support because it involved the use of commercial trade channels for surplus food distribution. Needy families would be authorized to purchase orange stamps approximately equal to their average food expenditures. These stamps could then be used to purchase, at regular prices, any type of food. For each dollar of orange stamps purchased, households would receive 50 cents worth of blue bonus stamps. Blue stamps could be used only for those foods declared in surplus by the secretary of agriculture.

The food stamp plan actually had no statutory authority of its own but was rationalized as a surplus disposal method under Section 32. This fact stands as a constant reminder of the wide-ranging flexibility to make policy used by the executive branch during the Roosevelt administration.

The early food stamp plan received widespread support. Both the Democratic and Republican party platforms called for continuation of the program in 1940. The food distribution industry supported the program with enthusiasm. In the years that followed, the program was put into effect in nearly half of the counties in the country. These counties contained almost two-thirds of the population of the United States, according to the 1940 census.

Secretary of Agriculture Claude Wickard ended the program in March 1943 by stating that the conditions that had brought it into existence—unmarketable food surpluses and widespread unemployment—no longer existed. The program ended largely without fanfare; the nation was preoccupied with war. But the seeds had been sown for later development of a nationwide food assistance effort based primarily on human need rather than on farm income support.

Food assistance program expansion.   After World War II, food assistance programs once again emphasized direct distribution of commodities that were in surplus. Section 32 provided a continuing source of funding. But in a land of abundance, some still went without adequate food.

President Kennedy, in his first executive order, issued a command to expand domestic food distribution. The executive order did not mention food stamps. But in early February 1961, he announced that food stamp pilot programs were to be initiated. This decision was to become the forerunner of the food stamp program, which was made part of permanent legislation in 1964.

Child nutrition programs were also expanded early in the decade. In 1961, the National School Lunch Act was amended to include authorizations for special assistance cash subsidies to the school lunch programs. The school breakfast program and broader nutrition aid for children from needy families were initiated in 1966.

Rapid expansion of food assistance efforts took place during the 1970s. This expansion was an outgrowth of a national effort to overcome poverty, which was given considerable importance by the widely publicized study entitled Hunger USA.[61] The expansion of food assistance programs had the following characteristics:

- Total food assistance spending increased from $1 billion in 1969 to nearly $19 billion in 1983, leveled out through 1989, and then increased again through 1996 to more than $37 billion.

- The number of persons receiving some sort of food assistance more than tripled.

- Targeted programs for pregnant women and infant children, the elderly, and children in day care centers, among others, were established.

- The primary emphasis shifted from essentially surplus disposal to mostly income supplement and improving the nutritional health of the nation's low-income people.

The modern food stamp program.   Since the early 1970s, the food stamp program has been America's primary food assistance program. In the 1980s, food stamp expenditures accounted for nearly 70 percent of all food assistance spending.

The food stamp program, like its predecessor, was designed to provide low-income households with the food-buying income necessary to purchase more nutritious diets through regular market channels. The total value of the monthly food stamp allotment is based on three factors: food costs, income, and family size. The basic guide used to allocate food stamps is that households should not have to spend more than 30 percent of their income to purchase a nutritionally adequate diet. Suppose, for example, that a family of four earns $750 per month and that USDA calculates that a low-cost nutritious diet (referred to as the thrifty food plan) would cost that household about $70 per week ($280 per month). That

---

[61]Citizens Board of Inquiry, *Hunger USA.*

household would receive about $55 per month in food stamps ($750 × 0.30 = $225; $250 − $225 = $55). A household that earns no income would receive the cost of the thrifty food plan (published monthly by USDA).

In 1996, participation in the food stamp program averaged 25.8 million people—down from a peak of 27.5 million in 1994. Benefits per recipient averaged about $73 per month.[62]

The food stamp program has been much debated. The issues that form the basis for the controversy relate primarily to the values and beliefs of the American middle class. Regardless of the evidence, many Americans are unwilling to believe that as many as 1 in 10 needs or deserves public food assistance. The following list includes the questions most frequently debated:

- **How should program eligibility be determined?** (Who deserves assistance?) Most people accept the fact of poverty in America. Whether all households with less than a poverty income deserve food assistance is another matter. Some suggest that work should be a condition for receiving stamps. But many recipients cannot work because they have to care for their children.

- **How much assistance is adequate?** The thrifty food plan, based on recommended daily allowances (RDA), serves as the basis for how much assistance should be provided. This plan is the least costly of four family food plans developed by USDA. It specifies the quantity of foods in 15 different food groups that families with different-age children might be expected to use to meet the recommended dietary allowances.[63]

- **Should there be a work requirement to receive food stamps?** One of the main features of the 1996 welfare reform package was a requirement that adult food stamp recipients under age 50 who have no dependents have three months to find at least 20 hours of work per week or lose their food stamps. Those who are jobless for more than 3 of 36 months likewise lose benefits. Even before this policy was implemented in December 1996, states began to seek exemptions for high unemployment counties. Even Republican congresspersons began to wonder if they were too tough on the poor.

- **Should food stamps be used to purchase items other than nutritious foods?** Most persons defend with a vengeance their right to make product purchases without government interference. Still, the most frequent criticisms of the food stamp program are that recipient households are seen purchasing foods that others cannot afford and that money that would have been spent on food is used to purchase cigarettes and liquor instead of more nutritious food items. The available data on food stamp use do not support

---

[62]Food and Consumer Service, *Some Food Stamp Facts*, internet document (Washington, D.C.: December 1996).

[63]Interestingly, the thrifty food plan was developed by USDA on court order after a finding that its predecessor, the economy food plan, did not provide sufficient daily allowances to maintain adequate diets and did not discriminate sufficiently among the sexes and ages of family members.

those criticisms, but such evidence has not reduced the popularity of the claims.[64]

■ **In what form should food stamp assistance be provided?** Until 1993 all food stamp assistance was provided in the form of stamps, which often became a substitute for money. Fraud involving food stamps had become a significant problem. A substitute for stamps is a magnetic/electronic strip on a plastic eligibility card, comparable to a credit card or ATM card. The food stamp card would operate like a bank debit card when used at a supermarket. Each food stamp recipient would have a card with a balance that would be debited at the time of each purchase. The items eligible to be purchased with food stamps would be identified in the supermarket UPC code with the price of the product.

■ **Do food stamp recipients have better nutrition?** In terms of the nutrients that the recipients are most likely to be deficient in, food stamps have been found to result in a statistically significant, but small, improvement in nutrition.[65] However, there is also evidence that as income increases, the effectiveness of food stamps in improving nutrition diminishes to zero. Food stamp recipients having a college education spend no more on food than nonrecipients having comparable education and income.[66] How can this be if they are on food stamps? *These recipients simply use food stamps in place of income that would have been spent on food. Food stamps thus free up income that can be used for nonfood purposes. In other words, giving food stamps, as opposed to giving cash, does not mean that either more food will be purchased or that the nutrition of the recipients will be improved.* Therefore, it is extremely difficult to restrict the use of food stamps or the use of the income freed up by food stamps. Likewise, the food distribution option frees up income that can be used by the consumer to purchase other items.

■ **Do food stamps benefit farmers?** To the extent that food stamps increase consumption, farmers benefit from food stamps. Research results suggest that increases in consumption range from 26 to 50 cents for each dollar of food stamp expenditures.[67] Most of increased consumption would go to producers of cereal grains, beef, pork, dairy products, and poultry.

---

[64]Sylvia Lane, "Poverty, Food Selection and Human Nutrition," in *Agricultural-Food Policy Review*, AFPR-2 (Washington, D.C.: ESCS, USDA, September 1978), pp. 39–44; and Sylvia Lane, "Food Distribution and Food Stamp Program Effects on Nutritional Achievement of Low Income Households in Kern County, California," *Program Evaluation: Completed Studies* (Washington, D.C.: FNS, USDA, April 1977).

[65]Chavas and Keplinger, "Impact of Domestic Food Programs," p. 162; and J. Allen and K. Godson, "Food Consumption and Nutritional Status of Low Income Households," in *National Food Review* (Washington, D.C.: ERS, USDA, Spring 1985), pp. 27–31.

[66]Jean-Paul Chavas and M. L. Young, "Effects of the Food Stamp Program on Food Consumption in the Southern United States," *South. J. Agric. Econ.* (July 1982), p. 136.

[67]Chavas and Keplinger, "Impact of Domestic Food Programs"; Chavas and Young, "Effects of the Food Stamp Program"; William T. Boehm and Paul E. Nelson, "Food Expenditure Consequences of Welfare Reform," *Agricultural-Food Policy Review* (Washington, D.C.: ESCS, USDA, September 1978), p. 48; and James A. Zellner and Rosanna Mentzer Morrison, *How Do Government Programs and Policies Influence Consumer's Food Choices?* Agric. Inf. Bul. 553 (Washington, D.C.: ERS, USDA, November 1988), p. 1.

These conclusions lead to mixed emotions about the merits of the food stamp program as opposed to straight cash welfare payments. Farmers benefit, but they also question whether welfare programs foster a lack of individual initiative. Perhaps the greatest farmer benefit from food programs is not the increased consumption but the ability to bargain politically (horse trade) for farm programs. As noted in Chapters 3 and 4, farm bills are also food bills. This reality brings the farmer and urban interests together.

School Lunch Programs

Surplus disposal activities of the Federal Surplus Commodities Corporation, together with appropriated monies from Section 32, also helped give rise to federal child-feeding programs. At first, surplus commodities were dispensed using schools as the distribution points. This led to the spread of school lunch programs nationwide. In 1937, more than 3,800 schools received commodities for lunch programs serving 342,000 children daily. Five years later, 79,000 schools were serving surplus commodities to 5.3 million children.[68] But, as with the food stamp plan, food donations dropped sharply from 1942 to 1944, raising serious questions about the long-term stability of the school meal programs.

Domestic food assistance to the nation's school children was formally established when the National School Lunch Act was passed in 1946. As stated in the authorizing legislation, the objectives of the National School Lunch Program (NSLP) were to "safeguard the health and well-being of the Nation's children and to encourage the domestic consumption of nutritious agricultural commodities and other food." To do this, the federal government encouraged and assisted public and nonprofit private schools of high school grade and under to serve well-balanced lunches to children. This assistance has, over the years, included the following:

- A basic cash and donated food subsidy for all lunches; children from low-income families received an added subsidy. Initially, the emphasis was on foods purchased under Section 32 but, over time, cash subsidies became more important.
- Cash subsidies for school breakfast provided mostly to children of lower income families at reduced prices.
- Funds to partially reimburse states for undertaking the added administrative activities.
- Funds to help schools acquire food service equipment (discontinued in 1981).
- Funds for nutrition education programs and special development projects.
- Funds for expansion of the program to include breakfast food for eligible students.

---

[68]Agricultural Statistics (Washington, D.C.: USDA, 1952), p. 392.

From 1947 to 1996, federal expenditures for the NSLP increased from less than $100 million to more than $8.0 billion (see Figure 15.7). Over time, the federal government has been paying for an increasingly larger share of the total cost of the program. In addition, an increasing share of school lunches has been provided either free or at a reduced price. Free and reduced-price lunches are factors that have sharply increased the cost of the school lunch program. About 50 percent of students receive either free or reduced-price lunches.[69] In 1982 the Reagan administration proposed major cuts in the school lunch program. The stated objective of these cuts was to reduce subsidies to medium- and high-income families. These efforts resulted in stable school lunch expenditures only throughout the 1980s.

The overall attitude toward the school lunch program appears to be considerably more favorable than that toward the food stamp program. Yet several issues exist:

- The most controversial issue involves the nutritional quality of school lunch and breakfast. The influential public interest group, Public Voice for Food and Health Policy, made improvement of the nutritional quality of the school lunch program one of its major initiatives. The emphasis in its study involved the need to reduce fat.[70] Subsequently, it was made clear by USDA that the dietary guidelines would be implemented into the school lunch program by the end of 1994.[71] When Madigan's term ended abruptly with the election of President Clinton in November 1992, the president of Public Voice, Ellen Haas, took over as the undersecretary in charge of food assistance programs.

  In 1993, Senator Patrick Leahy (Dem.–Vt.) took the initiative to make the government dietary guidelines for children mandatory for federal feeding programs, including the school lunch program.[72] Simultaneously, Secretary Espy released a critical report indicating that 1 percent of the schools participating in the lunch program followed the dietary guidelines.[73] According to the report, school lunches exceeded the guidelines for fat by more than 25 percent, saturated fat by 50 percent, and sodium by nearly 100 percent. Moreover, those who ate school lunches were found to be consuming significantly more fat than the children who got their lunches elsewhere.[74]

  Haas took on the challenge of reducing fat in school lunch with a vengeance as more fruits and vegetables appeared on the lunch menu. She even proposed purchasing only organically produced food for the school lunch program.

[69]M. Matsumoto, "Domestic Food Assistance Costs," 1991.

[70]Jeffrey Shotland, *What's For Lunch?* (Washington, D.C.: Public Voice for Food and Health Policy, September 1989).

[71]Edward Madigan, "Remarks to the American School Food Service Association," *Selected Speeches and News Releases*(Washington, D.C.: USDA, July 23–29, 1992), p. 2.

[72]*Food and Fiber Letter* (Washington, D.C.: Sparks Companies, October 18, 1993), p. 2.

[73]*Food and Fiber Letter* (Washington, D.C.: Sparks Companies, November 1, 1993), p. 4.

[74]Mathematica Policy Research, Inc., *The School Nutrition Dietary Assessment Study* (Washington, D.C.: Nutrition Research Education Service/USDA, 1993).

■ Pressure continues from producer interests to utilize the school lunch program for surplus commodity disposal. School lunch accounts for more than 4 percent of the demand for milk. During the mid-1980s, school lunch was used as an outlet for surplus cheese and beef purchased to reduce the price-depressing effect of the dairy termination program. When the surplus disappeared, the commodities dried up. Again in 1996 when milk and beef prices fell, USDA purchases for school lunch and other nutrition programs were initiated. Schools resist the pressure to use the program for surplus disposal, which dictates what they serve in lunches. Many would rather receive all cash with the flexibility to buy whatever foods meet the specifications of their dieticians. This, incidentally, makes it more difficult for USDA to control the content and source of food purchased.

■ The minimum nutrient requirements for the program have been specified by the USDA. These requirements have always been controversial and politically sensitive. A 1981 cost-cutting move to get ketchup classified as a serving of vegetable in the school lunch program was one of the factors resulting in the dismissal of the head of USDA's food assistance programs.

■ The acceptability of lunches varies widely. In the 1970s, an effort had been made to serve a larger proportion of foods that children like to eat, such as the "All-American Meal" of hamburger, french fries, tomatoes, lettuce, and milk. However, meal acceptability still remains highly variable. In addition, the "All-American Meal" failed to meet the nutritional standards set by Ellen Haas.

■ The extent of the federal subsidy in school lunch programs became a major issue in the early 1980s as the Reagan administration attempted to reduce costs. Shifting a higher proportion of the costs to the children raised questions as to how many would drop out of the program and thus receive inadequate nutrition. Because the evidence clearly indicates that school lunch improves nutrition,[75] support for the program continues to be particularly strong.

Women, Infant, Children Program

The Women, Infant, Children Supplemental Food Program (WIC) is targeted toward mothers with children who are already participants in other welfare programs. *WIC integrates health care, nutrition education, food distribution, and food stamps into a comprehensive health and nutrition program.* In the nutrition program, emphasis is placed on providing high-quality protein to pregnant and nursing mothers and young children.

---

[75]David W. Price, Donald A. West, Genevieve E. Scheier, and Dorothy Z. Price, "Food Delivery Programs and Other Factors Affecting Nutrient Intake of Children," *Am. J. Agric. Econ.* (November 1978), pp. 609–618.

WIC has been demonstrated to be one of the most successful food assistance programs.[76] A North Carolina study found increased birthweight of babies, reduced incidence of anemia, and improved nutrient intake by the participants.[77] A Harvard University study estimated that up to $3 in hospital costs are saved for every $1 spent on the prenatal component of WIC.[78] Chavas and Keplinger list crucial factors that contribute to the WIC program's success in improving nutrition:[79]

- Targeting the program to pregnant and breast-feeding women, and infants and children under four years old
- Identifying the nutritional deficiencies of participants
- Providing selected food items in addition to food stamps
- Combining nutrition education with food assistance

WIC does not have as good an image as its record of meeting its objectives would imply. The reason probably lies in the notion that the program fosters unwed mothers and a welfare-dependent lifestyle. It has also run counter to conservative political philosophies (see box).

Welfare Reform

Major questions about the form of public assistance have been raised since the 1970s. As the food programs grew in number and cost, so did the number of critics. Welfare rights advocates argued for fewer restrictions on expenditures so that recipients could make purchasing decisions like other citizens. Taxpayers complained that the programs were too costly, too complicated, and not sufficiently restrictive.

The food assistance programs have been a major part of welfare reform proposals. Several proposals have been made to cash-out all programs of special assistance for one cash payment. That is, the basic notion is that all welfare programs would be consolidated into a single cash assistance program. School lunch assistance would also be in the form of cash as opposed to the combination of cash and commodities. Such proposal are often referred to as "cashing out."

Welfare reform raises at least three important policy issues for agriculture:

- *A simple cash transfer system is less effective in enhancing farm income than programs that target the aid to food purchases directly.* When people are given the choice, the tendency is to spend 10 to 20 cents out of each additional dollar on food. If the food aid is direct, there is less choice. Farmers gain more for each dollar's worth of food stamps, although the gain is not as great as sometimes suggested.

---

[76]Carol Tucker Foreman, "Human Nutrition and Food Policy," in *Agricultural-Food Policy Review,* p. 19.

[77]Benjamin Senauer, "Food Programs and Nutritional Intake: What Evidence," in *Agricultural-Food Policy Review,* (Washington, D.C.: ESCS/USDA September 1978), p. 41.

[78]Michael Reese *et al.,* "Life Below the Poverty Line," *Newsweek* (April 5, 1982), p. 25.

[79]See footnote 66.

Laundering

Publication of the National WIC Evaluation was delayed because of changes in the study—including redesigning the study and replacing the principal investigator after two years, overly optimistic time frames, protracted USDA review times, and printing problems. USDA deleted the original concluding chapter and executive summaries and replaced them with its compendium of results because, officials told GAO, the research team's conclusions portrayed the WIC program more favorably than justified by the data.

   In contrast, the GAO found that the original executive summary used appropriate methodology, was accurately presented, and reported the study's main conclusions: that WIC improves the diet of pregnant women and children, adds to maternal weight gain, increases the use of prenatal care, and reduces preterm deliveries.

Source: General Accounting Office, *The National WIC Evaluation: Reporting and Follow-Up Issues,* GAO/RCED-90-3 (Washington, D.C.: U.S. Congress, December 1989).

- *Targeted food assistance programs are more efficient in improving the nutritional status of the recipient.* Taxpayers concerned about the nutritional status of the poor are able to ensure themselves of increased food intake if the programs restrict the recipients' choice. Cash-out debates in the Congress almost always include discussions that focus on whether children of the poor would really be helped if adult recipients used the cash to purchase alcohol and buy lottery tickets. That is largely an uncontrollable issue unless food programs are to be highly targeted and controlled, as in the case of WIC.
- *Food assistance programs have generally been part of major farm legislation and appropriations.* They provide a reason for urban congresspersons to vote for farm programs and for rural congresspersons to vote for welfare programs. Urban support for traditional farm programs could be considerably reduced without food assistance programs. USDA would be reduced to about one-third its size.

Agreement on the political significance of moving the food assistance programs out of USDA is far from unanimous. One point of view suggests a transfer of food assistance out of USDA because appropriations for food assistance programs are competitive with farm program appropriations. Others argue that to remove those programs from USDA would leave the department as a farm advocacy organization with a very small budget, which would ultimately erode the broad base of support required for USDA to remain a separate cabinet-level department.

   When the welfare reform issue is further considered by the Congress, the position of the agriculture establishment will be most interesting. Each organization will have to evaluate the impact not only on its membership but also on its effectiveness as an organization. The perceptiveness of farm organization leaders in weighing these impacts could ultimately also affect their positions within the organizations!

THE PERVASIVENESS OF FOOD POLICY ISSUES

Everyone can and does get excited about food policy—farmers, processors, retailers, consumers, and policymakers. Everyone must decide what foods to consume and in what quantity they should be consumed. No one debates this point. The crucial question is the role that government should play in attempting to influence what people eat. It is clearly having a greater influence—much greater than many realize. The role of government in improving nutrition will be around long after farm programs are abandoned—if that ever happens!

ADDITIONAL READINGS

1. The most important contribution to an understanding of food and nutrition issues is contained in Ben Senauer, Elaine Asp, and Jean Kinsey, *Food Trends and the Changing Consumer* (Washington, D.C.: Eagan Press, 1991). It is required reading for any student of the food industry and food policy.

2. The best current source of information on food policy issues and programs is the *National Food Review*, a publication issued three times a year by ERS, USDA, Washington, D.C.

3. A series of articles in *Agricultural-Food Policy Review*, AFPR-2 (Washington, D.C.: ESCS, USDA, September 1978), provides a good review of the issues in food assistance programs.

4. One of the most comprehensive statements on food safety policy was given by Carol Foreman before the Outlook Committee on Agriculture, U.S. House of Representatives, on September 16, 1980.

5. The most interesting and informative statement on pesticide policy is by Carol M. Browner, Richard Rominger, and David A. Kessler, Testimony before the Subcommittee on Department Operations and Nutrition, Committee on Agriculture (Washington, D.C.: U.S. House of Representatives, September 22, 1993).

6. F. J. Cronin *et al.*, "Developing a Food Guidance System to Supplement Dietary Guidelines," *J. Nutr. Educ.* (1987), pp. 281–301, is an excellent article on the technical development of the second edition of the dietary goals and guidelines.

7. John M. Antle, *Choice and Efficiency in Food Safety Policy* (Washington, D.C.: AEI Press, 1995) is an excellent reference on the economics of food safety.

16 RURAL DEVELOPMENT POLICY

I'd rather wake up in the middle of nowhere than in any city on the earth.
—*Steve McQueen*

Rural development policy[1] involves government programs that focus on the living and employment conditions of rural America. From the signing of the Declaration of Independence through the Great Depression, much of the major legislation enacted by Congress focused on rural areas because of the contribution of natural resources to overall productivity and the constraints placed on growth by distance. This legislation was designed to achieve settlement and to build a transportation system and a communication infrastructure. It helped to build a land-grant university system that had special responsibilities in the discovery and application of technology to solve rural problems.

In almost every farm bill since the Great Depression, one or more titles (major sections) of the farm bills have dealt with rural development. Many of these titles were highly effective, such as those designed to make electricity available to farm families throughout the United States. Others suffered a fate of authorizations without adequate appropriations to satisfy the bills' objectives. That is, new initiatives designed to deal with particular rural problems were enacted into law, but appropriations required to implement them effectively did not follow. This reflects a lack of commitment and consensus on rural policy. The following erroneous assumptions about rural policy were among those made:

---

[1]Much of the content of this chapter is attributable to Dennis U. Fisher, who has devoted his professional career to researching and working with rural businesses and policymakers to help them learn about and understand rural problems.

- Rural problems could be solved by farm programs designed to raise farm income.
- Legislation developed and applied to urban areas could automatically also help rural areas.
- The cost of providing public services in rural areas was less than that in urban areas.

In contrast with farm policy, many of the policies and programs designed to help rural communities and businesses are not primarily federal. Most are state and local policies and programs. To the extent that the federal government is involved, programs are often shared with urban areas. The federal government's role, to an important extent, is as a facilitator of change; to provide new ideas and technical assistance to rural leaders.

This is the spirit in which the Fund for Rural America contained in the 1996 farm bill was created. This fund was created to develop practical new ideas for business development and for solving rural problems. Most of the fund will be spent on research and technical assistance in the contemporary higher risk agriculture and rural environment. Hopefully the initial $100 million of appropriations for fiscal year 1996–97 will be only a beginning. This amount of money is only a drop in the bucket in terms of the magnitude of the rural problem.

Although rural America has an image of being a highly desirable place to raise a family, living conditions and employment opportunities have tended to fall behind those of urban areas. As a result, the best educated and most mobile rural youth have migrated to urban America. This process of selective outmigration took its toll on rural America. The ability to address and solve rural problems persistently declined except in those areas adjacent to metropolitan centers that received the spillover benefits of urban growth.

In certain respects, the problems of rural America are similar to those of central cities: deteriorating infrastructure, losing businesses that relocate, and declining average income levels relative to those of the rest of the country. In other respects, rural problems are more complex than urban ones because of the long distances required to deliver services, a sparse population, and the related problems of providing equity in public services.

SCOPE OF RURAL DEVELOPMENT POLICY

To many, rural development policy focuses on infrastructure and business development. In other words, the policy is designed to attract industry and jobs to rural areas. Infrastructure and business development are important dimensions, but the scope of rural development policy is considerably broader. Equally important to people is human capital policy—rural education, rural health policy, and rural

poverty policy. This chapter utilizes this broader perspective on rural development policy.

RATIONALE FOR GOVERNMENT INTERVENTION

Over the years, many specific circumstances have resulted in government becoming
involved in rural areas. These circumstances fall into four general problem areas:

- **Access.** Historically, a primary function of rural programs has involved providing access to rural areas and linking urban areas. These policies provided farmers and businesses located in rural areas access to production
  resources and to markets; the development of transportation systems is the
  most obvious example. Improved transportation systems have not helped all
  rural communities, however. Being bypassed by an interstate highway could
  equally have caused a community's death.
- **Public goods and market failure.** The access rationale is closely related to the
  concepts of public goods and market failure. Without government intervention, certain goods and services would not be available in rural areas. Electricity
  and telephone service would not have been provided nearly so soon were it not
  for the Rural Electric Administration (REA). It may never have been profitable
  to provide electrical or telephone service to much of the rural West. In a modern context, rural communication systems need to be modernized to allow
  rural businesses to compete with their urban counterparts.
- **Equity.** Access, public goods, and market failure are related to issues of equity. To what extent does the government have an obligation to provide equity in service to rural residents? The equity issues cover a very wide range of
  concerns beyond transportation and communication. Rural health services,
  education, and social services lag those in urban areas in quality, quantity,
  and access in varying degrees.
- **Information.** Relatively speaking, rural America has always lagged in entering the information age, whether it is access to telephone service, cellular
  telephone service, cable television, or computer communication. Yet communication has become increasingly important to business success.
  Government has been pressured constantly to help rural areas maintain
  competitive communication technology.

STATUS OF RURAL AMERICA

One of the prerequisites for addressing rural problems is having a clear understanding of its demographics and the nature of its economy. Misconceptions of the
nature of the rural economy abound. There is even disagreement over the definition of rural. Throughout this chapter, **rural** means outside of metropolitan areas.

A *metropolitan area* is a county containing a city of 50,000 or more people, several small cities totaling 50,000 or more people, or a total area population of at least 100,000.[2] The terms *rural* and *nonmetro* are used synonymously in this chapter, just as are the terms *urban* and *metro*.

Swanson notes that two false assumptions have guided rural public policy[3]:

■ The first is a pervasive tendency to associate rural economies and community well-being with farming.

■ The second is that with the possible exception of farmers, rural people are faring relatively well.

Farm/Rural

Farm organizations, the media, and even the agricultural economics profession foster the perception that agriculture and rural are synonymous. The visual perception as one drives across rural America is that most of the area is devoted to agriculture. However, the importance of agriculture in rural areas varies widely and depends on how agriculture is defined.

Promoters of *agriculture*—including farm organizations and those who counsel undergraduate students to major in agriculture—define it very broadly to include retail food stores and perhaps Wal-Mart. Using a more narrow, and perhaps more reasonable definition, farming, agricultural services, forestry, and fisheries account for approximately 10 percent of the rural jobs. Manufacturing, retail trade services, and government are all larger rural employers than is agriculture.[4]

Agriculturally dependent counties are defined as those counties having 20 percent or more of their income coming from agriculture over a five-year period. Only 20 percent of all rural counties have an economic base that depends on agriculture and only 7 percent of the total U.S. population lives in those counties.[5] Other sectors accounted for a higher percentage of income than agriculture in a number of these countries. Yet if your home happens to be in one of these counties, you perceive agriculture as being quite important.

Relative to their employment earnings, rural residents depend more on transfer payments than do urban residents. That is, although in absolute dollar terms per capita transfers in rural and urban areas are about the same, they account for a larger share of rural income because rural incomes are lower.[6] **Transfer payments** are defined *as cash or goods for which no work is currently performed*. Examples include

[2]*Rural Conditions and Trends* (Washington, D.C.: ERS/USDA, Spring 1990), p. 23.

[3]Louis E. Swanson, "The Rural Development Dilemma," *Resources* (Washington, D.C.: Resources for the Future, Summer 1989), pp. 14–16.

[4]*Rural Conditions and Trends*, p. 11.

[5]Kenneth Deavers, *Rural Vision—Rural Reality* (Madison, Wis.: University of Wisconsin, Department of Agricultural Economics, April 20, 1990), p. 3.

[6]Robert A. Hoppe and Linda M. Ghelfi, "Nonmetro Areas Depend on Transfers," in *Rural Development Perspectives* (Washington, D.C.: ERS/USDA, May 1990), pp. 22–24.

social security, workers' compensation, Medicare, Aid to Families with Dependent Children, unemployment insurance, and veterans programs. Farm subsidies are not considered transfer payments because they are classified as income from current farming activities.

The relatively stable nature of transfer payments adds an element of stability to the rural economy, thus offsetting some of the instability of agriculture. Transfer payments, dividends, interest, and rent are particularly important in rural areas that are havens for retirement. An amazing 475 of these retirement destination counties—three-fourths of the number of farm-dependent counties—exist in rural areas. Retirees bring with them not only the stability of transfer payments but also, frequently, a nest egg of income from private sector retirement programs and from the accumulation of net worth.

Rural People Are Not Doing Well

The second common misconception is that rural people, other than farmers, are doing quite well. As a general rule, *farmers are doing better* than the rest of the rural population. This is true for three reasons:

- Approximately 64 percent of commercial farmers' net income (88 percent of all farms) comes from off-farm jobs (see Table 9.1). In addition to farmers themselves having off-farm jobs, nearly as many farm wives work as urban wives.
- Farm households' average net worth is about twice that of all U.S. households. Farming requires substantial assets. Farm debt, generally, is low relative to the assets.
- Rural unemployment is higher than nonfarm unemployment, a phenomenon that began in the 1980s. Prior to the 1980s, rural unemployment was generally lower than urban unemployment, except in the South.

Having said this, there is great diversity in rural America. For example, in 1993, 14 percent of all U.S. households had incomes under $10,000 and 29 percent had incomes over $50,000. The equivalent for farm households was 19 percent with under $10,000 incomes and 25 percent with over $50,000 incomes.[7] As usual, generalizations based on averages are dangerous.

Rural earnings per job have lagged urban earnings because nonmetro jobs generally require less skill. In every category of employment, nonmetro workers earn less. In the manufacturing sectors, Deavers attributes this phenomenon to the relative absence of complex manufacturing activities in rural areas.[8]

[7]Robert A. Hoppe, *Structural and Financial Characteristics of U.S. Farms, 1993,* Agric. Int. Bul. 728 (Washington, D.C.: ERS/USDA, October 1996).
[8]Deavers, *Rural Vision,* p. 26.

As a result of lower earnings, the incidence of rural poverty is higher than that of urban poverty, although not quite as high as central city poverty. That is, while urban poverty has run about 11 to 12 percent of the households, rural poverty has tended to run in the 15 to 16 percent range, and central city poverty has been running about 17 percent.[9] Rural poverty tends to be particularly high for blacks, native Americans, other minority groups, families headed by women, and unrelated individuals living together.

The lack of job opportunities and the declining economic conditions have led to persistent outmigration from rural America. The nonmetro population growth rate is about one-half of the metro growth rate. Approximately half of the rural counties are losing population. These counties tend to be centered in the Great Plains and the Corn Belt—agriculture's heartland (Figure 16.1). Even worse, a higher percentage of the best educated leave. The disconcerting fact is that while individuals having less than a high school education are migrating into rural America, those with some college, a college degree, or more education are leaving. This reality does not bode well for the future of rural America!

RURAL DEVELOPMENT ISSUES

Some very important and far-reaching issues arise from the economic and demographic trends discussed. The solutions to the following problems have much to do with the appropriate course for future rural development policy.[10]

Population Concentration versus Dispersion

Rural counties represent 90 percent of the U.S. land area and about 25 percent of the population. Therefore, 75 percent of the population live on 10 percent of the land area. It is hardly debatable that the major U.S. cities have problems. Many of those problems result from congestion—too many people living in too small an area.

It is not the character of the U.S. economic and political system to tell people where to live. But the United States once had population distribution policies including significant economic incentives encouraging people to locate in specific areas. The most important of these was the Homestead Act and related legislation used to encourage settlement of the West. Out of this legislation arose the development of railroads, highways, farm-to-market roads, mail service, rural electrification, and rural telephones.

As the West was settled, the focus of these policies changed from encouraging people to live in specific areas to providing services to people where they live.

---

[9]*Rural Conditions and Trends*, p. 17.

[10]Much of the content of this section represents the conclusions of a set of regional rural development workshops held with rural leaders to look for solutions to complex rural problems. They are also published in and expanded on in Ronald D. Knutson and Dennis U. Fisher, *Focus on the Future: Options* in Sue H. Jones, ed., *Developing a New National Rural Policy* (College Station, Tex.: Texas A&M University and Aspen Institute, Fall 1988).

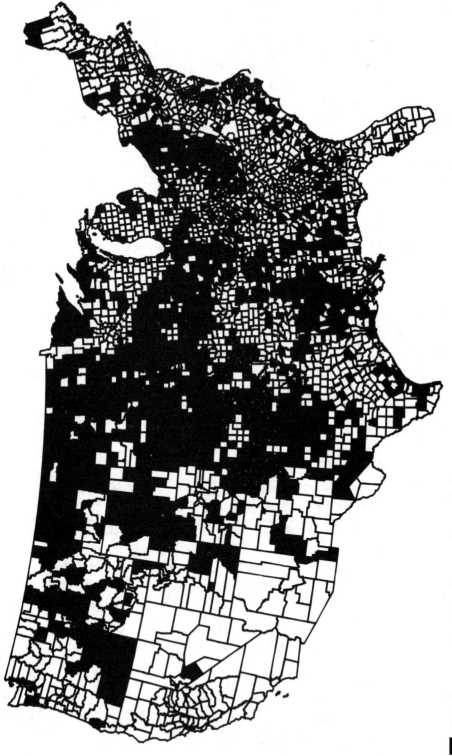

■ Counties experienced a decrease in population

□ Counties experienced an increase or no change in population

FIGURE 16.1 Counties with a falling population, 1980–1991.

By their very nature, policies that encourage location of infrastructure in specific locations have a *de facto* population distribution objective even if proponents of such policies do not recognize it.

For much of U.S. history since its settlement, educated rural youth have had economic incentives to move from rural areas. Whether that persistent migration is good for the United States as a whole is at issue. To what degree should the U.S. population become concentrated in a small area? Should anything be done to reverse the trend? If so, what should be done?

Diseconomies of Distance

Economic theory often ignores the costs of distance, information, communication, and population dispersity with the same quality of social services. Twentieth-century policymakers recognized these extra costs involved in establishing massive programs to provide electricity, telephone service, and roads to rural areas. This same recognition has not been applied, however, to social services. Often, the assumption has been that it costs less to deliver social services in rural areas. It also now appears to be assumed that the modernization and maintenance of roads and communication systems in rural areas cost no more per capita than in urban areas. Since the advent of the interstate highway systems, bridges in rural areas have been allowed to deteriorate, and many rural communities are unable to receive fax or computerized electronic exchange of information because of the lack of public investments in new communication technology.

Outmigration aggravates these problems. Many of the costs involved in providing public services are fixed. Even the cost to maintain a doctor in a rural area is fixed because a certain minimum salary must be provided to attract his or her services. Either that, or rural residents must pay the cost of transportation to obtain medical services at an urban center. Likewise, many of the costs of quality education are fixed. But as population and, therefore, the number of taxpayers decline, the ability to pay for those services decreases.

Equal Access to Public Services

For decades, many rural residents accepted the reality that living in rural America meant sacrificing access to public services that were readily available in the cities. Party-line telephone systems, poor-quality schools, and the lack of ready access to quality medical care were accepted as one of the costs of rural living. Economic decline, outmigration, and increased awareness of urban amenities have made rural residents less willing to pay the price of lower quality public services. Those who have the resources and have alternative opportunities continue to leave. In the process, rural representation in the Congress and state legislatures declines, making public solutions to rural problems less politically feasible. With an agriculture committee, but no rural committee, the rural population has depended on the agriculture committees for its programs. Legitimate danger exists that policy and program inequities may be creating a permanently disadvantaged rural society, a situation that may be perilously close at hand in many rural areas.

Programs Targeting Rural Areas

Few government programs, aside from farm subsidies, are specifically targeted, designed, or directed to solve rural problems. Instead, federal programs designed to aid economic development are national in scope with no rural–urban differentiation. Because of the uniqueness of rural areas, at least in terms of distance traveled and sparse population, national programs may not be as effective in solving rural problems as they are in addressing comparable urban issues. For example, in urban America, individuals receiving job training can generally find employment in the city where they live. This may not be the case in rural America, however, because the retrained worker may have to incur substantial moving costs to find a job. Likewise, rural schools have unique problems in scale of operation, the distance students need to travel to get to school, and the ability to offer enrichment courses in areas such as foreign languages and the basic sciences.

Targeting does not necessarily mean an increase in government expenditures, but it does require the recognition that rural problems are unique and cannot be forced into the bureaucratic blueprint for urban programs. It also means that public employees must be willing to travel to, and often live in, rural areas to provide the same quality of service as in urban areas. In other instances, public investments may be required, for example, to restructure rural health delivery systems or provide distance learning concepts to rural schools.

Safety Net for People or Places

Decisions must be made regarding the objectives of rural policy. One of the most basic decisions involves whether rural policy is designed to help rural people adjust to change or to preserve rural communities. What is best for rural people may not be best for rural communities. For example, with contemporary population dispersion and infrastructure policies, some communities may have little or no chance to attract new business ventures. The interests of some of their residents may be served best by retraining and possibly relocating to larger rural community business and service centers offering better employment opportunities, yet such retraining may not be in the best long-run interest of community survival.

The trade-offs between helping people versus places is difficult. Communities must make conscious decisions to find their niche. Every rural community cannot expect to become a growth center. The result is a need to help communities plan for the future, find their niche, and make the most of their situation. Likewise, strong support for education suggests a basic belief in helping youth achieve their capacity even though the consequence may be increased migration from rural areas. The alternative may be for government to support or perpetuate business that exploits low-cost rural labor. In certain respects, low-cost government loans to rural businesses have had this effect, as have low-cost loans to farmers.

Subsidies to Business with Locational Disadvantages

Distance may imply locational advantage or disadvantage. A retail business located in a rural community tends to have a locational advantage in serving its population. On the other hand, a business located in a rural community but designed to serve a broader national or international market may have a substantial locational disadvantage. Potential rural businesses may seek special concessions, subsidies, or protection to locate in rural areas and to continue their operations. The incentives for such subsidies are often substantial, and they can result in permanent subsidies to inefficient operations. In some cases, communities gave so many subsidies to attract business that, even though they were successful in recruiting these businesses, they were eventually worse off. The same has been true for competition for industry among states.

Physical versus Social Infrastructure

Substantial conflict exists between the advocates of greater emphasis on physical infrastructure and job creation versus those who would place more emphasis on social infrastructure designed to improve the capacity and flexibility of people. In the past, much rural development emphasis has been placed on physical infrastructure, that is, improvements in rural road systems, water treatment facilities, industrial parks, and retraining. A politician can more readily point to success in obtaining a grant to build a physical facility or road than to design a new social program.

Solving rural problems requires concern for the long-run needs of people. Strong rural education systems have the potential to make the local population more flexible, thus attracting business and reducing the need for publicly supported retraining programs. The dilemma is that rural communities end up paying for the cost of education, but their brightest and best educated citizens tend to move to cities, which realize the benefits of their contributions. Deavers concludes that this implies the need for a larger share of support for rural education from the federal government.[11] The long-run key, of course, is to improve the opportunities and quality of rural life enough for these citizens to earn advanced degrees and come back to contribute to rural growth and development.

DEFINITION OF RURAL DEVELOPMENT POLICY

As the preceding discussion implies, there is substantial debate over the appropriate scope of rural development policy. Traditionally, federal and state policies

---

[11]Deavers, *Rural Vision*, p. 16.

have emphasized infrastructure and technical assistance in planning for development. Social needs have tended to be handled by massive federal programs utilizing the same basic approach in rural and urban areas. While originally a local responsibility, schools have, in some states, received substantially increased state support. However, federal support is insignificant.

Pressures for federal rural development policies to break out of the infrastructure mode are increasing. Yet significant political and budget constraints fortified by organizational momentum support the infrastructure mode. Clearly, the problems of rural America include the lack of consistent quality education and social services as well as deteriorating infrastructure.

POLITICS OF RURAL AMERICA

From a political perspective, a very logical question arises. Why is it that farmers, representing 2 percent of the population, have received more support in seeking political solutions to their problems than rural residents at 25 percent of the population? There are at least three answers to this question[12]:

- Rural interests are fragmented into six interest groups: infrastructure, health, business development, education, poverty, and agriculture. Of these, the infrastructure and agriculture lobbies are clearly dominant, quite well organized, and politically astute. Rural interest groups have been the most effective when they mutually support one another, such as those seeking to extend electrical and telephone services to rural areas. Rural health, education, and poverty can be bundled together as human resources, but these interests have not tended to react as a unit. Instead, they have tended to act independently, seek only local or state solutions, or identify with national program counterparts that serve both urban and rural areas. Of these three interest groups, rural health interests have had some success in organizing and removing some of the inequities through the Rural Health Caucus of the Congress. Rural education has a small program targeted to rural schools. Rural business development programs tend to be dispersed among several agencies, each with its own constituency. Business development interests have enhanced their success somewhat by riding on the coattails of the infrastructure lobby.
- Rural interests that have depended on national programs to solve their problems have invariably found themselves in competition with the virtually insatiable competitive demands from the vast majority of the population that live in urban areas. With the political alignment of the House of Representatives favoring urban areas, rural interests are at a decided disadvantage.[13] In addition, the largest number of the most accessible recipients

---

[12]This issue is discussed in Ronald D. Knutson and Dennis U. Fisher "Rural Development Policy: Fragmentation Moving Toward Consensus," *Choices* (2nd Quarter 1989), pp. 8–11.

[13]For a discussion of the shift in political power in Congress, see Chapter 3.

of the benefits of national programs are located in or near urban areas. Therefore, those located in the rural hinterlands have a very difficult time getting the attention of the delivery mechanism for national programs from which they might benefit. Moreover, because most of the recipients and programs are urban, administrative procedures and priorities tend to be developed to serve their needs, not considering the uniqueness of the rural constituents.

- The only congressional committees with a true rural constituency are the House and Senate agriculture committees. All of the other committees are dominated by senators and congresspersons with primarily urban concerns. Moreover, the jurisdiction of the agriculture committees is limited by the rules of the respective bodies. For example, the agriculture committees can clearly deal with farm subsidy issues, but they are very limited in their ability to handle a bill dealing with rural health, education, or poverty. As a result, the interests in rural health have formed a caucus designed to obtain bipartisan support for health issues of concern to rural areas. Interestingly, the rural health caucus has a broader constituency base than the rural caucus, which tends to be more closely tied to agriculture issues. However, even with such organizational initiatives, agriculture committee leadership tends to be leading the charge. What they can accomplish is limited by their votes, their seniority, and their ability to horse trade—which could result in a reduction of their influence for another agricultural or rural constituency.

Equally important, the agriculture lobby has not supported rural programs that do not have a clear benefit to farmers—such as the case for rural electric or telephone programs. This lack of agriculture support is particularly evident in the case of social programs such as poverty programs. Interestingly, it has also extended to rural education and rural business development. In the case of rural health, there is evidence of a somewhat broader base of general farm organization support. However, the more powerful commodity organizations seldom take a position on such social and human resource issues. Aside from their basically conservative character, such organizations may be concerned about the budget trade-offs that could potentially reduce their programs in return for expanded social programs.

Even in the case of targeting or tailoring general program provisions to rural areas, agricultural interests may be concerned about confrontations with urban interests that could backfire in creating urban opposition to farm programs. Yet the members of farm organizations must live with the consequences in terms of lower educational quality, declining health care services, and a declining rural economy. Alternatively, perhaps the farmers who live in rural areas do not perceive these social issues as being problems or as directly affecting them. They may subscribe to the notion that those living in rural areas must endure certain hardships. Alternatively, perhaps they do not view inequities in education and social services as problems. Even if this is true for farmers, there is evidence that it is not true for the majority of rural residents.[14]

RURAL DEVELOPMENT POLICY OPTIONS

Despite these fragmentation and organizational issues, rural interests have indicated great concern about the broad rural development policy agenda. This section briefly summarizes some of the major options and consequences for each general area.

Business and Job Development

The policy goal in business and job development is to increase income and employment in rural areas. Many look on this area of emphasis as being the most direct means to rural development. The general policy options for achieving this goal include the following:

- **Industrial recruiting** is designed to attract new or relocating businesses to rural areas. This is the policy option most favored and initiated by rural communities. Ironically, it also has the lowest probability of success because of vigorous competition for new industry and the advantages of industries locating near urban areas. The types of businesses that have tended to locate in rural areas are those related to agriculture, those desiring to take advantage of low-cost rural labor, and those having substantial externalities that particularly affect the air.

- **Grants and loans** may be utilized as a tool for industrial recruiting. The grants or outright gifts generally are in kind such as a building, water systems, water treatment, solid waste disposal, or site location in an industrial park. Tax breaks may also be given, and loans may involve interest subsidies. These incentives are designed to offset any of the perceived disadvantages of locating business in rural areas. The dangers are that the costs of these subsidies may be greater than the benefits, that the industry may be subsidy dependent (i.e., never able to operate without a subsidy or remaining only as long as the subsidy continues).

- **Business retention and expansion** strategies emphasize helping existing rural businesses compete and expand. These strategies involve making the local community business environment conducive to growth by working with local businesspersons. Local businesses are contacted regularly to determine their condition and future plans and obstacles to remaining in the community. Public and private groups and agencies from the city council to

---

[14]This value-laden statement is based on the results of four regional rural development workshops and subsequent state development workshops suggesting substantial rural population concern about these issues. See Knutson and Fisher, "Rural Development Policy."

the chamber of commerce are involved in removing local impediments and assisting businesspeople with difficulties outside their immediate control.

One of the major problems of rural businesses is that the consumers do not always buy their goods and services locally. Therefore, business retention and expansion strategies in part involve creating a shopping environment in which residents of the local community and surrounding area will buy locally. Programs such as Main Street, which are designed to improve the attractiveness of the community, have this as an objective. In addition, assistance may be provided to help management and/or employees acquire new skills designed to modernize operations.

- **Business incubators** involve an assortment of activities to foster new business ventures using local resources. The most common form of the incubator includes subsidized rent in a building designed to accommodate several retail, service, and/or light manufacturing businesses. Business counseling, accounting, and central office needs such as receptionist, secretarial, copying, and fax services are usually provided at reduced rates. The idea is to provide services collectively that are necessary for a start-up business but that may not be available or may be prohibitively expensive in a rural area. The goal is to graduate or move the business out of the incubator after it has completed the early stages of development (five years). This part of the incubator concept has never worked well. Incubators without walls have been tried in some areas. All of the same provisions except a physical facility are included.

The businesses brought into the incubator are usually in their early stages and are operated by local people. However, entrepreneurs from outside the community sometimes have been recruited to enter incubators. This may be the type of support needed to attract a young person back to the rural community or to give a local person with a good idea the basis to implement it.

Education

The goal of rural education is to improve the level of literacy, skills, capacity, and leadership ability within the local community. This involves more than primary and secondary education, although rural schools are a key component. Rural schools may well serve as the focal point to develop rural leadership and training programs because, for many communities, the school teachers and administrators are the largest organized body that have the required expertise. However, if the leadership development and training role is to be assigned to rural schools, they must be provided the human and financial resources to implement this task effectively. The general policy options for achieving the rural education goal include the following:

- **Financing schools** traditionally has been viewed as a local responsibility through taxing the value of real estate. However, most states give farmers a tax break by valuing farmland on its agriculture income-producing value rather than its market value. The agriculture use value in Texas was 16.7 percent of the market value in 1995.[15] While this tax break was installed to discourage the conversion of farmland to urban development, it also substantially reduces the amount of revenue available to finance rural education. It likewise shifts the burden of rural education to rural businesses that are not farm and to residents of rural communities. These higher tax rates for nonfarmers makes it less attractive for businesses to locate in rural communities. Some states have been forced to step in and provide subsidies to rural schools or move property tax revenue from rich districts to poorer rural districts.

- **Literacy** is an essential ingredient for success. Dropping out of school before receiving a high school diploma is more common in rural than in urban areas. Regionally, it is more common in the South. General Equivalency Diploma (GED) programs provide an opportunity for people to receive a high school diploma as a step toward societal and job literacy. The need for literacy receives extensive lip service, but financial commitment to literacy programs has been minimal. Local schools are the logical provider of this service.

- **Job place literacy** is a program to develop individual worker knowledge and skills for employment. Job place literacy may be accomplished either through on-the-job training or formal training programs. Publicly supported job training is accomplished under the Job Training and Partnership Act (JTPA) supervised by the U.S. Department of Labor. Training may be offered under JTPA grants to high schools, councils of government, community colleges, vocational schools, or private industry. Initiation of a JTPA program requires local initiative and identifiable jobs focused on particular employers.

- **Local school teachers and administrators** have very important responsibilities in rural communities. Frequently, the faculty of a rural school represents the broadest base of knowledge and training that exists in the community; schools also are frequently the largest employer in the community. Their faculties, therefore, can play a leadership role considerably broader than educating the youth. This broader role extends to developing leadership programs, providing job training, and assisting in community planning. Few communities take full advantage of this expertise.

Declining populations put considerable pressure on the ability of the local community to finance a quality education program. Viewed in this broad context of responsibilities with teachers and administrators contributing to development and training, maintaining small rural schools may be justified. Unless there is

---

[15]David P. Ernstes, *et al., The Effects of Differential Assessment on Tax Incidence in Texas Counties* (College Station, Tex.: Texas A&M University, Agricultural and Food Policy Center Working Paper 96–7, November 1996), p. 15.

local school support, however, the inevitable result is consolidation of small schools and, for most rural communities, loss of the local school.

- **Leadership** is a most valued but lacking resource in rural communities. Leadership often is the key factor that makes the difference between economic growth and decline. Leadership development and training, therefore, are vital components of a rural community's education strategy and development strategy. Unfortunately, leadership is often stifled by the local community power structure, which may itself lack enlightened leadership. It therefore becomes very important to the development process that the power base for the community be overtly open to fostering new leadership patterns.
- **Community colleges and/or vocational technical schools** cannot exist in every rural community. At the same time, it is important that every community have access to the services offered by these educational institutions. Often local leaders can initiate and complement the educational programs offered by a community college or vocational school by offering adult education classes under the guidance and supervision of the college or vocational school. Such programs may involve teaching computer skills and/or leadership skills or initiating job training programs.

Rural Health

On a per capita basis, the cost of medical care rises as its complexity increases and as rural population declines. Many of the costs involved in providing medical care are fixed. Accordingly, it should not be surprising that rural communities are losing hospitals and doctors. Access to rural medical services is becoming a major problem—perhaps the major rural problem, as indicated by the formation of the Rural Health Caucus in the Congress. The Clinton administration put health care high on the agenda of needed policy reform. However, without specifically targeting unique rural health care problems, it is unlikely that they will be solved. The following are general policy options concerning rural health issues.

- **Limiting medical malpractice liability** places maximums on the damages awarded in suits brought against medical doctors practicing in rural areas. This option addresses the reality that rural doctors face extra risks because they may lack facilities, equipment, and specialized medical services. Particular problems have been encountered in the areas of obstetrics and other specialized practices. The consequences of limiting malpractice include retaining health practitioners in less-populated rural areas and maintaining a wide range of medical services while protecting the patrons' interests. The end result is to improve rural living conditions.
- **Restructuring rural hospitals and clinics** involves planning and developing a financially viable rural health care system. The need for restructuring

results from continuous improvements in medical equipment and rising costs combined with a declining rural population. Many of the rural hospitals that were constructed shortly after World War II have not only become obsolete but are also larger than needed to provide adequate rural medical services in light of increased access to regional or major medical facilities located in metropolitan areas. In fact, much of the debate over the need for rural hospitals and clinics arises from judgments regarding the distribution of medical services between central hospital facilities in urban areas and rural hospitals or clinics. For example, should rural medical facilities be limited to emergencies, obstetrics, surgery, and recovery and rehabilitation? These issues need to be addressed in the planning process, perhaps on a basis that is broader than that for the local community. The consequences of resolving the issues of the structure of rural hospitals and clinics include increased ability of rural communities to provide health services, increased cooperation between rural health care systems and major medical centers, improved use of rural medical facilities, and improved ability to retain medical doctors in rural areas.

- **Emergency medical services (EMS and 911 services)** are needed that provide qualified and timely medical care in the event of an emergency. The importance of EMS arises from medical research findings indicating that time and care in responding to medical emergencies are keys to survival. Private ambulance services have largely collapsed because of issues of legal liability and the qualifications of the emergency personnel. Increasingly, EMS services are being provided by, or in cooperation with, fire departments. A persistence of accidents in rural areas, frequently involving urban residents, has increased support for improved EMS often including 911 and lifeline helicopter services from rural to metro medical centers. The consequences of offering EMS services include improved access to and quality of medical services, improved emergency communication systems, and strengthened fire protection services (see also the discussion of hospital districts and medical service areas).

- **Rural doctors and related health care professionals,** are also needed. Providing these professionals involves financial assistance to practices in rural areas or to train doctors and dentists who make a commitment to practice in rural areas. The National Health Service Corporation (NHSC) was established for this purpose but has not solved the problem for remote rural areas. As a result, state and local governmental units have sought additional remedies, including salary supplements, facility subsidies, and education scholarships. The consequences for rural areas include improving the quantity and quality of health care and making rural areas a better place to live and work.

- **Hospital districts and medical service areas** are needed that include taxing entities and/or regulatory authorities whose objective is to provide and maintain quality health, medical, hospital, and emergency medical services over a given geographic area. These service areas are designed to spread the costs of these services over the general population to make them economically feasible. In qualifying for the status of a hospital district or medical service area, the hospital takes on the responsibility for indigent care. Hospital districts or areas are established under state and local authority. The number of districts and areas has expanded since the 1970s when hospitals and emergency services encountered financial problems. The consequences of creating hospital districts and medical service areas include maintaining rural medical services and reducing fragmentation within the health care system. Without strong leadership, the districts or areas can become resistant to change like any other institution.

Rural Poverty

Poverty program administration ranks second only to education in state and local government budgets and second to defense in the federal budget. The incidence of rural poverty varies widely regionally and is particularly prevalent in the South, but no region is exempt from it. Poverty programs tend to be mandated by the federal government but are carried out at the state and local levels.

Not surprising, because more than 75 percent of the incidence of poverty is in urban areas, programs are oriented to meet those needs. Rural areas have unique delivery problems associated with distance traveled and the more sparse recipient density. Rural residents are less likely to participate in poverty programs. Tailoring poverty programs to local conditions often depends, therefore, on the innovativeness of those responsible for their delivery. Often, the erroneous assumption is made that the cost and the human resources required per recipient for poverty program delivery is the same in urban as in rural areas. This assumption makes poverty program delivery considerably more difficult than in urban areas. With this in mind, some options follow for dealing with rural poverty conditions.

- **Income maintenance programs** provide financial assistance in time of need. The most common financial assistance program is Aid to Families with Dependent Children (AFDC), which is designed for the single parent.

The basic objectives of these programs are to provide a temporary financial base for families and individuals as they strive to achieve self-sufficiency and to encourage care of dependent children within their own homes. Unfortunately, eligibility standards and assistance levels vary widely from state to state, making it profitable for low-income families to seek out states with the most liberal standards. Variation results from the requirement that states bear a larger share of the program costs and from the differences in states' ability to pay. The consequences of the income maintenance option include increased income flow to recipients and to rural communities, and an increased cost burden on state and local governments. Benefits to the family must be weighed against the requirement that the spouse cannot be present; some consider this requirement to be a key factor in the incidence of broken homes, divorce, and family abandonment.

Work requirements in welfare programs, such as those contained in the 1996 welfare reform bill for recipients without dependents, place a larger burden on individuals located in high unemployment areas. To the extent that rural communities have higher unemployment it will be harder for recipients to find work.

- **Health care assistance programs** pay medical expenses for the poor. Recall from the previous discussion that one of the requirements for designation as a hospital district or medical service area is to provide health care to the indigent, including children, pregnant women, and the elderly. Medicaid greatly expanded these programs. The higher proportion of rural elderly people makes these programs particularly important. Existence of health care services in rural areas tends to be spotty. The major consequences of improved health care assistance include better living conditions for those needing health care, income transfers to rural areas, and large increases in budget costs relating to the rising cost of medical care.

- **Housing assistance programs** provide loans, subsidies, tax incentives, and grants to the poor for rental and owner-occupied housing. Provision for both rehabilitation of older homes and construction of new ones exists. This assistance allows the needy to stay where they are rather than to move or be homeless. The commitment to these programs has gone up and down like a roller coaster. Rural programs have been small relative to urban programs, and the Government Accounting Office has found overt program biases that favor families located near metropolitan areas.[16] The consequences of housing assistance include the provision of quality housing for at least a portion of the rural poor and an improved economic and living environment for rural areas.

- **Education assistance programs** to the poor provide loans and grants to help families and individuals bear the costs of primary, secondary, and college education. Head Start and Upward Bound are the largest of these programs paid for by state and federal funds. These programs are perhaps the most effective and adequately funded of all rural poverty programs, probably because most of the assistance flows through rural schools. It probably also has the largest long-run payoff. The consequences of educational assistance to the poor include the development of human resources, the achievement of higher levels of education, and the frequency with which children are set on a productive path out of poverty. However, the extent to which rural areas benefit depends on whether the graduating participants stay in rural America or migrate to the city.

Farm Labor

Farm labor issues have tremendous impacts on rural communities in areas where migrant workers dominate. Agencies impacted are schools, social services, and

---

[16]General Accounting Office, *Concentration of Home Loans,* GAO/RCED-93-57 (Washington, D.C.: U.S. Congress, 1993).

medical care. Unique systems for delivering these services to what is frequently a minority population have generally received little attention.

In 1995, the agricultural workforce included nearly 3.0 million full- and part-time employees, including farm operators and unpaid family workers.[17] It has been declining at a rate of about 2 percent annually as agriculture mechanization occurs. Of these, 954,000 were full- and part-time hired farm workers that were highly dependent on agriculture.[18] There were 146,000 migrant workers.

The migrant segment of farm workers has a long history of controversy from a policy perspective. From 1942 to 1964, the bracero program institutionalized farmers' reliance on foreign-born workers. The program was initiated during World War II to relieve the wartime labor shortage but continued long after the war was over. During most of the program, more than 90 percent of the migrant workforce was foreign.[19] At its peak in the 1970s, more than 450,000 braceros were employed on 50,000 farms in 38 states, but three-fourths were employed in Texas and California.

From 1965, when the bracero program was allowed to expire amid public recognition of its inequities, through 1986, farm labor issues periodically boiled with the apparently uncontrolled entrance and employment of illegal aliens. Limited enforcement activity focused on regulating the abuses of farm labor contractors, employment of minors, unemployment compensation, migrant labor housing, sanitation standards, occupational hazards (OSHA), efforts to unionize farm workers in California, and the impacts of periodic increases in the minimum wage.

Immigration Reform

The Immigration Reform and Control Act of 1986 deals broadly with the issue of roughly 3 million illegal aliens living in the United States, 350,000 of whom were employed in agriculture. It provided for most of these people to become U.S. citizens. It also made it illegal for employers, including farmers, to hire illegal aliens with the employer having the burden for obtaining proof of legal status.

The only legal loophole for employing imported agriculture labor became the so-called "H-2A program," a vehicle for importing nonimmigrant labor. "Workers with H-2A visas have only the right to earn wages for performing jobs during defined periods on farms duly certified by the Department of Labor. When the jobs end, they must leave."[20] However, the use of the H-2A program appears to have been limited to a small number of commodities such as sugarcane in the Southeast and apples in the Northeast. The requirements for certification became greater than the cost of hiring U.S. farm labor and/or mechanizing.

[17]*Agriculture Fact Book 1996* (Washington, D.C.: USDA, 1996), p. 23.
[18]*Farm Labor* (Washington, D.C.: NASS/USDA, November 1996), p. 2.
[19]Howard R. Rosenburg, *Emerging Outcomes in California Agriculture from the Immigration Reform and Control Act of 1986* (Davis, Calif.: University of California Agricultural Issues Center, February 1988), p. 5.
[20]Rosenburg, *Emerging Outcomes*, p. 8.

Minimum Wage

A 1966 amendment to the Fair Labor Standards Act extended the minimum wage to labor employed on farms that used more than 50 man-days during any calendar quarter of the preceding calendar year. A man-day was defined as any day in which a worker performs at least one hour of agricultural work. However, until 1978, the minimum wage for farm labor was maintained at a lower level than for nonfarm labor. Agriculture is not required to pay time and a half for overtime.

The minimum wage increased from $2.65 per hour in 1978 to $4.75 in 1996 and $5.15 in 1997. The basic rationale for the increase relates to the fact that most of the people who work at rates below the minimum wage earn poverty-level incomes. Many of the hired farm workers have low incomes and skills. Farmers justify paying low wages by suggesting that either they are paying the going wage or this is what the workers are worth.[21] The result is the creation of a cycle of low wages, low skills, and low education.

Farm interests have traditionally suggested that increases in the minimum wage result in greater incentives to mechanize and to use chemical weed controls and related labor-saving technologies. The result, they suggest, is increased unemployment. A substantial body of research supports this theory of substitution. This research generally shows that with all other variables held constant, for every 10 percent increase in the wage rate, the employment of hired farm labor declines about 2 percent in the short run (1 to 2 years) and 10 to 30 percent in the long run (5 to 10 years).[22] Thus, although substitution possibilities tend to be limited in the short run, incentives exist over time for the development and adoption of a wide range of new technologies that reduce labor requirements.

Right to Organize

While the power of U.S. labor unions declined during the 1980s, unionization of agricultural workers became a viable issue. The National Labor Relations Act (NLRA) is the basic federal statute that provides the framework within which employees organize and bargain collectively with employers who, in turn, are required to bargain in good faith. It prescribes unfair labor practices on the part of both unions and employers. It created the National Labor Relations Board (NLRB) to administer the act. Agricultural laborers are explicitly excluded from the definition of the term *employee* and are not subject to the provisions of the NLRA.

Most farm labor is not unionized. The major exception is the United Farm Workers, organized in the West by Cesar Chavez. Farmers have generally reacted

---

[21]Joseph D. Coffey, "National Labor Relations Legislation," *Am. J. Agric. Econ.* (December 1969), p. 1072.

[22]Ibid., p. 1067; Theodore P. Lianos, "Impact of Minimum Wages upon the Level and Composition of Agricultural Employment," *Am. J. Agric. Econ.* (August 1972), pp. 480-481; and Edward W. Tyrchniewicz and G. Edward Schuh, "Econometric Analysis of the Agricultural Labor Market," *Am. J. Agric. Econ.* (November 1969), p. 777.

negatively to the unionization activities of farm labor, and the American Farm Bureau Federation has been a consistent opponent to giving farm workers the right to organize. Farmers' objections to unionization pertain primarily to the following concerns:

- The impact of higher wages on their cost of production, with an inability to pass through those higher costs in the short run.
- The potential for strikes at harvest, having the potential to destroy a farmer's source of income. This threat is particularly serious for perishable crops for which a strike at harvest would be the most effective union bargaining tool.
- The potential for a **secondary boycott** involving *coercive pressure exerted on a third party*, such as a grocery chain, not to handle the products of farmers opposed to unionization. Such coercive pressure may include organized action of members not to buy groceries from noncooperating chains, encouraging other union members to do likewise, or making a general plea to consumers to buy only union products.

Although controversial, hired farm workers have tended to receive substantial public support. The California movement has received widespread support from church groups, student groups, civil rights groups, urban politicians, and organized labor. Even in agriculture, the National Farmers Union has been much more conciliatory in its attitude toward organized farm labor. It apparently believes that if all workers are given the right to organize and demand higher wages, farmers are in a better position to bargain for higher prices either through the political process or through the marketplace. The following are policy options for giving farm workers a set of rules within which organization and bargaining can take place:

- **Extending present provisions of the NLRA to agricultural workers** would bring agriculture under the same collective bargaining law, regulation, and court decisions as for other industries. It would prohibit the secondary boycott and would give the NLRB the authority to decide appropriate bargaining units and the access to farms for organizing, would specify the minimum size of farm operation covered, would provide for supervised elections and balloting procedures, and would supersede all state laws. It would not prevent strikes during critical periods of growing or harvesting.
- **Enacting special national labor relations legislation for agriculture** would consider the unique problems of agriculture in terms of preventing losses due to strikes and of compulsory arbitration, as well as the issue of secret ballots. Farmers have sought compulsory arbitration to allow an impartial third party to decide how a labor dispute should be settled. The arbitrator's decision would be binding on both parties. Organized labor, however, strongly opposes compulsory arbitration procedures. Even if such special legislation were enacted, many of the rulings already made under NLRA would likely be applied to the interpretation of the agriculture act.

■ **Enacting special state legislation for agriculture** would do the same as the
NLRA extension but could be tailored specifically to agriculture within a
particular state. The California agricultural labor relations law could probably be the model for any such legislation. The basic purposes of the
California law are to encourage and protect the right of farm workers to
organize, to select representatives for the purpose of bargaining, and to prohibit employers from interfering with those rights. The California law created a five-member Agricultural Labor Relations Board, which makes the ultimate rulings on the validity of elections and constitutes a court of appeals for
unfair labor practices cases. Although the California law prohibits secondary
boycotts, it does not extend to passing out leaflets, picketing, or labeling
products sold in supermarkets to draw attention to a dispute. Court interpretations of the California law would probably be considered in interpreting similar legislation in other states. The most controversial ruling of the
California board gave unions access to farms for the purpose of conducting
organizing efforts. Michigan and Ohio have each established a commission
that supervises private agreement arrangements and dispute settlements
between growers' associations, processors/packers, and labor organizations.

Each of these alternatives runs into many of the same consequences as those
that apply to the minimum wage. That is, higher wages and, in this case, the threat
of labor problems encourage mechanization and may lead to higher food prices.
On the positive side, unionization can result in a more orderly labor market, a
more dependable supply of workers, a better defined job structure, more consistent work rules, and generally a more productive workforce. This in turn can
reduce poverty, break the cycle of low wages that encourages low productivity,
and improve the health and education level of hired farm laborers. Some grower
organizations that bargain for the prices of fruits or vegetables have found that
their interests sometimes lie with the workers and have used union demands to
obtain concessions from processors.

Disputes over the rights of farm laborers to organize, the level of the minimum wage, and the employment of foreign laborers will continue. Yet their complexion will change as agriculture continues to industrialize and to move into a
higher technology era. In this era, the proportion of hired farm labor as a share of
farm employment will continue to increase.

The level of skill required of farm labor will also increase. More professional farm managers will be employed. Farm labor will have to be more adept
at using computers and increasingly complex farm machinery, performing
embryo transfers, injecting somatotropin, and adjusting the combination of
inputs to the more complex products of biotechnology. In other words, farm
labor will require back-breaking work and a higher level of skill—requiring
higher levels of education.

In this process of change, farm labor issues will parallel those developing in
the remainder of the economy. This should not be surprising as agriculture
becomes more like the rest of the economy.

Development Planning and Financing

With declining rural populations, some of the largest rural development issues involve planning for the future on a financially viable basis. Therefore, just as central cities are having financial problems, so are rural communities, counties, and related governmental subdivisions. Just as in the central cities, federal and state mandates have increased rural government costs, but neither the federal nor state governments have provided the funds to pay for the mandates. The options for addressing these issues and their consequences follow:

- **Development planning and technical assistance** activities recognize that although cities and counties may be the focal point for local political activity, rural economic activity often takes place on a broader, multicounty plane. Therefore, economic development districts and councils of governments have been established in most areas to deal with the broader planning functions in rural areas. These units function as a mechanism for securing loans and grants and as a basis for providing technical assistance in planning to communities and counties within the region. The effectiveness of these planning instruments depends on the extent of citizen involvement and the availability of funds, both of which have been highly variable. The consequences of effective multicounty planning include the development of a strategy that has a chance for success, adequate consideration of the required financial base, and an enhanced chance of achieving that financial base of support.

- **Self-financing** involves various means of raising local revenue including taxation, user fees, licensing, contracting, and privatization of local services. In addition to covering the costs of providing services, the objective is to maintain a rural infrastructure base conducive to development. Self-financing implies a pay-as-you-go approach, which may not always be possible. The alternative methods of self-financing have substantially different impacts on the various population segments. The main consequences of self-financing include avoiding debt service costs and delaying development initiatives until funds are available.

- **Public works loans and grants** have been provided in varying numbers by federal and state governments to finance expensive infrastructure. Specific targets for improved infrastructure have included roads, bridges, waste water treatment, water systems, solid waste disposal, fire protection, parks, and recreational facilities. Such initiatives are among the most difficult for rural communities to provide but are critical to economic development. Unfortunately, public sources of loans and grants are considerably less prevalent. The consequences of public works loans and grants include an improved basis for economic development now as opposed to the future, a basis for providing essential public services, and a means of implementing well-developed economic development plans.

■ **General obligation and revenue bonds** are debt financing instruments issued under state authority for specific purposes such as creating an infrastructure. Bonds may be issued either by the state and lent to local government units or issued directly by local government units. Such bonds provide an alternative source of debt financing. The consequences include the ability to proceed with the implementation of development planning and the requirement for debt service costs in the future.

TOWARD A CONCEPT OF A RURAL COALITION

This chapter noted previously that even though the agriculture committees are the focal point for rural development policy, farmers and farm organizations have had little interest in the contemporary rural development policy issues. Earlier in this book, we concluded that the prospect for increased government expenditures on farm programs is not particularly favorable. Moreover, the discussion of resource policy (Chapter 13) indicated that increasing demands are being made on farmers to reduce environmental externalities. Many of these same demands are being made on rural communities and related businesses in terms of solid waste disposal and water treatment.

Clearly, from the perspective of rural education, health care, communication, and infrastructure, all rural residents, including farmers, are affected by generally deteriorating conditions in the rural heartland. Even from a rural business perspective, many farmers depend on nonfarm businesses for a majority of their net income. With declining job numbers, farmers' off-farm job opportunities likewise decrease.

It makes sense, then, for all rural residents, including farmers, community political leaders, businesspersons, educators, health care workers, other employees, and retired folks to sit down together and decide what is best for them as a group—to develop their own policy agenda. The agenda would cover both agricultural policy and rural development policy—the combination of which might be termed *rural policy*. It would be advocated by a rural coalition representing about 25 percent of the U.S. population. Such a notion may be a pipe dream, but it is certainly one alternative for dealing with the common problems affecting rural America.

ADDITIONAL READINGS

1. Ronald D. Knutson and Dennis U. Fisher, *Focus on the Future: Options* in Sue H. Jones, ed., *Developing a New National Rural Policy* (College Station, Tex.: Texas A&M University and Aspen Institute, Fall 1988).

2. The most authoritative source on economic concepts related to rural development is by Ron Shaffer, *Community Economics* (Ames, Iowa: Iowa State University Press, 1989).

3. Dennis U. Fisher, Ronald D. Knutson, and Howard Ladewig, *Multicommunity Collaboration: An Evolving Rural Revitalization Strategy—Policy and Multicommunity Development* (College Station, Tex.: Texas A&M University, July 1992).

4. Dennis U. Fisher and Ronald D. Knutson, *Increasing Understanding of Public Problems and Policies 1989—Politics of Rural Development* (College Station, Tex.: Texas A&M University).

IMPACT OF AGRICULTURAL
AND FOOD POLICY
ON AGRIBUSINESS

The business of government is to keep government out of business—that is,
unless business needs government aid.

—Will Rogers

Agribusiness *as used in this chapter refers to those firms whose main business relates to providing food and fiber, but which are not primarily involved in production.* Agribusiness includes input suppliers such as agricultural lenders, fertilizer manufacturers and dealers, feed manufacturers, plant and animal breeders, seed companies, and dealers. Agribusiness also includes firms that transport, market, handle, process, distribute, wholesale, retail, and serve food—in essence, from the farm gate through the supermarket and/or fast-food outlet. Firms integrated into production are part of agribusiness. Thus, Tysons, the largest broiler producer and processor, is treated here as being an agribusiness firm (as opposed to a farmer) because of the importance of its input supply, processing, and marketing activities.

Most of the discussion in this book has related to the impact of agricultural and food policies on farmers, consumers, and taxpayers. Yet the farm value is only 21 percent of the retail price paid for food by consumers. The rest goes to the intermediaries. Actually most of this 21 percent does not stay with farmers. Figure 17.1 provides a different perspective on the distribution of the food and fiber sector gross domestic product. It indicates that after farmers have paid agribusiness input expenses, they retain only 6.4 percent to cover land rent, labor, and management. The input sector gets 32.4 percent. The processing and distribution sector gets 61.2 percent. In other words, out of the value of food and fiber at the retail level, more than 93 percent goes to agribusiness. From Chapter 8, you may recall that agribusiness employment totals 22.9 million.

Because of their prominence in agriculture, agribusiness firms are profoundly influenced by agricultural and food policy. This is evidenced by the

Total food and fiber sector GDP: $982.7 billion

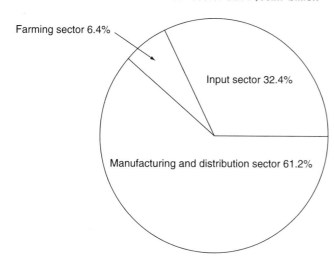

FIGURE 17.1
Relative importance of agribusiness as a component of GDP in the food and fiber sector, 1995.

Source: William Edmondson *et. al., Measuring the Economywide Effect of the Farm Sector: Two Methods* (Washington, D.C.: USDA/ERS, July, 1995), p. 35.

sophisticated network agribusiness firms have developed to influence farm and food policy decisions, as discussed in Chapter 4. Despite the importance of policy on agribusiness, little is written on the topic. Yet a large share of the readers of this book are likely to be or to become agribusiness employees.

Every agribusiness firm needs to have the ability to assess the impact of policy changes on its operations. This chapter is designed to provide insight into the nature of potential impacts of farm and food policy on agribusiness as well as some theoretical or analytical approaches to addressing the policy issues confronting business. The chapter first discusses the major agribusiness segments; then it specifies the policy goals of agribusiness. Subsequently, impacts are examined in terms of the main policy categories such as price and income policy, stocks policy, and so forth. Case examples are regularly utilized to illustrate particular points.

AGRIBUSINESS SEGMENTS

A reasonably common set of policy goals can be specified for agribusiness, but the priority that these goals receive varies. As a result, it is desirable first to define the agribusiness segments that are important from a policy perspective. For this purpose, five major agribusiness segments can be identified.

- *Input manufacturers* are the industrial firms that develop and manufacture the inputs used by farmers in the production of farm products. They often hold patents or copyrights on these inputs. Many of these firms are not exclusively dependent on agriculture for their sales but are multiproduct and even multinational petroleum, chemical, pharmaceutical, or machinery companies. These firms' most direct contact with agricultural and food policy probably relates to food safety and resource issues such as residues, endangered species, water quality, and soil erosion. However, the demand for the inputs that they manufacture is directly affected by farm programs that influence production.

- *Input suppliers* are the wholesale and retail distributors of farm supplies. One distinguishing feature among input suppliers is whether they are owned by farmers (predominantly as cooperatives) or by nonfarm investors. Farmer-owned input suppliers tend to be more sensitive to the needs and policy goals of farmers. Thus, while investor-owned fertilizer dealers often express strong political opposition to production controls, cooperative fertilizer dealers often are forced simply to bear the agony of reduced sales. Sometimes, however, cooperatives may be members of a trade association along with investor-owned import suppliers who lobby against farm programs that curb input sales.

- *Integrated agribusiness* is directly involved in input supply, sometimes manufacturing, farm production, marketing, and processing. Nonfarm-investor-owned companies such as Cargill, ConAgra, and Tysons, as well as some cooperatives such as Sunkist, Welch, and Farmland, have the characteristics of integrated agribusiness. In contrast with most farmers, integrated agribusinesses recognize the importance of controlling the production-marketing channel to their advantage. Their attitudes toward government tend to be influenced by the extent to which they can control or manage the policy tools. For example, while broiler integrators are highly opposed to government policies designed to manage grain supplies, fresh fruit and vegetable integrators (particularly cooperatives) embrace marketing orders as a market management tool.

- *Market intermediaries* cover a wide range of firms and functions that facilitate the sale of and/or physically handle farm products without changing their form. Market intermediaries profit from volume and price variations. For example, futures market brokers do not have much business when prices are resting on loan rates. Similarly, Cargill and Continental grain companies suffer when export volume is curtailed by high loan rates. In other words, market intermediaries are strong free-market advocates. They cover a wide range in size, from multinationals to independent auction or spot market operators.

- *Food processors and marketers* hold the franchise on the products that consumers buy. Included are the processors, manufacturers, wholesalers, retailers, and fast-food operators. Their most direct contract is with consumers. Their greatest policy interest, therefore, is in food policy, although farm and resource policies are also important concerns.

AGRIBUSINESS POLICY GOALS

The diversity of agribusiness is more apparent than the diversity of agriculture. Yet the business dimension of their operations makes agribusiness firms more inclined than farmers to have a common set of farm and food policy goals. In other words, agribusiness firms are more likely to approach policy issues from a strictly business perspective. This happens because business goals and economics take precedence over value judgment differences that tend to divide farmers and farm organizations. The following policy goals have been suggested for the agribusiness sector[1]:

- There is a desire for an *abundant supply of farm products that is readily available*. Agribusiness profits from volume. This is more typical of commodities than it is of branded products although even on branded inputs and food, product volume is a key to profit. Therefore, policies that reduce volume tend to be opposed by agribusiness.

- Agribusiness favors programs that *expand domestic and international markets*. Examples include foreign market development, check-off programs, foreign food aid, cooperator programs, food stamps, school lunch, and even export subsidies. It prefers programs that utilize commercial markets as opposed to government donation (food aid) programs. Farmers, on the other hand, prefer donation programs because more commodity is moved per dollar of government expenditures.

- Agribusiness firms want to *minimize regulation of their operations*. Agribusiness opposes programs that interfere with their marketing strategies. Firms want to maximize latitude for decision making. Government regulations either reduce firms' flexibility in making decisions or impose extra costs, and often they do both. Farmers and agribusiness tend to have similar attitudes toward regulatory issues—except in the arena of farm programs.

- Agribusiness firms prefer supply-and-demand *responsive markets*. In effect, this is saying that they prefer instability. Most large agribusiness firms have intelligence systems that make it possible for them to anticipate change, to take advantage of change, to hedge against it, or to adjust to it more rapidly than farmers. This is particularly true of market intermediaries and integrated agribusiness firms. It is less true of input manufacturers and input suppliers.

- Like other bureaucratic institutions, agribusiness *resists change*. Although, as a general rule, opposed to regulation, agribusinesses sometimes oppose deregulation. This is particularly true of "mature" industries that utilize regulation to protect against decline and change.

Agribusiness goals are not completely consistent. Preferring instability and resisting change are hardly consistent! Yet policy goals are seldom consistent. There are trade-offs. Priorities have to be set. Compromises have to be reached.

From these goals alone, an agribusiness policy agenda could be written. *It would be a market-oriented agenda with relatively low levels of government involvement.* But the world is not that simple. Some of the conflicts that arise become apparent in the discussion that follows.

---

[1]T. A. Stucker, J. B. Penn, and R. D. Knutson, "Agricultural-Food Policymaking: Process and Participants," *Agricultural-Food Policy Review*, AFPR-1 (Washington, D.C.: ERS, USDA, January 1977), pp. 1–11.

POLICY IMPACTS ON AGRIBUSINESS

Discussions of agribusiness policy impacts could be approached in several different ways. For example, five agribusiness segments could be discussed separately. However, on many—if not most—issues, agribusiness takes much the same policy position. The discussion that follows is in terms of the various policy areas discussed in this book with illustrative case examples in each area to indicate the complexity of the issues.

Trade Policy

Agribusiness firms tend to be free-trade oriented. They tend to be strong supporters of NAFTA and GATT. They strongly oppose any policies that would have government more involved in commodity trading. The reasons for this opposition are obvious:

- Free-trade policies expand trade volume. Commodity profits depend on volume and profits are probably easier to gain with lower price levels.
- Less government involvement leaves more functions to be performed by agribusiness. This is the case both on the seller (exporter) *and* buyer (importer) sides of the market.
- Free-trade policies reduce the market power of both producers and consumers. Therefore, relatively speaking, the agribusiness intermediaries are given more market power.
- Free trade increases price variability, thus allowing market intermediaries to generate higher returns due to hedging, price analysis, market information, and forecasting activities.

Where government is involved, agribusiness takes overt steps to be in a position to influence implementation decisions. The agribusiness strategy is one of having friendly elected and politically appointed individuals in positions of control, possessing insider knowledge, and influencing government decisions to their advantage.

GATT

Trade negotiations and GATT are often criticized for being ineffective. Quite clearly, trade negotiations and GATT procedures have been more effective for the industrial sector than for agriculture. Yet despite GATT's weaknesses, there have been many gains to producers and agribusiness from trade negotiations. Perhaps the most significant gain has been in the soybean trade with the European Union and Japan.

In the 1960–1961 Dillon Round of GATT, the United States and the European Union (EU) agreed not to impose any duties on soybeans and soybean meal for two reasons:

1. The European Union was highly deficient in vegetable proteins for balancing livestock and poultry feed rations. To a lesser extent, it was deficient in vegetable oils as well.
2. With the formation of the European Union in the late 1950s and the establishment of the Common Agricultural Policy (CAP), which utilized the variable levy as its main protective tool, the Union was under heavy pressure to show a degree of good faith toward GATT.

This EU soybean concession has had tremendous long-term value to U.S. farmers and agribusiness firms. EU soybean imports mushroomed as the poultry and hog industries expanded. In addition, the importance of soybeans in balancing rations became more apparent. The European Union imports soybeans both as whole beans and as soybean meal (Figure 17.2). EU soybean imports grew consistently from 1970 through 1980. The leveling, thereafter, of EU imports reflects EU policies subsidizing soybean production that were ruled to be a violation of GATT.

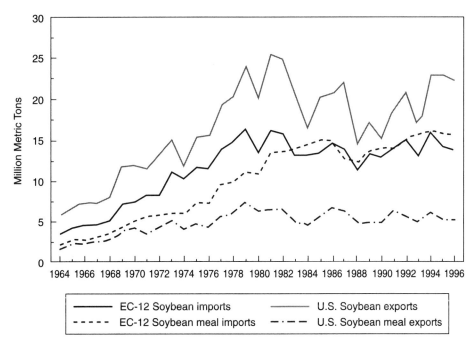

FIGURE 17.2 Trend in the imports of soybeans and soybean meal by the EU and in total U.S. exports, 1964–1996.

Source: PS&D View, electronic database (Washington, D.C.: USDA, 1996).

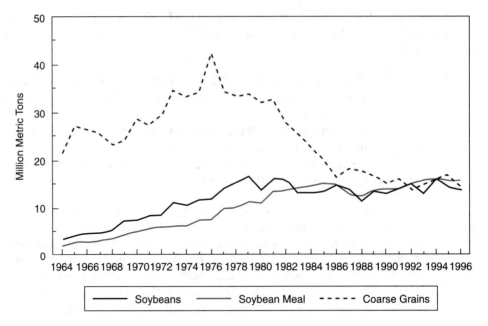

FIGURE 17.3   Trends in EU soybean, soybean meal, and coarse grain imports, 1964–1996.

Source: PS&D View, electronic database (Washington, D.C.: USDA, 1996).

The benefits to the United States of the EU exclusion of soybeans from the variable levy was very apparent when compared with its coarse grain imports. In coarse grain imports, where the European Union maintained a variable levy, the trend was negative. In the absence of a variable levy, the trend in soybean imports was clearly positive (Figure 17.3).

The Dillon Round decision on soybeans was not without controversy. EU farmers blamed soybean imports for displacing homegrown grains, reducing the consumption of olive oil, and increasing butter surpluses (a by-product of whole bean imports is soybean oil, which is used as a cooking oil and to make margarine). In the mid-1980s, the European Union threatened to place a tax on vegetable oil as a means of partially imposing a duty on soybeans. The United States strongly objected, with the effect of at least delaying the EU action.

In reaction to farmer unrest, the European Union installed a system of direct farmer payments for oilseed production not too much different from the U.S. target price system. These direct payments stimulated EU oilseed production, consequently reducing the need for soybean and soybean oil imports. The United States responded to the EU change in oilseed policy by filing a complaint with GATT. The subsequent ruling that the EU oilseed payments violated GATT became a central issue in the Uruguay Round. The European Union subsequently agreed to take action to reduce its soybean production.

In retrospect, there is probably no better testimonial on the importance of trade negotiations to agribusiness and farmers than the soybean experience. At the same time, the EU experience clearly indicates the constant pressures to impose protectionist policies.

Cartels and Marketing Boards

Agribusiness firms, led by the multinational grain companies, strongly oppose marketing boards. This opposition is so strong that it extends to attempting to prevent producer educational materials on marketing boards from being produced, distributed, and utilized.[2] The reasons are well explained by Schmitz in the situation where the board controls all marketing and sales[3]:

- *Marketing boards reduce the role of the private grain firms and the multinational grain companies.* These companies prefer to buy grain directly from farmers or from farmer-owned cooperatives, arrange for transportation to ports, maintain control of port-handling facilities, make sales directly to foreign customers, arrange for ocean freight, and handle international financial transactions. Under a marketing board system, these multinational companies buy grain from the board. In addition, a large proportion of board sales are government to government. Thus, the grain companies become strictly handlers performing largely transportation-related functions—often referred to as the logistics of the grain trade. Smaller private grain companies and grain cooperatives operate mainly as storage agents for the board. Generally, private companies and cooperatives do not buy grain from farmers—that is the responsibility of the board.

- *The objectives of the marketing board and the private companies directly conflict,* as illustrated in Figure 17.4. If the marketing board acts as a producer monopolist, it equates its marginal revenue with marginal cost (supply) producing quantity Q_b, selling at price P_b. On the other hand, a grain company with monopoly power in the seller's market and monopsony power in the producer market equates its marginal outlay for grain (*MO*) with its marginal revenue (*MR*), buying quantity Q_g, paying price P_p, and selling price P_g. In other words, the grain companies want to buy a smaller quantity, pay producers a lower price, and sell at a higher price than a producer-oriented marketing board.

- *The business of grain market intermediaries is reduced by marketing board activities.* The need for and role of the futures market under a marketing board is highly uncertain. With all grain purchased by the marketing board and with producers paid the average price generated by the board, there is little need for hedging at either the producer or grain handler levels. The volume of

---

[2]Mike Turner *et al., Who Will Market Your Grain?*, D-1057 (College Station, Tex.: Texas Agricultural Extension Service, March 1978) was the target of an agribusiness attempt to limit the use of marketing board educational materials.

[3]Andrew Schmitz *et al., Grain Export Cartels* (Cambridge, Mass.: Ballinger Publishing, 1981), pp. 38–48.

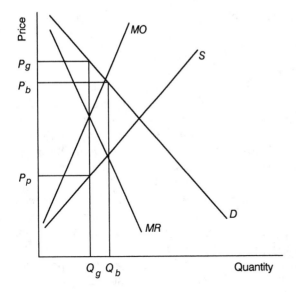

FIGURE 17.4
Objectives of marketing board and private companies conflict.

hedged trading is thereby reduced. Under these circumstances, the role and/or justification for the futures market is uncertain. Because the futures market is, in effect, a central pricing point or price discovery mechanism for the international grain market, a new world pricing point could be required.

■ *The producer marketing system would be radically changed.* There would be no producer marketing decision because marketing would be through the board. Storage decisions would likewise be at the discretion of the board. As indicated previously, cooperatives performing storage functions would become agents of the board.

Price and Income Supports

It is often asserted that no one benefits from farm programs other than producers. Similarly, it is often asserted that farm programs are mainly designed for crop producers and that livestock producers enjoy few benefits. Such generalizations are not always true. The following are two interesting cases in which agribusiness became a major beneficiary of crop policies. They also illustrate that the secondary effects of policy may be as important as the primary effects.

    Sugar policy.   Historically, the United States has been a net importer of sugar. U.S. sugar production has been maintained by setting a high support price and restricting imports sufficiently to achieve the support price level. However, this sugar policy has provided the umbrella for the development of a corn sweetener industry that has substantially reduced sugar consumption.

    Since the 1985 farm bill, the sugar price support has been effectively set at about $0.21 per pound while high-fructose corn sweeteners (HFCS) have tradi-

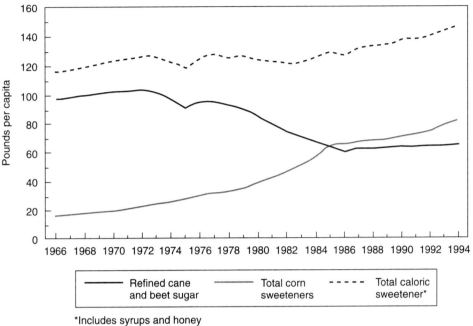

FIGURE 17.5   Trends in per capita sugar and corn sweetener consumption, 1966–1994.
Source: Food Consumption, Prices, and Expenditures, electronic database.
(Washington, D.C.: USDA/ERS, February 1996).

tionally been priced at $0.05 to $0.10 per pound (20 to 30 percent) less than sugar. HFCS is a nearly perfect substitute in some sugar uses. As a result, sugar consumption persistently declined, until the mid-1980s, while HFCS consumption persistently increased (Figure 17.5).

Interestingly, U.S. sugar producers have not been the ones hurt by this policy because their price has been supported. The lower sugar consumption has been largely at the expense of developing countries that export raw sugar to the United States. To maintain the sugar price at $0.21 per pound, import quotas were gradually reduced. The world sugar price has been in the $0.05 to $0.14 per pound range since 1980, so the United States, with its support price, has been a preferred market for developing countries.

U.S. sugar policy has had four fascinating effects on agribusiness:

- Sugar policy, in combination with ethanol subsidies, has built a corn sweetener industry involving major agribusiness firms with sales of about $1.5 billion in 1987. Without U.S. sugar policy, the modern corn sweetener industry would not exist.

- Noncaloric sweeteners have also enjoyed an umbrella of price protection. It may be argued that people consume noncaloric sweeteners for reasons other than price, but high sugar price supports have enabled competitive pricing of products such as diet soft drinks.

- Sugar refiners that process imported raw sugar have been going out of business.

- Imports of sugar-containing products have become a major competitive problem for U.S. processors of products such as candy.

For the same reasons that corn sweetener interests support high sugar price supports, corn interests would fight against lowering the sugar support price, converting the sugar program to a target price program, or eliminating the sugar program. Any of these actions would make sugar more price competitive with HFCS. The corn sweetener lobby, therefore, has become a prime advocate for the sugar program.

The 1996 farm bill presented an interesting problem for sweetener interests within the agribusiness community. Under the leadership of the Grain and Feed Dealers Association (GFDA) the agribusiness community lined up behind the Freedom to Farm decoupling concept. The GFDA strategy involved decoupling all commodities including lowering or eliminating the support price or loan rate. ADM, the largest manufacturer of HFCS, realized that a lowering or elimination of the support price for sugar would seriously reduce or eliminate HFCS's price advantage over sugar. This difference of opinion led to ADM's withdrawal from the agribusiness coalition. Sugar stood by itself as the only commodity retaining its high price support policy—a test to the political power associated with an alliance among sugar producers, corn producers, and HFCS manufacturers.

Price support loan.   The conditions under which the price support loan constitutes a floor price for commodities were discussed in Chapter 10. In general, the loan rate is a floor price when the government stands ready to accept forfeited commodities at the support price; when all producers are eligible for the loan (participate in the program); and when commodities are not released by USDA using other means. It stands to reason that if these conditions are fulfilled, agribusiness firms cannot attract commodities out of the hands of producers unless they are willing to pay the loan rate plus accumulated interest costs. Commodities may not, therefore, be available for export by the market intermediaries at prices that are competitive in world markets.

Even if the loan rate is below the world market price, price-supported commodities may not be available for export. Farmers' marketing decisions are influenced by expectations and risk. With no downside price risk (they can always forfeit at the loan rate), if farmers foresee a reasonable chance that the market might rise above the loan rate, they will tend to hold commodities off the market. This behavior also has adverse impacts on exporters. Therefore, the security of the loan itself can cause farmers not to make commodities available and thereby drive up the market price or deny the market the volume needed to satisfy demand at competitive prices. *It may be concluded that the loan rate does not have to be above the world market price for exports to be reduced.*

The effect of the loan program on producer behavior is an integral part of the residual supplier problem. The loan program puts export-dependent market intermediaries in a very difficult competitive position. However, multinational companies are not in as difficult a competitive position because they can trade the commodities of other countries.

Stocks Policy

One of the most complex and politically sensitive tasks of the government is that of managing commodity stocks. How stocks are managed has a tremendous impact on food security, price levels, price variability, commodity availability, and, therefore, on firm purchase, storage, and marketing decisions. For example, releasing CCC stocks depresses farm prices and discourages the accumulation of privately held stocks.

Releasing CCC stocks.   The 1996 farm bill directs the secretary of agriculture to minimize the accumulation of Commodity Credit Corporation (CCC) stocks. One of the secretary's most difficult decisions involves the release of government stocks. Charges of manipulation of prices and cheap food abound when the secretary places CCC stocks on the market and thereby suppresses commodity prices. A classic example occurred in 1987 after USDA made an agreement to lower the FOB port price sufficiently that the Soviet Union and China could buy U.S. grain at competitive world market prices. To make commodities available to U.S. exporting companies without substantial increases in the market price, payment-in-kind (PIK) certificates were used as subsidies to exporters of wheat to the Soviet Union. In addition, the secretary of agriculture took two actions:

- He established an auction market for CCC-owned wheat whereby the USDA would redeem the PIK certificates for wheat at a price that it deemed necessary to secure the Russian and Chinese business. The bidders were grain companies (presumably mostly exporters).
- Upon finding that there was not a sufficient surplus of PIK certificates, the secretary made 100 percent of the deficiency payments for wheat and barley in generic PIK certificates. In the past, PIK certificates had been used for only half of the payments.

The installation of a recourse loan would solve the nonrecourse farmer release problem experienced by agribusiness. Recourse loans have been implemented in honey and will be utilized in dairy beginning in year 2000 under the provisions of the 1996 farm bill. Expansion of the recourse concept could depend on what happens in dairy. Whether recourse loans are applied to crops could become a test of the unity and resolve of farm organizations versus agribusiness.

The farm organization lobby became irate at these actions, which, by releasing commodities from CCC stocks, had the effect of suppressing producer prices while directly benefiting a small number of exporters. Farm state congresspersons cited wheat market prices dropping as much as $0.39 per bushel from a market price of about $2.50. Prior to the USDA action, there were widespread predictions that the market price of wheat would need to rise sufficiently to cover interest on the CCC loan and thereby encourage farmers to release sufficient grain to sell the Soviets and Chinese. Then Chairman Daniel Glickman (Dem.–Kan.), of the House Wheat, Soybean, and Feed Grains Subcommittee, charged that by this action, the USDA was acting like a "grain board." Ironically, Secretary Glickman is now charged with making the release call so as to minimize stocks—an action that is almost certain to bring charges of market management by the USDA much like a grain board.

Holding stocks in other forms.   Government stocks depress market prices even when they are not released. The larger the stocks, the more depressed the price—up to the point at which the price simply rests on the loan rate. Why then does the government hold stocks in the form of commodities? Instead, why not hold them in the form of cash or land?

The money stock idea arose during the world food crisis in the early 1970s and at the World Food Conference in 1996. It was observed that the cost of storing commodities in strategic locations is so high that it would be cheaper to give countries needing commodities cash to buy grain as opposed to food aid. Yet if a country's ability to buy is enhanced when supplies are short, the market price is driven up even higher. The higher price tends to encourage production in future periods but also results in a more unstable price. In other words, *commodity stocks foster long-run supply and price stability, while a money reserve fosters instability.* Agribusiness firms would do better, in relative terms, under a money stocks policy because they are better able than most farmers at managing risk. However, both farmers and agribusiness would benefit.

Holding land out of production through the conservation reserve program (CRP) can also be viewed as a stocks policy. However, it has different effects than holding either commodity or money stocks. Stocks of land held out of production raise prices in the short run because supply is reduced. In the long run, however, land reserves also affect prices because of their potential for being put back into production. For agribusiness, land held in reserve is a two-edged sword: It reduces input sales and reduces volume of product marketed and/or stored. Thus, there is virtually unanimous agreement in agribusiness that CRP is bad policy, which is directly contrary to the thinking of most farmers.

Impacts of public stocks on private stocks.   Government involvement in stocks has a marked impact on agribusiness inventory management policies. Money is made and lost over changes in the value of inventories. Government stocks policy, therefore, influences privately held inventories during periods of changing prices.

Dairy price supports provide an excellent illustration of how the private and public sectors interact in holding stocks. Since 1949 the government has stood ready to purchase any manufactured dairy products (butter, cheese, and nonfat dry milk) offered to it at the support price. Government stocks could be bought by private firms at 10 percent over the support price. Manufactured dairy products are produced continuously. However, more milk is produced in the spring than in the fall. Milk prices may rise above the support level in the fall when milk supplies are short and may fall to the support level in the spring.

With 50 years of experience, butter, nonfat dry milk, and cheese manufacturers got used to milking the price support program. They let the government acquire and hold the stocks in the winter and spring while purchasing them back in the fall, if needed.

All of this experience will end in year 2000 when the price support program is replaced by a recourse loan. Purchasing stocks from the government will no longer be an option. Contract suppliers of cheese for McDonald's or Burger King will no longer be able to rely on the government for its reserve stocks. Likewise, there will be no floor on the market to shield the industry against a reduction in inventory value. Managing inventories and risk will become a new higher priority item on the agenda of the dairy industry.

Production Controls

The impact of government on agribusiness perhaps can be seen even more clearly in cases in which public decisions are made to control production. Input manufacturers, input suppliers, and market intermediaries experience almost immediate effects on their operations, with impacts often extending to the economic viability of rural communities. The 1980s provide an excellent environment to evaluate and, to a certain extent, quantify these impacts. Examples from land retirement and the dairy termination programs are used to illustrate the complexity of the effects of production control policies.

Land retirement programs. Recall that throughout the 1970s there were strong incentives to expand production. Most of the land that was retired by government programs in the 1960s (other than that planted to trees) was put back into production in the 1970s. The decline in exports during the early 1980s, combined with ever-increasing production, led to the potential for a large accumulation of government stocks. To forestall this development, large quantities of land were removed from production (Figure 17.6). The government implemented the original (PIK) program, which removed a record 83 million acres of land from production. Farm input sales, most of which were already in inventory, plummeted. For example, while from 1974 to 1981 fertilizer sales increased at an annual rate of 7.7 percent, in the period 1981 to 1985, they declined at a rate of 5.6 percent (Figure 17.7).[4] Fertilizer employee numbers declined at a rate of 5.5 percent annually, and

---

[4]Stan Daberkow, *Agricultural Input Industry Indicators in 1974–1985*, Agric. Inf. Bull. 534 (Washington, D.C.: ERS, USDA, November 1987).

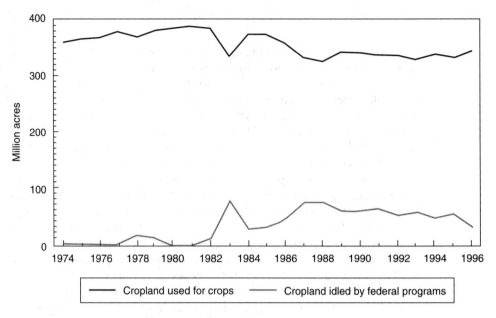

FIGURE 17.6   Total cropland and cropland idled by federal programs, 1945–1996.

Source: Data provided by the USDA/ERS, Natural Resources and Environment Division, Natural Resource Conservation and Management Branch (Washington D.C.: USDA/ERS, November 1996).

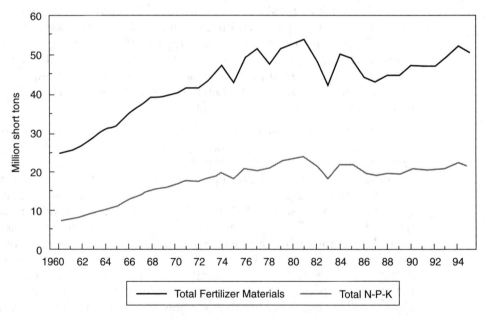

FIGURE 17.7   Fertilizer sales declined in response to increased land retirement during the 1980s.

Source: Data supplied by USDA/ERS, Natural Resources and Environment Division, Production, Management and Technology Branch (Washington, D.C.: USDA/ERS, December 1996).

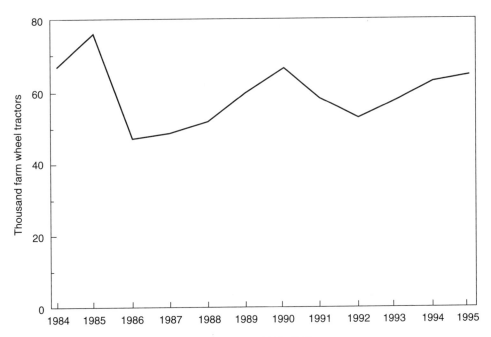

FIGURE 17.8   Tractor sales, over 40 horsepower, 1984–1995.

Sources: Economic Research Service, *Agricultural Resources: Inputs*, AR-29 (Washington, D.C.: USDA/ERS, February 1993), p. 31; and Economic Research Service, *AREI Updates: Farm Machinery Purchases*, Exports and Imports No. 5 (Washington, D.C.: USDA/ERS, June 1996) p. 2.

capital expenditures declined 16.2 percent annually. In farm machinery, tractor sales increased by 15.2 percent annually from 1974 to 1981 but decreased by 15.3 percent from 1981 to 1985.[5] Farmer cooperatives, such as Farmland Industries, went through years of downsizing as sales of fertilizer, petroleum, and other farm supplies declined. For machinery dealers such as International Harvester, PIK was the straw that broke the camel's back as tractor sales declined during subsequent years and International Harvester was acquired by Case, which was owned by Tenneco (Figure 17.8).

The impact of particular government programs and economic forces on agribusiness during the 1980s, in contrast with the 1960s, is very interesting. In the 1960s, the prevailing policy tools were acreage allotments and price supports. (There were no target prices.) In the presence of only allotments and price supports, a cutback in acreage had two economic effects:

- It reduced the quantity of land to be farmed, thus tending to reduce input sales.
- It had the potential to increase commodity prices, thus tending to increase input sales.

[5]Ibid., p. 13.

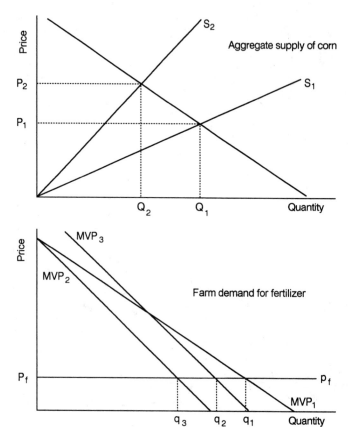

FIGURE 17.9
Effect of production controls on the supply of corn and on the demand for fertilizer with price supports.

The effects are seen more clearly in Figure 17.9. The marginal value product (MVP) curve represents the demand for fertilizers per acre of corn produced. MVP is the marginal product (the extra corn production resulting from an additional unit of fertilizer applied) times the price of corn.

Removing land from production in the 1960s reduced the marginal product of land because land is farmed more intensely due to the existence of diminishing returns. This shifts the MVP curve to the left from MVP_1 to MVP_2 (Figure 17.9). This has the effect of reducing the quantity of fertilizer demanded from Q_1 to Q_2 at fertilizer price P_f. However, if P_1 is greater than either the support price or the target price, production controls increase the price of corn from P_1 to P_2, which shifts the fertilizer demand curve back to the right from MVP_2 to MVP_3.[6] The result of the higher corn price is to increase fertilizer demand from Q_2 to Q_3. The net effect is a small reduction in the quantity of fertilizer demanded from Q_1 to Q_3. Throughout the 1960s, these opposing forces operated with no great harm to the

---

[6]This analysis assumes that the price support level is sufficiently low that the price of corn rises above the support level.

farm input industries. In fact, input demand grew as the proportion of purchased inputs used by farmers increased (see Figure 9.1).

In the 1980s, taking the same amount of land out of production had a much more adverse impact on agribusiness. The reason lies in a change in farm policy—the substitution of target prices for support prices. With the target price programs, any increases in producer prices that resulted from the large production cutbacks simply meant fewer government payments, not higher incomes (returns) to crop producers! The difference in economic effect of price support versus target price policy is more clearly illustrated in Figure 17.9. With the target price set substantially above the now-higher market price (P_1), removing land from production is not compensated for by an increase in producer returns. Therefore, while the shift in fertilizer demand occurs from MVP_1 to MVP_2, there is *no compensating shift to MVP_3* due to the increased market price *because the target price, not the market price, drives production decisions.*[7] The 1996 farm bill switches the focus of farmers' production decisions back on the market price. Thus input sellers are no longer shielded from the effects of CRP decisions.

Dairy termination program.   The indirect impacts of farm programs are often unanticipated and can become highly controversial. One of the best illustrations has been the impact of dairy programs on the beef industry. For years, the beef industry had little or no interest in domestic farm policy related to other commodities. The basic attitude of cattle growers tended to be that of "give us our beef import quotas and leave us alone." This was the case at least until the dairy diversion program of 1983 and the dairy termination program of 1986.

The impact of dairy programs on the beef industry results from about 15 percent of the beef and 25 percent of the hamburger meat being by-products of the dairy industry. Both the dairy diversion and termination programs had substantial impacts on beef supplies because they were explicitly designed to slaughter cows. This meant even more beef in the supermarket during a period of dairy herd liquidation.

From an economic perspective, the impact of these programs on beef prices is relatively easy to analyze.[8] For example, the dairy termination program bought dairy farmers out of production and required that their herds, composed of about 1.5 million head of cows, heifers, calves, and bulls, be slaughtered. Slaughter requirements were divided into three 6-month periods selected by the dairy farmers whose bids were accepted by the government. Economic theory suggests that

---

[7]This analysis assumes that no decoupling policies and no effective payment limits exist. A good exercise involves explaining why!

[8]To some economists, this analysis may be viewed as being overly simplistic. It is presented as an illustration of a general approach rather than a definitive answer to a relatively complex issue. For example, all red meat and even poultry might be considered when analyzing price impacts. The illustration is not designed to discourage more complete approaches, which would be expected to be utilized by agribusiness firms.

participating farmers would tend to slaughter at either the beginning or the end of the period. Farmers who were not covering their variable costs of production would tend to

- Sign up in the first slaughter period
- Slaughter at the beginning of the period.

Of 1.5 million dairy animals signed up for the program, 832,000 were to be slaughtered in the first period. Utilizing the expectation that 52 percent would go to market in the first two months of the first period, it was estimated that 63 million pounds of beef would be added to the supply. This was 5.6 percent of the total expected beef slaughter based on past experience and outlook information. With a beef price elasticity of demand of −0.31 at the farm level, a 5.6 percent increase in supply would be expected to lead to an 18 percent decline in the price of fed cattle by the following formula:

$$\frac{\text{Percentage increase in supply}}{\text{Elasticity of demand}} = \text{Percentage change in price}$$

$$\frac{5.6}{-0.31} = -18$$

A portion of the downward price pressure created by increased beef supplies was offset by mandated purchases of 400 million pounds of red meat by USDA for school lunch, export, and military commissaries located outside the United States. If these meat purchases were spread equally throughout the 17-month slaughter period, the quantity demanded would increase by 1.6 percent per month. This would reduce the expected price decline by 5 percent (1.6/0.31), to 13 percent during the first two months of the program.

The result of the actual drop in price, which was very close to that estimated, was a lawsuit by livestock producers designed to force increased purchases of beef by USDA to regulate the flow of beef to slaughter. As a result of this lawsuit, USDA purchases increased, although only minor steps were taken to redistribute slaughter.

Without question, the indirect impacts of dairy programs on livestock sharply escalated cattle grower's interest in domestic farm program issues. The subsequent Harkin bill proposal to more than double the price support level for feed grains and milk, and also impose mandatory production controls, led to the conclusion that beef prices would fall by more than 23 percent if the resulting increased slaughter of 2 million head were spread over a six-month period.[9]

[9]Ronald D. Knutson *et al.*, *Policy Alternatives for Modifying the 1985 Farm Bill* (College Station, Tex.: Texas A&M University, 1987), p. 75.

The dairy termination program had both positive and negative impacts on agribusiness. Meat packers slaughtering dairy animals were deluged with business, utilizing facilities to capacity. Similarly, the business of market intermediaries (brokers, auction markets, and cow jockeys) increased sharply. Milk cooperatives and processors located in milk-deficit regions having heavy participation in either the diversion or termination program scrambled for milk supplies. Intermarket movement of milk supplies increased sharply to meet local market needs. Milk cooperatives having commitments to supply the full milk needs of large processors (full supply contracts) were particularly hard pressed to meet their commitments.

The important long-run effect of the termination program is that it motivated the National Cattlemen's Beef Association (NCBA) to become involved in farm policy. From their interest in dairy policy NCBA developed an interest in crop policy, which affects the price of feed, the profits of feedlots, the price of fed cattle, the price of feeder cattle, and the price of replacement breeding stock.

Structure and Resource Policy

From settlement of the frontiers to the present, government has attempted to influence the structure of agriculture and facilitate technological progress. The agribusiness impacts of these policies are often overlooked. Examples are presented from cooperative and patent policy.

Cooperative policy. Cooperatives not only enjoy the benefits of limited antitrust exemptions, but also have the privilege under the price support program of taking out commodity loans on behalf of their member producers. Referred to as *Form G lending authority,* cooperatives can utilize this privilege if they have pooling contracts with their members. Under such a contract, the producer commits to market through the cooperative. The cooperative normally takes out the CCC loan as soon as the commodity is received from the member and provides the producer an advance, which may be some percentage of the loan rate. The cooperative is free to market the commodity just like the producer and must pay off the loan as soon as the commodity is sold. The producer receives the pool or average price whenever the pool is settled.

Some advantages to cooperatives from Form G lending authority follow:

- The commodity is controlled by the cooperative from the time it is received. The cooperative is free to sell the commodity in domestic or foreign markets whenever it is advantageous.
- The CCC loan interest rate may be less than the market rate.
- The cooperative enjoys the nonrecourse loan forfeiture privileges if the market price does not rise above the loan rate—just like the farmer.
- If there is a marketing loan, the cooperative is in a position to take advantage of its benefits by blending the proceeds with those received from product sales.

Form G lending authority has been used extensively by rice and cotton cooperatives, which were the original recipients of the privilege. Cooperatives are much stronger in these commodities than is typical of grain cooperatives. When, in the mid-1970s, Form G lending authority was extended to grain cooperatives, the multinational grain companies were sufficiently upset to challenge the USDA's authority in court. The multinationals apparently feared that the cooperatives' relative market positions would become stronger in grain. However, this has never materialized. Three alternative explanations follow for cooperatives' lack of success in exploiting Form G lending authority in grain to the advantage of producers:

- Grain farmers hold the right to market their grain independently more dearly than do rice and cotton producers.

- Management of grain cooperatives does not have the same level of marketing talent as does that of rice and cotton cooperatives. As a result, they would be unable to generate higher average levels of producer returns.

- Multinational grain companies have so many other advantages, such as being able to buy and sell the grains of countries throughout the world, that grain cooperatives were unable to compete.

Reality indicates that each of these factors probably plays a role in grain cooperatives' relative lack of use of Form G lending authority.

Research and patent policy.   Publicly sponsored research has been a tradition in U.S. agriculture, extending to the creation of the land-grant universities in each state. These universities, with a prime responsibility for agricultural research and technology transfer, are a driving force behind the progressiveness of American agriculture.

Agribusiness has been a primary beneficiary of agricultural research in the following ways:

- Input manufacturers have converted the products of agricultural research into new varieties, chemicals, and machinery. Agricultural extension has encouraged the adoption of this new technology whenever it proved advantageous.

- Food processors and marketers have enjoyed the benefits of higher quality, lower priced farm products.

- Input manufacturers and market intermediaries have been more competitive in foreign markets, where both the inputs and the farm products have been sold.

In the 1980s, patent rights were extended to the products of biotechnology. This expansion of patent rights allows input manufacturers and food processors to capture an increased share of the benefits from agricultural research and could change the structure of farming.[10] The benefits of patents are obvious in terms of higher profits (monopoly rents) for inputs or products having patent rights. In addition, clearly superior patented inputs or products might be used as a lever for gaining increased control of a commodity through contract or ownership integration. Commodities having the benefits of patent rights might be effectively converted into branded food products.

The advent of biotechnology and patenting resulted in a proliferation of biotech firms, many of which were new small enterprises and some of which were established firms such as Monsanto and DuPont. In the 1990s there was a consolidation of the biotech industry with the established firms buying up many of the smaller companies. High prices for the products of biotechnology, in part reflecting regulatory cost to prove efficacy and safety, led to questions regarding who was receiving the benefits of publicly supported research. Such questions could become the focal point of future controversy regarding the structure of agriculture and of public sector support for land-grant universities.

Food Policy

The impact of food programs and demand expansion programs on agribusiness is often overlooked. Yet whole industries have been built to serve the food needs created by programs such as school lunch. The food stamp program has a decidedly different impact on food processors and marketers than commodity distribution programs. In addition, domestic and international generic advertising and promotion programs create demand surges with direct agribusiness benefits.

School lunch. USDA is the largest purchaser of prepared foods in the United States through its school lunch program. Schools serve more than 25 million children each school day. All of these meals are, in varying degrees, subsidized by USDA. Arguably, the school lunch program fostered the food service sector of the food industry and contributed significantly to the development of the fast-food business.

Food processor and marketer impacts are materially different, depending on whether schools' food needs are subsidized by giving the schools commodities or cash. If schools are given commodities, USDA makes the decision about what is bought. In addition, commodity purchases are from national companies as opposed to local businesses. Schools have much more flexibility when they buy commodities themselves. Local processors and marketers often get the business. In addition, more value-added foods are generally purchased by schools. That is, although USDA has a tendency to buy a standard product such as turkey breasts or canned beans, schools buy frozen pizza or lettuce for a salad bar. USDA aims

---

[10]Universities also gain from the extension of patent rights to biotechnology because they receive increased private sector support from agribusiness and generally hold the patents with rights being sold to an input manufacturer or food processor.

for having the greatest impact on producer prices, while schools buy with greater consideration of student preferences.

School lunch also influences what people eat. Without school lunch, children would carry their own lunches made up largely of bread, luncheon meat, peanut butter, jelly, cheese, fresh fruit, and soft drinks. School lunch fosters the consumption of vastly increased quantities of dairy products (particularly milk and processed cheese), hamburgers, canned peaches, canned pears, beans, canned tomatoes, hams, macaroni, potatoes, fats, oils, eggs, turkeys, and hot dogs. A major change in policy brought on by USDA Undersecretary Haas for the Clinton administration involved reducing fat and increasing fruits and vegetables. These changes in policy are obviously designed to generate spillover effects on what the population eats.[11] But there is also a different distribution of farmer benefits both in terms of commodities and geographically.

Food stamps versus commodity distribution.   One of the historic controversies in USDA food programs is whether to distribute food stamps or commodities. The major difference between the two systems is in its food processor and marketer impacts. Commodity distribution bypasses food retailers. In addition, different food processors may supply USDA's commodity purchase programs rather than sell to food retailers. Secondary impacts of commodity distribution programs may extend to other commodities. For example, if butter is distributed, margarine sales are adversely affected.

Because commodity distribution programs displace commercial sales, food retailers have been ardent advocates of food stamps. Food stamps allow the poor to purchase not only more food but also more value-added products. The major beneficiaries of food stamps, therefore, are the food retailers and processors that produce branded, value-added products.

Advertising and Promotion

Enabling legislation at federal and state levels charges the USDA and related state agencies with supervising the collection and expenditure of about $1.0 billion in check-off funds used mostly for promotion and advertising.[12] Agribusiness is a beneficiary of these producer expenditures. The prime objective of advertising and promotion programs is to expand demand. This has the short-run effect of raising prices and the longer run effect of expanding supplies. Agribusiness benefits from the short-run price increase impact as much as farmers—perhaps even more. When demand is expanding, it is easier for food processors and marketers to pass on price increases. Profits, therefore, tend to be stronger. Some of the most profitable years for dairy processing plants were in the mid-1980s, when demand

---

[11]Students can readily identify with and discuss how school lunch influenced what they ate in high school and their current attitudes toward certain foods.

[12]Olan D. Forker and Ronald W. Ward, *Commodity Advertising* (New York, N.Y.: Lexington Books, 1993), p. 101.

for dairy products was rising while the milk price support level was being reduced. The reason lies in the combined effects of increased advertising financed by a milk producer check-off program and reduced milk prices. Interestingly, the reduced milk price effects on consumption served to provide the illusion that the advertising program was considerably more effective than it really was. However, the producer program has been sufficiently successful and the competition with soft drinks sufficiently keen that milk processors sought and obtained a comparable check-off program designed to complement that of producers.

Market intermediaries benefit from international demand expansion activities. The FAS cooperator program helps teach potential foreign buyers how to most effectively utilize U.S. agriculture products. This responsibility would otherwise fall largely on the exporter. Benefits are realized by both the farmer in terms of expanded demand and market intermediaries in terms of increased volume of product handled at reduced costs. The precise magnitude of these benefits is difficult to quantify but with about $55.8 billion in exports, it is not unreasonable to anticipate agribusiness benefits that exceed producer contributions!

STRIKING A BALANCE

Agribusiness involvement in policy is both offensive and defensive. Defensive involvement is designed to prevent adverse impacts on a firm's costs, volume, and profits. Offensive involvement is designed to capture a share of the benefits of government assistance that may be enacted primarily to help farmers and/or consumers. Little, if any, attention has been given to the distribution of costs and benefits. However, judging from the proliferation of agribusiness lobbying, the payoffs must be substantial from both offensive and defensive lobbying.

The future of agricultural policy is about as difficult to predict as the future of agriculture itself. Agriculture and food problems are moving targets. Surpluses and low prices are sometimes interrupted by deficits and high prices. Economic booms are interrupted by recessions and unemployment. Agricultural and food policy goals shift as priorities change.

The basic volatility of agriculture and the importance of food to the survival of humankind complicate the problems of farmers, agribusiness, consumers, and policymakers in arriving at a mutually acceptable agricultural and food policy. This initially requires an understanding of the policy options, their consequences, and their interrelationships. Secondarily, it requires a willingness on the part of those affected by agricultural and food policy to recognize each other's interests and seek compromise solutions. Striking a balance is not an easy task, but it is as important to agriculture as food is to life itself.

Index